Molecular Logic-based Computation

Monographs in Supramolecular Chemistry

Series Editors:
Philip Gale, *University of Southampton, UK*
Jonathan Steed, *Durham University, UK*

Titles in this Series:
 1: Cyclophanes
 2: Calixarenes
 3: Crown Ethers and Cryptands
 4: Container Molecules and Their Guests
 5: Membranes and Molecular Assemblies: The Synkinetic Approach
 6: Calixarenes Revisited
 7: Self-assembly in Supramolecular Systems
 8: Anion Receptor Chemistry
 9: Boronic Acids in Saccharide Recognition
10: Calixarenes: An Introduction, 2nd Edition
11: Polymeric and Self Assembled Hydrogels: From Fundamental Understanding to Applications
12: Molecular Logic-based Computation

How to obtain future titles on publication:
A standing order plan is available for this series. A standing order will bring delivery of each new volume immediately on publication.

For further information please contact:
Book Sales Department, Royal Society of Chemistry, Thomas Graham House, Science Park, Milton Road, Cambridge, CB4 0WF, UK
Telephone: +44 (0)1223 420066, Fax: +44 (0)1223 420247,
Email: booksales@rsc.org
Visit our website at http://www.rsc.org/Shop/Books/

Molecular Logic-based Computation

A Prasanna de Silva
School of Chemistry and Chemical Engineering, Queen's University Belfast, Belfast, UK
Email: a.desilva@qub.ac.uk

RSC Publishing

Monographs in Supramolecular Chemistry No. 12

ISBN: 978-1-84973-148-5
ISSN: 1368-8642

A catalogue record for this book is available from the British Library

© The Royal Society of Chemistry 2013

All rights reserved

Apart from fair dealing for the purposes of research for non-commercial purposes or for private study, criticism or review, as permitted under the Copyright, Designs and Patents Act 1988 and the Copyright and Related Rights Regulations 2003, this publication may not be reproduced, stored or transmitted, in any form or by any means, without the prior permission in writing of The Royal Society of Chemistry or the copyright owner, or in the case of reproduction in accordance with the terms of licences issued by the Copyright Licensing Agency in the UK, or in accordance with the terms of the licences issued by the appropriate Reproduction Rights Organization outside the UK. Enquiries concerning reproduction outside the terms stated here should be sent to The Royal Society of Chemistry at the address printed on this page.

The RSC is not responsible for individual opinions expressed in this work.

Published by The Royal Society of Chemistry,
Thomas Graham House, Science Park, Milton Road,
Cambridge CB4 0WF, UK

Registered Charity Number 207890

Visit our website at www.rsc.org/books

Printed in the United Kingdom by Henry Ling Limited, Dorchester, DT1 1HD, UK

Foreword

Chemistry is one of the oldest and most fascinating fields of science. It deals with molecules, peculiar entities that are extremely small (sub-nanometer dimension) and unbelievably numerous and diverse. Five million different molecular types have been discovered in Nature and at least fifty million more have been synthesized by chemists in their laboratories. Each type of molecule has its own composition, size, shape and structure, and therefore it is a unique "object". Some molecules are simple, others are extremely complex, as is the case for most natural molecules of biological interest. Indeed, in the past century chemistry became the science of molecules.

At the dawn of the twenty-first century, the most creative act in chemistry was thought to be the total synthesis of complex natural molecules, an activity defined as "a precise science and a fine art capable of penetrating the secrets of Nature". Today, total synthesis continues to be an important field of chemical research, but in the last 15 years quite new features of chemistry have emerged on the scene. Two of them are directly related to the topic discussed in this book: novel conceptual interpretations of old classes of chemical processes, and the marriage of the synthetic talent of chemists with a "device driven" creativity. This book, indeed, originates from novel interpretations of old processes and expands by illustrating the clever assembly of molecular building blocks to construct a variety of molecular devices capable of performing all kinds of logic functions. As a result, a new branch of chemistry, namely molecule-based computation, is established.

Light interacts with matter in a very specific way and therefore it can be employed for investigating the molecular world. Indeed, the old branch of chemistry called *spectroscopy* is extensively used by physical chemists to elucidate the structure of molecules and by analytical chemists to identify chemical species, define the composition of complex systems, and verify the purity of compounds. The interaction of light with matter is exploited by another traditional discipline of chemistry, namely *photochemistry*, for decomposing or converting a molecule into a different one. In the latter case, the reaction can often be reverted (*photochromism*).

Such old fields of chemistry, *i.e.* spectroscopy, photochemistry, and photochromism, can be viewed with new eyes in the frame of digital information science. Since each molecule is a very specific entity, its presence (1) and absence (0) can be associated with a bit of information. In such a vision, spectroscopy "reads", photochemistry "writes", and the back reaction of a photochromic system "erases" a bit of information.

These concepts can be extended to inputs other than photons. For example, we can "read" the presence of a molecule by using electrons (electrochemical analysis) or by adding a specific reactant (chemical analysis), and we can also "write" (and then "delete") information bits on molecules using electrons and chemicals. Photonic, electronic and chemionic inputs can be used in a sequence and combined, and the resulting outputs can be harvested by a variety of techniques. Suitable merging of inputs and outputs allows properly designed molecules to perform logic functions and to make complicated operations. Since molecules are so small, an enormous number of bits can theoretically be processed by a macroscopic chemical system. The outstanding signal-processing capacity of the molecular world is thoroughly illustrated and discussed in this excellent book.

Although input/output molecular-level processes are very common and have been known for a long time, their binary logic aspect has been recognized only recently. For example, there are hundreds of compounds whose fluorescence (optical output) is quenched by the addition of a quencher (chemical input); such processes were extensively used for sensing purposes, but their intrinsic nature of a NOT logic gate went unnoticed.

The possibility of developing an artificial molecular-scale computer was first mentioned in the well known Pimentel report "Opportunities in Chemistry" in 1985, and the leap of imagination to execute logic operations at the molecular level was made by Arieh Aviram in 1988. The field of molecular logic-based computation, however, began to develop only 5 years later when, in a paper published in *Nature*, A. P. de Silva and co-workers demonstrated the analogy between molecular switches and logic gates. In the following years the field of molecular logic gates has developed enormously in several laboratories and along many directions: the kind of input/output employed, their number, the construction of novel molecular and supramolecular systems suitably designed for logic-based computation, the complexity of the logic operations performed, the interested disciplines and, last but not least, the variety of applications.

This book is very well organized. It starts by illustrating the basic concepts of digital computation and explaining why molecules are such an outstanding substrate for this task. The following chapters describe a large number of examples, beginning with single input–single output logic gates and continuing with more complex operations, such as those involved in encoders, decoders, full-adders and full-subtractors. Separate chapters are dedicated to history-dependent systems, multi-level logic, and molecular quantum computing. At the end, a long chapter is dedicated to describing a variety of applications, particularly in the field of medicine, such as photodynamic therapy, intracellular computing, and measuring blood electrolytes.

The ultimate goal of molecular-based computation is the development of a computing machine based on molecules that can make modern silicon computers obsolete. In the previously mentioned report, Pimentel was explicit and optimistic about the development of molecular computers: "There are those who dismiss as far-fetched the idea of man-made molecular scale computers.... But since we know that molecular computers are routine accessories of all animals from ants to zebras, it would be prudent to change the question from whether there will be man-made counterparts to questions concerning when they will come into existence and who will be leading in their development. The when question will be answered on the basis of fundamental research in chemistry; the who question will depend on which countries commit the required resource and creativity to the search." Although the path to reaching such an ultimate goal is still very long, this book shows that important progress has been made since Pimentel's vision.

In conclusion, A. P. de Silva has admirably succeeded in introducing the important topic of molecular-based computation. The whole scientific community, and particularly young people, who are less burdened by impressions of what is impossible, will certainly find in this book many suggestions for their research.

Vincenzo Balzani
Bologna, January 2012

*For Innocence
and the Good Gone Before*

Preface

My grandfather was the teacher in our little Sri Lankan town. On some days, he would sit at his desk dressed in his best clothes and his gold-rimmed spectacles. The ceremonial oil lamp would be lit. At a carefully appointed time, a mother and her child would arrive bearing gifts. The child was in a starched white school uniform, her innocent eyes shining in the lamplight while the mother's face was warm with hope for her youngster's future. The child had a slate in one hand and a scribe in the other. My grandfather would take her writing hand in his and navigate her around the first character of our alphabet while saying it aloud. The child would repeat after him and enter the new world of learning.

We all learn – in schools, factories, bars and streets. We gather, store, process and transmit information in society. Molecular systems involved in our senses and within our brains allow all this to happen. Molecular systems allow living things of all kinds to handle information for the purpose of survival and growth. Nevertheless, the vital link between molecules and computation was not generally appreciated until a few decades ago. Semiconductor-based information technology had penetrated society at many levels. The interest in maintaining momentum of this revolution led to the consideration of molecules, among others, as possible information handlers. Such an overlap between the recent engineering-oriented revolution and the ancient biology-oriented success story is bound to be interesting.

George Boole's times in Ireland 150 years ago produced the logic ideas that provide the foundations of computation to this day. Chemically minded scientists are well placed to design and construct molecular systems which carry out Boolean logic operations. Such molecular logic gates can benefit from designers coming from backgrounds of various kinds. It is quite logical that a rich variety of approaches has resulted, making the field very interesting to be in. Workers in the field are regularly challenged to learn ideas which are well outside their comfort zones. For all these reasons, I felt that the time was ripe for a comprehensive monograph on this topic. The only book that comes even close, *Molecular Computation* (eds. T. Sienko, A. Adamatzky and M. Conrad, MIT Press, Cambridge, MA, 2001), is an edited collection of chapters on biochemical computing with no reference at all to Boolean logic concerning small molecules. However, a valuable chapter on the topic is available in V. Balzani, M. Venturi and A. Credi, *Molecular Devices and Machines*, Wiley-VCH, Weinheim, 2nd edn, 2008. It is a delight to report that the situation has changed for the better with the appearance of three books concerning this general area: E. Katz (ed.), *Molecular and Supramolecular Information Processing* Wiley-VCH, Weinheim, 2012; E. Katz (ed.), *Biomolecular Information Processing* Wiley-VCH, Weinheim, 2012; K. Szacilowski, *Infochemistry Wiley*, Chichester, 2012.

This book is my attempt to tell the story of the growth of this field. It also tries to gather up and classify the outputs of scientists from around the world as comprehensively as possible. However, the literature in the field is already so large that all the work cannot be discussed at the same depth in a book of this size. Some representative reports will be summarized either in the text or in annotated tables. The earliest papers outlining a concept will come in for special attention. Only experimental work will be discussed, since the field has suffered enough from speculation and hype. While apologizing in advance for any errors of omission and commission in these pages, I would appreciate being informed of these via e-mail (a.desilva@qub.ac.uk). One of my hopes is that workers in molecular devices and related fields will find this account a handy reference. Another hope is that this story will give fresh and bright minds a taste of what this field of molecular logic-based computation has produced in just under two decades. Then they will hopefully feel that the field deserves their own inputs to make it stronger still.

I remain grateful to many people and institutions for many helpful roles.

Hosts: University of Colombo, Sri Lanka and Queen's University Belfast, Northern Ireland.

Fellow labourers at the two host institutions: Dayasiri Rupasinghe, Shantha Samarasinghe, Amila Norbert, Annesley Peiris, Dayal de Costa, Nalin Goonesekera, Saliya de Silva, Thilak Fernando, Lalith Silva, Sisira de Silva, V. Edwin, Sydney Ramyalal, Ranjith Jayasekera, Aruna Dissanayake, Suram Patuwathavithana, Saman Sandanayake, Richard Bissell, Nimal Gunaratne, Aiden Bryan, Mark Lynch, Alan Patty, Graham Spence, Kemuel Nesbitt, Linda Daffy, Sean O'Callaghan, Glenn Maguire, Aine Kane, Colin McCoy, Thorri Gunnlaugsson, Terry Rice, Anne Goligher, Catherine McVeigh, Emma O'Hanlon, Pamela Maxwell, Isabelle Dixon, Jude Rademacher, Juliette Roiron, Nathan McClenaghan, Allen Huxley, Joanne Ferguson, Aoife O'Brien, Marc Schroeder, David Fox, Tom Moody, Gareth McClean, Sheenagh Weir, Gareth Brown, Sara Pagliari, Manel Querol, Bernadine McKinney, Bridgeen McCaughan, Seiichi Uchiyama, David Farrell, John Callan, Gregory Coen, Robert Boyd, David Magri, Boontana Wannalerse, Catherine Dobbin, Vinny Vance, Matthew West, Glenn Wright and Bernard McLaughlin.

Preface

Colleagues at the two host institutions: All the chemistry staff (past and present) working in the offices, lecture theatres, teaching laboratories, research laboratories, workshops, services, stores and corridors.

Collaborators: Jean-Philippe Soumillion, the late Jean-Louis Habib-Jiwan, Otto Wolfbeis, Jim Tusa, Hua-rui He, Marc Leiner, Stephane Content, Kaoru Iwai, Mark James, David Pears and Steven Magennis.

Short-term hosts: Universite Catholique de Louvain, Ecole Normale Superieur de Cachan, Universite de Bordeaux, Universite Louis Pasteur in Strasbourg, University of Peradeniya, Institute of Fundamental Studies in Kandy, Nara Women's University, Chulalongkorn University in Bangkok and East China University of Science & Technology in Shanghai.

Funding agencies: Department of Employment and Learning of Northern Ireland, Engineering and Physical Sciences Research Council, European Community, InvestNI, McClay Trust, Avecia Ltd, Procter and Gamble Company, Japan Society for the Promotion of Science and Roche-AVL Company.

Supporters: The Abeysekeras, Bastiampillais, Bells, Camerons, Corrys, de Soysas, de Silvas (Sri Lanka, Australia and USA), Dikkumburas, Ekanayakes, Fernandos, Gunaratnes, Gunnlaugssons, Guruges, Halahakoons, Jayaweeras (Sri Lanka and UK), Killorans, Koelmeyers, Liyanages, McIntyres, McNamaras, Naqvis, Namasivayams, Norberts, Pathiranas, Pereras, Ramasubbus, Rupasinghes, Senanayakes, Sridharans, Surendrakumars, Udeshis and the past/present members of the Niall McClean band.

Teachers: Tissa Silva, Indra de Silva, Errol Fernando, Vincent Arkley, R. S. Ramakrishna, K. Mahadeva, Satish Namasivayam, James and Jadwiga Grimshaw, Brian Walker, Ron Grigg and Fraser Stoddart.

A special 'thank you' is due to two people. 'Prof' Ron Grigg: if not for his foresight, the last 25 years would have been very different for me. Fraser Stoddart: if not for his early invitation and belief, this book probably would not have happened.

Finally, I owe Seiichi Uchiyama, Amila Norbert and David Magri for their proofreading, Bernard McLaughlin for his cover artistry, Vincenzo Balzani for graciously contributing a Foreword and Alice Toby-Brant and Janet Freshwater for keeping me on my toes. My thanks are due to you all, as well as to providence, luck and serendipity.

Contents

Chapter 1	A Little History	1
	1.1 Introduction	1
	1.2 Early Proposals for Molecular Logic	1
	1.3 Photochemical Approach to Molecular Logic and Computation	5
	References	7

Chapter 2	Chemistry and Computation	12
	2.1 Introduction	12
	2.2 Why Molecules?	15
	2.3 Breadth of Molecular Logic and Computing	18
	2.4 Indicators and Sensors	18
	2.5 The Digital–Analogue Relationship in Chemical Systems	19
	2.6 Molecular Device Characteristics	20
	2.7 Molecular Logic for the Spectrum of Disciplines	21
	References	21

Chapter 3	A Little Logic and Computation	24
	3.1 Introduction	24
	3.2 Truth Tables and Algebraic Expressions for Logic Gates	25
	3.2.1 Single Input–Single Output Devices	25
	3.2.2 Double Input–Single Output Devices	26

Monographs in Supramolecular Chemistry No. 12
Molecular Logic-based Computation
By A Prasanna de Silva
© The Royal Society of Chemistry 2013
Published by the Royal Society of Chemistry, www.rsc.org

	3.3 Logic Gates in Electronics	29
	3.4 Number Manipulation	31
	References	33

Chapter 4   A Little Photochemistry and Luminescence — 34

    4.1   Introduction — 34
    4.2   Excited States Involving Charge Transfer — 35
    4.3   Metal-centred (MC) Excited States — 37
    4.4   nπ* and ππ* Excited States — 37
    4.5   Photoinduced Electron Transfer (PET) — 38
    4.6   Electronic Energy Transfer (EET) — 41
    4.7   Excimers and Exciplexes — 41
    4.8   Vibrational Deexcitation and Excited State Intramolecular Proton Transfer (ESIPT) — 42
    4.9   Relationships between some of the Photochemical Principles used in Switching — 43
    References — 44

Chapter 5   Single Input–Single Output Systems — 50

    5.1   Introduction — 50
    5.2   YES — 50
        5.2.1   Electronic Input — 50
        5.2.2   Chemical Input — 51
        5.2.3   Temperature Input — 74
        5.2.4   Light Dose Input — 76
    5.3   Irreversible YES — 77
        5.3.1   Chemical Input — 78
        5.3.2   Light Dose Input — 83
    5.4   NOT — 83
        5.4.1   Electronic Input — 83
        5.4.2   Chemical Input — 83
        5.4.3   Temperature Input — 90
        5.4.4   Light Dose Input — 91
    5.5   Irreversible NOT — 92
        5.5.1   Anion Input — 93
        5.5.2   Oligonucleotide Input — 93
        5.5.3   Protein Input — 94
    5.6   PASS 1 — 95
    5.7   PASS 0 — 95
    References — 96

Chapter 6 Reconfigurable Single Input–Single Output Systems   **109**

    6.1 Introduction   109
    6.2 Nature of Inputs   109
    6.3 Output Observation Technique   111
    6.4 Observation Wavelength   112
    References   112

Chapter 7 Double Input–Single Output Systems   **114**

    7.1 Introduction   114
    7.2 AND   114
        7.2.1 Distinguishable and Separate Inputs   115
        7.2.2 Indistinguishable and Separate Inputs   129
        7.2.3 Distinguishable and Connected Inputs   132
        7.2.4 Indistinguishable and Connected Inputs   134
        7.2.5 Light Dose Input(s)   136
        7.2.6 Biopolymeric AND Gates   141
        7.2.7 AND Gates using Molecule-based Materials   143
    7.3 OR   145
        7.3.1 OR Gates using Molecule-based Materials   151
    7.4 NOR   152
    7.5 NAND   156
    7.6 INHIBIT   161
        7.6.1 INHIBIT Gates using Molecule-based Materials   167
    7.7 XOR   169
        7.7.1 Light Dose Input(s)   173
        7.7.2 XOR Gates using Molecule-based Materials   174
    7.8 XNOR   175
    7.9 IMPLICATION   176
    7.10 TRANSFER   179
    7.11 NOT TRANSFER   180
    7.12 PASS 0 and PASS 1   181
    References   181

Chapter 8 Reconfigurable Double Input–Single Output Systems   **195**

    8.1 Introduction   195
    8.2 Module Connectivity within Device   195
    8.3 Functional Group Connectivity within Input Array   197
    8.4 Functional Group Configuration within Input Array   198
    8.5 Nature of Inputs   199
    8.6 Output Observation Technique   202

8.7	Nature of Output (within a given Observation Technique)	204
	8.7.1 Observation Wavelength	204
8.8	Starting State of Device	205
8.9	Applied Voltage or Redox Reagents	207
References		208

Chapter 9 Double Input–Double Output Systems — 210

9.1	Introduction	210
9.2	Half-adder	210
9.3	Half-subtractor	219
9.4	1:2 Demultiplexer	223
9.5	Magnitude Comparator	223
9.6	Reversible Logic	224
References		225

Chapter 10 More Complex Systems — 228

10.1	Introduction	228
10.2	Three-input AND	230
	10.2.1 Three-input AND with Mixed Input Types	233
10.3	Three-input OR	233
10.4	Three-input NOR	235
10.5	Three-input INHIBIT	236
10.6	Three-input IMPLICATION	240
10.7	Three-input Enabled OR	240
10.8	Three-input Enabled NOR	242
10.9	Three-input Enabled IMPLICATION with Wavelength-Reconfigurability	244
10.10	Three-input Disabled OR with Wavelength-Reconfigurability	245
10.11	Three-input Disabled INHIBIT	247
10.12	Three-input Disabled XNOR	248
10.13	Three-input Disabled IMPLICATION	249
10.14	Three-input Inverted Enabled OR	253
10.15	2:1 Multiplexer and 1:2 Demultiplexer	253
10.16	Other 3-Input Systems	257
10.17	Four-input AND	257
10.18	Four-input Doubly Disabled AND	259
10.19	4-to-2 Encoder and 2-to-4 Decoder	260
10.20	Other Four-Input (and Higher) Systems	263

	10.21	Higher Arithmetic Systems		267
		10.21.1 Combined Half-adder and Half-subtractor		267
		10.21.2 Full-adder		270
		10.21.3 Combined Full-adder and Full-subtractor		272
	10.22	Gaming Systems: Tic-tac-toe		275
	References			276

Chapter 11 History-dependent Systems 285

11.1	Introduction	285
11.2	R–S Latch	285
11.3	D Latch	290
11.4	Molecular Keypad Lock	292
References		298

Chapter 12 Multi-level Logic 302

12.1	Introduction	302
12.2	'Off–on–off' Switching Systems	302
12.3	Other Variants	308
References		312

Chapter 13 Quantum Aspects 316

13.1	Introduction	316
13.2	Nuclear Magnetic Resonance Spectroscopy Approach	316
13.3	Electronic Absorption and Emission Spectroscopy Approach	318
	13.3.1 Internal Charge Transfer (ICT) systems	318
	13.3.2 Electronic Energy Transfer (EET) Systems	326
	13.3.3 Excimer and Exciplex Systems	330
13.4	Raman Spectroscopy Approach	331
References		333

Chapter 14 Applications 336

14.1	Introduction	336
14.2	Optical Sensing based on YES and NOT Logic and Superpositions Thereof	337
	14.2.1 Tracking Species and Properties within Cells and in Tissue	337
	14.2.2 Measuring Blood Electrolytes	343
	14.2.3 Monitoring Air Pressure on Aerofoils	344

	14.2.4	Sensing Marine Toxins	345
	14.2.5	Monitoring Nuclear Waste Components	346
	14.2.6	Screening for Catalysts	348
	14.2.7	Detecting Chemical Warfare Agents	348
14.3	Improved Sensing		349
	14.3.1	Improved Sensing with AND Logic	349
	14.3.2	Improved Sensing with Superposed AND, INHIBIT and TRANSFER Logic	353
	14.3.3	Near-simultaneous Monitoring of Multiple Species with XOR Logic	354
14.4	Identification of Small Objects in Populations		356
14.5	Improved Medical Diagnostics		359
14.6	Improved Therapy		364
14.7	Photodynamic Therapy		366
	14.7.1	Targeted Photodynamic Therapy	367
14.8	Intracellular Computation		370
14.9	Conclusion		374
References			375

List of Abbreviations and Glossary   **384**

Subject Index   **387**

CHAPTER 1

A Little History

1.1 Introduction

The physicist and percussionist Feynman is widely credited with envisioning atom assembly as a way of building ultra-small devices.[1,2] While this vision can be realized by atom manipulation with scanning probe microscope tips,[3] it is clear that chemists have a treasure chest of methods to arrange atoms in intricate arrays to produce molecules on a kilogram scale. Once synthesized, these molecules can be investigated one at a time – at the single molecule level – if necessary. More commonly, these molecules will be studied in large populations of many billions.

1.2 Early Proposals for Molecular Logic

Perhaps the best known early proposal for molecular logic and computation was announced in 1988 by Aviram[4,5] and is outlined in Figure 1.1. Given his affiliation with IBM, it was natural that the system **1** used electric voltages for both the inputs and the outputs. The vision was as follows. A π-system$_1$, an oligothiophene of *ca.* 25 nm length would be fixed via thiol terminals to two gold[6] electrodes (electrode$_1$ and electrode$_2$). In this electrically neutral state, the oligothiophene would be non-conducting.[7–10] The middle of the oligothiophene contains a spiro linkage so that another π-system can be positioned orthogonally. The latter π-system$_2$ (different[5] from or identical[4] to the first oligothiophene) would be held in its radical cation form which is electrically conducting.[7–10] The spiro linkage would hinder electron transfer between the two π-systems so that the first oligothiophene remains electrically neutral. However if a strong electric field is applied to the spiro linkage via two orthogonal switching electrodes aimed at it, theoretical calculations[11,12] suggested that electron transfer would occur between the two π-systems. The first oligothiophene would then pass into its radical cation state which is an electron

Monographs in Supramolecular Chemistry No. 12
Molecular Logic-based Computation
By A Prasanna de Silva

Published by the Royal Society of Chemistry, www.rsc.org

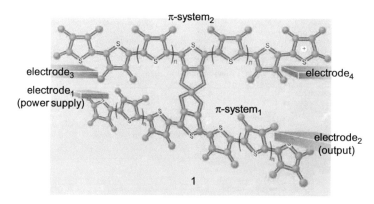

Figure 1.1 Artist's impression of a proposed molecular electronic switch **1**. Another pair of switching electrodes which is aimed at the centre of **1** is not shown. Adapted from L. F. Lindoy, *Nature*, 1993, **364**, 17 with permission from Nature Publishing Group.

conductor, as mentioned earlier. Thus a voltage applied at one gold electrode$_1$ would appear at the other (output) gold electrode$_2$ in response to a voltage input applied across two switching electrodes aimed at the centre of **1**, which are not shown in Figure 1.1. In other words, an electric field-induced insulator-to-conductor transition at the molecular scale would serve as the switching mechanism in this device. The above argument does not need electrode$_3$ and electrode$_4$ but they are required[4] for building NOT logic gates, which are essential for general computing. Wiring of multiple copies of these switches was then expected to yield various logic functions.

As we go to press 24 years later, Aviram's proposal remains a proposal. Much effort has been expended by Tour, for example, to synthesize close relatives of **1** and many other derivatives in resourceful ways.[13,14] Tour and his collaborators Allara, Weiss and Reed also performed pioneering two-electrode experiments on molecules simpler than **1**,[15-18] some of these results turning out to be controversial.[19-21] Nevertheless, he concluded that it was too difficult to focus more than two metal electrodes selectively onto a small molecule such as **1** in order to perform the crucial test.[22,23] Fabricating three-electrode devices based on single molecules remains a difficult art.[24,25] Logic gates based on graphene and (bundled or single) carbon nanotubes have also been constructed.[26,27]

Though not proposed for molecular logic-based computation as such, Aviram and Ratner's suggestion (made in 1974) for a molecular diode/rectifier[28] reached a happier conclusion. An electron donor π-system linked via a σ-bridge to an electron acceptor π-system such as **2** was Aviram's and Ratner's original suggestion. After a long odyssey and many controversies concerning the nature of the metal electrode–organic molecule interface in determining current output–voltage input profiles,[29] the rectification behaviour of monolayers of, for example, **3** was demonstrated.[30-32]

Aviram's baton has been picked up by several others to develop different approaches to molecular electronic logic and computation. These have focused on molecules different from **1**, e.g. rotaxanes,[33] catenanes,[34] carbon nanotubes[26,27] and graphene.[26,27]

This electronic approach to molecular logic and computation was and is very popular, perhaps because of its direct connection with the burgeoning semiconductor computer industry and also because it engaged the concepts and techniques of chemists, physicists and engineers in equal measure. Naturally, there have been many good discussions in the literature.[35–37] Well-resourced national programmes were launched in several countries. The USA operation occurred in two waves, with the first centred in the early 1980s[38,39] and the second starting in the late 1990s and continuing to this day.[40] The entire field of molecular logic-based computation has benefited as a result, but also (undeservedly) shared in the trauma of exposed hype at various times.

Hole-burning spectroscopy was another early approach which involved discussions on molecular computation. In the 1980s, Wild's laboratory demonstrated computing functions such as elementary addition on molecular substrates.[41] This was a photochemical method. A laser picked out a few molecules in a specific characteristic microenvironment in a rigid medium on the basis of their characteristic absorption lines, to the accuracy of 0.0001 cm^{-1}. These molecules were converted after photoselection to their tautomeric forms, which would persist. However, the logic functions were not intrinsic to the molecules but arose from optical images which were impressed upon them, for instance.[41] The molecules served as an information storage medium, which is a highly sensitive one at appropriately low temperatures. So this approach does not come within the scope of this book, even though we need to pay homage to this very elegant science (Figure 1.2).[42–45]

Chemists were developing a separate strand of thought after reflecting on biology. This was mentioned in the Pimentel report on science presented to the US House of Congress,[46] which observed that research in molecular computation should be possible given that the brain is a pinnacle of such activity. However we must not forget that each cell possesses vital computational skills of its own.

As far as we are aware, the first claim of an intrinsically molecular-level logic experiment appeared in the conference literature.[47] However, these reports concerning porphyrin molecules have never crossed over into the refereed primary literature in sufficient detail to allow corroboration via the usual scientific process. Additionally, we have been unable to locate any work which followed up the original claims. Birge's NAND gate molecules **4** and **5** were designed to contain two input chromophores to which light can be directed. A retinal Schiff base derivative was used as the output chromophore, whose

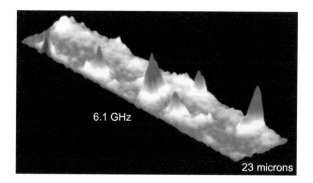

Figure 1.2 Fluorescence excitation of single perylene molecules dispersed in polyethylene film as a function of the probe laser frequency (range 6.1 GHz) and position (range 23 μm).
Reprinted from W. E. Moerner and T. Basche, *Angew. Chem. Int. Ed.*, 1993, **32**, 457 with permission from Wiley-VCH.

absorption band was expected to shift upon electric charging of the porphyrin serving as a charge integrator. The input chromophores were stated to be selectively excitable although their structures do not differ significantly in **4**. The electric charging upon excitation of the input chromophores was considered to arise from photoinduced electron transfers (PET, see section 4.5) in **4** and from internal charge transfer (ICT, see section 4.2) in the excited state of **5**. Proper mechanistic evaluation is difficult in the absence of detailed information. Birge's later statements to the press[48] do not give us much hope concerning these specific cases. However, the subsequent studies by Wasielewski,[49] Gust[50] and Andréasson[51] are distinctly different and employ clear mechanistic designs. These represent progress along this general avenue.

5; R = (structure: NMe₂-C₆H₄-CH=CH-CH=CH-CH=CH-C₆H₄-NO₂)

R' = (structure: NMe₂-C₆H₄-CH=CH-C₆H₄-NO₂)

1.3 Photochemical Approach to Molecular Logic and Computation

It was in 1993 that we published the first general and practical approach to intrinsically molecular logic.[52] This approach, based on ion-driven luminescent signalling systems, grew steadily and without fuss as an activity among chemists and molecular physicists initially. This philosophy forms a substantial fraction of the molecular logic gates currently known. Its use of high-school chemistry fundamentals in combination with basic computer science has appealed to the younger generations of scientists from various backgrounds. The past 19 years[50,51,53–75] have shown how molecules can begin to perform some of the computational functions achieved in semiconductor technology.[76–80] and begin to apply these functions in biological contexts. Notably, the recognition[52] that two-way communication with molecular logic devices can be achieved with chemical species and light signals launched progress down this avenue. It is important to note that Fromherz[81,82] had also seen a connection between pH-dependent fluorescence phenomena and electronic devices.

Small molecules, with their large diversity in structure and properties, have yielded more and more logic functions including small- and medium-scale serial and parallel integration. Arithmetic systems such as half/full adders/subtractors are appearing steadily. We note that different logic functions are being achieved by different molecular structures, rather than by differently connecting copies of a parent structure, which is the semiconductor computing paradigm.[4,76–80] Criticism of the photochemical approach, usually from a semiconductor electronic computing viewpoint,[22,83] has eased in recent years as the biological paradigm – at cellular and organism levels – has become appreciated as an equally valid computing scheme.[84,85] The photochemical approach[52] has also demonstrated strategies seen only in modern semiconductor computing, e.g. concerning reversibly or irreversibly reconfigurable logic systems. Some of these are simultaneously and multiply reconfigurable.[86] The latter cases are superposed logic systems, which are unheard of in conventional semiconductor-based computing and are close to what is dreamt of by the quantum computing community,[87] who are suffering their own traumas concerning previous hype.

6   Chapter 1

Figure 1.3 (a) and (b) Approximate world maps of the sources of molecular logic devices and cases which are interpretable as such. Only the names of corresponding authors from the literature are given. The simplest cases of single-input, single-output binary logic devices are not included. The author will be happy to receive evidence of errors and omissions so that future versions of the maps can be improved.

This basic photochemical approach,[52] though conceived for small molecules, is now being applied to large biomolecules with much success[67] and thus drawing in the molecular biology community. The information handling capability of DNA has never been in doubt,[88,89] but now it is clear that the humblest molecules, synthetic or natural, are achieving similar information handling all around us and even inside.

Adleman's oligonucleotide approach to molecular computation,[90,91] which does not directly involve Boolean logic, caused quite a stir in 1994 owing to the promise of solving problems considered too hard for solution by current semiconductor computers. The natural celebrity of DNA and the opportunity for molecular biologists to get involved in information science also aided its popularity. While this topic lies outside the scope of this book, it is important to note that some subsequent examples of this approach have involved Boolean logical arguments.[92,93] Currently the progress appears to have been tempered following Ellington's critique of this approach.[94,95] This progress is chronicled in a series of conference proceedings.[96] It is telling that the title of the conferences has been modified since 2009 from 'International Meeting on DNA Computing' to 'International Meeting on DNA Computing and Molecular Programming', reflecting the growing influence of molecules other than DNA in computing. Another important trend in recent years is the application of the concepts of small-molecule Boolean logic[52] to DNA and RNA,[97–101] so that there is plenty of momentum but not along the original Adleman avenue. The structural capability of DNA has also permitted a tiling approach to molecular logic and computation.[102]

The discussion above has shown that, as small as they are, molecules have many logic and computing tricks up their sleeves. A world map showing the laboratories involved in this enterprise (Figure 1.3) brings us almost up to the state of the present day.

References

1. R. P. Feynman, *Eng. Sci.*, 1960, **23**, 22. Also available at www.zyvex.com/nanotech/feynman.html.
2. P. Rodgers, *Nature Nanotechnol.*, 2009, **4**, 781.
3. R. Wiesendanger, *Scanning Probe Microscopy and Spectroscopy*, Cambridge University Press, Cambridge, 1994.
4. A. Aviram, *J. Am. Chem. Soc.*, 1988, **110**, 5687.
5. A. Aviram, *Mol. Cryst. Liq. Cryst.*, 1993, **234**, 13.
6. A. Kumar, N. L. Abbott, E. Kim, H. A. Biebuyck and G. M. Whitesides, *Acc. Chem. Res.*, 1995, **28**, 219.
7. H. Shirakawa, *Angew. Chem. Int. Ed.*, 2001, **40**, 2575.
8. A. G. MacDiarmid, *Angew. Chem. Int. Ed.*, 2001, **40**, 2581.
9. A. J. Heeger, *Angew. Chem. Int. Ed.*, 2001, **40**, 2591.
10. K. K. Kanazawa, A. F. Diaz, W. D. Gill, P. M. Grant, G. B. Street and G. P. Gardini, *Synth. Met.*, 1980, **1**, 329.

11. J. R. Reimers, A. Bilic, Z. L. Cai, M. Dahlbom, N. A. Lambropoulos, G. C. Solomon, M. J. Crossley and N. S. Hush, *Aust. J. Chem.*, 2004, **57**, 1133.
12. A. Farazdel, M. Dupuis, E. Clementi and A. Aviram, *J. Am. Chem. Soc.*, 1990, **112**, 4206.
13. J. M. Tour, *Chem. Rev.*, 1996, **96**, 537.
14. J. M. Tour, *J. Org. Chem.*, 2009, **74**, 7885.
15. Z. J. Donhauser, B. A. Mantooth, K. F. Kelly, L. A. Bumm, J. D. Monnell, J. J. Stapleton, D. W. Price, A. M. Rawlett, D. L. Allara, J. M. Tour and P. S. Weiss, *Science*, 2001, **292**, 2303.
16. L. A. Bumm, J. J. Arnold, M. T. Cygan, T. D. Dunbar, T. P. Burgin, L. Jones II, D. L. Allara, J. M. Tour and P. S. Weiss, *Science*, 1996, **271**, 1705.
17. J. Chen, M. A. Reed, A. M. Rawlett and J. M. Tour, *Science*, 1999, **286**, 1550.
18. J. He, Q. Fu, S. Lindsay, J. W. Ciszek and J. M. Tour, *J. Am. Chem. Soc.*, 2006, **128**, 14828.
19. J. L. He, B. Chen, A. K. Flatt, J. J. Stephenson, C. D. Doyle and J. M. Tour, *Nature Mater.*, 2006, **5**, 63.
20. R. F. Service, *Science*, 2003, **302**, 556.
21. P. S. Weiss, *Science*, 2004, **303**, 1137.
22. J. M. Tour, *Acc. Chem. Res.*, 2000, **33**, 791.
23. J. M. Tour, in *Stimulating Concepts in Chemistry*, ed. F. Vogtle, J. F. Stoddart and M. Shibasaki, Wiley-VCH, Weinheim, 2000, p. 237.
24. A. Bachtold, P. Hadley, T. Nakanishi and C. Dekker, *Science*, 2001, **294**, 1317.
25. H. Song, Y. Kim, Y. H. Jang, H. Jeong, M. A. Reed and T. Lee, *Nature*, 2009, **462**, 1039.
26. P. Avouris, Z. H. Chen and V. Perebeinos, *Nature Nanotechnol.*, 2007, **2**, 605.
27. A. K. Geim and K. S. Novoselov, *Nature Mater.*, 2007, **6**, 183.
28. A. Aviram and M. L. Ratner, *Chem. Phys. Lett.*, 1974, **29**, 277.
29. R. M. Metzger, *J. Mater. Chem.*, 2008, **18**, 4364.
30. A. S. Martin, J. R. Sambles and G. J. Ashwell, *Phys. Rev. Lett.*, 1993, **70**, 218.
31. A. Stabel, P. Herwig, K. Mullen and J. P. Rabe, *Angew. Chem. Int. Ed. Eng.*, 1995, **34**, 1609.
32. R. M. Metzger, B. Chen, U. Hopfner, M. V. Lakshmikantham, D. Vuillaume, T. Kawai, X. Wu, H. Tachibana, T. V. Hughes, H. Sakurai, J. W. Baldwin, C. Hosch, M. P. Cava, L. Brehmer and G. J. Ashwell, *J. Am. Chem. Soc.*, 1997, **119**, 10455.
33. C. P. Collier, E. W. Wong, M. Belohradsky, F. M. Raymo, J. F. Stoddart, P. J. Kuekes, R. S. Williams and J. R. Heath, *Science*, 1999, **285**, 391.
34. C. P. Collier, G. Mattersteig, E. W. Wong, Y. Luo, K. Beverly, J. Sampaio, F. M. Raymo, J. F. Stoddart and J. R. Heath, *Science*, 2000, **289**, 1172.

35. J. Jortner and M. Ratner (ed.), *Molecular Electronics*, Blackwell, Oxford, 1997.
36. J. A. Schwarz, C. Contescu and K. Putyera (ed.), *Encyclopedia of Nanoscience and Nanotechnology*, Dekker, New York, 2004.
37. P. Rodgers (ed.), *Nanoscience and Nanotechnology*, World Scientific and MacMillan, Singapore and London, 2010.
38. F. L. Carter (ed.), *Molecular Electronic Devices*, Dekker, New York, 1982.
39. F. L. Carter, R. E. Siatkowski and H. Wohltjen (ed.), *Molecular Electronic Devices*, Elsevier, Amsterdam, 1988.
40. www.nano.gov
41. U. P. Wild, S. Bernet, B. Kohler and A. Renn, *Pure Appl. Chem.*, 1992, **64**, 1335.
42. W. E. Moerner and T. Basche, *Angew. Chem. Int. Ed.*, 1993, **32**, 457.
43. R. Jankowiak, J. M. Hayes and G. J. Small, *Chem. Rev.*, 1993, **93**, 1471.
44. R. Ao, L. Kummerl and D. Haarer, *Adv. Mater.*, 1995, **7**, 495.
45. J. Friedrich, in *Encyclopedia of Spectroscopy and Spectrometry*, ed. J. C. Lindon, G. E. Tranter and J. L. Holmes, Academic, San Diego, 2000, p. 826.
46. G. C. Pimentel, *Opportunities in Chemistry*, National Research Council, National Academy Press, Washington, DC, 1985.
47. R. R. Birge, in *Nanotechnology; Research and Perspectives*, ed. B. C. Crandall and B. C. Lewis, MIT Press, Cambridge, MA, 1992, p. 156.
48. P. Ball, *Nature*, 2000, **406**, 118.
49. A. S. Lukas, P. J. Bushard and M. R. Wasielewski, *J. Am. Chem. Soc.*, 2001, **123**, 2440.
50. D. Gust, T. A. Moore and A. L. Moore, *Chem. Commun.*, 2006, 1169.
51. J. Andréasson and U. Pischel, *Chem. Soc. Rev.*, 2010, **39**, 174.
52. A. P. de Silva, H. Q. N. Gunaratne and C. P. McCoy, *Nature*, 1993, **364**, 42.
53. A. P. de Silva, N. D. McClenaghan and C. P. McCoy, in *Electron Transfer in Chemistry*, ed. V. Balzani, Wiley-VCH, Weinheim, 2001, vol. 5, p. 156.
54. A. P. de Silva, N. D. McClenaghan and C. P. McCoy, in *Molecular Switches*, ed. B. L. Feringa, Wiley-VCH, Weinheim, 2001, p. 339.
55. F. M. Raymo, *Adv. Mater.*, 2002, **14**, 401.
56. V. Balzani, A. Credi and M. Venturi, *ChemPhysChem*, 2003, **4**, 49.
57. A. P. de Silva and N. D. McClenaghan, *Chem. Eur. J.*, 2004, **10**, 574.
58. V. Balzani, A. Credi and M. Venturi, *Nano Today*, 2007, **2**, 18.
59. A. P. de Silva and S. Uchiyama, *Nature Nanotechnol.*, 2007, **2**, 399.
60. R. Ballardini, P. Ceroni, A. Credi, M. T. Gandolfi, M. Maestri, M. Semeraro, M. Venturi and V. Balzani, *Adv. Func. Mater.*, 2007, **17**, 740.
61. U. Pischel, *Angew. Chem. Int. Ed.*, 2007, **46**, 4026.
62. A. Credi, *Angew. Chem. Int. Ed.*, 2007, **46**, 5472.
63. V. Balzani, A. Credi and M. Venturi, *Chem. Eur. J.*, 2008, **14**, 26.
64. K. Szacilowski, *Chem. Rev.*, 2008, **108**, 3481.

65. V. Balzani, M. Venturi and A. Credi, *Molecular Devices and Machines*, Wiley-VCH, Weinheim, 2nd edn, 2008.
66. Y. Benenson, *Mol. Biosyst.*, 2009, **5**, 675.
67. E. Katz and V. Privman, *Chem. Soc. Rev.*, 2010, **39**, 1835.
68. M. Amelia, L. Zou and A. Credi, *Coord. Chem. Rev.*, 2010, **254**, 2267.
69. U. Pischel, *Aust. J. Chem.*, 2010, **63**, 148.
70. P. Ceroni, A. Credi, M. Venturi and V. Balzani, *Photochem. Photobiol. Sci.*, 2010, **9**, 1561.
71. H. Tian, *Angew. Chem. Int. Ed.*, 2010, **49**, 4710.
72. S. Minko, E. Katz, M. Motornov, I. Tokarev and M. Pita, *J. Comput. Theor. Nanosci.*, 2011, **8**, 356.
73. A. P. de Silva, *Chem. Asian J.*, 2011, **6**, 750.
74. A. P. de Silva, T. P. Vance, B. Wannalerse and M. E. S. West, in *Molecular Switches*, ed. B. L. Feringa and W. R. Browne, Wiley-VCH, Weinheim, 2nd edn, 2011, p. 669.
75. G. de Ruiter and M. van der Boom, *Acc. Chem. Res.*, 2011, **44**, 563.
76. A. P. Malvino and J. A. Brown, *Digital Computer Electronics*, Glencoe, Lake Forest, 3rd edn, 1993.
77. J. Millman and A. Grabel, *Microelectronics*, McGraw-Hill, New York, 2nd edn, 1988.
78. J. R. Gregg, *Ones and Zeros*, IEEE Press, New York, 1998.
79. A. L. Sedra and K. C. Smith, *Microelectronic Circuits*, Oxford University Press, Oxford, 5th edn, 2003.
80. C. Maxfield, *From Bebop to Boolean Boogie*, Newnes, Oxford, 2009.
81. P. Fromherz, *Chem. Phys. Lett.*, 1974, **26**, 221.
82. P. Fromherz and W. Arden, *J. Am. Chem. Soc.*, 1980, **102**, 6211.
83. R. S. Williams, *quoted in ref.*, **48**.
84. A. Arkin and J. Ross, *Biophys. J.*, 1994, **67**, 560.
85. D. Bray, *Nature*, 2002, **376**, 307.
86. A. P. de Silva and N. D. McClenaghan, *Chem. Eur. J.*, 2002, **8**, 4935.
87. M. A. Nielsen and I. L. Chuang, *Quantum Computation and Quantum Information*, Cambridge University Press, Cambridge, 2000.
88. G. L. Zubay, W. W. Parsons and D. E. Vance, *Principles of Biochemistry*, Dubuque, Iowa, 1995.
89. J. M. Berg, J. L. Tymoczko and L. Stryer, *Biochemistry*, Freeman, New York, 6th edn, 2006.
90. L. M. Adleman, *Science*, 1994, **266**, 1021.
91. L. M. Adleman, *Sci. Am.*, 1998, **279**, 54.
92. K. Sakamoto, H. Gouzu, K. Komiya, D. Kiga, S. Yokoyama, T. Yokomori and M. Hagiya, *Science*, 2000, **288**, 1223.
93. K. Komiya, K. Sakamoto, A. Kameda, M. Yamamoto, A. Ohuchi, D. Kiga, S. Yokoyama and M. Hagiya, *BioSystems*, 2006, **83**, 18.
94. J. C. Cox, D. S. Cohen and A. D. Ellington, *Trends Biotechnol.*, 1999, **17**, 151.
95. X. Chen and A. D. Ellington, *Curr. Opin. Biotechnol.*, 2010, **21**, 392.

96. *DNA Computing: 13th International Meeting on DNA Computing, DNA13*, Memphis TN, USA, ed. M. H. Garzon and H. Yan, Springer, Berlin, 2008.
97. B. Yurke, A. P. Mills and S. L. Cheng, *Biosystems*, 1999, **52**, 165.
98. M. N. Stojanovic and D. Stefanovic, *Nature Biotechnol.*, 2003, **21**, 1069.
99. G. Seelig, D. Soloveichik, D. Y. Zhang and E. Winfree, *Science*, 2006, **314**, 1585.
100. K. Rinaudo, L. Bleris, R. Maddamsetti, S. Subramanian, R. Weiss and Y. Benenson, *Nature Biotechnol.*, 2007, **25**, 795.
101. B. M. Frezza, S. L. Cockroft and M. R. Ghadiri, *J. Am. Chem. Soc.*, 2007, **129**, 14875.
102. C. D. Mao, T. H. LaBean, J. H. Reif and N. C. Seeman, *Nature*, 2000, **407**, 493.

CHAPTER 2
Chemistry and Computation

2.1 Introduction

First of all, this book is not about computational chemistry. That is the subject that depends on computer programs based on quantum theory[1,2] running on semiconductor-based hardware to provide information about atoms, molecules, materials and reactions.[3] The more recent availability of density functional methods[4] has made such calculations more accessible and useful to experimental chemists. Computational chemistry also includes methods which only require hand calculation to produce powerful insights.[5]

On the other hand, this book is about molecules and chemical systems which possess innate ability to compute, at least in a rudimentary way, like machines based on semiconductor transistors (or magnetic relays or mechanical abacuses in the past) or like people. Although the field is only 19 years old, it already has applications (see Chapter 14) which are uniquely useful.

Chemists have grown up with the slogan 'Chemistry is the central science'. If chemistry meshes more with the spectrum of disciplines, the more central a science it will be. Meshing means, at a minimum, to engage with the concepts of that discipline and, more hopefully, to solve problems in that field. Given the atomic and molecular basis of living matter, it is not surprising that biology and medicine have received large contributions from chemistry. Even civil, mechanical and electronic engineering benefit from chemistry-based materials development. On the other hand, mathematics and computer science have not benefited from chemistry as much. Of course, chemistry has contributed to the material aspects of computer science, *e.g.* the production of pure semiconductors followed by controlled doping. An exception is the chemical manifestation of topology.[6,7] Molecular logic and computation can help fill this gap.

It is clear that the modern information technology revolution[8] has permeated society. However, there is a much older and more vital information revolution. This is the one operating within our genes, cells, nerves and brains.[9] Chemical

---

Monographs in Supramolecular Chemistry No. 12
Molecular Logic-based Computation
By A Prasanna de Silva
© The Royal Society of Chemistry 2013
Published by the Royal Society of Chemistry, www.rsc.org

processes start and drive it, even though its complexities take us into the realms of biology. As the semiconductor-based revolution gradually deals with ever-smaller features, molecules become ever more attractive as information processors. Indeed, small molecules easily go to small vital spaces where semiconductor devices fear to tread. It therefore becomes a responsibility for chemists to explore the information-processing capabilities of molecules.

This should not be difficult, because chemists have been exposed to molecular information processing since high school. Many chemistry experiments involve the exposure of a compound to a reagent and/or heat. Similar operations, perhaps with less quantitation, occur in kitchens around the world several times each day. Physical organic chemists[10,11] consider the key compound to be the substrate. The progress of the reaction is seen by the change of some visible property to that of the product, such as the colour. This response of the compound to the stimulus can be appreciated in a different way[12] (Figure 2.1). To borrow a computer scientist's language, an input is applied to a molecular device so that an output will result.

Of course, the outputs of many simple semiconductor logic devices are continually controlled by the inputs. Therefore, reversible chemical reactions are the best suited to be interpreted in a computational manner. Otherwise such a molecular device would have the output frozen in one state even though the inputs undergo subsequent changes. However, there is another class of simple semiconductor logic devices which can be locked in a given output state. Some of these have been addressed by chemists,[13,14] and we will discuss these in Chapter 11.

Electronic logic devices are characterized by easy application of inputs and equally easy observation of the outputs. We need to have similarly convenient operation of inputs/outputs to and from the molecular logic devices. Chemical inputs are easy to apply in a laboratory context. Optical responses (absorbance, emission intensity, emission anisotropy, ellipticity signals in circular dichroism spectra) serve as outputs since these are the most convenient for quick observation with inexpensive equipment or, sometimes, none at all. Of these, absorbance in the visible region is particularly convenient and visually striking. Emission intensity in the visible region is perhaps even more arresting, nearly as convenient and, furthermore, is detectable at the single molecule level.[15–19]

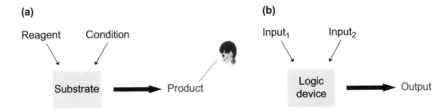

Figure 2.1 (a) Chemical and (b) computational aspects of stimulus–response situations. Adapted from A. P. de Silva and S. Uchiyama, *Nature Nanotechnol.*, 2007, **2**, 399 with permission from Nature Publishing Group.

Of course, other spectroscopic signals arising from products of equilibria (nuclear magnetic resonance – NMR, Raman...) are also usable in this way. All of these circumvent problems of wiring between the molecular and macro worlds. Molecule-based materials also fit here even though they cannot result in single-molecule logic behaviour. Nevertheless, single-molecule logic is not essential for some imagined applications.

Optical inputs, *e.g.* light dose, which cause photochemical reactions, are also useful. 'High' and 'low' signal levels can be taken as binary 1 and 0 respectively (or the opposite) whatever the inputs or outputs under discussion. There are also molecular situations in which electric voltages are used as input and output signals. Much effort has been expended to miniaturize existing electronic logic devices where the semiconductors themselves have been replaced by molecules or at least molecular ensembles.[20,21] Indeed, many other chemical phenomena can be co-opted into molecular logic schemes[12] (Figure 2.2).

While it is true that computation became a household phenomenon through the success of the semiconductor industry, let us not forget that computational ideas were available for a long time before semiconductor transistors and even magnetic relays were invented.[22] The recognition of computational concepts in

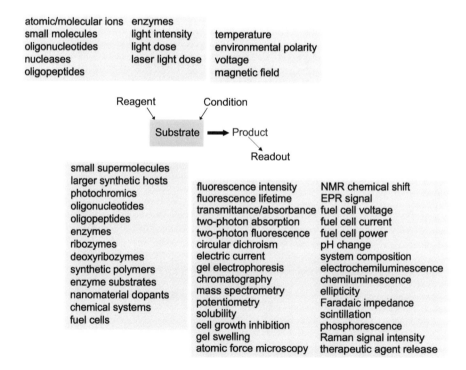

Figure 2.2 The range of inputs, outputs and molecular devices employed in logic operations.
Adapted from A. P. de Silva and S. Uchiyama, *Nature Nanotechnol.*, 2007, **2**, 399 with permission from Nature Publishing Group.

biology, including human behaviour, is a much more recent event.[23–28] Consider for a moment all the computations required inside our heads to recognize someone who approaches us. Many questions are posed about that person's face. Within milliseconds, the answers are obtained by inspection and compared with stored information. If we focus down to the molecular level, oligonucleotides are being regulated in various ways to express proteins. Equally importantly, proteins of various kinds are receiving various chemical inputs and outputting product molecules. Occasionally, a modification of the protein itself becomes the output. Such molecular computational operations keep us alive and well.

2.2 Why Molecules?

Although semiconductors are essential for current computing devices, this book aims to demonstrate that computation with molecules is no less realistic. Semiconductor-based computing depends on electrically wiring multiple copies of a fundamental unit, a transistor or a diode, in various patterns. The same model has been proposed for molecular computation,[29] but no such fundamental unit has yet emerged in a practical sense. In contrast, this book develops the idea that different molecules can be constructed to perform different computational tasks. Although the properties of semiconductors can be adjusted by changing particle size,[30] or by doping bulk materials, the entire repertoire of chemical reactions discovered over the history of synthetic chemistry allows us to access a wide diversity of molecular structures. Since this history has given us a systematic, rational and design-based process, we can have a high degree of control of molecular properties. Molecular devices therefore become the hardware and the design ideas serve as the software in order to produce an algorithmic output. Molecules or supermolecules[31] carrying several modules are especially useful for information handling because, for example, one module might capture a chemical input and another could then be triggered to give an observable output.

Current electronic computing depends critically on integration *via* connectivity, *i.e.* the 0 V or 5 V output from one device is used to supply the input to the next.[32] This is quantitative input–output homogeneity. Therefore we need to consider the issue of connectivity and the relationship between inputs and outputs when nanosized devices such as molecules are under discussion. As mentioned above, an early proposal of molecular electronic logic suggested using wires to connect single diode molecules in different combinations.[29] Connectivity is also essential in biological information processing because cascades down a stream are involved, where molecules are passed from one enzyme to another. However, making the output from one molecular logic gate into the input to the next gate is not so straightforward, because the inputs and outputs can be quite different in nature.

For example, ionic inputs and fluorescence output were involved in the first molecular logic gate.[33] In fact, the first molecular logic devices were unequivocally demonstrable because their inputs and outputs were so different qualitatively that there was no danger of the output feeding back into the input channel and causing a short circuit. When inputs and outputs are quantitatively homogeneous, they must be kept separate by running them along conduits. Such conduits or wires obviously add to the complexity of the complete device. Controversies have also arisen concerning the metal wiring of molecular electronic gates.[34]

Some solutions to the problem of molecular logic device integration, such as functional integration and qualitative input-output homogeneity, will be discussed in later chapters on specific types of logic gates. Notably, most of the molecular logic gates which are described in this book arise from the connectivity of various atoms in different patterns informed by suitable designs.

It is important to note that synthetic molecular computation needs to play to its strengths and operate in situations where semiconductor computing shows weaknesses. In other words, the semiconductor computing model need not be followed in every detail by designers of computing molecules. Although one aim may be to make molecular substitutes for semiconductors within existing logic architectures,[20,21] we are free to follow conceptually different paths. A crucial example is the integration of synthetic logic systems with biological entities, such as living cells, where semiconductor devices lose potency because of issues of size, safety and material incompatibility. Living cells process information concerning internal or external conditions.[23–28] In that case, intracellular species are strictly controlled in terms of concentration, spatial position and time. Problems arise when this condition breaks down in some way, *e.g.* when the concentration deviates from the normal. Then, a synthetic molecular device can respond to logical combinations of these 'high' or 'low' concentrations to release a therapeutic agent[35–37] or to generate a light signal for diagnostics[38] (see Chapter 14).

What about the issue of speed? The rates of diffusion and binding of chemical species limit some molecular logic devices to millisecond time scales. However, this is sufficient to deal with biological processes of similar or longer duration.[28] As mentioned above, the far faster semiconductor devices cannot easily operate in these environments. Thus, speed does not necessarily lead to success, as Aesop illustrated with hares and tortoises.[39] All-optical molecular logic devices are liberated from the diffusion limit and can operate in the nanosecond regime.[40]

Molecular devices, being far smaller than the space-scale that humans operate in, require to be addressed and be communicated with. It is possible to communicate electrically with semiconductor logic devices via metallic wires. This approach is rare in the molecular logic context,[20,21] where most examples are currently handled in different ways. For instance, chemical input signals are applied to molecular devices *via* diffusion in solution. The fluorescence output signal from the molecules easily reaches our eyes.

This human–molecule interface is routinely used in the form of sensors in cell biology. As mentioned at various points in this book, sensors are forms of YES or NOT logic gates. A fine example is Tsien's **1** which measures Ca^{2+} levels,[41] with millisecond and micrometre resolution, within living cells *via* the fluorescence intensity at 510 nm under excitation at 335 nm. More of these compounds will come up for discussion in Chapter 14. Related sensors are observable at the level of single molecules.[15–19] Though not directly visible, levels of H^+ and other chemicals can be similarly deduced from nanometric spaces near membranes by careful positioning of molecular devices employing hydrophobic effects.[42,43] The H^+ concentrations near more complex membranes are critical in bioenergetics.[44] Single molecule addressing[15–19] is not required for some of these purposes if the devices are all in similar environments because they will report the same average information. Of course, the statistics of variability will not be available in that case.

1

We need to also consider the stability of molecular computational devices. Organic molecules are not noted for their longevity. This is in stark contrast to the robustness of inorganic semiconductor materials. Whenever a molecular device is monitored by ultraviolet–visible absorption spectroscopy or even fluorescence spectroscopy, excited states are invariably produced. Since the energies of these excited states are comparable to the strengths of chemical bonds, this can lead to slow photochemical destruction even in the most carefully chosen cases, causing a limit on robustness. Exposure to atmospheric O_2 only adds to the destruction. However, even the ancients noted that life is prone to decay. So the lack of robustness is shared by biology and its building blocks, organic molecules. Computing molecules and cellular systems are naturally suited to each other according to this parameter as well.[28] Their interfacing is therefore a profitable opportunity. Fluorescent molecular logic systems can build on the real-life success of the sensors discussed in the previous paragraph.

However, if we needed to extend the useful readable life of molecular devices, spectroscopies such as optical rotatory dispersion (ORD) and circular dichroism (CD) are available provided that chirality is built-in to the device.[45] A large part of ORD spectra occurs outside the electronic absorption band because absorption is a limited feature on the general dispersion, stretching over a wide wavelength range. Indeed, optical properties such as the refractive index are examples of this dispersion. Thus ORD spectra can be monitored in their long-wavelength wing without populating excited states, thus avoiding the photo-destruction discussed above.

So we see that the common engineering objections raised against molecular computation do not cause problems when the relevance and importance of the biological and chemical arena is acknowledged. Indeed, active molecular devices capable of information processing are not only realistic but useful as well. It is worth noting that such devices were one of the aims of nanotechnology,[46,47] as originally proposed.

2.3 Breadth of Molecular Logic and Computing

Figure 2.2 classifies known molecular logic systems according to various chemical or physical parameters. A range of substrates, reagents, reaction conditions and readout modes can be discerned. Biological entities contain many chemicals which can naturally serve as inputs to synthetic devices embedded therein. Electrical inputs to molecular logic gates can be supplied along wires by conventional semiconductor devices.[8] On the other hand, optical inputs can be delivered remotely. Some of these cases featured in Figure 2.2 need bulk materials, such as electrodes, gels, chromatography-type media, reactors or cuvet assemblies. Although these approaches cannot be used for single molecule detection, this level of sensitivity is not required for many studies and applications. In fact, only a few fluorescent logic gates have been examined at the single molecule level[15–19] under carefully chosen conditions.

2.4 Indicators and Sensors

Historically, chemists have used the term 'indicator' to describe molecules which give optical responses to indicate the conclusion of a titration.[48] Subsequently, the term was also applied to molecules that responded optically to variations in pH or redox state.[48] Engineers have used the term 'sensor' to describe larger devices (nanometric and above) which do similar things. Thoughtful analytical chemists appreciated the molecular interfaces to these larger devices, such as fibre-optic equipment.[49] Thus the indicator was considered to be the front-end of the sensor device. Nowadays, molecules can operate autonomously inside cells with no physical connection to an engineering device. The molecules themselves are the devices.[50] So we see that the engineering idea of 'sensor' can be applied to molecules on their own, since they are sensitive to, and measure properties of, their immediate environment. Then we can see how the temperature and pressure sensors of engineering can have molecular counterparts, *i.e.* molecular thermometers and barometers.

Further generalization[51] shows that, at their binary extremes, 'off–on' sensors are YES logic gates and that 'on–off' sensors are NOT logic gates. Boolean logic can be an umbrella for these and many other phenomena, provided that their analogue response regions are appreciated.[52–56] The connection between sensing and logic will be cemented in the next section.

2.5 The Digital–Analogue Relationship in Chemical Systems

The stimulus–response curve for a large population of molecules is shown in Figure 2.3. The case of the response level increasing with an increasing stimulus level is pictured but the opposite situation is handled equally by the following discussion. Usually, the stimulus level is shown in a logarithmic scale whereas the response level is on a linear scale. As we learned in high school, mass action of molecule populations lies behind the sigmoidal shape of this curve. Generally, this arises from chemical equilibria involving various binding interactions. For simple cases like the dependence of light absorption or emission of a sensory dye on proton concentration, as found in pH indicators,[48] the curve form is well described by the Henderson–Hasselbalch equation (2.1).[57,58]

$$\log[(I_{max} - I)/(I - I_{min})] = \log(\text{stimulus level}) - \log\beta \quad (2.1)$$

where the response level I varies between a maximum (I_{max}) and a minimum (I_{min}) and β is the binding constant for the equilibrium of the sensory dye and the stimulating species. The stimulant concentration where the response is halfway between the minimum and maximum values corresponds to β. Equation (2.1) is stated in a logarithmic form because the log(stimulus level) is frequently stated as –p(stimulus), *e.g.* $-\log(H^+) = pH$.

Indicator or sensor studies are usually carried out in the region of the curve where a small change in the stimulus level causes a corresponding measurable change in the response. From a sensing viewpoint, such quantitation can be seen to be the embellishment of a qualitative judgement of the presence or absence of a species. From a computing viewpoint, this is the analogue region of the curve which can be developed into fuzzy logic operations.[53,54] The extremes of the curve possess two plateaux whose intensities are largely

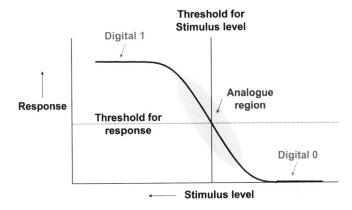

Figure 2.3 Stimulus–response curve for a molecular device.

independent of the stimulus level, at least locally. These extremes clearly display the bistable nature of the system. From a computing standpoint, these plateaux can be viewed as the 'high' (binary 1) and 'low' (binary 0) responses respectively. These together become the digital region. It is important to note that the analogue region would not exist unless the two plateaux differed significantly in their 'high' and 'low' values. In other words, the discrete binary states form the foundation of the analogue sensing region. If we think chemically, the two plateaux arise from two discrete states of a molecule (target-bound and target-free), *i.e.* the binary nature persists.

As mentioned above, the analogue experiments are conducted by making small changes in the stimulus level and measuring the small changes of the response. Digital experiments, such as those establishing logic truth tables, are done by making large excursions in the stimulus level and measuring the (hopefully) large changes of response. The mid-point on the curve, which allowed estimation of β, is also convenient to define the threshold levels for the stimulus level and for the response level (Figure 2.3). When these thresholds are substantially exceeded or undershot the corresponding 'high' or 'low' levels are attained. Approach to within 9% of the maximum or minimum response level (or, more than 91% or less than 9% of whole response level) is considered to be attainment of that level. In fact, the classical indicator range of 2 pH units arises in this way by applying equation 1 to pH equilibria.[48]

Further quantitation of digital switching studies is carried out by venturing into the analogue region so that the β value can be determined. This, as well as the I_{max} and I_{min} values, defines the stimulus–response characteristics of the system. When the molecular system is considered as a device from a computing viewpoint, these values form its input–output characteristics. So we see how closely sensing and switching are related – they are entwined within the same fundamental stimulus–response curve of chemical systems. Furthermore, if these molecules are considered individually, as in modern single molecule studies,[15–19] each one would switch sharply between the 'high' and 'low' levels as the stimulus level is changed. Again, the binary foundations are obvious. Finally, the use of sensing data by managers to give 'go/no go' or 'well/ill' decisions takes us into the binary digital domain again. So, binary logic is present at the beginning, end and throughout sensing operations.

2.6 Molecular Device Characteristics

The stimulus–response curve for the molecular device (Figure 2.3) is therefore a characteristic of its guest-binding behaviour. When we consider the stimulus species as the input and the response is taken as the output, input–output characteristics can also be seen in the current–voltage curves of electronic devices. Non-linearity is key to both systems. There are differences too. Electronic devices have to be circuits where an in-line power supply keeps the current flowing in a particular direction. If the power supply is a battery, as is the case in many mobile applications, the chemical potential stored inside drives the current. The physical electronic symbols, or even the wiring diagrams, in Chapter 3 may not look like circuits but there are always earth connections, for example, which complete circuits through the earth, which are not shown for simplicity's sake. On the other hand, many molecular systems are composed of cascades down a chemical potential gradient. There can be small circuits in some enzyme pathways where some intermediate species are cycled, but there is usually no circuit overall. In some all-optical molecular systems concerned with electronic energy transfer (EET), the cascades occur along an excited state energy gradient.

2.7 Molecular Logic for the Spectrum of Disciplines

Boolean logic is an entire framework for unifying and categorizing input–output or stimulus–response patterns. Though Boolean logic was long-practised by physicists, engineers and mathematicians, now chemists and biochemists have access to molecular phenomena which can benefit from the organization according to this formal framework. Then we can see the relationships between known phenomena and also see where the gaps lie. Some of these phenomena will have specific chemical descriptions but now they can be described in a general way so that non-chemists may take an interest. For instance, elements of cooperativity[59] and co-agonist[60] phenomena can be viewed as AND logic operations. Learning and using the language of other disciplines is a way of taking chemical ideas to non-chemist hearts and minds. Engineers stand to gain from this transfer. The small size and ubiquity of molecules in life processes can help to carry Boolean ideas into biological and medical situations where conventional engineering devices cannot go.

References

1. E. Heilbronner and H. Bock, *The HMO-Model and its Applications*, Verlag Chemie, Weinheim, 1976.
2. G. C. Schatz and M. A. Ratner, *Quantum Mechanics in Chemistry*, Dover, New York, 2002.
3. C. J. Cramer, *Essentials of Computational Chemistry*, Wiley, New York, 2002.
4. W. Koch and M. C. Holthausen, *A Chemist's Guide to Density Functional Theory*, Wiley-VCH, Weinheim, 2nd edn, 2002.
5. M. J. S. Dewar and R. C. Dougherty, *The PMO Theory of Organic Chemistry*, Plenum, New York, 1975.
6. E. Flapan, *When Topology Meets Chemistry*, Cambridge University Press, Cambridge, 2000.
7. Eds. J.-P. Sauvage and C. Dietrich-Buchecker (ed.), *Molecular Catenanes, Rotaxanes and Knots*, VCH, Weinheim, 1999.
8. A. P. Malvino and J. A. Brown, *Digital Computer Electronics*, Glencoe, Lake Forest, 3rd edn, 1993.

9. E. R. Kandel, J. H. Schwartz and T. M. Jessell (ed.) *Principles of Neural Science*, Elsevier, New York, 3rd edn, 1991.
10. L. P. Hammett, *Physical Organic Chemistry*, McGraw-Hill, New York, 2nd edn, 1970.
11. E. V. Anslyn and D. A. Dougherty, *Modern Physical Organic Chemistry*, University Science Books, Mill Valley, CA, 2006.
12. A. P. de Silva and S. Uchiyama, *Nature Nanotechnol.*, 2007, **2**, 399.
13. F. M. Raymo, *Adv. Mater.*, 2002, **14**, 401.
14. D. Margulies, C. E. Felder, G. Melman and A. Shanzer, *J. Am. Chem. Soc.*, 2007, **129**, 347.
15. S. Brasselet and W. E. Moerner, *Single Molec.*, 2000, **1**, 17.
16. L. Zang, R. C. Liu, M. W. Holman, K. T. Nguyen and D. M. Adams, *J. Am. Chem. Soc.*, 2002, **124**, 10640.
17. M. W. Holman and D. M. Adams, *ChemPhysChem*, 2004, **5**, 1831.
18. R. Ameloot, M. Roeffaers, M. Baruah, G. De Cremer, B. Sels, D. De Vos and J. Hofkens, *Photochem. Photobiol. Sci.*, 2009, **8**, 453.
19. M. Elstner, K. Weisshart, K. Müllen and A. Schiller, *J. Am. Chem. Soc.*, 2012, **134**, 8098.
20. C. P. Collier, E. W. Wong, M. Belohradsky, F. M. Raymo, J. F. Stoddart, P. J. Kuekes, R. S. Williams and J. R. Heath, *Science*, 1999, **285**, 391.
21. C. P. Collier, G. Mattersteig, E. W. Wong, Y. Luo, K. Beverly, J. Sampaio, F. M. Raymo, J. F. Stoddart and J. R. Heath, *Science*, 2000, **289**, 1172.
22. G. Boole, *An Investigation of the Laws of Thought*, Dover, New York, 1958.
23. J. Monod, *Chance and Necessity*, Knopf, New York, 1971.
24. D. Bray, *Nature*, 1995, **376**, 307.
25. K. A. Williams, *Nature*, 2000, **403**, 112.
26. R. A. L. Jones, *Nature Nanotechnol.*, 2009, **4**, 207.
27. J. Hasty, D. McMillen and J. J. Collins, *Nature*, 2002, **420**, 224.
28. S. Istrail, S. Ben-Tabou De-Leon and E. H. Davidson, *Developmental Biol*, 2007, **310**, 187.
29. A. Aviram, *J. Am. Chem. Soc.*, 1988, **110**, 5687.
30. X. Michalet, F. F. Pinaud, L. A. Bentolila, J. M. Tsay, S. Doose, J. J. Li, G. Sundaresan, A. M. Wu, S. S. Gambhir and S. Weiss, *Science*, 2005, **307**, 538.
31. J.-M. Lehn, *Supramolecular Chemistry*, VCH, Weinheim, 1995.
32. E. Hughes, *Electrical Technology*, Longman, Burnt Mill, 6th edn, 1990.
33. A. P. de Silva, H. Q. N. Gunaratne and C. P. McCoy, *Nature*, 1993, **364**, 42.
34. J. L. He, B. Chen, A. K. Flatt, J. J. Stephenson, C. D. Doyle and J. M. Tour, *Nature Mater.*, 2006, **5**, 63.
35. R. J. Amir, M. Popkov, R. A. Lerner, C. F. Barbas III and D. Shabat, *Angew. Chem. Int. Ed.*, 2005, **44**, 4378.
36. E. Shapiro and Y. Benenson, *Sci. Am.*, 2006, **294**(5), 44.
37. J. Macdonald, D. Stefanovic and M. N. Stojanovic, *Sci. Am.*, 2008, **299**(5), 84.
38. D. C. Magri, G. J. Brown, G. D. McClean and A. P. de Silva, *J. Am. Chem. Soc.*, 2006, **128**, 4950.

39. O. Temple and R. Temple, *Aesop, The Complete Fables*, Penguin Classics, New York, 1998.
40. A. S. Lukas, P. J. Bushard and M. R. Wasielewski, *J. Am. Chem. Soc.*, 2001, **123**, 2440.
41. G. Grynkiewicz, M. Poenie and R. Y. Tsien, *J. Biol. Chem.*, 1985, **260**, 3440.
42. S. Uchiyama, G. D. McClean, K. Iwai and A. P. de Silva, *J. Am. Chem. Soc.*, 2005, **127**, 8920.
43. S. Uchiyama, K. Iwai and A. P. de Silva, *Angew. Chem. Int. Ed.*, 2008, **47**, 4667.
44. F. M. Harold, *The Vital Force*, Freeman, New York, 1986.
45. Y. C. Zhou, D. Q. Zhang, Y. Z. Zhang, Y. L. Tang and D. B. Zhu, *J. Org. Chem.*, 2005, **70**, 6164.
46. K. E. Drexler, *Engines of Creation*, Anchor Press, New York, 1986.
47. M. A. Ratner and D. Ratner, *Nanotechnology: A Gentle Introduction to the Next Big Idea*, Prentice-Hall, Upper Saddle River, NJ, 2003.
48. E. Bishop (ed.) *Indicators*, Pergamon, Oxford, 1972.
49. K. Cammann, G. G. Guilbault, E. A. H. Hall, R. Kellner and O. S. Wolfbeis, *Proceedings, Cambridge Workshop on Chemical Sensors and Biosensors*, Cambridge University Press, New York, 1996.
50. V. Balzani, M. Venturi and A. Credi, *Molecular Devices and Machines*, Wiley-VCH, Weinheim, 2nd edn, 2008.
51. A. P. de Silva, D. B. Fox and T. S. Moody, in *Stimulating Concepts in Chemistry*, ed. F. Vogtle, J. F. Stoddart and M. Shibasaki, Wiley-VCH, Weinheim, 2000, p. 307.
52. K. P. Zauner and M. Conrad, *Biotechnol. Prog.*, 2001, **17**, 553.
53. P. L. Gentili, *Chem. Phys. A*, 2007, **336**, 64.
54. P. L. Gentili, *J. Phys. Chem. A*, 2008, **112**, 11992.
55. V. Privman, M. A. Arugula, J. Halamek, M. Pita and E. Katz, *J. Phys. Chem. B*, 2009, **113**, 5301.
56. D. Melnikov, G. Strack, M. Pita, V. Privman and E. Katz, *J. Phys. Chem. B*, 2009, **113**, 10472.
57. K. A. Connors, *Binding Constants*, Wiley, New York, 1987.
58. R. A. Bissell, E. Calle, A. P. de Silva, S. A. de Silva, H. Q. N. Gunaratne, J.-L. Habib-Jiwan, S. L. A. Peiris, R. A. D. D. Rupasinghe, T. K. S. D. Samarasinghe, K. R. A. S. Sandanayake and J.-P. Soumillion, *J. Chem. Soc. Perkin Trans.*, 1992, **2**, 1559.
59. A. Whitty, *Nature Chem. Biol.*, 2008, **4**, 435.
60. G. L. Collingridge and J. C. Watkins (ed.), *The NMDA Receptor*, Oxford University Press, Oxford, 2nd edn, 1995.

CHAPTER 3
A Little Logic and Computation

3.1 Introduction

Given the rich textbook literature available on general computer science written by experts,[1–3] only a few thumbnail sketches will be offered from a chemist's viewpoint which outline essential background for what follows. Chemists who take an interest in computer science might find it beneficial to learn some of its language, foundations and ideas.

The year is 1849. The British Queen Victoria launches three Queen's Colleges of Ireland in the cities of Belfast, Galway and Cork. A bright young English mathematician applies for a Professorial position in the Colleges. Belfast and Galway lose him but Cork and the world gain. The mathematician is George Boole. He writes a major work in 1853,[4] expanding and strengthening a shorter piece that he had written earlier,[5] within a few weeks of conceiving the idea that logic could be placed on a mathematical basis.

Logic is essential to sensible conversation because it builds up an argument in a stepwise fashion. It is critical to the conduct of business by thoughtful politicians, lawyers, civic leaders and, of course, scientists and students. The Ancients, particularly Aristotle, set up logic as a classical discipline. However it fell to Boole to analyse everyday language in terms of algebraic variables, relationships and the quantities 0 and 1. Variables could be x, y, ... while relationships could be 'and', 'or', ... The quantities 0 and 1 were the only numbers used within this concept. The assignment of truth as 1 and falsehood as 0 rationalizes the absence of higher decimal numbers. This is the binary basis for many aspects of human information transfer. Information can be handled in these binary units, or bits.

Monographs in Supramolecular Chemistry No. 12
Molecular Logic-based Computation
By A Prasanna de Silva
© The Royal Society of Chemistry 2013
Published by the Royal Society of Chemistry, www.rsc.org

A Little Logic and Computation

3.2 Truth Tables and Algebraic Expressions for Logic Gates

Following Boole's early death in 1864 after exposure to the Cork rain, the baton was taken up by many others to develop various aspects of Boolean algebra.[6-9] The many possible relationships among variables were summarized in terms of number patterns. Truth tables show these patterns particularly clearly. These assumed added value nearly a century later when logic became important in information handling related to the growing discipline of computer science. This should not be surprising because logic was already important in information handling within human language. Because 0 and 1 are the only quantities involved, we can quickly determine how many relationships are possible for a given situation. The variables of interest to computer science/engineering are the data inputs that enter a device and the outputs which leave it. Because these devices control a flow (of information) through them, they are appropriately called gates.

3.2.1 Single Input–Single Output Devices

Discussion of single input–single output devices is a good place to start. The input can exist in only two states *i.e.* 0 or 1. For each of these input states, the device can produce a corresponding output in either of two states, 0 or 1 again. So the total number of input–output patterns is four (2^2) and no more. Let us logically generate these four patterns (Table 3.1) as follows. The input state will be incremented starting at 0. So the first row of input states will go 0, 1 as we move to the right. The four distinguishable output patterns corresponding to this input pattern are given in the next four rows. The output pattern in the second line of Table 3.1 has only 0's. Then we will try patterns containing just a single 1, starting by inserting it at the left. When that digit 1 has reached the end of the row, we have exhausted all of those possibilities. Finally we use two 1's, which take up both the output states available in the last row. The four output patterns that we generated in Table 3.1 can be named PASS 0, NOT, YES and PASS 1 respectively. The PASS 0 logic gate does what it says and only produces a constant 0 output whatever the input state may be. A NOT gate stubbornly says 'no' to any input state that it is confronted with and produces the inverse. On the contrary, a YES gate says 'yes' to the input state in a sycophantic manner and follows the input state exactly. Hence an input string of 01

Table 3.1 Truth tables for single input–single output logic gates.

Input	0	1
Output PASS 0	0	0
Output NOT	1	0
Output YES	0	1
Output PASS 1	1	1

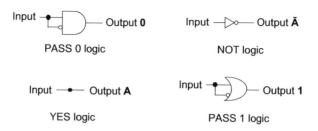

Figure 3.1 The four single-input logic gates. Input is symbolized by the variable A for algebraic purposes throughout. The black spot represents an electrical connection throughout. The symbol for NOT logic can be represented by an open circle when used in association with other gates.

produces an output string of 01. For this reason, a YES gate is also sometimes called EQUIVALENCE or IDENTITY. A PASS 1 gate is also true to its label and only passes a 1 as output whatever the input state is. It is also worth noting that all four logic gates are of equal complexity, from this analysis conducted in the Boolean spirit, even though different values may get attached to each gate at various points in our story.

From an algebraic viewpoint, the outputs of PASS 0, NOT, YES and PASS 1 would be 0, $\bar{A}$, A and 1 respectively, for an input variable A (Figure 3.1). Some readers may find the algebra approach to be the most elegant even when relatively complex gate arrays are being considered. The simplest array to handle a given logic problem can be deduced, either by applying various rules of combinational logic[1-3,6-9] or by resorting to graphical tools such as Karnaugh maps.[1-3,6-9] We will use the latter to check gate arrays in Chapter 10 and to simplify them further where necessary. Software is available for minimizing large gate arrays.[10]

3.2.2 Double Input–Single Output Devices

Next stop: double input–single output devices. Since there are two inputs to consider, each of them can independently exist in any one of two states, 0 or 1. Therefore, the full status of all inputs is represented by both the states of $input_1$ and $input_2$. A maximum of four (2^2) states of inputs can be imagined. For each of these, the output state of the device can be 0 or 1. So the total number of input–output patterns is 16 (4^2). Following the logical development in the previous paragraph, we can work out each of the 16 patterns for double-input, single-output gates (Table 3.2). Figure 3.2 provides additional information about these gates. First, we will develop the full set of input patterns in the first two lines of Table 3.2.

A reasonable start would be to make the pair of $input_1$, $input_2$ a pair of 0's. Then we can generate pairs containing just a single 1, starting by inserting it at the top. When that digit 1 has reached the end of the short column (in blue) representing the two inputs, we have exhausted all of those possibilities. Then we consider two 1's, which take up both the states available in the last short column of inputs (in blue). Once all the input patterns have been found, we can logically develop all 16 output patterns, starting with four 0's in the third line of

A Little Logic and Computation   27

Table 3.2 Truth tables for double input–single output logic gates.

Input$_1$	0	1	0	1
Input$_2$	0	0	1	1
Output PASS 0	0	0	0	0
Output NOR	1	0	0	0
Output INHIBIT(input$_2$)	0	1	0	0
Output INHIBIT(input$_1$)	0	0	1	0
Output AND	0	0	0	1
Output NOT TRANSFER(input$_2$)	1	1	0	0
Output NOT TRANSFER(input$_1$)	1	0	1	0
Output XNOR	1	0	0	1
Output XOR	0	1	1	0
Output TRANSFER(input$_1$)	0	1	0	1
Output TRANSFER(input$_2$)	0	0	1	1
Output NAND	1	1	1	0
Output Reverse IMPLICATION	1	1	0	1
Output IMPLICATION	1	0	1	1
Output OR	0	1	1	1
Output PASS 1	1	1	1	1

Table 3.2. Next, the four cases with a single 1, moving one place to the right in each of these patterns. Following these are six cases with two 1's and after these we have four patterns containing three 1's. As we proceed down the last set, the 0 digit can be seen to move leftwards gradually. Table 3.2 is rounded off with the single pattern of four 1's.

A set of commonly used names for these 16 logic gates is given in Table 3.2. The two gates with constant output, PASS 0 and PASS 1, form the two ends of the table. Among the others, the commutative cases, *i.e.* where swapping the two inputs around does not alter the output pattern, tend to be more popular. The OR gate returns an output of 1 if either input$_1$ or input$_2$ sends in a 1 digit. NOR, shortened from NOT OR, logic therefore produces the opposite number pattern. AND logic produces an output of 1 only if both input$_1$ and input$_2$ are in the state of 1. The NOT AND gate (NAND gate) inverts the pattern of AND logic. Exclusive OR (abbreviated to XOR) logic outputs a digit 1 only if input$_1$ and input$_2$ are in different states. Again XNOR (exclusive NOT OR) logic gives the pattern inverted from that of XOR, *i.e.* it outputs a digit 1 only if input$_1$ and input$_2$ are the same or coincident. For the latter reason, XNOR is also called COINCIDENCE logic. Four pairs of non-commutative logic gates remain. For example, TRANSFER(input$_1$) logic transfers the state of input$_1$ to the output while ignoring input$_2$. In other words, it is a single-input YES gate operating on input$_1$ while the input$_2$ line is left unconnected within the device (Figure 3.2). Understandably, the NOT TRANSFER(input$_1$) gate produces the inverted pattern from that of TRANSFER(input$_1$). Then we have INHIBIT(input$_1$) where the 1 state of input$_1$ disables the output, *i.e.* produces a 0 state. Finally, IMPLICATION(input$_1 \Rightarrow$ input$_2$) is related inversely to INHIBIT(input$_1$), *i.e.* it refers to the IF THEN operation. The algebraic representation of the output from each gate, given an input$_1$ variable A and input$_2$ variable B, is

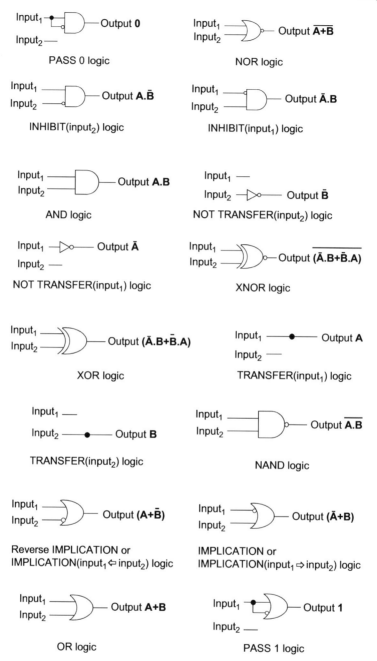

Figure 3.2 The 16 double-input logic gates. $input_1$ and $input_2$ are symbolized by variables A and B for algebraic purposes throughout.

given in Figure 3.2, *e.g.* the outputs from the AND and OR gates are A.B and A + B respectively. Analogous to what we noted under single-input, single-output logic, all these 16 double-input, single-output logic gates are of equal complexity in the Boolean sense. Tables 3.1 and 3.2 will be pulled apart into separate truth tables for the 20 logic gates in subsequent chapters as we deal with molecular versions.

Naturally, many other input and output sets, *e.g.* for triple input–single output devices, can be developed in an analogous manner where necessary.

3.3 Logic Gates in Electronics

The mathematical ideas put forward by Boole have been crystallized into electronic devices within modern computing machines of all kinds. These are the semiconductor-based logic gates which operate with electric voltage, say 5 V, as the binary 1 state and 0 V as the binary 0 state for both the inputs and the outputs. So now we have 'high' and 'low' levels of a physical property standing for Boole's truth and falsehood respectively. It can be noted that the moment we desert mathematics for natural science and engineering phenomena, we come to face to face with fluctuation of properties and errors occurring during their measurement. However, semiconductor-based logic gates are digital electronic systems where the voltage may vary between 3 and 7 V (at least in one common convention) and still be considered as the binary 1 state. Similarly, the binary 0 state will stand for all voltages below 2 V. This leads to a certain robustness of operation because only the 'no man's land' voltage range of 2–3 V will confuse the device. More fundamentally, the binary system, as opposed to a ternary or higher order system, is also maximally robust towards error accumulation during serial operation or during transmission over distances. Herein lies one reason for the success of the Boolean mathematical idea within modern computers.[11]

Chemists frequently employ pictures, cartoons and formulae to convey information without succumbing to the monopoly of text. We have already seen how logic ideas can be presented in terms of truth tables or algebraic expressions. Some readers might find a pictorial representation to be the most appealing of all. In order to represent logic gates pictorially, we will follow the general symbolism employed in modern textbooks on computer science and engineering.[1–3] Figures 3.1 and 3.2 consider single-input, single-output and double-input, single-output gates respectively. Wires are represented by lines. YES logic is considered to be trivial by computer engineers (though not from a molecular viewpoint) and could be any point on a wire. Symbols for single-input NOT, double-input AND and double-input OR gates are fundamental for drawing purposes. These are combined in various ways to develop all other logic gates (and their symbols), which is called the 'sum of products' approach;[1–3,6–9] the name becomes clear when we inspect their Boolean algebraic symbols. However, the XOR gate receives a special symbol to reduce wiring complexity (Figure 3.3). This combinatorial approach is used by

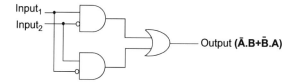

Figure 3.3 The array of NOT, OR and AND gates corresponding to double-input XOR logic.

engineers to build gate arrays of blinding complexity which power the world's semiconductor-based information processors.

Let us look at Figure 3.1 now with an awareness of its electronic possibilities. Even though the non-switching single-input PASS 0 logic is shown here as a specifically wired combination of NOT and AND gates, it behaves as if the output is permanently connected to the earth. Practically, a simpler wiring pattern can produce the required logic behaviour in such a case. A PASS 0 device can have its output directly wired to the earth while the input line is left unconnected. In the case of single-input PASS 1 logic, it is as if the output is permanently connected to the power supply. PASS 1 logic behaviour can be arranged by leaving the input line unconnected whereas the output is directly wired to the power supply. We note that electronic devices require a defined voltage at an output terminal, *e.g.* the power supply voltage (5 V say) or the earth (0 V). An unconnected output terminal on a device would have an undefined voltage and would thus be meaningless in this context.

It is also useful to look within a couple of these electronic logic gates. Although current gates are fashioned from transistors, Figures 3.4 and 3.5 show how AND and OR gates can be assembled in a minimal fashion from even simpler components such as diodes.[12] The latter devices are 'one-way streets' for passage of electric current. More specifically, these possess local electric fields at the interface between n-type and p-type semiconductors (where n- and p- represent the negative and positive majority charge carriers within, say, bulk silicon). These electric fields are the traffic controllers of electric current which can only be supported by a driving voltage applied in one direction.

Let us consider Figure 3.4, representing the AND gate, in some detail. The power supply voltage is identical to the 'high' levels of $input_1$ and $input_2$, *i.e.* 5 V. If $input_1$ is 'low' (0 V, *i.e.* earth) there is 5 V dropping across $diode_1$ (right to left) because the power supply is connected to its right-hand side. Such a positive voltage drop causes the diode to conduct an electric current (limited by the resistor) to the earth via the $input_1$ connection. The current running through the resistor will drop the voltage to nearly 0 V at the junction. This is sensed by the output and it will read 0 V ('low'). The same output state applies if $input_2$ is 'low'. On the other hand if both $input_1$ and $input_2$ are 'high' (5 V), both diodes have 0 V drop across them. So they both cease to conduct. Now the output will clearly sense the voltage (5 V) of the power supply, from which essentially no current is being drawn. Thus the output is in the 'high' state.

A Little Logic and Computation   31

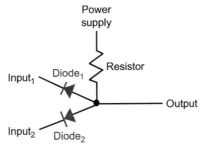

Figure 3.4 Electrical circuit diagram for a diode–diode–logic AND gate. The electrical circuit is completed through appropriate earth connections throughout.

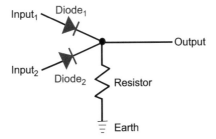

Figure 3.5 Electrical circuit diagram for a diode–diode–logic OR gate.

The previous analysis reveals the integration of several components that is needed for even a rather simple double-input, single-output AND gate, whose symbol in Figure 3.2 draws a veil over all these inner workings.

When we inspect Figures 3.1 and 3.2, it needs to be noted that electronic versions show integrations of 'component' gates, *i.e.* the symbol for NOR logic is constructed by connecting the output of an OR gate to the input of a NOT gate, even though the mathematical discussion showed that NOR and OR logic arise at the same level of complexity within the Boolean analysis. Thus, it should not be a surprise that some gate arrays in Figures 3.1 and 3.2 and elsewhere in this book, even after maximum simplification, may appear rather complex for the equivalent algebraic expression.

3.4 Number Manipulation

Each of us learned our numbers on our mother's knee and then in school. Implicitly we were introduced to the ascending hierarchy of numbers of 0, 1, 2 .... Importantly, our molecule-based brain learned this information for application throughout life. A bit later, we were introduced to the way to sum and carry digits, for example when we tried to add 6 to 7 and we ran out of fingers. Thirteen arose when one of the 10's was carried forward (to the left) and three 1's remained. The latter is the sum digit. Though conventional semiconductor-based calculators or computers deal with only 0 and 1, the concept of sum and carry digits remains. The half-adder is the logic device that accomplishes this by running two binary digits into a parallel array of AND and XOR gates where the carry digit is outputted from the AND gate and the sum digit emerges from the XOR gate (Figure 3.6). The truth table for the half-adder (Table 3.3) can now be interpreted as a composite of AND and XOR logic functions. We note that the symbols '0' and '1' now have the usual mathematical meaning, *i.e.* as integers, besides referring to the states 'low' and 'high' in an engineering sense or the states 'true' and 'false' in a logical sense respectively.

Let us consider each row of Table 3.3 by looking for the mathematical relationship between the numbers represented by the inputs and outputs, by reading the output string as $Output_1 Output_2$. Since zeros to the left of a number are valueless, let us also add a zero to the left of each input number. Inspection shows that the following relationships hold true, when we remember that 10 in the binary number system is the decimal number 2 and that 01 in the binary number system is the decimal number 1.

$$00 + 00 = 00 \quad \text{or} \quad 0 + 0 = 0 \tag{3.1}$$

$$00 + 01 = 01 \quad \text{or} \quad 0 + 1 = 1 \tag{3.2}$$

$$01 + 00 = 01 \quad \text{or} \quad 1 + 0 = 1 \tag{3.3}$$

$$01 + 01 = 10 \quad \text{or} \quad 1 + 1 = 2 \tag{3.4}$$

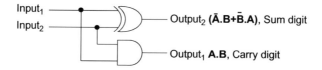

Figure 3.6 The array of XOR and AND gates corresponding to the half-adder.

Table 3.3 Truth table for the half-adder.

$Input_1$	$Input_2$	$Output_1$	$Output_2$
0	0	0	0
0	1	0	1
1	0	0	1
1	1	1	0

The pairs of equations (3.1)–(3.4) demonstrate number addition in the binary and decimal systems, where the former is what Table 3.3 teaches us and the latter is what we learned from our parents. We will defer further discussion about arithmetic until we reach Chapter 9.

References

1. J. Millman and A. Grabel, *Microelectronics*, McGraw-Hill, New York, 2nd edn, 1988.
2. A. P. Malvino and J. A. Brown, *Digital Computer Electronics*, Glencoe, New York, 3rd edn, 1993.
3. A. L. Sedra and K. C. Smith, *Microelectronic Circuits*, Oxford University Press, Oxford, 5th edn, 2003.
4. G. Boole, *An Investigation of the Laws of Thought*, Dover, New York, 1958.
5. G. Boole, *The Mathematical Analysis of Logic*, G. Bell, London, 1847.
6. M. Ben-Ari, *Mathematical Logic for Computer Science*, Prentice-Hall, Hemel Hempstead, 1993.
7. J. R. Gregg, *Ones and Zeros*, IEEE Press, New York, 1998.
8. C. Maxfield, *From Bebop to Boolean Boogie*, Newnes, Oxford, 2009.
9. F. M. Brown, *Boolean Reasoning*, Dover, New York, 2nd edn, 2003.
10. R. K. Brayton, G. D. Hachtel, C. T. McMullen and A. L. Sangiovanni-Vincentelli, *Logic Minimization Algorithms for VLSI Synthesis*, Kluwer, Dordrecht, 1984.
11. R. W. Keyes, *Rev. Mod. Phys.*, 1989, **61**, 279.
12. E. Hughes, *Electrical Technology*, Longman, Burnt Mill, 6th edn, 1990.

CHAPTER 4
A Little Photochemistry and Luminescence

4.1 Introduction

The field of molecular logic-based computation has been blessed with the continual addition of experimental techniques from across the sciences (see Figure 2.2). However, the bulk of the examples have benefited from strategies based on light absorption and fluorescence, as well as other forms of luminescence. This is because such optical techniques allow us to see what we are doing and to show others what is being done in the molecular domain. Such human–molecule communication is vital, especially when non-scientists' hearts and minds need to be won. The fact that single molecules are now detectable via their fluorescence[1] is a reminder of how sensitively light signals can be detected by modern instruments and even the human eye. Absorption and emission spectrometers are workhorses of analytical laboratories and are widely available. They tend to be rather inexpensive too, especially if cut-down versions of the spectrometers are chosen for dedicated application. When nanoscale systems of biology or materials science are being interrogated with fluorescent molecules, the degree of perturbation of the host is therefore minimal. On the other hand, the excited state of the fluorescent guest molecule is easily perturbed by its immediate environment and hence the fluorescence signal can be an effective reporter. Since the excited state lifetimes of fluorescent molecules are typically rather short (*e.g.* 10 ns), the time-resolution of such interrogations is sufficient for even the fastest bioprocesses. Fluorescence lifetime measurements allow self-calibrating results which are immune to several experimental variables. Normal and confocal fluorescence microscopy permits routine examination of small biorelevant spaces.[2] The space-resolution of such experiments is usually controlled not by the molecular size (*ca.* 1 nm) but by the

---

Monographs in Supramolecular Chemistry No. 12
Molecular Logic-based Computation
By A Prasanna de Silva
© The Royal Society of Chemistry 2013
Published by the Royal Society of Chemistry, www.rsc.org

classical Abbe limit (*ca.* the wavelength of light). Recent years, however, have seen this limit being rolled back with the advent of super-resolution techniques,[3,4] at least with the sacrifice of time-resolution.

So it becomes beneficial for us to look briefly at the photochemical foundations[5–43] of absorption and fluorescence spectroscopies from a molecular switching viewpoint. Then we can design molecules, quantitatively and predictively in some cases, to carry out switching operations which can then be built up into logic systems. Remarkably, a relatively small number of principles[9,28,33,44,45] make up these photochemical foundations.

4.2 Excited States Involving Charge Transfer

A fluorophore can be directly integrated with a receptor[46] (Figure 4.1) so that their orbitals overlap. Partial charge transfer within an excited fluorophore is a consequence of the mixed atomic composition of most molecular π-systems. Internal charge transfer (ICT) is most clearly seen in 'push–pull' π-systems which have electron-donating and electron-withdrawing groups at opposite ends.[28,33,47,48] Such excited states have very different electric dipole moments when compared with their corresponding ground states. The energies of these ICT excited states can be perturbed by monopolar or dipolar species (Figure 4.2), to produce shifts in emission or absorption wavelengths.[49,50] Spectral

Figure 4.1 The integrated 'fluorophore–receptor' format of fluorescent ICT systems.

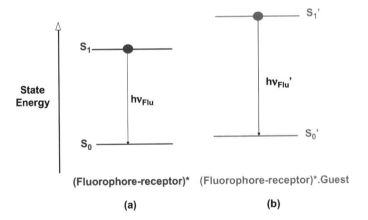

Figure 4.2 State energy diagrams of the 'fluorophore–receptor' system in (a) the guest-free situation and (b) the guest-bound situation.

shifts allow analysis by calculating the ratio of intensities at two separated wavelengths, which is a useful analytical method when complex matrices, e.g. intracellular environments,[51–54] are examined. Many fluorescent pH indicators[55] are now understandable as ICT systems, e.g. **1**. A special case of dipolar species perturbing ICT excited states is seen in the spectral red-shifts induced by polar solvents,[6,28,33,41] where a conventional receptor is not required. These dipolar species can even take the form of neighbouring amino acid residues in the tertiary structure of a protein when they influence the emission spectrum of an unnatural amino acid **2**, which has been incorporated at a chosen specific site.[56]

Intensity changes are not predicted directly by the ICT model and whatever is seen experimentally is usually moderate or minor. So order-of-magnitude changes of the emission quantum yield are found in ICT-type systems only if a bond torsion path is available to deconjugate donor and acceptor segments of the fluorophore to complete the charge separation.[57] These twisted internal charge transfer (TICT) excited states have features that are equivalent in many ways to an electron transfer discussed in section 4.5 under PET systems.[46] Fluorescence switching 'off' of **3**[58] is selective for Hg^{2+} in this way. The boron-dipyrromethine fluorophore and the 8-hydroxyquinoline receptor in **3** are nearly orthogonal.

Charge transfer (CT) phenomena also occur when the two partners are separate π-systems if they can take up a sandwich configuration.[59] While there is a classical literature on this aspect,[60] our emphasis is on its exploitation within supramolecular systems.[61] The simplest expression of CT behaviour is the emergence of a new red-shifted absorption band. This happens at 404 nm, for instance, when tryptophan enters the cavity of **4**.[62]

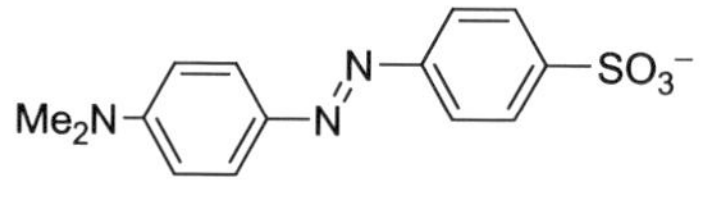

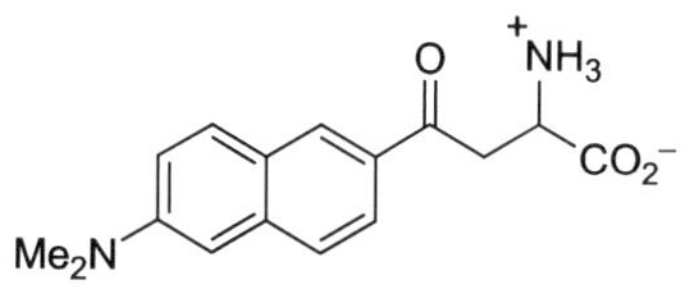

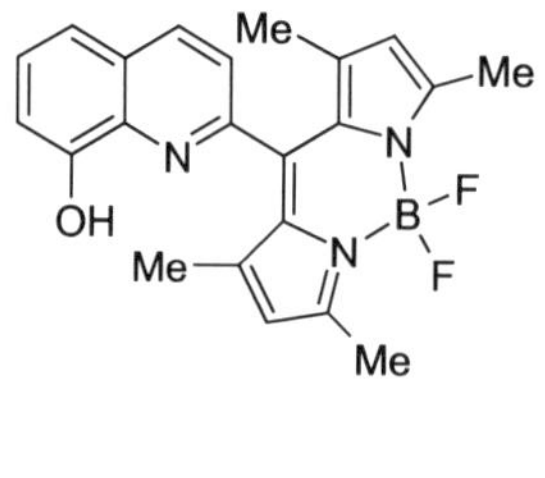

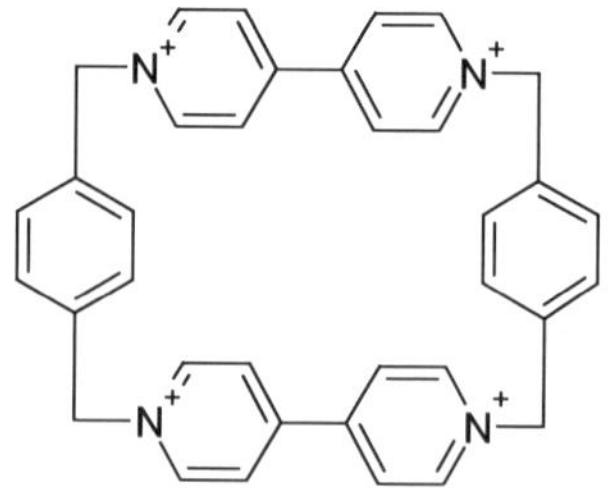

ICT-type phenomena within metal-containing systems are most commonly seen in low oxidation state metal centres with π-acid ligands. These form the metal-to-ligand charge transfer (MLCT) category of excited states.[63,64]

Reference works on inorganic photochemistry[12,22,23] are available for those wishing to study this topic further. Polypyridyl-Ru(II) and -Re(I), with their metal-to-ligand charge transfer (MLCT) excited states, are the most popular examples.[12]

4.3 Metal-centred (MC) Excited States

These excited states are most clearly seen in lanthanide complexes[65–70] whose atomic luminescence has the appeal of rather long lifetimes and line-like emissions, which allow easy extraction from contaminant fluorescence signals. A nice example is Gunnlaugsson's **5** which succeeds in imaging microcracks in bone[71] (see section 14.2.1 for further discussion of related cases). Sames' group reveal a case where an esterase causes luminescence enhancement of **6** by a factor of 9 when it hydrolyses to **7**.[72] Since the organic chromophore of these compounds need to be excited, its triplet energy level determines the efficiency of electronic energy transfer (EET) to populate the lanthanide centre before emission. The degree of ICT within the chromophore is greater in the case of **6** owing to the extra electron withdrawing power of the ester, *c.f.* **7**. Presumably this will preferentially lower the excited state energy of **6**. Hence, better sensitization of emission is seen with the hydrolysis product **7**. The corresponding Eu^{3+} complexes show no such bias owing to their far lower excited state energies, allowing their use for calibration of the emission change discussed above.

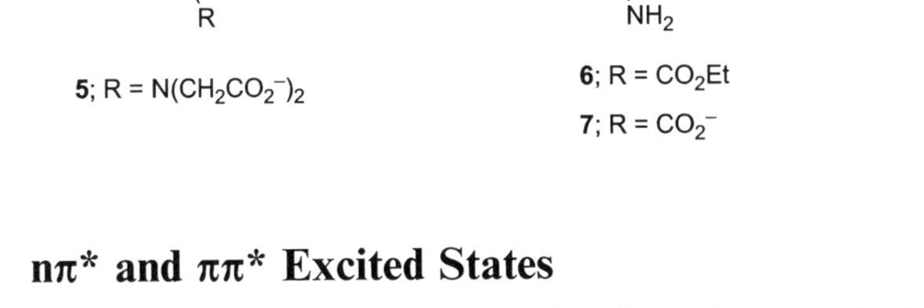

5; R = $N(CH_2CO_2^-)_2$

6; R = CO_2Et

7; R = CO_2^-

4.4 nπ* and ππ* Excited States

Switching of the lowest excited state-type within a fluorophore can also be very useful. Fluorophores whose lowest singlet excited states are of the nπ* type are usually poorly emissive under ambient conditions. Such states have a small radiative rate constant because of the poor overlap in space between n and π* orbitals. Perhaps more importantly, such singlet states are also accompanied by a ππ* triplet state at slightly lower energy, which allows rapid intersystem crossing according to El-Sayed's rules.[44] A designed binding of the n electron pair can raise the energy of the nπ* singlet to be above that of the more emissive ππ* singlet state so that the emission is enhanced. The emission of **8** at 515 nm is enhanced 8-fold by Pb^{2+} in this way.[73]

8; n = 3

4.5 Photoinduced Electron Transfer (PET)

An electron can be fully transferred from/to an excited fluorophore to/from a suitable electron system of a partner[21,30,74,75] such as a receptor for a chosen guest. 'Fluorophore–spacer–receptor' systems (Figure 4.3) are of this type,[45,46,76–80] where the spacer is typically a short aliphatic unit but it can be virtual or even absent.[81] The nature of the spacer can control the path of PET,[82,83] and its length governs the rate of PET.[84,85] The frontier orbital representation of the thermodynamic situation[86] is shown in Figure 4.4a. The kinetic situation is also understandable.[87,88] Redox potentials[89] can be used to choose components so that the receptor will transfer an electron to/from the excited fluorophore. If electrochemical data are unavailable or unreliable, the frontier orbital energies obtained by quantum mechanical calculation[90–105] can be used towards the same end. This means that fluorescence is suppressed. Binding of a guest species can change the redox potential of the receptor so that the electron transfer is no longer thermodynamically feasible, and the excited state energy of the fluorophore is therefore released as a humanly visible emission (Figure 4.4b). Guest-induced fluorescence enhancement arises in this way. The quantitative design and the predictability of many observable parameters are notable. All absorption spectral data and the fluorescence wavelength remain essentially unchanged except when ICT fluorophores are employed.[106] Even then, the changes are relatively minor. Guest-induced

Figure 4.3 The 'fluorophore–spacer–receptor' format of fluorescent PET systems.

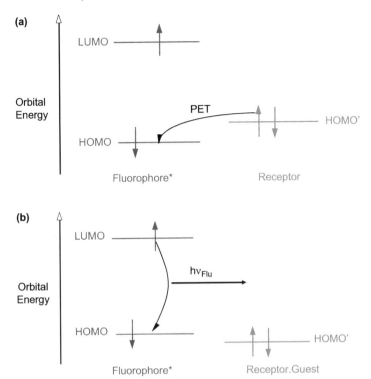

Figure 4.4 Molecular orbital energy diagrams of the frontier orbitals of the fluorophore and the receptor within the 'fluorophore–spacer–receptor' system in (a) the guest-free situation and (b) the guest-bound situation. The electron occupation in the excited fluorophore is shown.

fluorescence reduction is similarly designable. Furthermore, various forms of luminescence, such as phosphorescence, can also be accommodated. Maps (Figure 4.5a and 4.5b) summarize the worldwide activity in these systems.

Nagamura's experiments with **9**,[107,108] with its heteroaromatic fluorophore, methylene spacer and azacrown receptor for Sr^{2+}, support these ideas. Laser flash photolysis[108] of **9**[107] directly shows PET from the azacrown nitrogen to the fluorophore. Supporting evidence is available in the immeasurably fast fluorescence decay seen for **9** which increases to 230 ps upon binding Sr^{2+}. Transient absorption spectra of the fluorophore component **10** show the excited singlet state characterized by bands at 525 and 750 nm. This evolves to the excited triplet state (band at 460 nm). A similar but faster evolution is seen for **9**, along with the production of the fluorophore radical anion (band at 610 nm). Similar observations of fluorophore radical anions arising from PET systems are available.[105] Binding of Sr^{2+} to **9** slows and weakens these phenomena.

Figure 4.5 Approximate world maps of sources of fluorescent PET sensors. Only the names of corresponding authors from the literature are given. The author will be happy to receive evidence of errors and omissions so that future versions of the map can be improved.

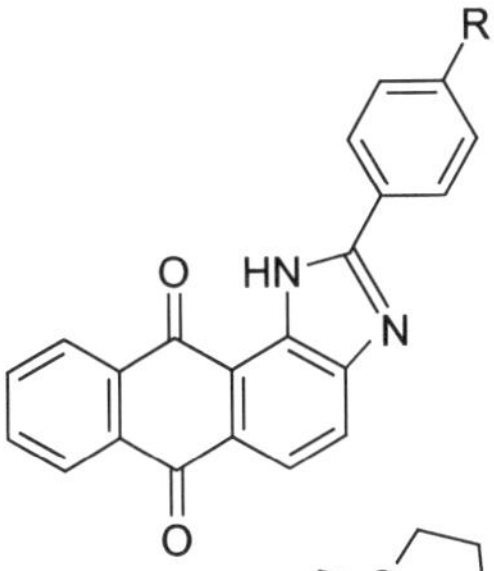

9; R = [structure: N with two CH2CH2-O-CH2CH2-O groups forming aza-crown]

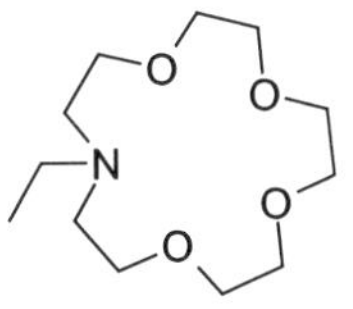

10; R = t-Bu

4.6 Electronic Energy Transfer (EET)

EET was touched upon in section 4.3, but now it takes centre stage. Electronic energy transfer[28,33,44,109] requires two partners – an energy donor and an energy acceptor – so that energy can run downhill. Since each excited state can emit its own signature, we have a two-colour output system where the intensities at the two wavelengths can be ratioed.[52–54,110] EET does not require electric concepts at the simplest phenomenological level. EET systems have a set of advantages and disadvantages complementary to those based on ICT and PET. Indeed EET is widely used in biochemical circles under the acronym FRET (Forster/fluorescence resonance energy transfer) even though resonance is only one of three distinct mechanisms of EET. Inter-partner distances of around 5 nm are particularly amenable to EET-based experiments concerning larger biomolecules. Of course, EET efficiency drops off with increasing distance. The amount of overlap between the absorption spectrum of the acceptor and the emission spectrum of the donor also controls the efficiency of EET. Since the 5 nm distance is rather larger than those commonly encountered in the smaller systems of chemistry, strong sensory/switching outputs are obtained mainly in chemical systems where full association/dissociation occurs so that the two relevant distances become <1 nm and >100 nm (on average, for a 10^{-6} M solution) respectively. Particularly strong examples using EET for logic purposes are available from Stojanovic's laboratory[111] and will be discussed in several following chapters.

4.7 Excimers and Exciplexes

Excited state association processes[44] have been known for over 50 years.[6] Excited states possess half-filled orbitals which can interact attractively with other ground-state orbitals. This leads to π–π overlap and these delocalized states show up in red-shifted and broad emission signatures. The latter are due to the instability of the corresponding ground states, *i.e.* extremely short lifetimes. Heisenberg's uncertainty principle then leads to an ill-defined energy. These features are well illustrated by Prodi's use of **11** to detect Zn^{2+} by induction of an excimer emission.[112] Excimers and exciplexes are particularly useful for our purposes when in intramolecular or pseudo-intramolecular formats. Then the design requires geometric flexibility of the spacer between the lumophore and its partner. Indeed the flexibility requirement for the passage of the extended conformation to the folded counterpart to allow excimer formation leads to a useful dependence on solvent viscosity. Excimers, which arise from two identical copies of a fluorophore, are better suited for this role because of their lack of electric effects compared with exciplexes. The latter, which are formed from an excited fluorophore and another electron-rich or -poor partner, behave as frustrated PET systems. These show substantial charge separation which leads to a dependence on solvent polarity.

11

4.8 Vibrational Deexcitation and Excited State Intramolecular Proton Transfer (ESIPT)

Structural rigidity has always been considered in textbooks to be an important requirement for efficient fluorescence.[9] While this remains true, our emphasis here is on its detrimental effect on emission efficiency because of the vibrations offered by bound water molecules. In organic structures, the water could be bound to an ICT pole, thereby opening a route to a strengthening or weakening of the hydrogen bond upon excitation. Strong coupling of this vibrator to the fluorophore can lead to almost complete quenching of emission.[113] This would correspond to a Born–Oppenheimer hole.[10,25] Even stronger coupling of water oscillators to lumophores is seen with lanthanide ions, notably Eu^{3+}.[65,66,114] Parker's **12** is such an example, with bicarbonate-induced displacement of water oscillators leading to 2.3-fold luminescence enhancement at 618 nm.[115] In some organic systems, an intramolecular phenolic O–H can take the place of the water vibrator with much more potency. Some of these couple so well that an excited state intramolecular proton transfer (ESIPT) is found. Most of these lead to quenching of the emission, even though new emission bands can also arise.[41,116] A designed disruption of this fluorophore–vibrator coupling can lead to an effective mechanism for emission enhancement.

A Little Photochemistry and Luminescence 43

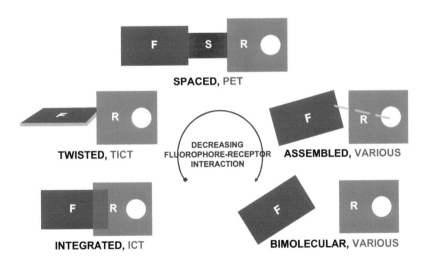

12; R = N-alanyl

4.9 Relationships between some of the Photochemical Principles used in Switching

The previous sections have outlined how various photochemical principles give rise to the switching of fluorescence signals. One way of appreciating the relationships between some of these principles is given in Figure 4.6.[46] Module connectivity – the geometric relationship between the fluorophore and the spacer – is shown here. The connectivity goes from being tightly integrated to none at all, *via* intermediate situations such as twisted, spaced and assembled cases. The major photochemical principle involved in each case is also noted in Figure 4.6. Some switch designs mix several photochemical mechanisms. For instance, a fluorescent PET system may employ an ICT fluorophore.[106]

Figure 4.6 The spectrum of fluorophore–receptor interactions within fluorescent switching systems. The photochemical principle involved in each situation is also mentioned.
Redrawn from R. A. Bissell, A. P. de Silva, H. Q. N. Gunaratne, P. L. M. Lynch, G. E. M. Maguire and K. R. A. S. Sandanayake, *Chem. Soc. Rev.*, 1992, **21**, 187, with permission from the Royal Society of Chemistry.

Before concluding, it is necessary to note that structural rigidification,[117] paramagnetism and heavy atom effects[44] are also occasionally used in some switch designs.

An example of an assembled switching system is the fluorophore **13** and Cu^{2+}-receptor **14**.[118] Hydrophobic effects drive detergent molecules such as **15** to form micellar aggregates in water.[119] These effects also push **13** and **14** to locate in the micelles. Electrostatic attraction also aids this push. Even though Cu^{2+} will face repulsion from the micelle surface, it will be captured by the polar ligating groups of **14**. A deprotonation of the amide unit is also involved at that point. Now the Cu^{2+} centre is rather close to the fluorophore and quenches the fluorescence via EET and PET.

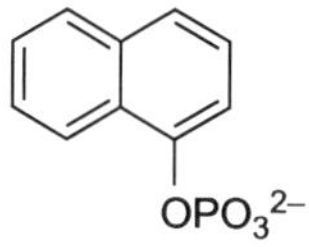

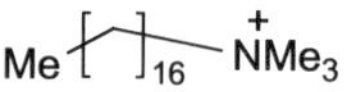

References

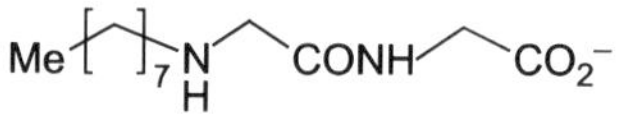

1. C. Gell, D. Brockwell and D. A. Smith, *Handbook of Single Molecule Spectroscopy*, Oxford University Press, Oxford, 2006.
2. D. L. Taylor and Y. L. Wang (ed.), *Fluorescence Microscopy of Living Cells in Culture. Parts A and B*, Academic, San Diego, 1989.
3. I. Amato, *Chem. Eng. News*, 2006, **84**(36), 49.
4. M. Heilemann, *J. Biotechnol.*, 2010, **149**, 243.
5. H. H. Jaffe and M. Orchin, *Theory and Applications of Ultraviolet Spectroscopy*, Wiley, New York, 1962.
6. J. B. Birks, *Photophysics of Aromatic Molecules*, Wiley, London, 1970.
7. R. P. Wayne, *Principles and Applications of Photochemistry*, Oxford University Press, Oxford, 1988.
8. B. M. Krasovitskii and B. M. Bolotin, *Organic Luminescent Materials*, VCH, Weinheim, 1989.
9. G. G. Guilbault, *Practical Fluorescence*, Dekker, New York, 2nd edn, 1990.
10. J. Michl and V. Bonačić-Koutecký, *Electronic Aspects of Organic Photochemistry*, Wiley, New York, 1990.
11. R. J. Hurtubise, *Phosphorimetry. Theory, Instrumentation and Applications*, VCH, Cambridge, 1990.
12. V. Balzani and F. Scandola, *Supramolecular Photochemistry*, Ellis-Horwood, Chichester, 1991.

13. V. Ramamurthy (ed.), *Photochemistry in Organized and Constrained Media*, VCH, New York, 1991.
14. O. S. Wolfbeis (ed.), *Fibre Optic Chemical Sensors and Biosensors*, CRC Press, Boca Raton, vols 1–2, 1991.
15. D. L. Wise and L. B. Wingard (ed.), *Biosensors with Fiberoptics*, Humana Press, Clifton, 1991.
16. I. Hemmila, *Applications of Fluorescence in Immunoassays*, Wiley, New York, 1991.
17. J. Kopecky, *Photochemistry. A Visual Approach*, VCH, New York, 1992.
18. K. Kalyanasundaram, *Photochemistry of Polypyridine and Porphyrin Complexes*, Academic, London, 1992.
19. A. W. Czarnik (ed.), *Fluorescent Chemosensors of Ion and Molecule Recognition*, American Chemical Society, Washington DC, 1993.
20. S. G. Schulman (ed.), *Molecular Luminescence Spectroscopy*, Wiley, New York, Parts 1–3, 1985–1993.
21. G. J. Kavarnos, *Fundamentals of Photoinduced Electron Transfer*, VCH, Weinheim, New York, 1993.
22. O. Horvath and K. L. Stevenson, *Charge Transfer Photochemistry of Coordination Compounds*, VCH, New York, 1993.
23. D. M. Roundhill, *Photochemistry and Photophysics of Metal Complexes*, Plenum, New York, 1994.
24. J. R. Lakowicz (ed.), *Topics in Fluorescence Spectroscopy*, Plenum, New York, vols 1–4, 1991–1994.
25. M. Klessinger and J. Michl, *Excited States and Photochemistry of Organic Molecules*, VCH, New York, 1995.
26. J.-P. Desvergne and A. W. Czarnik (ed.), *Chemosensors of Ion and Molecule Recognition*, Kluwer, Dordrecht, 1997.
27. W. Rettig, B. Strehmel, S. Schrader and H. Seifert (ed.), *Applied Fluorescence in Chemistry, Biology and Medicine*, Springer, Berlin, 1999.
28. B. Valeur and M. N. Berberan-Santos, *Molecular Fluorescence*, Wiley-VCH, Weinheim, 2nd edn, 2012.
29. B. Valeur and J.-C. Brochon (ed.), *New Trends in Fluorescence Spectroscopy*, Springer, Berlin, 2001.
30. V. Balzani (ed.), *Electron Transfer in Chemistry*, Wiley-VCH, Weinheim, vols 1–5, 2001.
31. V. Ramamurthy and K. S. Schanze (ed.), *Molecular and Supramolecular Photochemistry*, Dekker, New York, vols 1–10, 1997–2003.
32. H. S. Nalwa (ed.), *Handbook of Photochemistry and Photobiology*, American Scientific Publishers, Stevenson Ranch, CA, 2003.
33. J. R. Lakowicz, *Principles of Fluorescence Spectroscopy*, Springer, New York, 3rd edn, 2006.
34. M. Montalti, A. Credi, L. Prodi and M. T. Gandolfi (ed.), *Handbook of Photochemistry*, CRC Press, Boca Raton, FL, 3rd edn, 2006.
35. J. R. Lakowicz (ed.), *Topics in Fluorescence Spectroscopy*, Plenum, New York and Springer, Berlin, vols 1–11, 1992–2006.

36. J. R. Albani, *Principles and Applications of Fluorescence Spectroscopy*, Wiley-Blackwell, 2007.
37. B. L. Feringa and W. R. Browne (ed.), *Molecular Switches,* Wiley-VCH, Weinheim, 2nd edn, 2011.
38. V. Balzani, A. Credi and M. Venturi, *Molecular Devices and Machines*, VCH, Weinheim, 2nd edn, 2008.
39. O. S. Wolfbeis (ed.), *Springer Series on Fluorescence*, Springer, Berlin, vols 1–6, 2000–2008.
40. E. M. Goldys (ed.), *Fluorescence Applications in Biotechnology and Life Sciences*, Wiley-Blackwell, London, 2009.
41. A. P. Demchenko, *Introduction to Fluorescence Sensing*, Springer, New York, 2009.
42. C. D. Geddes and J. R. Lakowicz (ed.), *Reviews in Fluorescence*, Kluwer, New York, vols 1–5, 2004–2010.
43. Invitrogen, *The Molecular Probes Handbook*, 11th edn (www.invitrogen.com/site/us/en/home/References/Molecular-Probes-The-Handbook.html).
44. N. J. Turro, V. Ramamurthy and J. C. Scaiano, *Modern Molecular Photochemistry of Organic Molecules*, University Science Books, Mill Valley, CA, 2010.
45. A. P. de Silva, H. Q. N. Gunaratne, T. Gunnlaugsson, A. J. M. Huxley, C. P. McCoy, J. T. Rademacher and T. E. Rice, *Chem. Rev.*, 1997, **97**, 1515.
46. R. A. Bissell, A. P. de Silva, H. Q. N. Gunaratne, P. L. M. Lynch, G. E. M. Maguire and K. R. A. S. Sandanayake, *Chem. Soc. Rev.*, 1992, **21**, 187.
47. B. Valeur, in *Molecular Luminescence Spectroscopy*, ed. S. G. Schulman, Wiley, New York, 1993, Part 3, p. 25.
48. B. Valeur, in *Topics in Fluorescence Spectroscopy. Probe Design and Chemical Sensing*, ed. J. R. Lakowicz, Plenum, New York, 1994, vol. 4, p. 21.
49. M. M. Martin, P. Plaza, Y. H. Meyer, F. Badaoui, J. Bourson, J. P. Lefebvre and B. Valeur, *J. Phys. Chem.*, 1996, **100**, 6879.
50. J.-F. Létard, R. Lapouyade and W. Rettig, *Pure Appl. Chem.*, 1993, **65**, 1705.
51. R. Y. Tsien, *Chem. Eng. News*, 1994, **72**(29), 34.
52. A. P. Demchenko, *Lab Chip*, 2005, **5**, 1210.
53. A. P. Demchenko, *Trends Biotechnol.*, 2005, **23**, 456.
54. A. P. Demchenko, *Anal. Biochem.*, 2005, **343**, 1.
55. G. F. Kirkbright, in *Indicators*, ed. E. Bishop, Pergamon, Oxford, 1972, p. 685.
56. B. E. Cohen, T. B. McAnaney, E. S. Park, Y. N. Jan, S. G. Boxer and L. Y. Jan, *Science*, 2002, **296**, 1700.
57. W. Rettig, *Top. Curr. Chem.*, 1994, **169**, 253.
58. S. Y. Moon, N. R. Cha, Y. H. Kim and S. K. Chang, *J. Org. Chem.*, 2004, **69**, 181.
59. C. J. Bender, *Chem. Soc. Rev.*, 1986, **15**, 317.

60. R. Forster, *Organic Charge Transfer Complexes*, Academic, London, 1969.
61. V. Balzani, A. Credi, F. M. Raymo and J. F. Stoddart, *Angew. Chem. Int. Ed.*, 2000, **39**, 3349.
62. T. T. Goodnow, M. V. Reddington, J. F. Stoddart and A. E. Kaifer, *J. Am. Chem. Soc.*, 1991, **113**, 4335.
63. G. A. Crosby, *Acc. Chem. Res.*, 1975, **8**, 231.
64. Q. Zhao, F. Y. Li and C. H. Huang, *Chem. Soc. Rev.*, 2010, **39**, 3007.
65. F. H. Richardson, *Chem. Rev.*, 1982, **82**, 541.
66. D. Parker and J. A. G. Williams, *J. Chem. Soc. Dalton Trans.*, 1996, 3613.
67. C. P. Montgomery, B. S. Murray, E. J. New, R. Pal and D. Parker, *Acc. Chem. Res.*, 2009, **42**, 925.
68. A. P. de Silva, H. Q. N. Gunaratne and T. E. Rice, *Angew. Chem. Int. Ed. Engl.*, 1996, **35**, 2116.
69. A. P. de Silva, H. Q. N. Gunaratne, T. E. Rice and S. Stewart, *Chem. Commun.*, 1997, 1891.
70. K. Hanaoka, K. Kikuchi, H. Kojima, Y. Urano and T. Nagano, *Angew. Chem. Int. Ed.*, 2003, **42**, 2996.
71. B. McMahon, P. Mauer, C. P. McCoy, T. C. Lee and T. Gunnlaugsson, *J. Am. Chem. Soc.*, 2009, **131**, 17542.
72. M. S. Tremblay, M. Halim and D. Sames, *J. Am. Chem. Soc.*, 2007, **129**, 7570.
73. M. Kadarkaraisamy and A. G. Sykes, *Inorg. Chem.*, 2006, **45**, 779.
74. M. A. Fox and M. Chanon, ed., *Photoinduced Electron Transfer*, Elsevier, Amsterdam, 1988.
75. J. Mattay (ed.), *Photoinduced Electron Transfer*, in *Top. Curr. Chem.*, 1990, **156**, **158**; 1991, **159**; 1992, **163**; 1993, **168**; 1994, **169**.
76. A. J. Bryan, A. P. de Silva, S. A. de Silva, R. A. D. D. Rupasinghe and K. R. A. S. Sandanayake, *Biosensors*, 1989, **4**, 169.
77. R. A. Bissell, A. P. de Silva, H. Q. N. Gunaratne, P. L. M. Lynch, G. E. M. Maguire, C. P. McCoy and K. R. A. S. Sandanayake, *Top. Curr. Chem.*, 1993, **168**, 223.
78. A. P. de Silva, D. B. Fox, T. S. Moody and S. M. Weir, *Pure Appl. Chem.*, 2001, **73**, 503.
79. A. P. de Silva, T. S. Moody and G. D. Wright, *Analyst*, 2009, **134**, 2385.
80. C. Lodeiro, J. L. Capelo, J. C. Mejuto, E. Oliveira, H. M. Santos, B. Pedras and C. Nunez, *Chem. Soc. Rev.*, 2010, **39**, 2948.
81. A. P. de Silva, C. M. Dobbin, T. P. Vance and B. Wannalerse, *Chem. Commun.*, 2009, 1386.
82. M. N. Paddon-Row, in *Stimulating Concepts in Chemistry*, ed. F. Vogtle, J. F. Stoddart and M. Shibasaki, Wiley-VCH, Weinheim, 2000, p. 267.
83. J. W. Verhoeven, *Pure Appl. Chem.*, 1990, **62**, 1585.
84. M. Onoda, S. Uchiyama, T. Santa and K. Imai, *Luminescence*, 2002, **17**, 1.
85. J. C. Beeson, M. A. Huston, D. A. Pollard, T. K. Venkatachalam and A. W. Czarnik, *J. Fluoresc.*, 1993, **3**, 65.

86. A. Weller, *Pure Appl. Chem.*, 1968, **16**, 115.
87. R. A. Marcus, *J. Chem. Phys.*, 1956, **24**, 966.
88. R. A. Marcus, *J. Chem. Phys.*, 1956, **24**, 979.
89. H. Siegerman, in *Techniques of Electroorganic Synthesis. Part II*, ed. N. L. Weinberg, Wiley, New York, 1975.
90. K. Rurack, J. L. Bricks, A. Kachkovski and U. Resch, *J. Fluoresc.*, 1997, **7**, 63S.
91. K. Rurack, J. L. Bricks, B. Schulz, M. Maus and U. Resch-Genger, *J. Phys. Chem. A*, 2000, **104**, 6171.
92. S. Uchiyama, T. Santa and K. Imai, *Analyst*, 2000, **125**, 1839.
93. C. J. Fahrni, L. C. Yang and D. G. VanDerveer, *J. Am. Chem. Soc.*, 2003, **125**, 3799.
94. A. Chatterjee, T. M. Suzuki, Y. Takahashi and D. A. P. Tanaka, *Chem. Eur. J*, 2003, **9**, 3920.
95. Y. Gabe, Y. Urano, K. Kikuchi, H. Kojima and T. Nagano, *J. Am. Chem. Soc.*, 2004, **126**, 3357.
96. P. Yan, M. W. Holman, P. Robustelli, A. Chowdhury, F. I. Ishak and D. M. Adams, *J. Phys. Chem. B*, 2005, **109**, 130.
97. M. E. McCarroll, Y. Shi, S. Harris, S. Puli, I. Kimaru, R. S. Xu, L. C. Wang and D. J. Dyer, *J. Phys. Chem. B*, 2006, **110**, 22991.
98. J. Cody, S. Mandal, L. C. Yang and C. J. Fahrni, *J. Am. Chem. Soc.*, 2008, **130**, 13023.
99. Y. Ooyama, H. Egawa and K. Yoshida, *Dyes Pigm.*, 2009, **82**, 58.
100. X. Zhang, L. N. Chi, S. M. Ji, Y. B. Wu, P. Song, K. L. Han, H. M. Guo, T. D. James and J. Z. Zhao, *J. Am. Chem. Soc.*, 2009, **131**, 17452.
101. I. D. Petsalakis, I. S. K. Kerkines, N. N. Lathiotakis and G. Theodorakopoulos, *Chem. Phys. Lett.*, 2009, **474**, 278.
102. T. Kowalczyk, Z. L. Lin and T. Van Voorhis, *J. Phys. Chem. A*, 2010, **114**, 10427.
103. A. Samanta, B. K. Paul, S. Jana and N. Guchhait, *Photochem. Photobiol.*, 2010, **86**, 1022.
104. S. J. Lee, H. C. Chen, Z. Q. You, K. L. Liu, T. J. Chow, I. C. Chen and C. P. Hsu, *Mol. Phys.*, 2010, **108**, 2775.
105. L. Flamigni, B. Ventura, A. Barbieri, H. Langhals, F. Wetzel, K. Fuchs and A. Walter, *Chem. Eur. J.*, 2010, **16**, 13406.
106. A. P. de Silva, H. Q. N. Gunaratne, P. L. M. Lynch, A. L. Patty and G. L. Spence, *J. Chem. Soc. Perkin Trans.*, 1993, **2**, 1611.
107. K. Yoshida, T. Mori, S. Watanabe, H. Kawai and T. Nagamura, *J. Chem. Soc. Perkin Trans.*, 1999, **2**, 393.
108. H. Kawai, T. Nagamura, T. Mori and K. Yoshida, *J. Phys. Chem. A*, 1999, **103**, 660.
109. B. Valeur, in *Fluorescent Biomolecules: Methodologies and Applications*, ed. D. M. Jameson and G. D. Reinhart, Plenum, New York, 1989, p. 269.
110. R. Guliyev, A. Coskun and E. U. Akkaya, *J. Am. Chem. Soc.*, 2009, **131**, 9007.
111. J. Macdonald, D. Stefanovic and M. N. Stojanovic, *Sci. Am.*, 2008, **299**, 84.

112. L. Prodi, R. Ballardini, M. T. Gandolfi and R. Roverai, *J. Photochem. Photobiol. A Chem.*, 2000, **136**, 49.
113. M. D. P. de Costa, A. P. de Silva and S. T. Pathirana, *Can. J. Chem.*, 1987, **65**, 1416.
114. D. Parker, *Chem. Soc. Rev.*, 2004, **33**, 156.
115. Y. Bretonniere, M. J. Cann, D. Parker and R. Slater, *Org. Biomol. Chem.*, 2004, **2**, 1624.
116. W. Kloppfer, *Adv. Photochem.*, 1977, **10**, 311.
117. H. S. Jung, K. C. Ko, J. H. Lee, S. H. Kim, S. Bhuniya, J. Y. Lee, Y. M. Kim, S. J. Kim and J. S. Kim, *Inorg. Chem.*, 2010, **49**, 8552.
118. P. Grandini, F. Mancin, P. Tecilla, P. Scrimin and U. Tonellato, *Angew. Chem. Int. Engl.*, 1999, **38**, 3061.
119. Y. Moroi, *Micelles: Theoretical and Applied Aspects*, Springer, Berlin, 1992.

CHAPTER 5

Single Input–Single Output Systems

5.1 Introduction

Now that we are armed with the background information given in the previous four chapters, we are ready to commence our journey. Table 3.1 in Chapter 3 collected the truth tables of all the possible single-output patterns when triggered by a single input. The four logic gates that emerged in this way are now discussed in turn with regard to their molecular implementation.

5.2 YES

YES logic requires the output to follow the input. Many of the chemical cases to be discussed below represent 'off–on' switching of the output of some description, *i.e.* the application of a 'high' input signal enhances the output signal.

5.2.1 Electronic Input

Getting electrons in and out of a molecule (or a small population of them) while preserving its integrity[1] is not trivial. Much effort is being expended to examining suitable interfaces to the molecule(s).[2–5] An all-electronic YES gate simply corresponds to any point along a wire. Single molecular electronic implementation of a wire[6] is accomplished by planting **1** *via* its thiol moiety at a grain boundary of a self-assembled monolayer of **2** on a gold surface and then contacting the distal terminus with the tip of a scanning tunnelling microscope.[7]

Monographs in Supramolecular Chemistry No. 12
Molecular Logic-based Computation
By A Prasanna de Silva

Published by the Royal Society of Chemistry, www.rsc.org

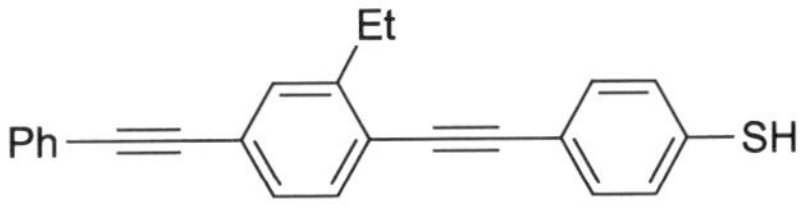

5.2.2 Chemical Input

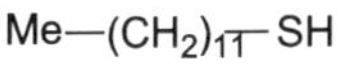

Cases with non-electronic input can be viewed as sensors or switches when the input species interacts reversibly with the molecular device. If appropriate codings are used, any bistable system with an easily recognizable physical output would fall into this category. There are many examples of such behaviour,[8–27] most of which have been developed for other applications by their original authors and therefore have not been recognized in this logical context. In order to keep this chapter within reasonable bounds, only a selection are summarized in Table 5.1 and only a representative number are discussed in the text. Many of the molecular examples presented here can be immobilized on materials such as mesoporous silica,[28] if desired.

5.2.2.1 Cation Input

The classical literature of absorptiometric and fluorescent reagents for protons and metal ions in analytical chemistry is a particularly rich source[8–10] of examples of this type.

5.2.2.1.1 Proton Input. Protons may be the simplest of chemical species, but they can be a convenient proving ground for many a concept. For instance, individual fluorescent sensor molecules at work[65] can be observed by confocal microscopy. Adams' **3**[66] is essentially a 'fluorophore–spacer–receptor' system. PET from the anilinic unit to the perylenetetracarboximide leads to negligible fluorescence. Protonation of the anilinic unit gives a strong fluorescence enhancement (Figure 5.1). A few more H^+-driven cases are discussed below. Several other interesting cases are summarized in Table 5.1.

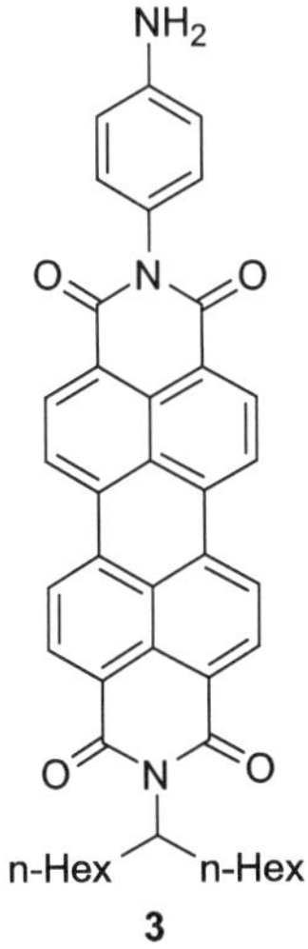

Table 5.1 Collected data for some molecular YES logic gates.

Device[a]	Input[b] (hi,lo)[c]	Output[d] (hi,lo)[e]	Power[f]	Characteristics[g]
4[9] in H_2O	$H^+(10^{-2},10^{-7})$	$Abs_{blue}(>10,1)$	$h\nu_{blue}$	$\log\beta_{H+}$ 3.1[29]
5[30] in MeOH:H_2O	$H^+(10^{-2},10^{-9})$	$Flu_{480}(19,1)$	$h\nu_{385}$ logε 4.3	$\log\beta_{H+}$ 4.1
"	$Na^+(10^{-1},0)$	$Flu_{480}(9,1)$	"	$\log\beta_{Na+}$ 2.8
6[31] in H_2O	$H^+(10^{-2},10^{-9})$	$Flu_{489}(120,1)$	$h\nu_{388}$ logε 4.4	$\log\beta_{H+}$ 5.7,3.0
"	$Ca^{2+}(10^{-3},0)$	$Flu_{489}(92,1)$	"	$\log\beta_{Ca2+}$ 6.5
7[32] in MeCN:CH_2Cl_2	$H^+(10^{-5},0)$	$Flu_{337}(7,1)$	$h\nu_{275}$ logε 4.7	$\log\beta_{H+}$ 7.4,7.0[h]
8[34] in Dioxan[j]	$Li^+(10^{-1.4},0)$	$Abs_{586}(23,1)$	$h\nu_{586}$ logε 4.6	$\log\beta_{Li+}$ 8.3[i]
9[35] in MeCN	$Li^+(10^{-3},0)$	$Flu_{432}(3,1)$	$h\nu_{381}$ logε 3.9	$\log\beta_{Li+}$ 5.1
10[36] in MeCN	$Li^+(10^{-3.3},0)$	$Flu_{337}(9,1)$	$h\nu_{280}$	$\log\beta_{Li+}$ 5.4
11[37] in MeOH	$Na^+(10^{-2},0)$	$Flu_{449}(15,1)$	$h\nu_{368}$ logε 3.9	$\log\beta_{Na+}$ 3.0
12[38] in MeOH	$Na^+(10^{-2},0)$	$Flu_{532}(7,1)$	$h\nu_{461}$ logε 4.3	
13[39] in MeOH	$Na^+(10^{-1.3},0)$	$Flu508(37,1)$	$h\nu_{508}$ logε 5.1	$\log\beta_{Na+}$ 2.5
14[40] in H_2O	$Na^+(10^{0.5},0)$	$Flu_{525}(5,1)$	$h\nu_{506}$ logε 4.8	$\log\beta_{Na+}$ 0.4[k]
15[41] in H_2O	$K^+(10^{-0.8},0)$	$Flu_{500}(2.5,1)$	$h\nu_{424}$	$\log\beta_{K+}$ 0.6
16[42] in MeCN	$Mg^{2+}(10^{-4},0)$	$Flu_{454}(200,1)$	$h\nu_{365}$ logε 4.2	$\log\beta_{Mg2+}$ 5.3
17[43,44] in H_2O	$Ca^{2+}(10^{-3},0)$	$Flu_{575}(24,1)$	$h\nu_{547}$ logε 4.9	$\log\beta_{Ca2+}$ 6.1
18[45] in H_2O	$Ca^{2+}(10^{-4.4},0)$	$Flu_{782}(3.4,1)$	$h\nu_{760}$ logε 5.3	$\log\beta_{Ca2+}$ 6.6
19[46] in H_2O[l]	$Ca^{2+}(10^{-3},0)$	$Flu_{598}(32,1)$	$h\nu_{579}$ logε 5.0	$\log\beta_{Ca2+}$ 4.5
20[47] in H_2O	$Zn^{2+}(10^{-2.8},0)$	$Flu_{545}(50,1)$	$h\nu_{442}$ logε 4.3	$\log\beta_{Zn2+}$ 3.9
"	$H^+(10^{-2},10^{-12})$	$Flu_{545}(100,1)$	"	$\log\beta_{H+}$ 3.1
21[48] in H_2O	$Zn^{2+}(10^{-2.2},0)$	$Flu_{468}(12,1)$[m]	$h\nu_{381}$ logε 4.2	$\log\beta_{Zn2+}$ 3.8
"	$Cd^{2+}(10^{-2.2},0)$	$Flu_{506}(16,1)$[m]	"	$\log\beta_{Cd2+}$ 4.2
22[49] in H_2O	$Zn^{2+}(10^{-8},0)$	$Flu_{529}(2.3,1)$	$h\nu_{511}$ logε 4.9	$\log\beta_{Zn2+}$ 9.1
23[50] in MeCN:H_2O	$Zn^{2+}(10^{-7},0)$	$Flu_{514}(22,1)$	$h\nu_{360}$ logε 4.9	$\log\beta_{Zn2+}$ 8.2
24[51] in H_2O	$Zn^{2+}(10^{-8},0)$	$Flu_{304}(13,1)$	$h\nu_{270}$ logε 3.9	
25[52] in H_2O	$Cd^{2+}(10^{-4.4},0)$	$Flu_{454}(19,1)$	$h\nu_{356}$ logε 4.0	$\log\beta_{Cd2+}$ 4.6
26[53] in Dioxan:H_2O	$Hg^{2+}(10^{-4.7},0)$	$Flu_{536}(8,1)$	$h\nu_{492}$ logε 4.4	
27[54] in MeOH:H_2O	$Hg^{2+}(10^{-4.4},0)$	$Flu_{476}(12,1)$	$h\nu_{423}$ logε 4.7	$\log\beta_{Hg2+}$ >10
28[55] in H_2O	$Hg^{2+}(10^{-5.5},0)$	$Flu_{524}(5.4,1)$	$h\nu_{500}$ logε 4.8	$\log\beta_{Hg2+}$ 6.4

Table 5.1 (*Continued*)

Device[a]	Input[b](hi,lo)[c]	Output[d](hi,lo)[e]	Power[f]	Characteristics[g]
29[56] in EtOH:H$_2$O	Hg^{2+}(10$^{-4.8}$,0)	Flu$_{556}$(17,1)	hν_{415} logε 4.3	logβ_{Hg2+} 5.2
30[57] in MeCN	Hg^{2+}(10^{-3},0)	Flu$_{440}$(2.5,1)	h$\nu_{\sim 300}$ logε 4.1	logβ_{Hg2+} 5.7
31[58,59] in MeOH	Cu$^+$(10$^{-4.9}$,0)	Flu$_{436}$(50,1)	hν_{350}	
32[60] in H$_2$O	Cu^{2+}(10$^{-5.3}$,0)	Flu$_{770}$(10,1)	hν_{745}	logβ_{Cu2+} 5.6
33[61] in DMF	Ni^{2+}(10^{-4},0)	Flu$_{609}$(49,1)	hν_{440}	logβ_{Ni2+} 9.4[n]
34[61] in DMF	Fe^{3+}(10^{-5},0)	Flu$_{609}$(138,1)	hν_{440}	
35[62] in MeOH	Fe^{3+}(10^{-4},0)	Flu$_{510}$(>500,1)	hν_{470} logε 4.4	logβ_{Fe3+} 4.8
36[63] in THF	PdCl$_2$(10$^{-5.3}$,0)	Flu$_{414}$(2.5,1)	hν_{375} logε 4.3	logβ_{PdCl2} 5.4
R-37[64] in MeOH:H$_2$O	Glucose^o(10^{-1},0)	Flu$_{358}$(4.0,1)	hν_{289}	log$\beta_{Glucose}$ 3.3
"	Glucose^p(10^{-1},0)	Flu$_{358}$(2.0,1)	"	log$\beta_{Glucose}$ 3.1

[a]The device is taken as a given molecular structure in a given medium.
[b]Chemical concentration unless noted otherwise.
[c]High (hi) and low (lo) input values. All concentrations are in M units.
[d]Output is given as the intensity of fluorescence (or the absorbance) at a given wavelength. All wavelengths are in nm units.
[e]High and low output values. The 'low' output value observed is scaled to unity.
[f]Power supply is light at a given wavelength. The extinction coefficient (ε) of the device at that wavelength is given in units of M^{-1}cm^{-1}.
[g]Characteristics of device operation regarding inputs include chemical binding constants (β) in units of M^{-1}. Characteristics of device operation regarding outputs include emission quantum yields (ϕ).
[h]Exciplex, excimer and EET interactions are also present. Similar switching in a non-dendrimeric case is available.[33]
[i]Value for model receptor in wet CHCl$_3$.
[j]With 20% (v/v) H$_2$O.
[k]Quantum chemical calculations and cyclic voltammetry data are also available.
[l]The structure also carries a ω-azidoalkyl chain branching from the central diaryloxyethane unit of the Ca^{2+} receptor and a chloro instead of the methyl substituent on the aryl unit.
[m]Exciplex emission.
[n]2:1 (**33**:Ni^{2+}) stoichiometry found by fluorescence tiration.
[o]D-enantiomer.
[p]L-enantiomer.

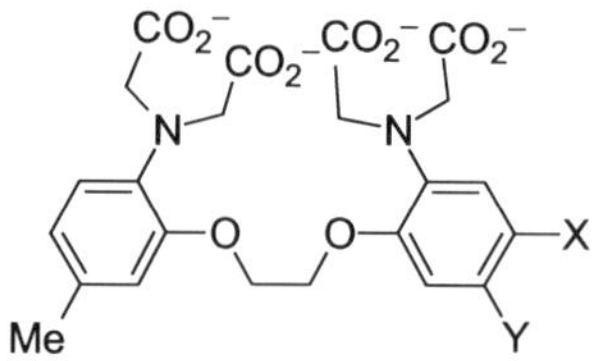

6; X = H, Y =

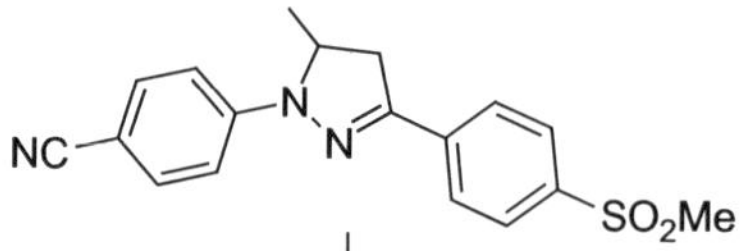

17; X = H, Y =

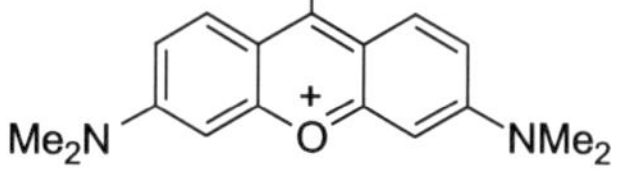

18; Y = H, X =

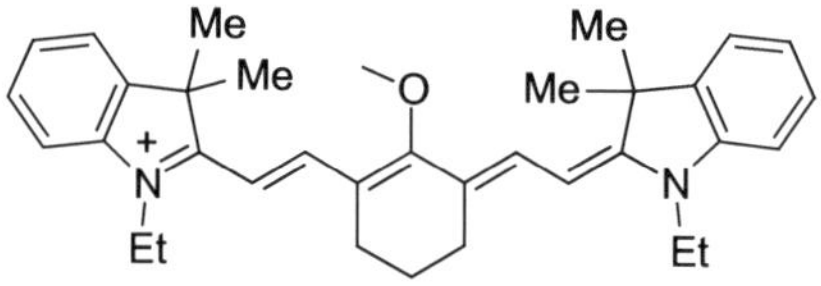

19; X = H, Y =

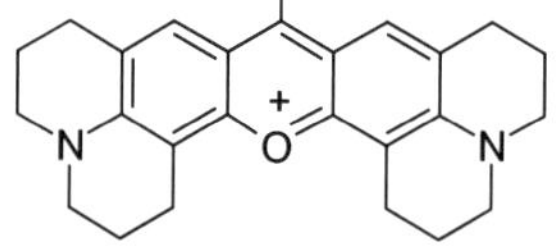

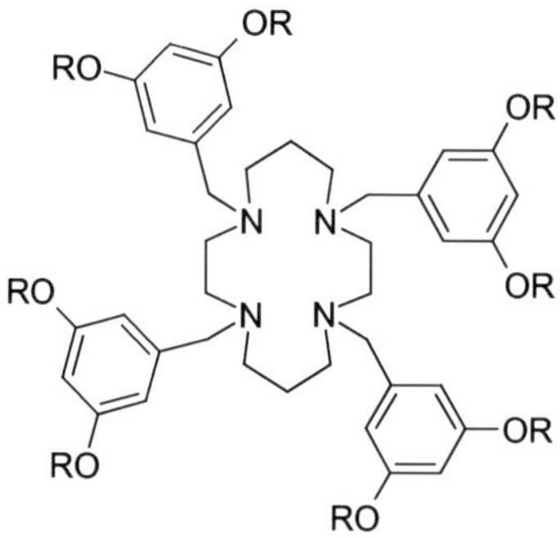

7, R = 2-naphthylmethyl

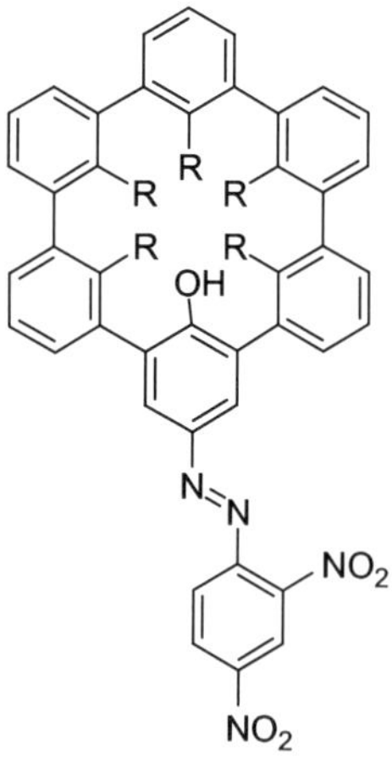

8; R = OMe

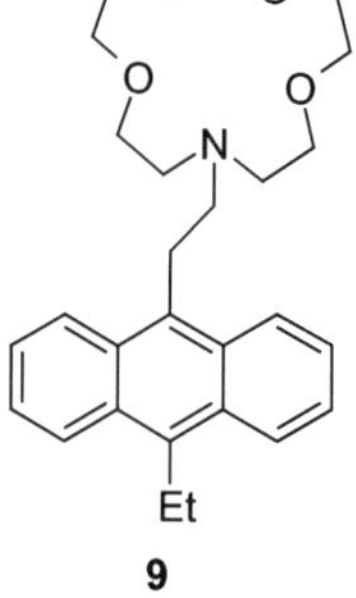

9

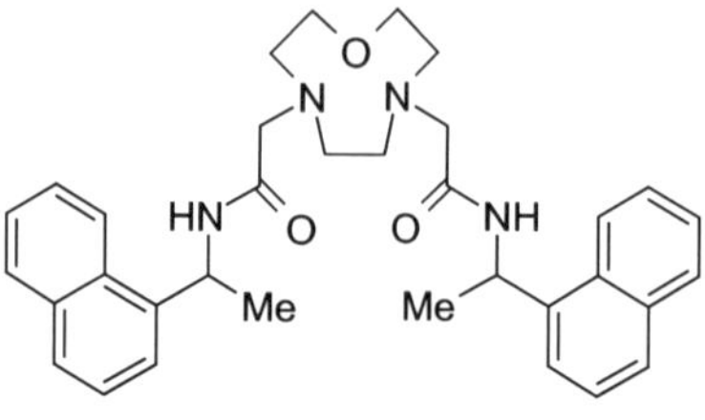

10

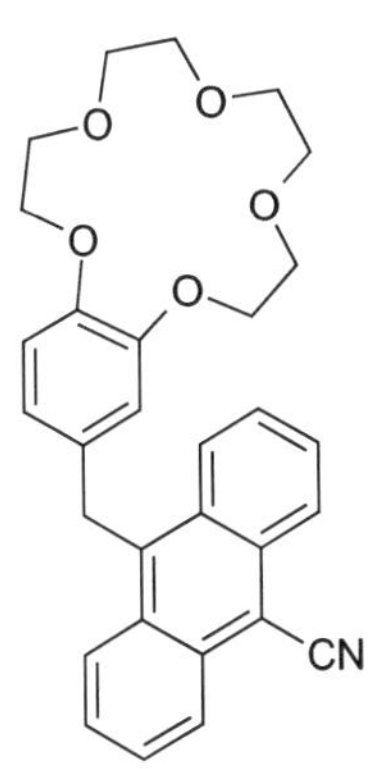

11

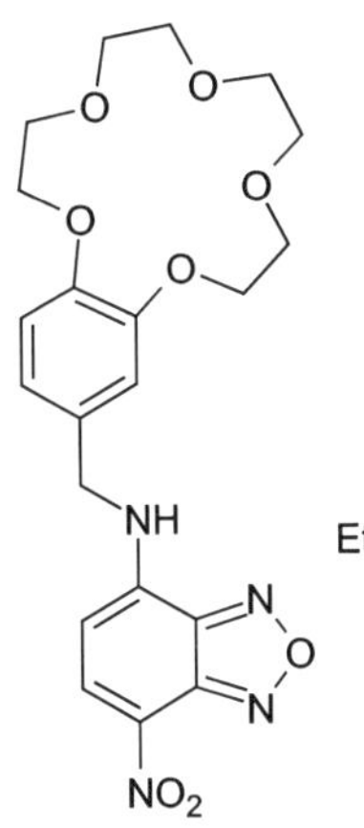

12

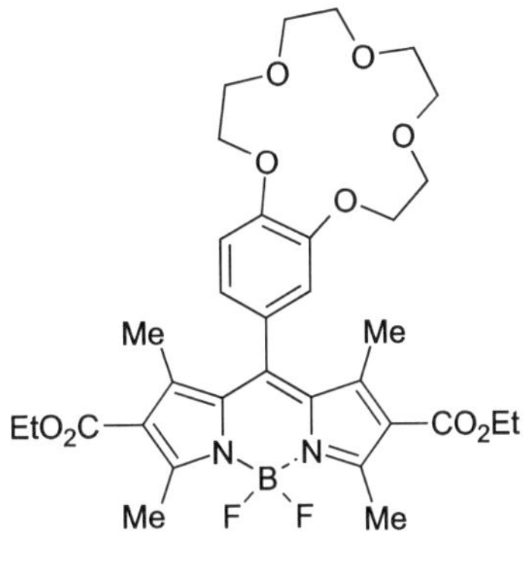

13

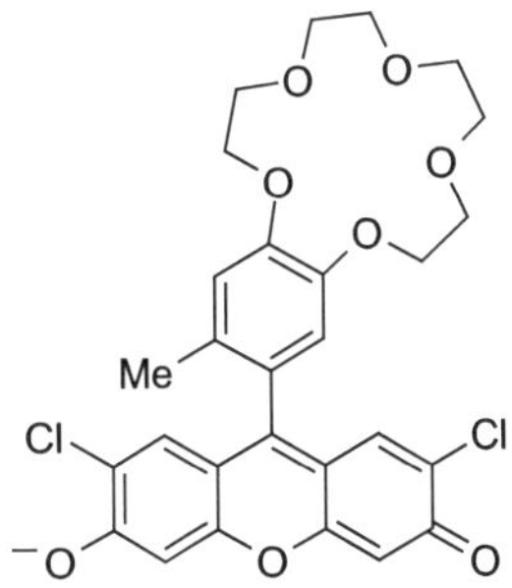

14

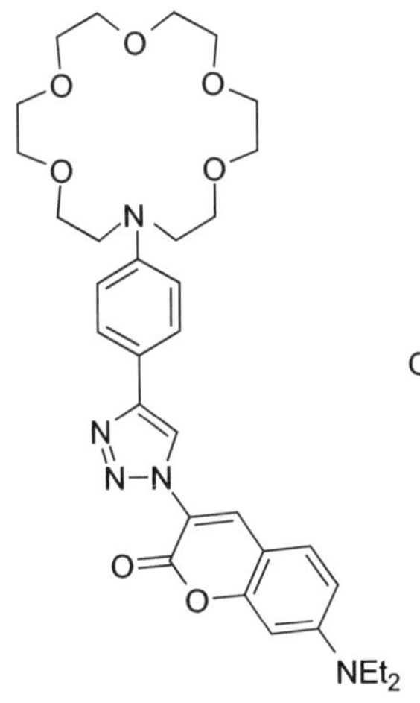

15

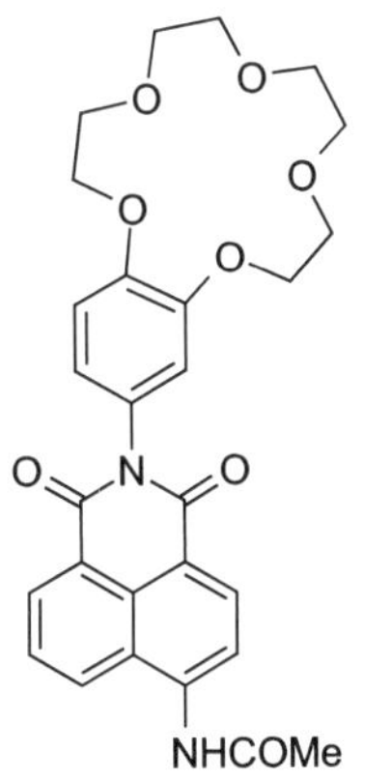

16

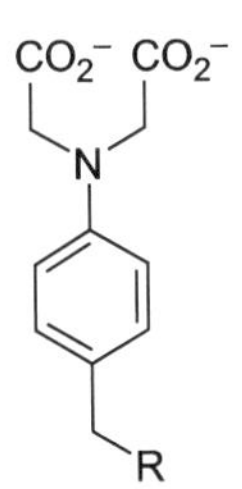

20, R =

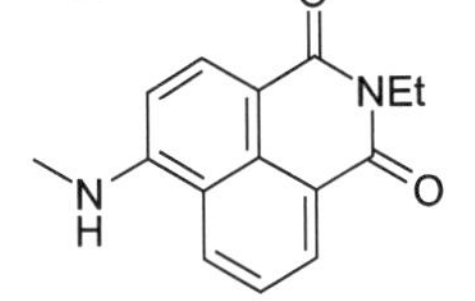

21; R = 9-anthryl

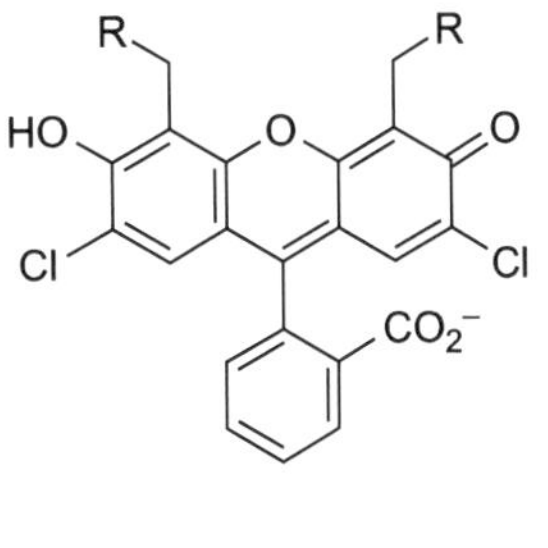

22; R = bis(2-pyridylmethyl)amino

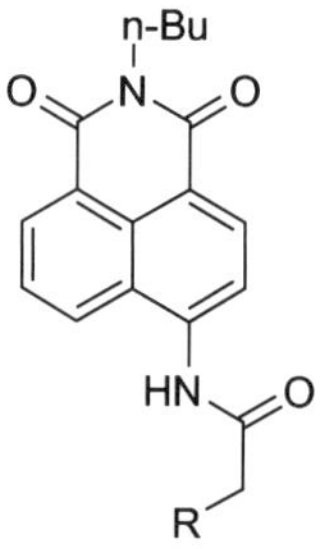

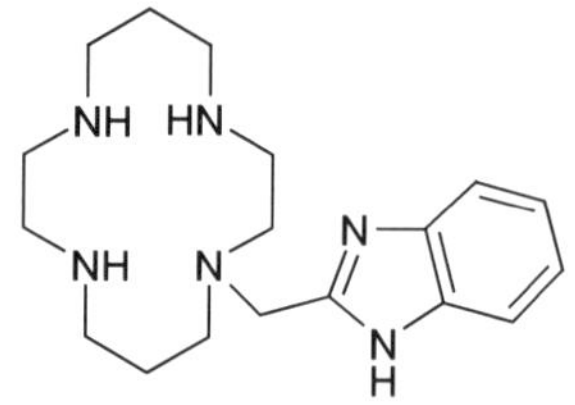

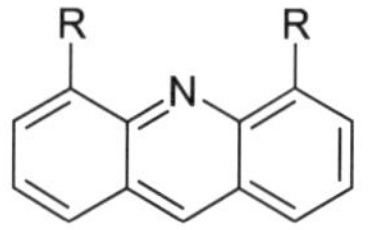

23; R = bis(2-pyridylmethyl)amino

25; R = CH$_2$N(CH$_2$CH$_2$OH)$_2$

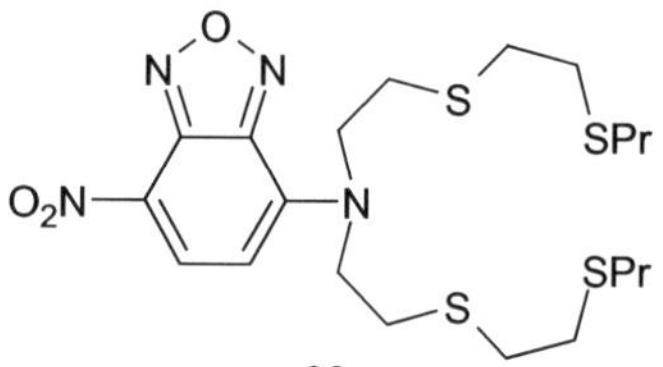

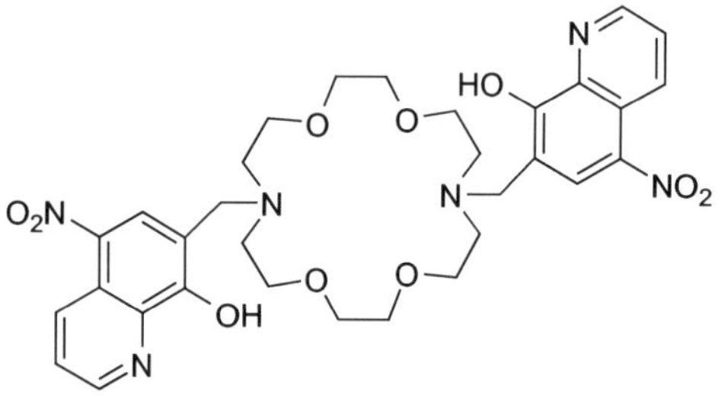

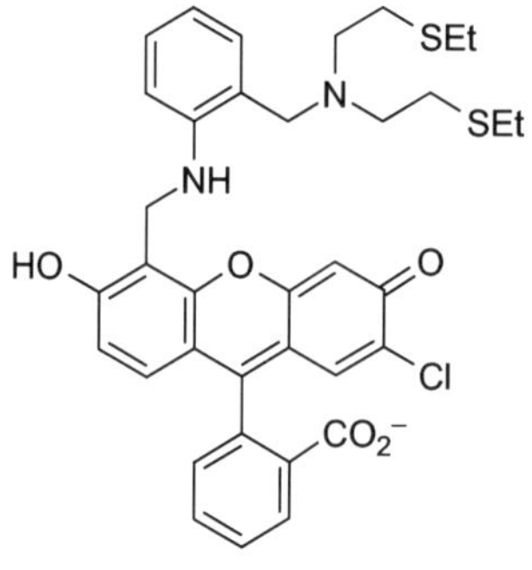

29; R =

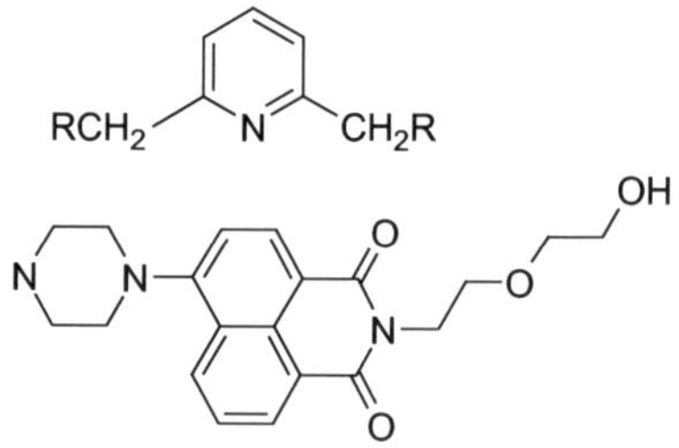

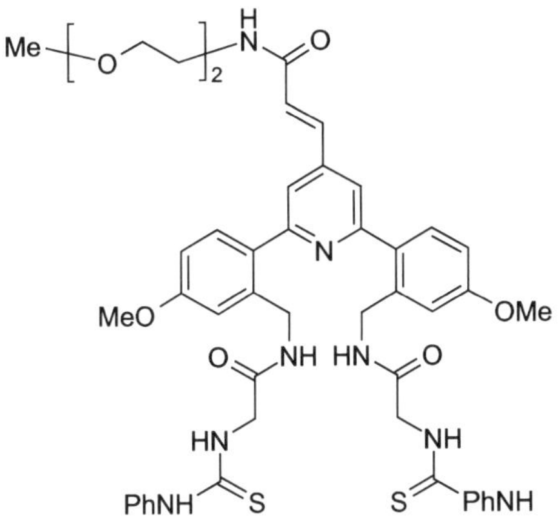

30

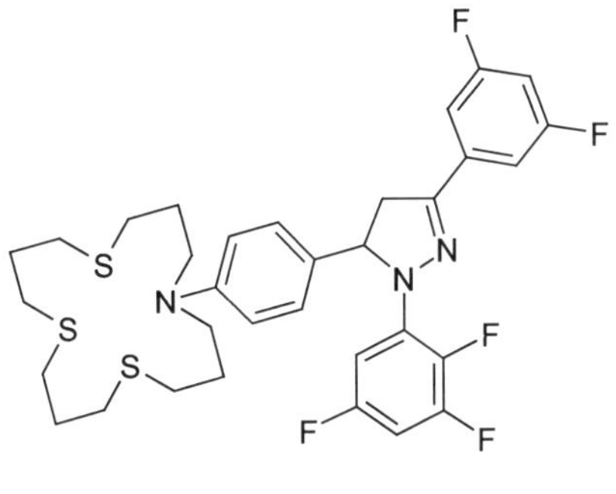

31

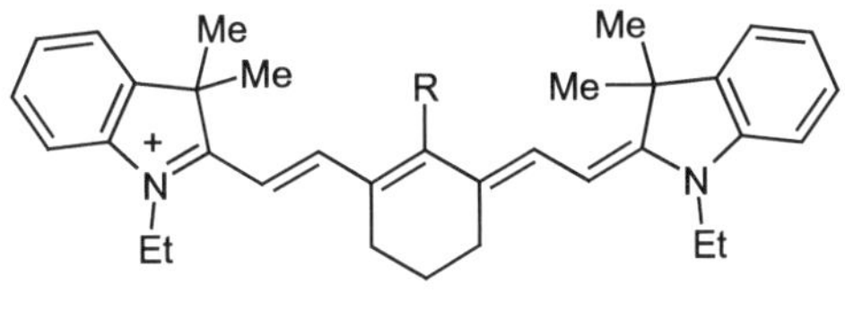

32; R = N(CH₂CONHOH)₂

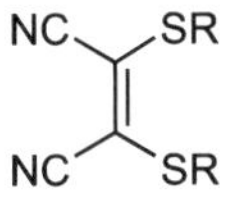

36; R = 9-anthrylmethyl

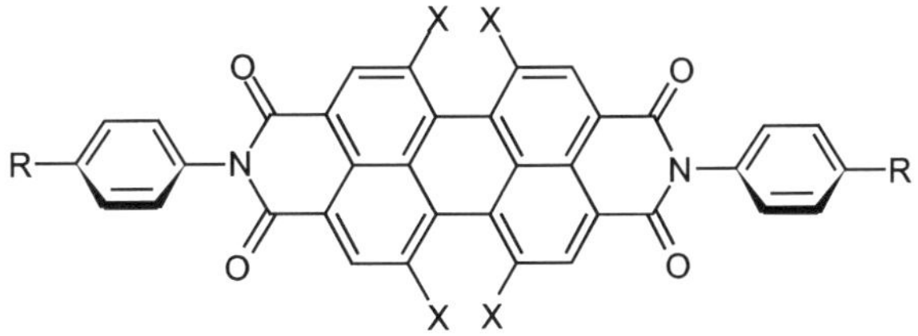

33; X = 4-t-Butylphenoxy, R = bis(2'-pyridylmethyl)amino

34; X = 4-t-Butylphenoxy, R = bis(2'-pyridylmethyl)aminoethylamino

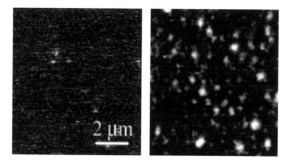

Figure 5.1 Single-molecule fluorescence scanning confocal micrographs of **3** on quartz before (left) and after (right) to HCl–dioxan vapour. Excitation wavelength 488 nm.
Adapted from L. Zang, R. Liu, M. W. Holman, K. T. Nguyen and D. M. Adams, *J. Am. Chem. Soc.*, 2002, **124**, 10640 with permission from the American Chemical Society.

High levels of H^+ trigger strong fluorescence from **38**[67,68] whereas there is almost no emission if H^+ concentrations are kept low. PET occurs from the amine lone electron pair to the fluorophore in the latter situation. Arrival of high enough levels of H^+ blocks the lone electron pair of the amine and lets fluorescence reassert itself (Table 5.2).[69] A notable feature is the switching of the fluorescence output right across the range of observation wavelengths (Figure 5.2). Systems built from relatives of **38** can also be designed to produce the fluorescence switching action in a less discontinuous manner. While this deviates from the usual requirement for logic behaviour, the region for analogue measurement (section 2.5) is extended (Figure 5.3).[70–72] Similar extensions are achievable by compounds carrying two different types of fluorophore, which mixes PET, ICT and EET processes.[73]

Single Input–Single Output Systems

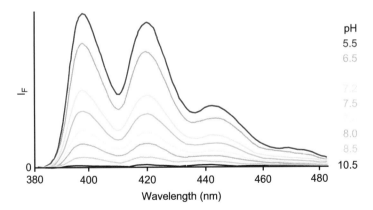

38

A heterocyclic fluorophore is visible in Fahrni's **39**,[74,75] which switches 'on' its emission when the aniline receptor picks up H^+. The structure and function of **39** is largely subsumed within **5**[30] and **6**[31] (Table 5.1), designed to respond principally to Na^+ and Ca^{2+} respectively. Both **5** and **6** have been tested for

Table 5.2 Truth table for **38**.[67]

Input H^+	Output Fluorescence[a]
low (10^{-10} M)	low (0.002)
high (10^{-4} M)	high (0.41)

[a]Quantum yields in methanol : water (1 : 4, v/v), λ_{exc} 366 nm, λ_{em} 398, 420, 444 nm.

Figure 5.2 Fluorescence intensity of **38** in methanol : water (1:4, v/v) as a function of pH. Excitation wavelength 366 nm.
Adapted from data in R. A. Bissell, E. Calle, A. P. de Silva, S. A. de Silva, H. Q. N. Gunaratne, J.-L. Habib-Jiwan, S. L. A. Peiris, R. A. D. D. Rupasinghe, T. K. S. D. Samarasinghe, K. R. A. S. Sandanayake and J.-P. Soumillion, *J. Chem. Soc. Perkin Trans. 2*, 1992, 1559 with permission from the Royal Society of Chemistry.

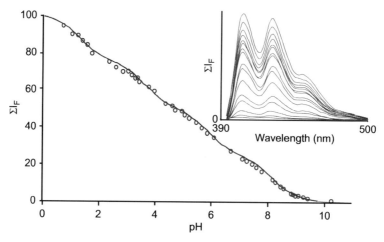

Figure 5.3 The summed fluorescence intensity–pH profile of a set of four compounds related to **38** (open circles) in methanol : water (1:4, v/v). The full line is the optimum theoretical curve. The set of emission spectra is shown in the inset. Emission wavelength 427 nm. Excitation wavelength 372 nm.
Adapted from A. P. de Silva, S. S. K. de Silva, N. C. W. Goonesekera, H. Q. N. Gunaratne, P. L. M. Lynch, K. R. Nesbitt, S. T. Patuwathavithana and N. L. D. S. Ramyalal, *J. Am. Chem. Soc.*, 2007, **129**, 3050 with permission from the American Chemical Society.

their response to H^+ during selectivity studies. Indeed, strong fluorescence switching 'on' is seen. PET from the aniline unit to the pyrazoline fluorophore carrying electron-withdrawing groups such as CN was first appreciated by Pragst and co-workers.[76]

39

Tian's copolymer **40**[77] includes the 'fluorophore–spacer–receptor' system where H^+ binding to the tertiary aliphatic amine inhibits PET and switches 'on' the emission of the naphthalimide unit.[78–80] The H^+ input is provided by a photogenerated acid, from precursors such as triphenylsulfonium salts, to achieve 'light writing' of fluorescent images.

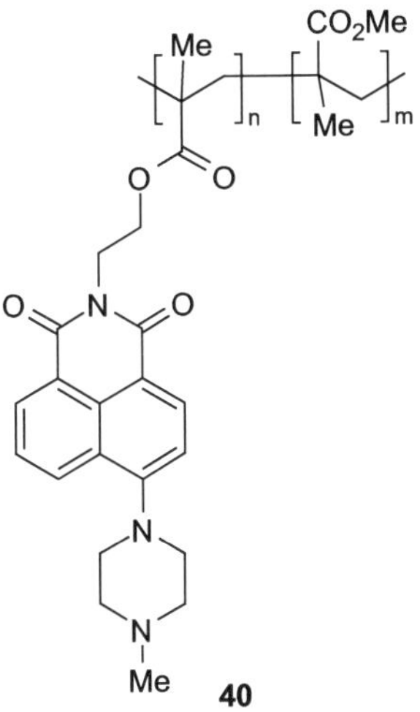

Single Input–Single Output Systems   61

5.2.2.1.2 Light Metal Ion Input. An early illustration of this type would be the anthracene-appended azacrown ether **41**[81] which responds to a K$^+$ input with a fluorescence quantum yield enhancement (FE) factor of 47 due to PET suppression within the 'fluorophore–spacer–receptor' system.[82] Cases carrying the components of **41** in a polymer chain also respond in the same way but with smaller FE values.[83,84] Similar behaviour, but stimulated by Ca^{2+} input, is seen for **6**[31] and **42**.[85] A selection of cases which build on this approach is included in Table 5.1. Some of these have virtual spacers in that the fluorophore and the receptor modules are held orthogonal across a carbon–carbon bond.[44,86] YES logic can also be built into rotaxanes by hiding a squaraine fluorophore[87] inside a cage with the aid of Na$^+$.

A unique example **43**[88] deals with scintillation as an output. When ionizing radiation (α, β or γ) is absorbed by an aromatic solvent, electron–hole pairs are produced. These recombine to produce high-energy excited states of the solvent. EET from the solvent to a suitable solute eventually gives the lowest excited state. PET deactivates excited **43**[88] as usual. When K$^+$ enters the azacrown receptor of **43**, PET is arrested[81] so that scintillation increases strongly.

Lanthanide-based PET systems showing strong luminescence switching are still uncommon.[89–92] Gunnlaugsson and Leonard[93] use their experience[94] to build useful examples for alkali cations under simulated physiological conditions, *e.g.* **44** produces an FE of 40 for Na$^+$. The corresponding diaza-18-crown-6 derivative targets K$^+$ nicely. The latter target is also tackled with good selectivity by Thibon and Pierre,[95] who exploit K$^+$-π interactions within **45.**K$^+$.

The Na$^+$-sensing data of **12**,[38] due to Toyo'oka, is summarized in Table 5.1 but needs further discussion here. The 4-amino-7-nitro-2,1,3-benzoxadiazole fluorophore (amino-NBD) is rather electron-poor which enables it to engage in PET from the benzo-15-crown-5 ether. This fluorophore also shows an emission quantum yield which decreases rapidly as the environmental polarity is increased, hence deserving a mention in section 5.4.2.4 on NOT logic. There is a substantial literature, both classical and modern, on emission/absorption intensity and wavelength of dyes being separately controlled by environmental parameters such as dipolarity/polarity, hydrogen bond acidity and viscosity.[96–103]

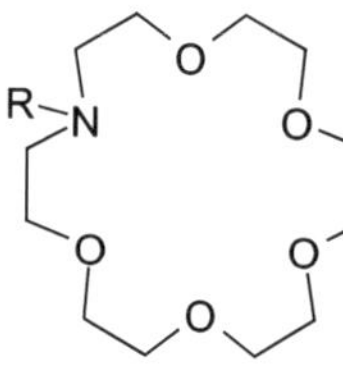

41; R = 9-anthrylmethyl

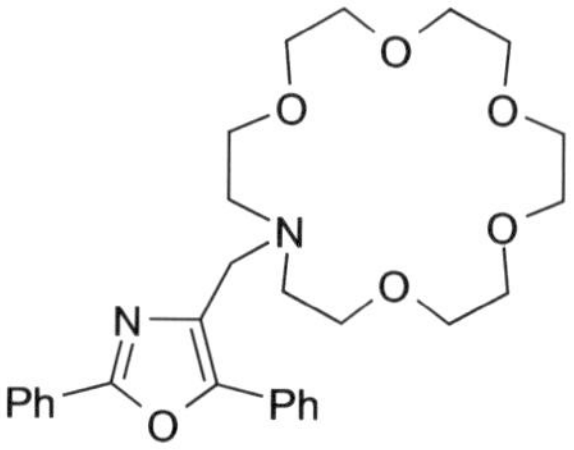

43

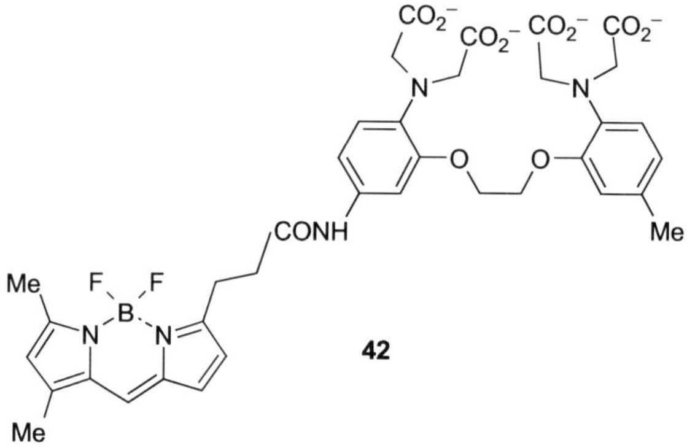

42

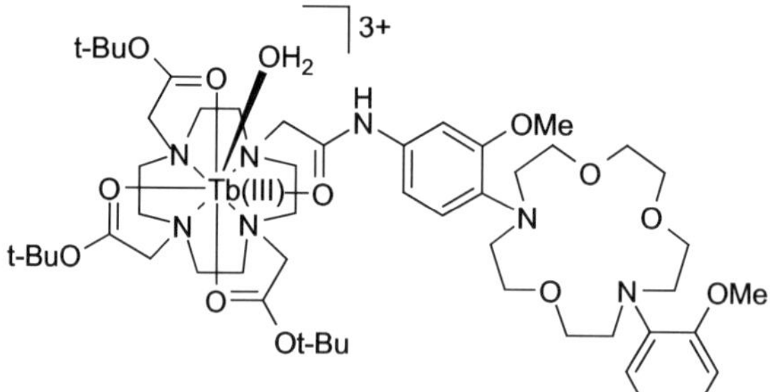

44

The two compounds **46** and **47** are isomers which differ in the amide orientation and the anthryl linking point. Their fluorescence response follows YES logic with respect to the smaller Mg^{2+} and the larger Ca^{2+} respectively. Qin's **46**[104] starts as a weakly emitting excimer (at 415 nm, $\phi_F = 0.020$) owing to its flexible polyether motif but binds a Mg^{2+} each at opposite ends of the dicarboximido polyether receptor. This produces a stretched structure so that monomer emission at 380 nm increases ($\phi_F = 0.088$). In contrast, Nakamura's **47**[105] shows Ca^{2+}-induced suppression of a TICT excited state[106–108] by rigidification of the carbonyl group and the oligooxyethylene chain. A monofluorophore relative of **47** is **48**,[109] which also shows a Ca^{2+}-induced FE of 37 for the same reason. A polyether diester receptor is discernible in **49**,[110] where PET minimizes fluorescence. Unusually, the input of Ca^{2+} produces an intramolecular exciplex emission in acetonitrile because of the receptor wrapping around Ca^{2+} to bring the terminal π-systems close.

For a glimpse of other mechanisms underlying YES gates, let us consider the Ca^{2+}-driven fluorescence switching of **50**.[111] The weak fluorescence of **50** is due to the non-planar xanthone possessing an nπ* singlet excited state. Ca^{2+} binding to the ring of oxygens raises the n-π* state energy above that of the ππ* state. This state-switching causes the fluorescence enhancement,[112,113] as discussed in section 4.4. Isomer **51** serves as a control, because its carbonyl oxygen is not pointing into the receptor ring.

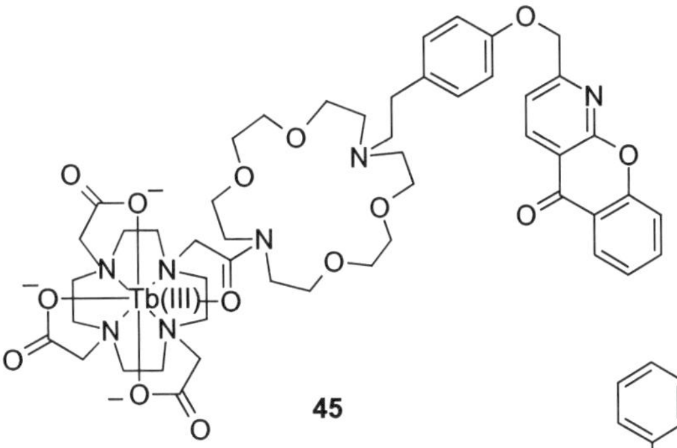

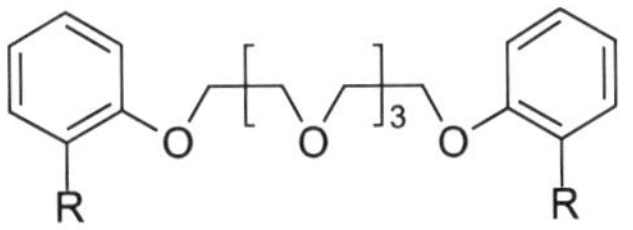

46; R = CONH-2-anthryl
47; R = NHCO-9-anthryl

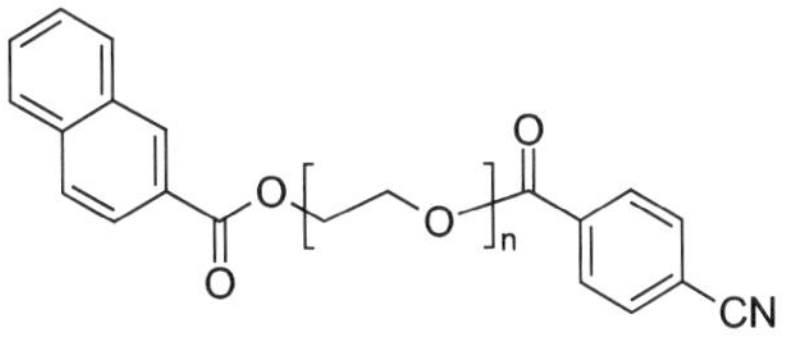

45

49

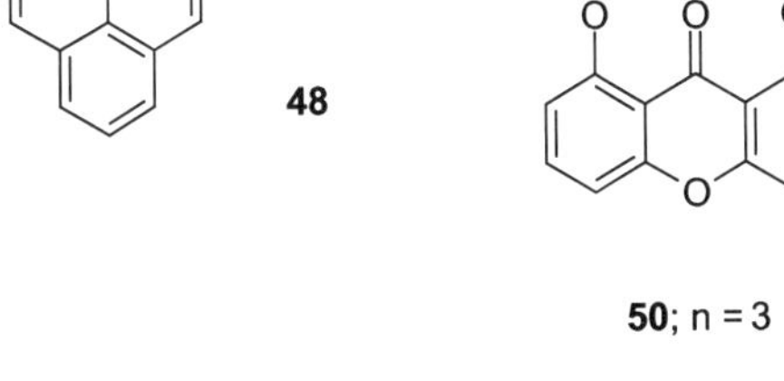

48

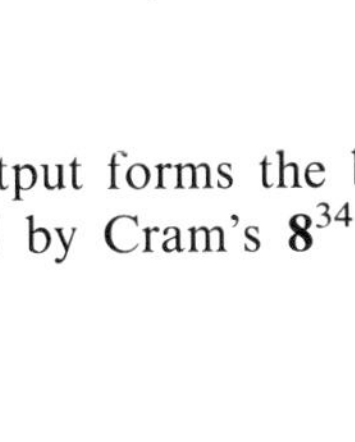

50; n = 3

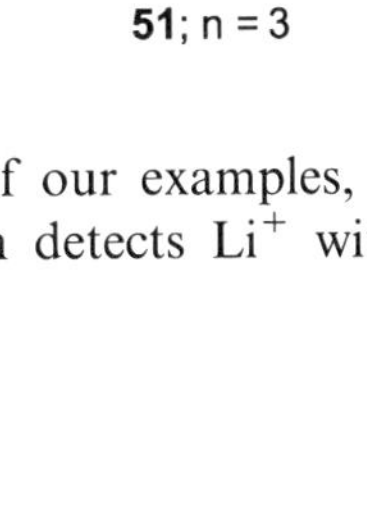

51; n = 3

Though fluorescence output forms the bulk of our examples, absorbance output is nicely illustrated by Cram's **8**[34] which detects Li^+ with exquisite sensitivity. An amount as small as 10^{-8} M is indicated with the development of an absorption band at 586 nm. Binding of Li^+ into the spherand receptor causes H^+ ejection from the phenol group.[114] It is the well-delocalized π-system of the Li^+-paired phenolate that produces this intense absorption.

5.2.2.1.3 Heavy Metal Ion Input. Being a closed-shell ion, Zn^{2+} tends to avoid many paths that deactivate molecular excited states, which were summarized in Chapter 4. Unsurprisingly, there are many examples of Zn^{2+}-driven fluorescent YES gates with excellent FE values, most of which are based on PET designs.[18,22,51,115–119] One of the earliest of these with good selectivity is de Silva's **52** with an anthracene fluorophore and a di(2-pyridyl)amine receptor, where the amine provides the electron donor site.[120] Lippard's team adapt this approach extensively for Zn^{2+} imaging in cell neurophysiology,[116] so that some of these cases would not be out of place in Chapter 14 concerning applications. For instance **22**[49,121,122] has FE = 2.3 upon 1 : 1 Zn^{2+} binding. 1 : 1 Stoichiometry is seen from the binding isotherm, in spite of the presence of two identical receptor regions and an X-ray structure of the 2 : 1 complex. It appears that the first Zn^{2+} displaces an H^+ from one of the protonated benzylic amines at pH 7. Unusually, H^+ seems poorer than Zn^{2+} at suppressing PET in this case. Cases like this serve as a reminder that bireceptor PET systems can qualify as 2-input AND gates (see section 7.2.2) only if there is experimental evidence that two binding events are necessary for the major switching phenomenon to be observed.[123]

52; R = bis(2-pyridylmethyl)amino

While fluorescent PET switches/sensors have many advantageous features, they cannot usually be monitored at two wavelengths to yield different responses. Such ratiometric observations are very suitable for intracellular use.[124] Woodroofe and Lippard[125] tackle this problem with **22** bound *via* a carboxamide linkage to a Zn^{2+}-insensitive, blue-emitting fluorophore **53** (λ_{Flu} = 488 nm). Non-specific esterases in the cytosol cleave the ester linkage to yield two fluorescent molecules which are monitored at separate wavelengths. Using two unconnected fluorophores to supply two colour channels[126] requires checking of the co-localization of the two fluorophores. Unimolecular ratiometric sensors exploit long spacers[127] or carefully chosen fluorophore pairs[128] to minimize EET between them.

Another simple but effective receptor for Zn^{2+} which has been built into fluorescent switches is *N*-phenyliminodiacetate, as illustrated by Gunnlaugsson's PET-based **20**.[47]

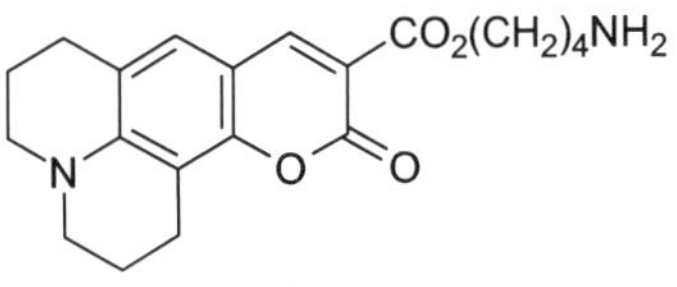

53

In contrast, Licchelli's **54** carries a pyridyl imine receptor and a naphthalimide fluorophore.[129] Since the latter is prone to excimer formation,[130,131] Zn^{2+} induces this output as multiple copies of **54** are brought together by the metal ion. Cd^{2+}, a hard target to detect selectively, is handled by Qian's YES gate **55**[132] which gives FE = 195 and operates nicely inside cells (Figure 5.4). A PET mechanism is indicated although a clear spacer unit is not discernible. A similar situation arises for Guchhait's **56**[133] which gives not only a Zn^{2+}-induced FE = 143 but also a Cu^{2+}-induced FE = 49 and similar values for other open-shell metal ions. Another case of a Cu^{2+}-induced fluorescence enhancement which is beaten by Zn^{2+} is due to Yoon and Spring.[134] A $n\pi^*$–$\pi\pi^*$ state inversion or an ESIPT process is also plausible for **56** due to the involvement of an imine electron pair or the phenolic group respectively. A PET-type process also applies when a fluorescent quantum dot[135] is employed instead of an organic fluorophore. The capping agent on the quantum dot can be any of several polyazamacrocycles, which also serve as a receptor for Zn^{2+}, and Zn^{2+}-induced FE values of up to 9 are seen.[136]

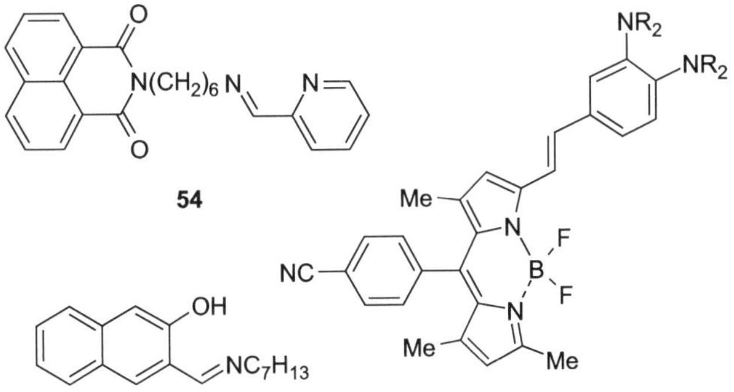

54

56   **55**; R = $CH_2CONHCH_2CH_2OH$

The need to understand Hg^{2+} toxicity has led to the development of fluorescent YES gates driven by this heavy metal cation.[18,137] The soft cation Hg^{2+} binds to Nolan and Lippard's **28**[55] via the sulfurs and the neighbouring nitrogens. The fluorescence response occurs by suppression of the PET from the aniline moiety to the fluorophore. Though Hg^{2+} is normally redox-active enough to introduce a PET channel of its own, S-coordination mitigates this.[138] However, this case also responds to H^+ in neutral water ($pK_a = 7.1$) owing to the presence of the aliphatic amine unit, which is responsible for moderating the

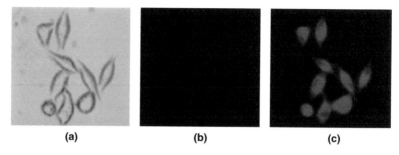

Figure 5.4 Micrographs of HeLa cells containing **55**: (a) bright-field image; (b) fluorescence image, observed at 570 nm and excited at 550 nm; (c) same as (b) but with 5 mM Cd^{2+} added.
Reprinted from T. Cheng, Y. F. Xu, S. Zhang, W. P. Zhu, X. H. Qian and L. Duan, *J. Am. Chem. Soc.*, 2008, **130**, 16160 with permission from the American Chemical Society.

FE value. An environmentally useful example is Chang's **57**[139] which produces a large fluorescence enhancement when driven by Hg^{2+} in neutral water. The tetrathiamonoaza-15-crown-5 ether receptor targets Hg^{2+} and the virtually spaced PET system[44] is reminiscent of a case due to Rurack *et al.*[138]

57

There are cases which distinguish themselves by employing a receptor devoid of sulfur, *e.g.* Qian's **29**,[56] or by Hg^{2+}-induced conformational changes instead of PET suppression, *e.g.* Mello and Finney's **30**,[57] or by using an ESIPT mechanism, *e.g.* Savage *et al.*'s **27**.[54] X-ray crystallographic evidence shows that the diazacrown unit in **27**[54] is not involved in binding Hg^{2+},[140] but soaks up H^+ liberated from the hydroxyquinolines when they bind Hg^{2+}.

Cu^{2+} is notorious for its fluorescence quenching ability, but Tang and coworkers arrange a Cu^{2+}-induced FE of 10 due to PET suppression in **32**.[60] The advantage of near-infra red (IR) excitation and emission of such PET systems[141] for use in tissues is demonstrated (Figure 5.5). Pd^{2+} (in the form of $PdCl_2$) is another hard target, but Holdt has a nice set of studies with **36**[63] and relatives[142] where PET from an anthracene fluorophore to the dithiomaleonitrile receptor is suppressed to achieve FE values as high as 250.[143]

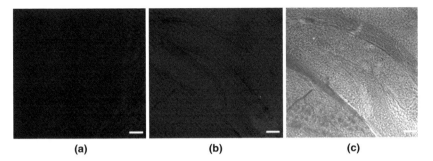

Figure 5.5 Confocal fluorescence micrographs of Cu^{2+} at a depth of 90 μm in rat hippocampal slices (a) incubated with **32**; (b) incubated with **32** and Cu^{2+}. (c) Bright field image. Emission wavelength 750–800 nm. Scale bar 100 μm. Adapted from P. Li, X. Duan, Z. Z. Chen, Y. Liu, T. Xie, L. B. Fang, X. R. Li, M. Yin and B. Tang, *Chem. Commun.*, 2011, **47**, 7755 with permission from the Royal Society of Chemistry.

5.2.2.2 Membrane-bounded Cation Input

Though tiny in size, this section concerns the vital matter of examining molecular switching in nanospaces.[102] The matter is indeed vital, because membrane-bounded nanospaces are involved in cellular energy-handling and in nerve signalling.[144] There are a few cases of cation-driven YES gates which only succeed in certain charge-neutral detergent micelle media. Ba^{2+} and Zn^{2+} are selectively signalled by fluorescent PET systems **58**[145] (FE = 6.5) at pH 10 and **59**[146] (FE = 25) at pH 7 respectively. Since these molecular devices are membrane-bound, they respond only to those ions which are adjacent to the membrane. The less polar micellar environment[147] strengthens cation-binding, but also decreases local ion densities.[148]

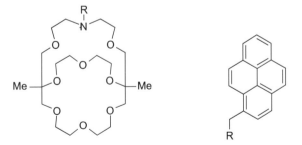

58; R = 1-pyrenylmethyl   **59**; R = bis(2-pyridylmethyl)amino

5.2.2.3 Anion Input

Fluorescent and absorptiometric systems to detect anionic species selectively[149–153] are still less common than those targeting cations. A few representative cases of small anions are featured below while larger organic counterparts will be held back for the next section.

We can take the current at an applied potential of -0.35 (vs. Fc^+/Fc) V (during cyclic voltammetry) as the output from **60**[154] as the input F^- is applied. The magnitude of this current increases by a factor of 4.6, illustrating YES logic, because of a F^--induced 190 mV cathodic shift in the reduction potential of the ferrocene unit. Coincidentally, a FE value of 12 is found in the emission of the naphthalene component. Example **60** is just one of several from Tarraga and Molina's laboratories.[155]

Now we follow up with some fluorescence-based examples. $CH_3CO_2^-$-binding decreases the electron deficiency of the thiouronium unit in **61**[156] and hence discourages PET from the fluorophore, which leads to an FE value of 4.1. CO_3^{2-} produces a FE of 53 from **62**,[157] though there is some time-dependence. A thiadiazole derivative which is the oxidation product of **62**, with fewer hydrogen-bond acidic N–H groups, shows a smaller FE value of 12, which indicates the importance of hydrogen-bonding in the complexation of CO_3^{2-}. A significant CO_3^{2-}-induced blue-shift is also seen. CO_3^{2-}-induced rigidification and even aggregation may be behind the fluorescence enhancement.

Next we consider an assembly of a fluorophore **63** coordinated to a $diCu^{2+}$-cryptand **64**.[158] EET from the fluorophore **63** to the ligated Cu^{2+} centres means that fluorescence is quenched. Owing to its more favourable bite angle, HCO_3^- displaces the carboxylate **63** away from the quenching Cu^{2+} centres, thus switching **63**'s fluorescence back 'on'. A related case is Ballester's squaramide-based macrocycle **65**,[159] which binds tetrahedral SO_4^{2-} even in water via hydrogen bonds and electrostatics. The receptor **65** is then paired with an anionic fluorophore to produce a non-emissive assembly. When SO_4^{2-} is added, it competes with the fluorophore for pairing to **65**. The released fluorophore emits strongly.

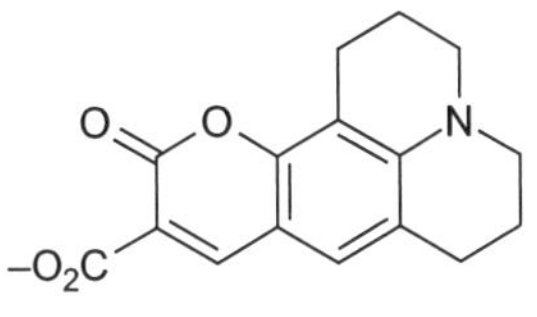

5.2.2.4 Organic Input

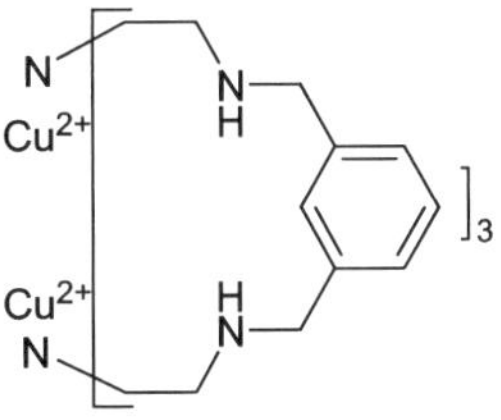

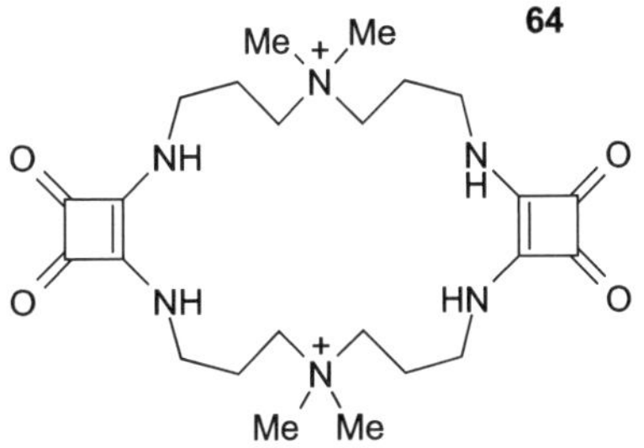

Large targets are harder for sensor/switch designers to handle because the major fraction of chemical recognition is currently done with endoreceptors[160] which, almost by definition, aim for smaller guests. Some representative cases of devices employing organic molecular inputs of various types appear below or in later sections.

Caffeine is handled by Waldvogel's receptor **66** in CH_2Cl_2.[161] Mixing with a caffeine analogue (**67**) containing an EET acceptor will receive the excitation energy of **66**,[162] quenching the fluorescence of the complex **66.67**. This complex serves as our device. Addition of caffeine displaces[163] **67** to remove the EET channel so that the fluorescence of **66** is recovered. Organic phosphines are handled by the **68**.Ag^+ complex,[164] which shows PET-like charge transfer[165] from the anthracene group to Ag^+, resulting in weak fluorescence. If an organic phosphine is applied as input to **68**.Ag^+, **68** is displaced. Free **68** is strongly fluorescent because the nitrogen electron pairs mix with the P(III) centres to suppress PET to anthracene.

When complexed to phenylazopyridine (PAP), **69**[166] is poorly emissive due to PET from the porphyrin–Zn(II) unit to PAP being faster than EET to the porphyrin free base component. If dimethylaminopyridine is employed as an input, it displaces PAP to occupy the axial position on Zn(II) to recover the porphyrin free base fluorescence.[167] A displacement is also involved in a PET-derived, non-emissive assembly which arises from the trianionic fluorophore **70** within the cavity of tetracationic receptor **71**.[168,169] Guanosine triphosphate (GTP) binds even more strongly to **71** than to **70** owing to a combination of electrostatics and π-stacking. Displacement of **70** in this way produces strong fluorescence (FE = 150) (Figure 5.6).

66

67

68

69; R = mesityl

70

71

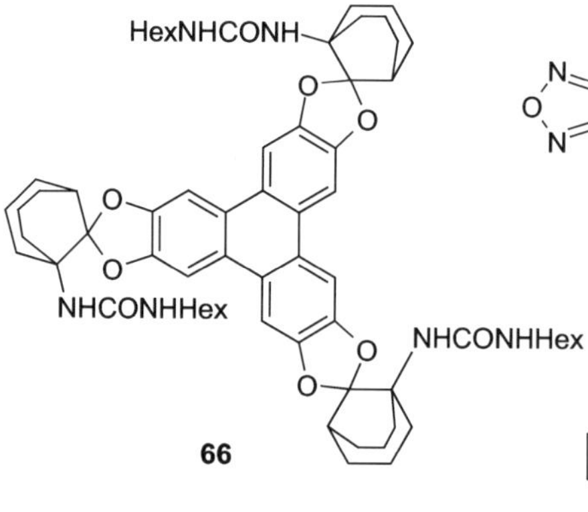

Devices which operate on organic inputs in an enantioselective fashion[170,171] are of particular value, *e.g.* Pu's (+)-**72**[172] which is silent towards *L*-*N*-boc-phenylglycine but responds to the *D*-enantiomer (FE = 5.7) albeit in organic solvent. H^+-transfer from the acid to the amines of **72** to prevent PET is involved but ESIPT from the naphthols to the amines can also play a role. However, the classic enantioselective device is Shinkai's **37**.[64] *D*-Glucose binds to the phenylboronic acid to increase its Lewis acidity so as to engage the amine group close by. PET from the amine to the fluorophore is arrested in this way, giving FE = 4.0, in the case of the *R*-enantiomer of **37**. *L*-Glucose only gives

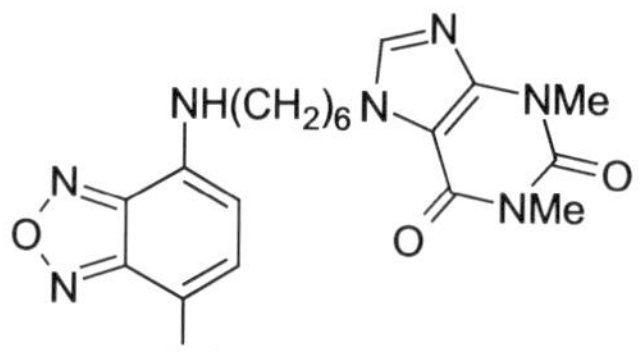

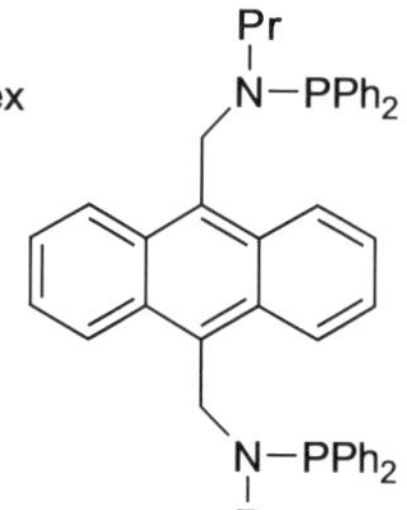

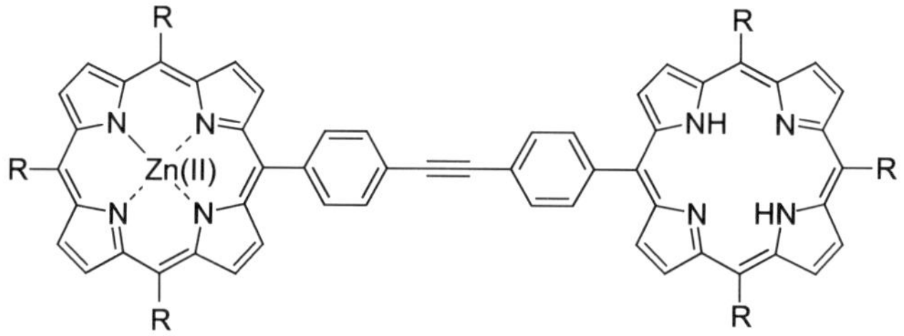

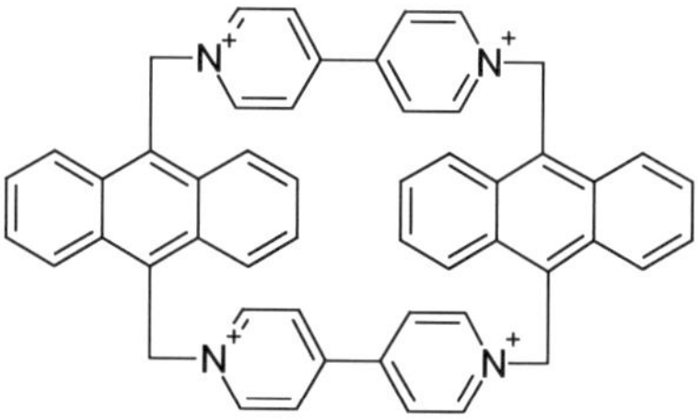

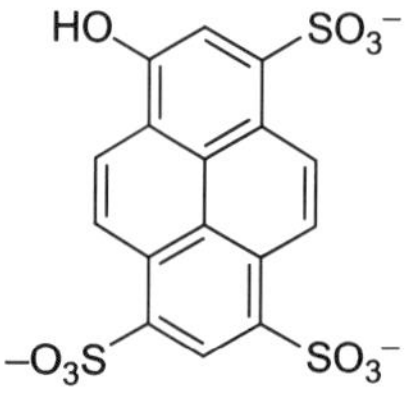

Figure 5.6 Fluorescence of the following compounds in water (left to right): **70**, **70**+**71**, **70**+**71**+cytosine triphosphate, **70**+**71**+adenosine triphosphate and **70**+**71**+guanosine triphosphate. Excitation wavelength 367 nm.
Adapted from P. P. Neelakandan, M. Hariharan and D. Ramaiah, *J. Am. Chem. Soc.*, 2006, **128**, 11334 with permission from the American Chemical Society.

FE = 2.0. Conversely, the *R*-enantiomer of **37** produces FE values of 2.0 and 3.7 respectively with *D*- and *L*-glucose. Indeed, sugars are perhaps the most popular organic input to fluorescent PET-based devices. Amine-capped fluorescent quantum dots[135] also respond in a YES-logical manner to glucose when the latter binds to a viologen **73** to pull it off the quantum dot so that a PET process is stopped.[173]

72

73

5.2.2.5 Oligonucleotide Input

Now we consider oligonucleotides on their own, which are rather large molecular inputs. We begin with a 110-mer ribonucleic acid (RNA). Sparano and

Koide used *in vitro* selection[174–176] to pick out the particular aptamer which binds the *N*-arylheterocyclic motif in **74** from a library of 10^{13} RNA sequences.[177] Component **74** plays the role of exoreceptor within the device **75**.[160] In fact, **75** has PET occurring from the **74** units to the fluorescein fluorophore. The aptamer engages the extended conformer of **75**, so that the **74** units alone are bound. Electron transfer slows down in the extended conformer owing to the increased distance between PET components, which leads to fluorescence enhancement (FE = 13).

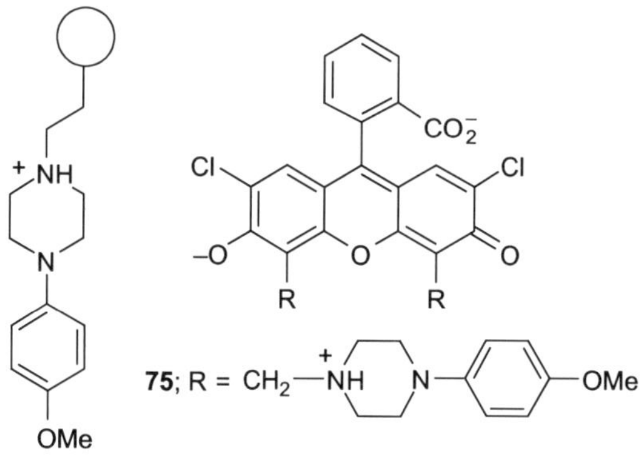

An oligonucleotide hairpin with a large loop can be straightened out by hybridizing it with a complementary strand. This straightening destroys an EET process which exists between fluorophore **76** and quencher **77** anchored at the two ends of the hairpin.[178,179] Therefore the complementary strand serves as the input to the hairpin YES gate, with the fluorescence of **76** serving as output.

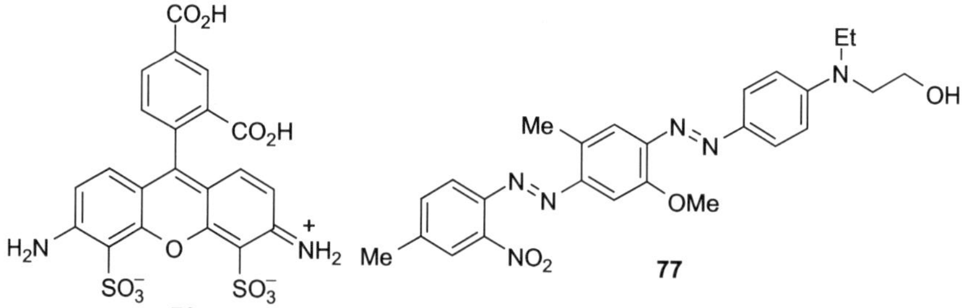

5.2.2.6 Protein Input

The biochemistry community have always had to wrestle with large targets. So the earliest examples of luminescent switching for targets as large as proteins are to be found there (*e.g.* serum albumin,[180] see section 5.4.2.4). Here, we consider an illustration of a large protein input driving a YES gate with fluorescence output. The extreme length and the high anionic charge of **78** contribute to the efficient quenching of its fluorescence by cationic **79** *via* PET

Single Input–Single Output Systems 73

from the fluorophore to the pyridinium centre. However **79** also contains a biotin moiety which is captured by avidin. Therefore avidin competes for **79** and displaces[163] **78** so that the latter's fluorescence is recovered.[181]

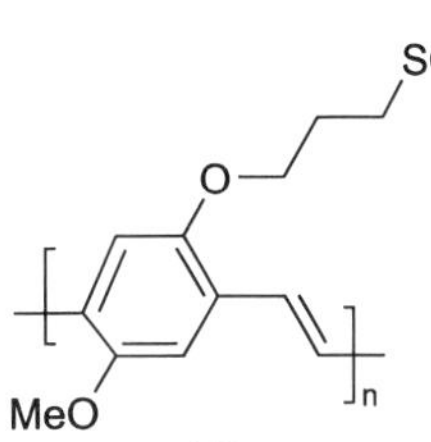

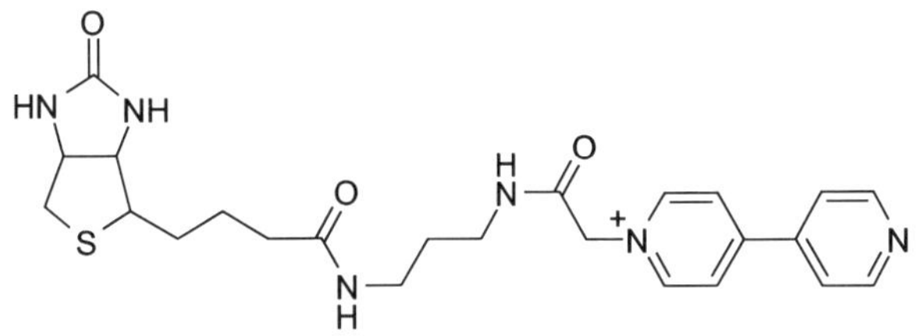

5.2.2.7 Redox Input

Electrochemical properties such as redox potential can also serve as inputs to molecular YES logic gates because redox indicators have a rather long history.[9,182] However, many of these systems[183–188] are stable in either the oxidized or reduced forms in the absence of applied potentials. Therefore these are better suited to discussion in Chapter 11.

5.2.2.8 Polarity Input

Solvent polarity can also serve as an input and **80** produces a fluorescence output according to YES logic.[189] Changing the solvent from hexane to methanol:water (1:4, v/v) produces a jump in the fluorescence quantum yield. In line with older data on simple coumarins,[190] **80** shows an $n\pi$–$\pi\pi^*$ state inversion with increasing polarity and the $\pi\pi^*$ state becomes the lower energy form.[189]

An water-induced FE value of 7.5 is seen for Harima's **81** in acetonitrile.[191] Water levels as low as 0.001 weight% are detectable in this way and can be improved further.[192] This low weight percentage corresponds to 4.4×10^{-4} M H_2O. Given that the concentration of **81** is 2.2×10^{-5} M, at most 20 molecules of H_2O could make up the average hydration shell of **81**. Even though this may not represent bulk water, the polarity and proticity of this shell encourage **81** to pass into the zwitterionic form *via* H^+ transfer from the acid group to the amine unit. PET from the amine to the fluorophore is therefore arrested. This method succeeds provided that the organic solvent is not basic enough to prevent protonation of the amine.

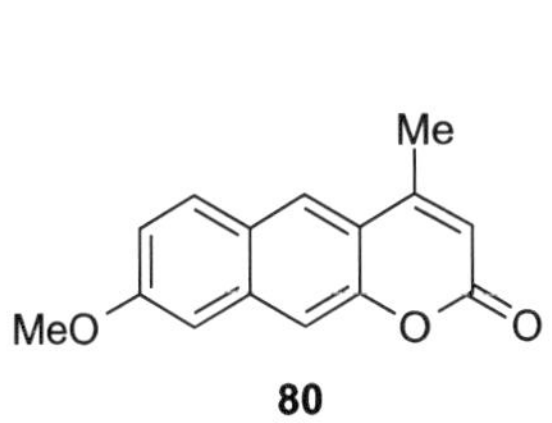

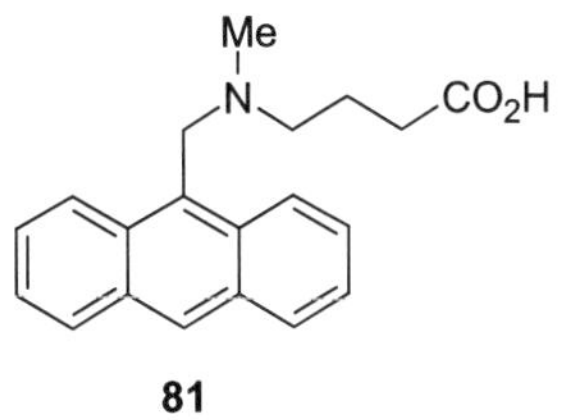

5.2.3 Temperature Input

It is also possible to use physical parameters such as temperature as an input. Fluorescent molecular thermometers[193,194] can be developed by combining fluorescent monomers and polymers that undergo a temperature-induced phase transition, to produce copolymers with a high fluorescence quantum yield[195] and long fluorescence lifetime[196] above a threshold temperature. The thermometers are composed of thermosensitive acrylamide polymers labelled with less than 1% of a benzofurazan-containing fluorophore **82**. The benzofurazan is sensitive to solvent polarity. With increasing temperature, there is a decrease in the microenvironment polarity near the main chains of the copolymer as evidenced by the location of the maximum emission wavelength. As the dielectric constant of the medium increases from that of isobutanol to that of water, the maximum emission wavelength undergoes a red shift by 15 nm. These polymers are remarkably sensitive temperature probes, with the fluorescence increasing substantially per 1 °C increase upon passing a sharp threshold temperature (Figure 5.7). Unlike many previous examples,[193,194] these fluorescent thermometers switch 'on' with an increase in temperature. The sharpness of the phase-transition particularly favours the description of **82** as a temperature-driven YES logic action.

The properties of fluorescent molecular thermometers such as **82** can be modulated by polymerizing three rather than only two co-monomers.[197] The introduction of two kinds of acrylamide unit in the copolymer makes the temperature ranges different from those of the original polymers while preserving the sensitivity and reproducibility. This approach allows for facile modification of the operating temperatures in aqueous solution to be centered at any value between 18 and 54 °C (Figure 5.8).

N_4-Coordinated Ni(II) displays the spin crossover equilibrium.[198] Coordinating anions like CN^- favour the blue high-spin state, whereas passive anions like ClO_4^- choose the yellow low-spin state. The latter condition is also favoured by higher temperatures. The availability of two states with different optical and magnetic properties allows fluorescent detection of the equilibrium position *via* interaction of the fluorophore and the Ni(II) centre. Indeed, the fluorescence of **83** increases with increasing temperature in spite of growing

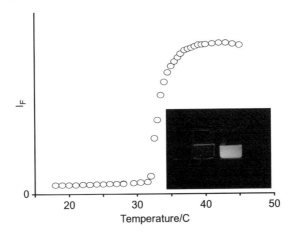

Figure 5.7 Fluorescence intensity of **82** in water as a function of temperature. Emission wavelength 533 nm. Excitation wavelength 444 nm. Inset: Photograph of **82** in water at 20 °C (left) and 40 °C (right).
Adapted from S. Uchiyama, Y. Matsumura, A. P. de Silva and K. Iwai, *Anal. Chem.*, 2003, **75**, 5926 with permission from the American Chemical Society.

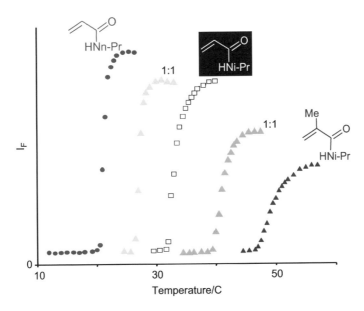

Figure 5.8 Fluorescence intensity of several polymers and copolymers of the monomers shown (related to polymer **82**) in water as a function of temperature. The monomer feed ratios are also given where appropriate. Emission wavelength 533 nm. Excitation wavelength 444 nm.
Adapted from S. Uchiyama, Y. Matsumura, A. P. de Silva and K. Iwai, *Anal. Chem.*, 2004, **76**, 1793 with permission from the American Chemical Society.

spectral overlap between the fluorophore emission and the low-spin state absorption. The latter parameter controls the efficiency of the EET mechanism underlying the fluorophore–Ni(II) centre interaction. It is likely that other factors, *e.g.* differences in magnetism between the two spin states, govern the emission–temperature profile.

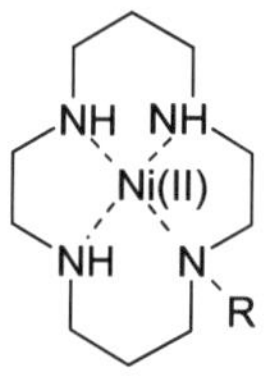

83; R = 2-naphthylmethyl

Reversible formation of covalent bonds is another route to molecular thermometers, if care is taken to maintain steady-state conditions. A temperature-dependent Diels–Alder reaction mixture is a particular example.[199] Reactions of this type reverse at higher temperatures. The two pyrene-based components of system **84** + furan + **85** have very different fluorescence yields: 0.02 (**84**) and 0.15 (**85**). PET from pyrene to the electron-poor maleimide ($E_{red} = -1.0$ V *vs.* sce) is responsible for the weak fluorescence of **84**. The Diels–Alder product **85** does not possess such a suitable electron acceptor.

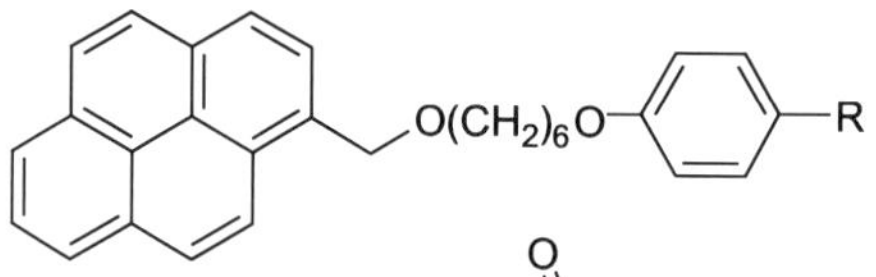

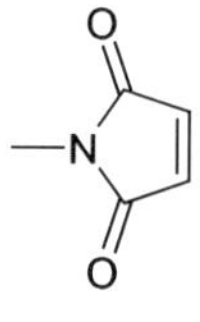

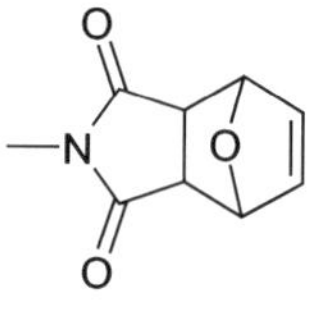

5.2.4 Light Dose Input

The commonest all-optical input–output systems[200–202] are photochromics[203–205] where an absorbed light dose forms an isomeric product state whose absorption lies at a longer wavelength. Thus the light dose input leads to an absorbance output. However, a large fraction of these photochromics will hold a product state for a period of time after the light dose has expired. Such latching leads to a memory effect which needs to be removed in discussions of combinatorial logic. Practically, such memory erasure can be done by resetting the starting state for the testing of each condition of a truth table. Such resetting can be done by applying a new dose of light which is absorbed by the product state or by waiting until thermal relaxation is complete in suitable cases. A number of examples involving photoisomerization to be featured in this book will have latching aspects to their behaviour which should ideally be treated in this way except when history-dependent systems are being considered (Chapter 11). Fluorescent versions of photochromics have also been available for 40 years,[206] though these do not appear to be reversible either thermally or photochemically. However, fluorophores with relatively long wavelengths can be built into hydrogen-bonded assemblies with photochromics. The fluorescence output of these assemblies can be modulated reversibly.[207]

However, we feature a case which involves very short timescales. The fast all-optical switch **86** is a 'donor–bridge–acceptor' system where selective excitation of the bridge module opens up a new channel for electron recombination between the terminal units.[208] In fact the recombination rate jumps up by a factor of 7000. The relatively long-lived fully charge-separated radical ion pair state involving the terminal units is created by exciting the porphyrin donor.[209] This actually occurs *via* a charge shift from the intermediate 'donor$^{•+}$–bridge$^{•-}$–acceptor' state. Transient absorption is the output.

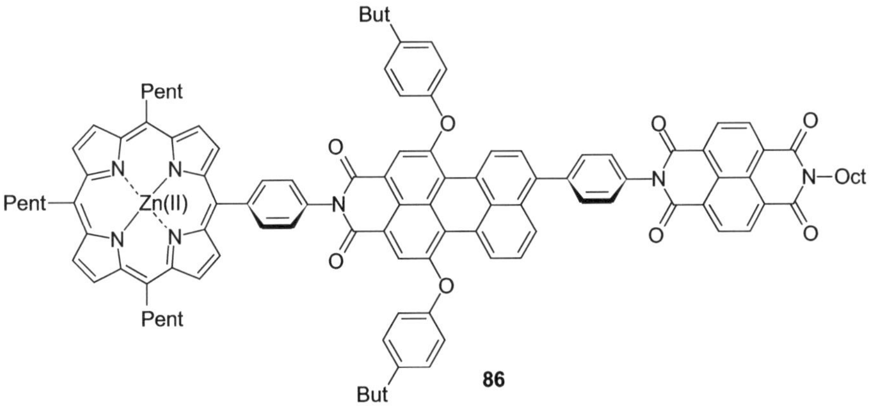

5.3 Irreversible YES

Supramolecular interactions are ideal for designing molecular logic devices because the device–input interactions are reversible, *i.e.* the device can be reset easily. Thus the device is available for use nearly continuously. However, molecular logic can also embrace irreversible interactions if we are willing to entertain once-only computations. Some of these arise in medical diagnosis contexts (Chapter 14). While time-varying data cannot be accurately obtained in this way, the approximate data can be valuable. This is especially true in situations where suitable supramolecular interactions are currently unavailable. In general, irreversible interactions lead to a memory aspect. Chapter 11 describes cases which can be reset in some way because they correspond to certain well-characterized semiconductor devices. Others are collected here.

5.3.1 Chemical Input

Chemical reagents[210]/chemdosimeters[211] produce easily visible outputs after binding irreversibly to their targets. The classical analytical chemistry literature[10] contains many examples of this type.

5.3.1.1 Cation Input

An early case is Chae and Czarnik's **87**[211] whose Hg^{2+}-induced hydrolysis produces strongly emissive **88** once the sulfurs are excised. Other cases which respond similarly to Hg^{2+} in water[212] and in organic solution are available.[213–215]

5.3.1.2 Organic Input

A classical design is the creation of a push–pull π-electron system with extensive delocalization. The output can be absorbance at long enough wavelengths where the starting material does not absorb. Strong fluorescence would also be excitable. An example is thiol-driven **89** where the fluorine is displaced by sulfur.[216–218] Similar cases driven by amines are also known.[219]

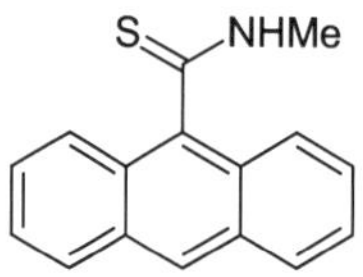

87

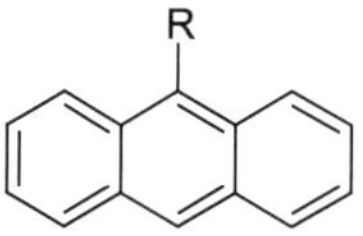

88; R = CO_2H
90; R = $CH_2NH(CH_2)_2OH$
91; R = CH_2NHMe

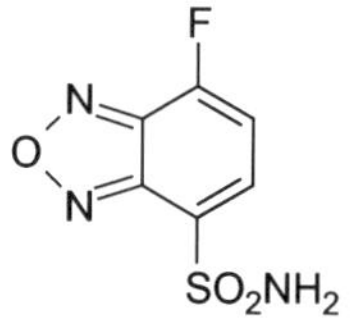

89

Ubiquitous CO_2 and toxic methyl isocyanate react with PET donor amino groups in **90**[220] and **91**[221] respectively, so that PET is prevented. The amino groups are converted to carbamate and urea functionalities respectively, leading to FE values of 2 and 100. Carboxylic acid-driven **92**[222] and acetic anhydride-driven **93**[223] are rather similar in design, where amides are produced from the amino groups. Conversion of an amine to a quaternary ammonium group is involved in fluorescent reagents for iodomethane and other alkylating agents.[224,225] Phosphine[226] and sulfide[227] PET donors can be used instead of amines. These can be converted to their oxides by reacting with hydroperoxides and peracids, so that PET is stopped. FE values as high as 30 are caused by cumyl hydroperoxide acting on **94**.[226] PET-suppression is also involved in Michael addition of thiols to enones **95** and **96** for instance.[210,228–230] Furthermore, nerve gases can be detected by fluorescent PET systems[231,232] and relatives.[233]

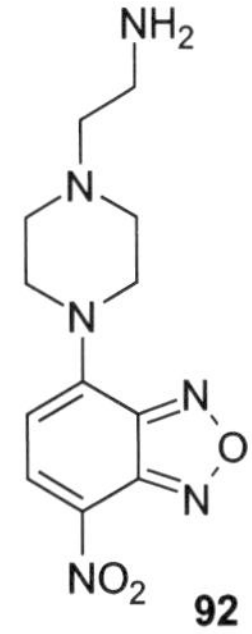

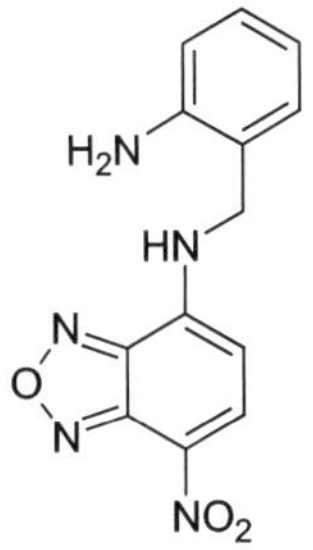

Acetone serves as the input to **97**[234] so that the aldol reaction which produces **98** takes the PET acceptor benzaldehyde out. From literature data,[235,236] we calculate that $\Delta G_{PET} = -0.6$ eV for **97**. An FE value of 78 is the result in neutral aqueous solution. There are older cases where a PET acceptor benzoic acid is taken out to cause fluorescence enhancement.[237] On the other hand, the PET-donor 1,2-diaminobenzene within **99**[238] is taken out with NO by conversion into a benzotriazole (FE = 12). The cyanine fluorophore allows excitation in the red. Monitoring NO in vital processes with fluorescent probes is an important area of research.[239–241]

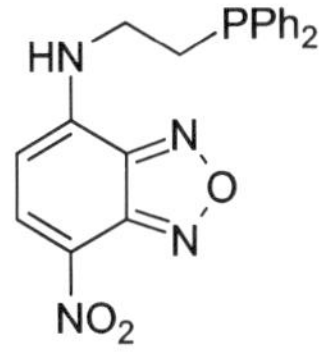

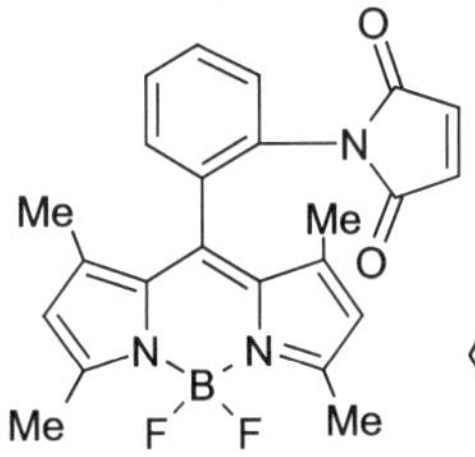

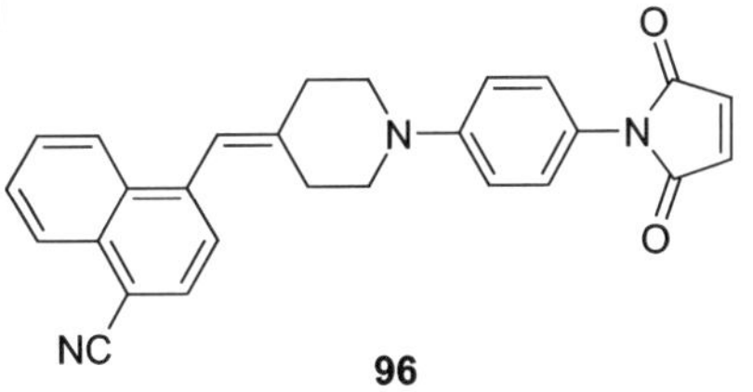

D'Souza's **100**[242] switches 'on' fluorescence in the presence of hydroquinone. Mechanistically, this is based on hydroquinone–benzoquinone interaction *via* hydrogen bonding and charge transfer.

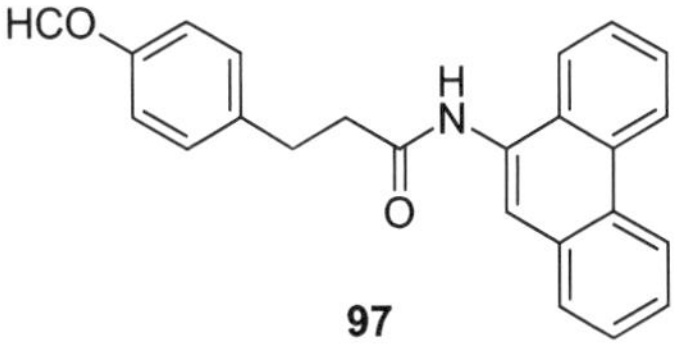

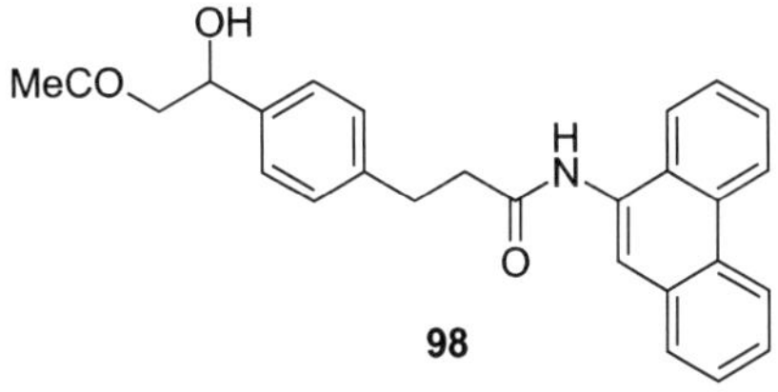

100; R = benzo-1,4-quinon-2-yl
128; R = 1,4-hydroquinon-2-yl

5.3.1.3 Oligonucleotide Input

DNA-based computing, but without direct use of Boolean logic functions, has been known since 1994.[243] Stojanovic and Stefanovic develop deoxyribozyme-based logic gates[244] by applying small molecule logic[245] to DNA. Deoxyribozymes are DNA-based enzymes, many of which require Mg^{2+} for their activity. The self-complexed stem–loop on the right hand side of **101** can be opened by offering it a stronger hybridization opportunity with input oligonucleotide **102**. Then the substrate **103** can bind across the bottom motif[246] and be stretched to the point that hydrolysis occurs at the ribonucleotide rA. Once fragmented in this way, the fluorescein unit is separated from the EET acceptor tetramethylrhodamine at the other end of **103**. Thus when excited at 480 nm, fluorescein emission at 525 nm emerges as the 'high' output. If input **102** is not applied, fluorescein emission is quenched by EET. A YES logic operation is thus satisfied. This fine application of molecular beacons[179] leads to various logic applications which will be featured in later sections. It must be noted, however, that this approach has an in-built irreversibility, unlike electronic or photoionic approaches, because a chemical hydrolysis reaction is involved. Furthermore, NOT logic would be the result if the fluorescence acceptor (tetramethylrhodamine) is used as our output. This illustrates wavelength reconfigurable logic (Chapters 6 and 13). The need to use a ribonucleotide in this approach can be a problem owing to their ease of hydrolysis, but this can be avoided[247] by using oligonucleotides with entirely deoxysugars and a Cu^{2+}-dependent DNA-cleaving deoxyribozyme.[248]

The approach in the previous paragraph can be extended to produce greater diversity of gates in the following way.[249] Let us remove the stem of a suitable deoxyribozyme and attach two runs of bases to produce two separate strands, *e.g.* **104** and **105**. These strands cannot stay together because they lack complementary runs of sufficient length. However, **104** and **105** can be brought together by offering the input strand **106** (to regenerate a stem–loop) and also the substrate **107**. The symbol BHQ1 in structure **107** stands for esterified **77**. Then the substrate will be hydrolyzed as before and recover the fluorescein emission. While this individual case illustrates YES logic, the road is clear for production of many other gates by using libraries of the various strands.

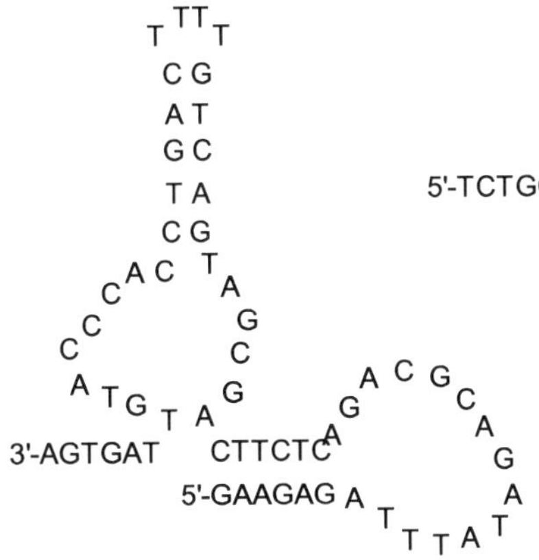

Single Input–Single Output Systems

81

5'-TCTGCGTCTATAAAT-3'
102

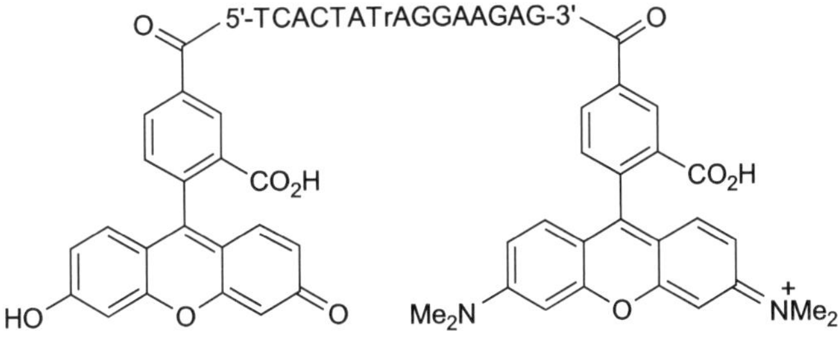

101

103

5'-CGAATCCTGAGTCTACAAATACCTA-3' **106**

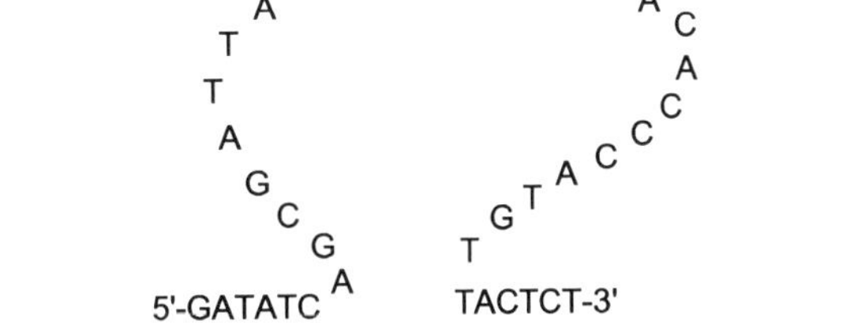

104 **105**

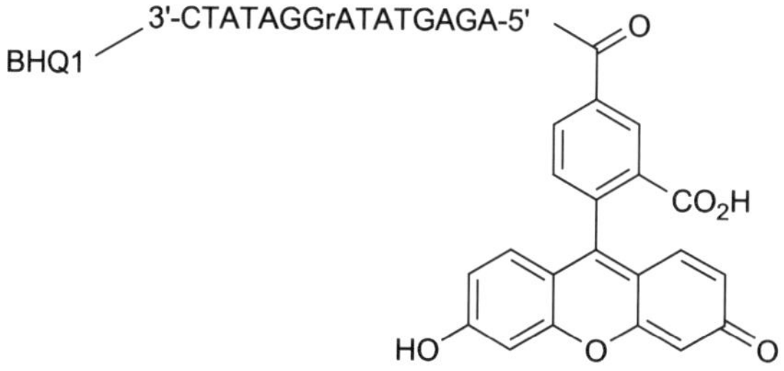

107

5.3.1.4 Protein Input

Protein stains are common,[27,250] but improvements are always welcome. Large blue-shifts in the emission (63 nm) and absorption (118 nm), a large protein-induced FE value of 50, small size of reagent and no change in the charge of protein upon derivatization, are all delivered by Wolfbeis' **108** when faced with human serum albumin as input.[251] The last two characteristics are important in order to preserve a single spot in electrophoretograms, especially under conditions of multiple labelling. Pyrylium ion **108** reacts with protonated lysine units of proteins at pH 7.2 to produce the corresponding pyridinium ion with a different degree of ICT. While **108** shows protein-driven YES logic in its emission output, its absorbance shows wavelength-reconfigurable logic which would fit in Chapters 6 and 13.

Many enzyme assays generate absorbance or fluorescence outputs.[252] One case is where a protein kinase serve as input to pyrene-appended peptides which contain a serine as well as several arginine residues.[253] The latter create a cationic and hydrophobic region. The peptide is exposed to an anionic dye before the kinase is presented so that the non-covalently bound dye quenches the pyrene fluorescence *via* EET. When the kinase arrives, the serine is phosphorylated. This allows a phosphoserine binding domain to capture the peptide so avidly that the dye is sloughed off. Strong fluorescence of pyrene is the result. Another important case is a protease assay. When the terminals of suitable peptides are derivatized with **109** and **110** as EET donor and acceptor respectively, the fluorescence output is low. When the correct protease is applied, the peptide is hydrolysed and the freed **109** emits strongly.[254]

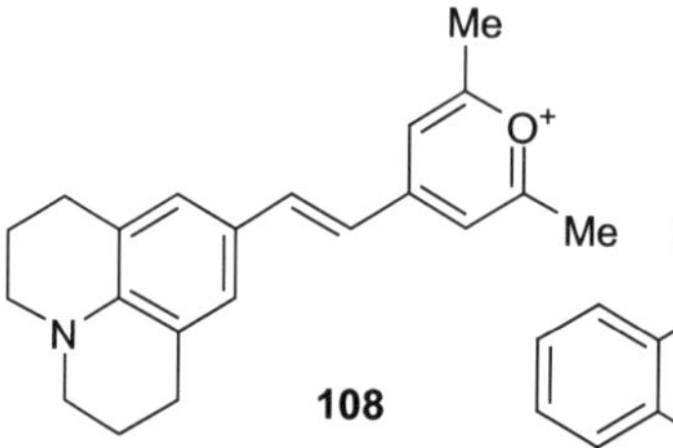

108 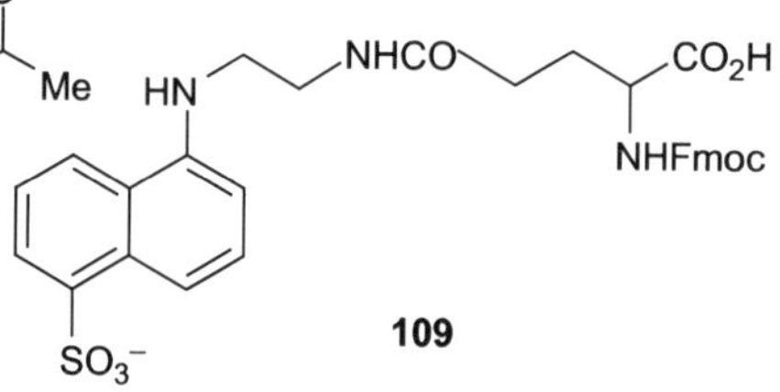   **109**

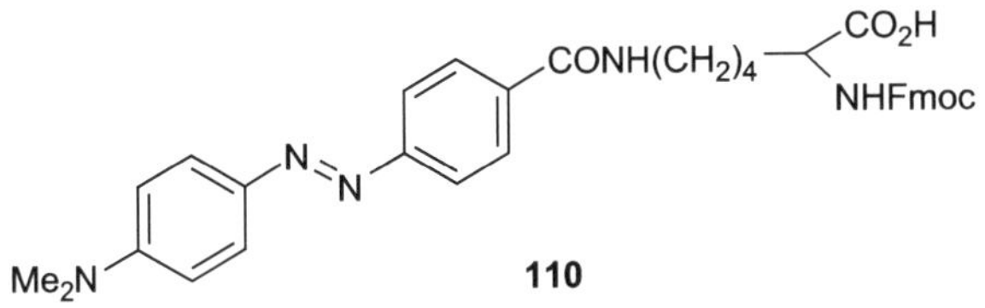

Single Input–Single Output Systems

110

5.3.2 Light Dose Input

Photochemical formation of dyes represents light dose-driven YES gates when their absorbance or fluorescence[206] is monitored. For instance, irradiation of **111** at 254 nm leads to growth of absorbance at 337 nm because the more-delocalized **112** is formed by dehydrohalogenation.[255]

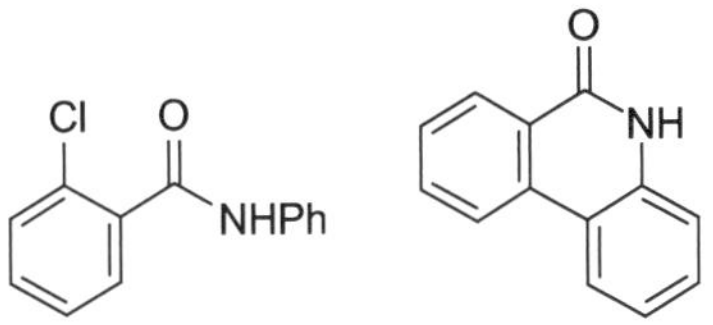

111 **112**

5.4 NOT

NOT logic naturally has the output behaving opposite to the input. In contrast to YES gates, the non-electronic NOT logic devices can be discerned in many sensors and reagents whose absorbance or fluorescence responds to the input in an 'on-off' manner. In other words, the application of a 'high' input signal causes the output signal to fall.

5.4.1 Electronic Input

The inversion of a voltage signal in the electronic case receives a molecular implementation within a single bundle of carbon nanotubes.[256] Avouris' team treats part of this intrinsically *p*-type bundle with potassium metal to leave it *n*-doped. This forms the twin hearts of connected, but complementary, field-effect transistors, which produces NOT logic behaviour in the same way as currently seen in common semiconductor devices (Figure 5.9). While this NOT gate is not at molecular scale, subsequent studies apply similar ideas and techniques to a single carbon nanotube.[257,258] Quinquethiophene-based monolayers self-assembled on silicon give rise to NOT logic in a similar way.[259]

5.4.2 Chemical Input

As indicated in section 5.4, a convenient source of chemically driven NOT gates is the sensor literature, particularly the analytical fluorescence literature.[10] This is to be expected because the quenching of fluorescence is a common phenomenon.

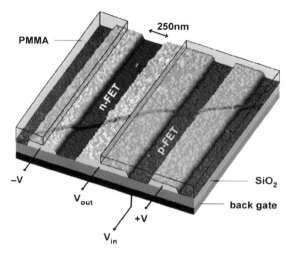

Figure 5.9 NOT logic gate constructed from a carbon nanotube bundle. Various voltages applied to the device are shown.
Reproduced from V. Derycke, R. Martel, J. Appenzeller and P. Avouris, *Nano Lett.*, 2001, **1**, 453 with permission from the American Chemical Society.

5.4.2.1 Cation Input

Gate **113**[260] is a 'fluorophore–spacer–receptor' PET system like **38** but with opposite characteristics. Gate **113** contains a pyridine receptor for H^+ which becomes more reducible when bound whereas **38** has an amine receptor which becomes less oxidizable upon protonation. Therefore **113** launches fluorescence-quenching PET from the fluorophore to the pyridine receptor only when H^+ arrives. Details for this and several other cases are given in Table 5.3.

The fluorescence of borondipyrromethine derivative **114**[261] is selective (FE = 0.005) for Ca^{2+}. Ca^{2+} inclusion in the calixcrown allows H^+ ejection from the proximal phenol[271] to produce a phenolate PET donor.[272]

Another way of designing fluorescent NOT logic gates is to arrange PET involving the input guest species itself. Redox-active metal ions such as Hg^{2+} and Cu^{2+} are usable in this way, though the latter metal also introduces an EET channel owing to its low-lying d–d excited state.[273,274] Several examples of this type are summarized in Table 5.3.

Langmuir–Blodgett films of **119**,[266] as well as assemblies of its fluorophore and receptor components, also show Cu^{2+}-driven NOT logic action regarding their fluorescence output. Similar behaviour is found when related components are assembled in micelles[275,276] or on silica particles.[277] An attraction of the self-assembled approach is the flexible choice of modules that can be made in a combinatorial manner without the inconvenience of organic synthesis. However, some synthesis cannot be avoided if modules are to be endowed with suitable functionality. Besides the normal Cu^{2+}-induced PET/EET effects, metal-induced deprotonation of the sulfonamide fluorophore also governs the fluorescence quenching.

Single Input–Single Output Systems

Table 5.3 Collected data for some molecular NOT logic gates.

Device^a	Input^b (hi,lo)^c	Output^d (lo,hi)^e	Power^f	Characteristics^g
113[260] in MeOH:H$_2$O	H$^+$ (10^{-2}, 10^{-7})	Flu$_{337}$(0.02,1)	hν_{348} logε 4.2	logβ_{H+} 4.5
114[261] in MeOH:H$_2$O	Ca^{2+} ($10^{-4.6}$,0)	Flu$_{507}$(0.005,1)	hν_{480}	logβ_{Ca2+} 5.7
115[262] in H$_2$O	Hg^{2+} (10^{-5},0)	Flu$_{381}$(0.17,1)	hν_{311}	
116[263] in H$_2$O	Hg^{2+} (10^{-4},0)	Flu$_{416}$(0.05,1)	hν_{310} logε 3.9^h	
117[264] in Dioxan–H$_2$O^i	Hg^{2+} ($10^{-3.3}$,0)	Flu$_{524}$(0.016,1)	hν_{480} logε 4.2	logβ_{Hg2+} 4.9
118[265] in H$_2$O–DMF	Cu^{2+} ($10^{-3.9}$,0)	Flu$_{558}$(0.02,1)	hν_{485}	logβ_{Cu2+} 6.3
119[266] in H$_2$O	Cu^{2+} (10^{-5},0)	Flu$_{508}$(0.35,1)	hν_{350}	
120[267] in MeCN-H$_2$O	MoO$_4^{2-}$ (–,0)	Lum$_{615}$(0.3,1)	hν_{460}	
121[268,269] in DMSO	CH$_3$CO$_2^-$ (10^{-2},0)	Flu$_{419}$(0.35,1)	hν_{370}	log$\beta_{CH3CO2-}$ 2.2
122[270] in MeCN	CH$_3$CO$_2^-$ (10^{-3},0)	Flu$_{422}$(0.02,1)	hν_{265} logε 4.4	log$\beta_{CH3CO2-}$ 5.2

^{a}The device is taken as a given molecular structure in a given medium.
^{b}Chemical concentration unless noted otherwise.
^{c}High (hi) and low (lo) input values. All concentrations are in M units.
^{d}Output is given as the intensity of fluorescence (or the absorbance) at a given wavelength. All wavelengths are in nm units.
^{e}High and low output values. The 'high' output value observed is scaled to unity.
^{f}Power supply is light at a given wavelength. The extinction coefficient (ε) of the device at that wavelength is given in units of M^{-1} cm^{-1}.
^{g}Characteristics of device operation regarding inputs include chemical binding constants (β) in units of M^{-1} Characteristics of device operation regarding outputs include emission quantum yields (ϕ).
^{h}At 370 nm.
^{i}The structure **117** is given in Chapter 4 as **3**.

5.4.2.2 Anion Input

A natural siderophore serves as a selective receptor for MoO$_4^{2-}$ within device **120**.[267] When the input species is received at pH 5.7, a PET process occurs from the catecholate component to the Ru(II) lumophore[278] to quench the luminescence by a factor of 3.3 (LE = 0.3).

A thiourea receptor within **121**[268,269] and a urea receptor within **122**[270] bind CH$_3$CO$_2^-$, resulting in the quenching of fluorescence. This is due to enhancement of PET from the receptor to the fluorophore by the anionic charge.[279] Though a formal covalent spacer is not present in **122**, the planes of phenanthroline and urea units are tilted by 35° in its X-ray crystal structure, so that PET occurring across a virtual spacer[44,106] can be imagined. So we have acetate-driven NOT logic in both these cases. The binding constants and FE values can be increased by using an *N*-amidothiourea receptor instead of thiourea.[280]

The pyridine-2,6-dicarboxamide receptor within **123**[281] binds F$^-$, though not selectively. In this case, the luminescence quenching (LE = 0.03) can be assigned to a Born–Oppenheimer hole created at the N...H...F bond.[98,282,283] The luminescence quenching response of the Zn^{2+} complex of **124**[284] to $H_2PO_4^{2-}$ (LE = 0.12) seems to be similar. The F^--induced red shift in the emission of **123** is caused by stabilization of the Re(I) MLCT excited state by the extra anionic charge on the amidopyridine units being pumped onto the metal centre.

113

114; R =

115

116

118; R = $(CH_2)_4NH(CH_2)_2NH_2$

119

121; R = 9-anthrylmethyl

122

120

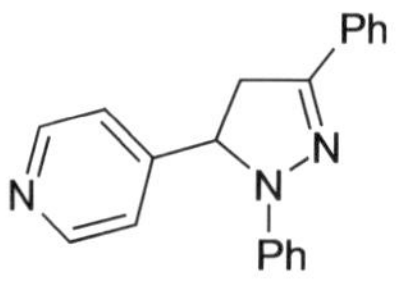

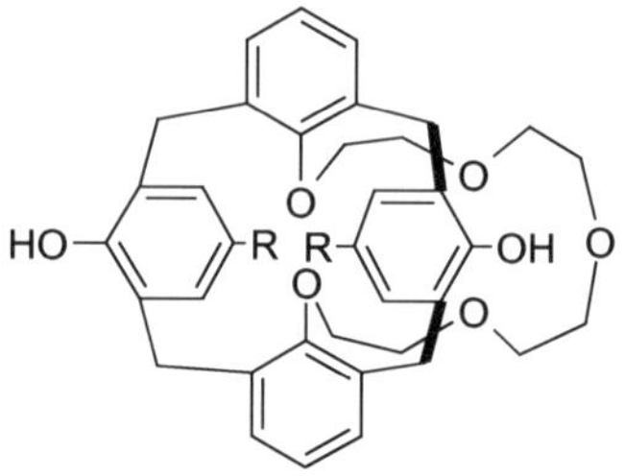

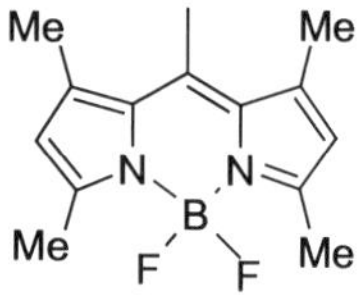

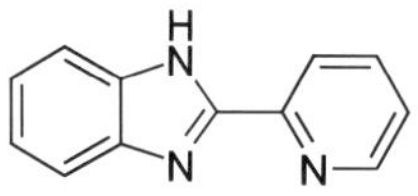

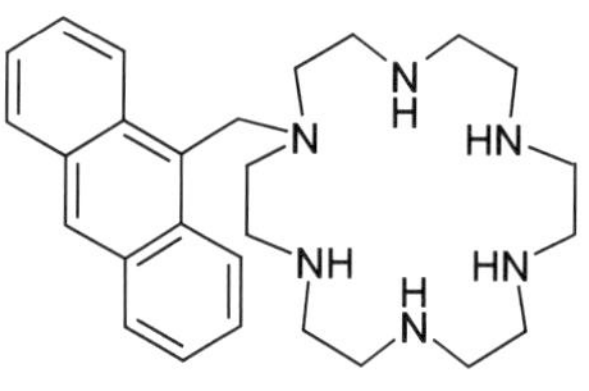

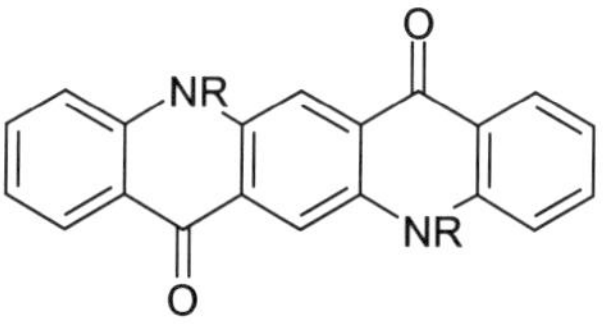

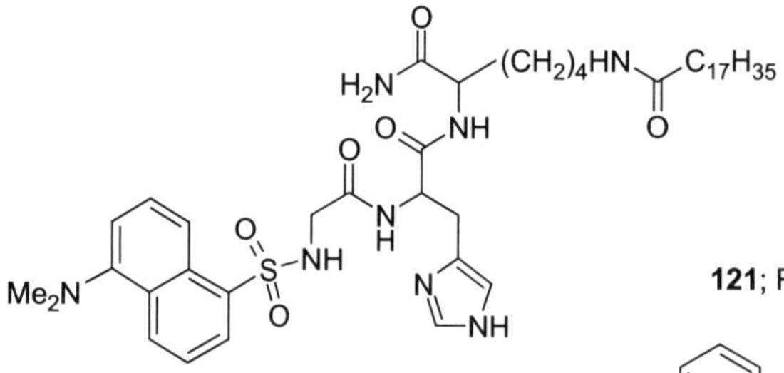

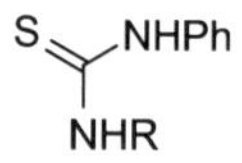

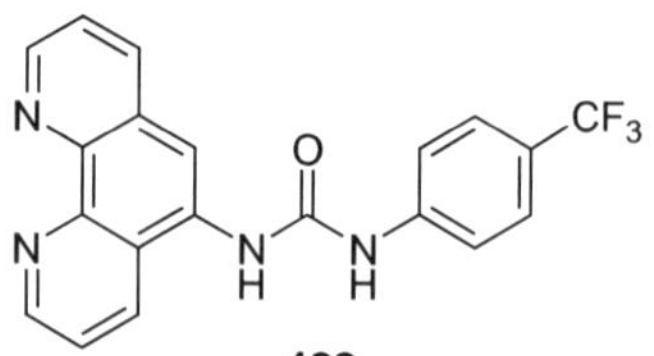

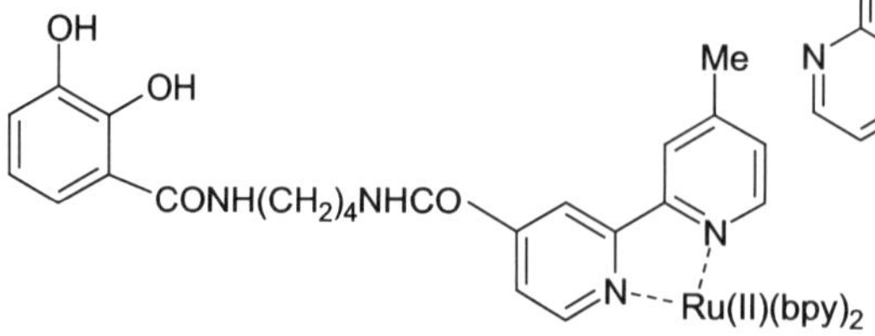

Single Input–Single Output Systems

A guanosine triphosphate (GTP)-induced FE of 0.16 is seen for **125**,[285] possibly due to PET from GTP to the electron-deficient anthracene fluorophore and also due to hydrogen bonding from the bound GTP to the anthracene π-cloud.[286]

The intercalation complex **126.127** is brightly emissive because the fluorophore **126** is protected from water.[287] A new intercaland serves as the input by displacing **126** into bulk water so that fluorescence is quenched.[288,289]

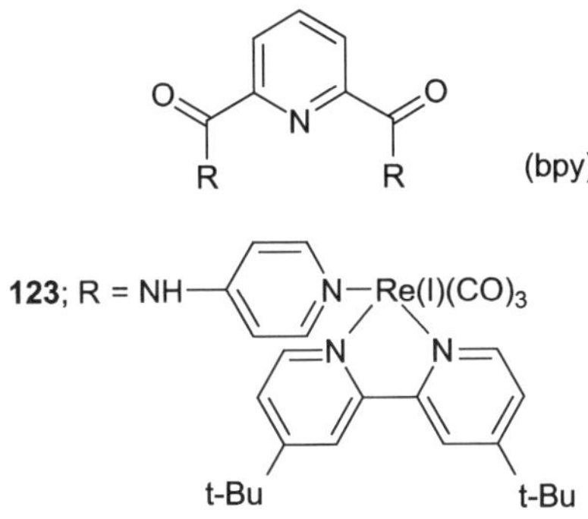

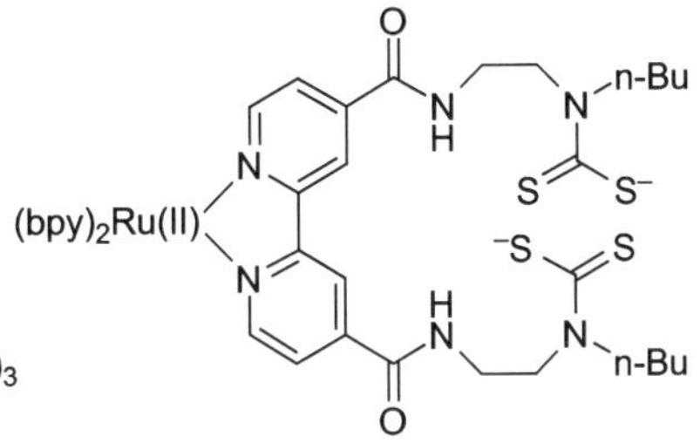

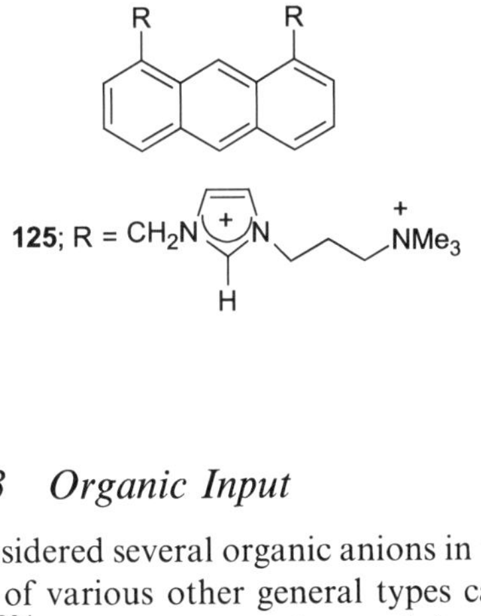

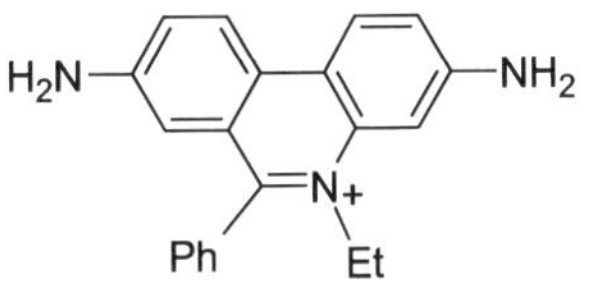

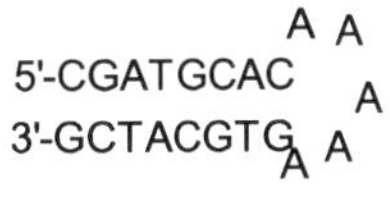

5.4.2.3 Organic Input

We considered several organic anions in the previous section. However, organic species of various other general types can serve as inputs to NOT logic devices,[290–294] and these are featured here. For instance, the fluorescence of D'Souza's **128**[290] is quenched by exploiting hydrogen-bonding and π-π stacking capabilities of benzoquinone (**129**) to pair with the hydroquinone moiety of **128** (the structure **128** is given next to its cousin **100**). Just as the electron deficiency of **129** led to **128**'s NOT logic behaviour, viologens quench the fluorescence of dialkoxynaphthalene units within larger structures.[291] Czarnik's **130**[292,293] interacts with catechol derivatives to produce **131**. It is probable that **131** fits the format of a twisted PET system with a virtual C_0 spacer. So it is understandable why the fluorescence of **130** is sharply quenched upon transformation to **131**.

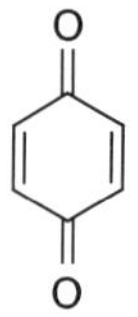

The complex between **132** and a boronic acid is brightly emissive because ESIPT[295] is prevented by conversion to a boronate ester.[296] Addition of fructose displaces **132** so that fluorescence is killed off.[296] Another case involving a competition occurs when thiols ligate to cyclen-bound Zn^{2+} within **133**.[297] The metal centre also carries a bound imide by a combination of coordinative and hydrogen-bonding interactions.[298] The imide is poorly emissive but the fluorescence is improved after ligation to the Zn^{2+} centre.[297,299] An $n\pi^*-\pi\pi^*$ state inversion is involved here (Chapter 4) because the imide is deprotonated during the binding. Application of the glutathione input to **133** causes the imide to be displaced in the neutral form so that fluorescence at 470 nm (λ_{ex} = 355 nm) is weakened (FE = 0.47).

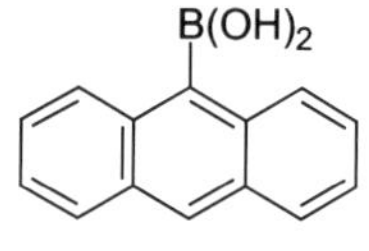

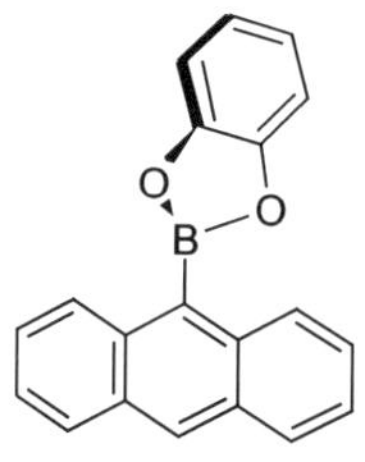

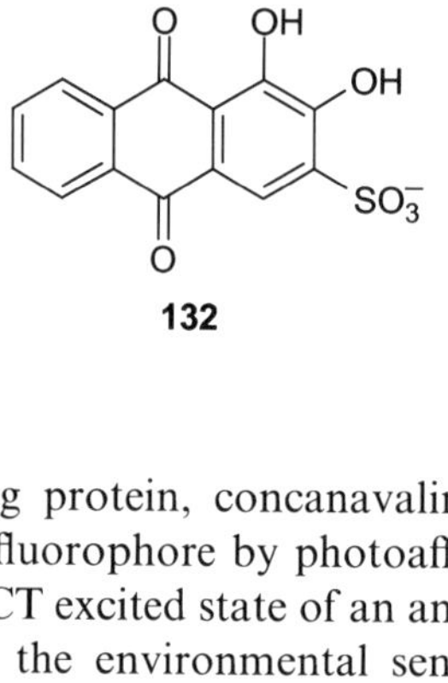

A sugar-binding protein, concanavalin A (Con A) is derivatized with a polarity-sensitive fluorophore by photoaffinity labelling and subsequent modification.[300] The ICT excited state of an aminonaphthalene sulfonate derivative is responsible for the environmental sensitivity. Application of methyl-α-D-mannoside causes the active site of Con A to be occupied and pushes the fluorophore to a more polar environment so that the fluorescence intensity decreases (and red shifts).

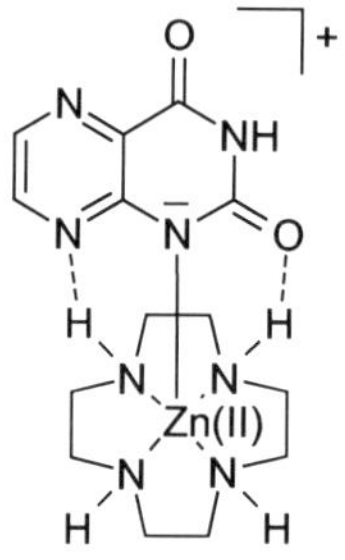

Addition of **134** causes the uv–visible absorbance of **135**[301] at 386 nm to drop by a factor of 1.4, as well as a red shift of 11 nm. This is caused by π–π stacking besides N–H–N hydrogen bonding, even though the complexation also relies on N–H–O hydrogen bonding. The fluorescence output at 434 nm is also perturbed similarly (FE = 0.66), though a PET process from the naphthalene to the acridine probably also contributes to this. Significant enantioselectivity is another asset of this system. Enantioselective quenching of fluorescence is also seen when chiral amino alcohols serve as the input to dendrimer **136**.[302] Excited state proton transfer from the binaphthol core to the amino alcohol is responsible for the quenching.

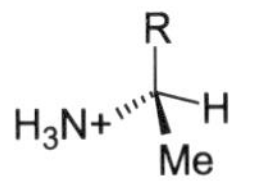

134; R = 1-naphthyl

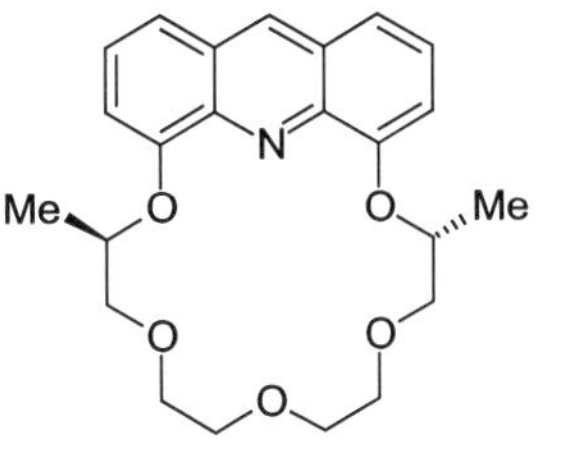

135

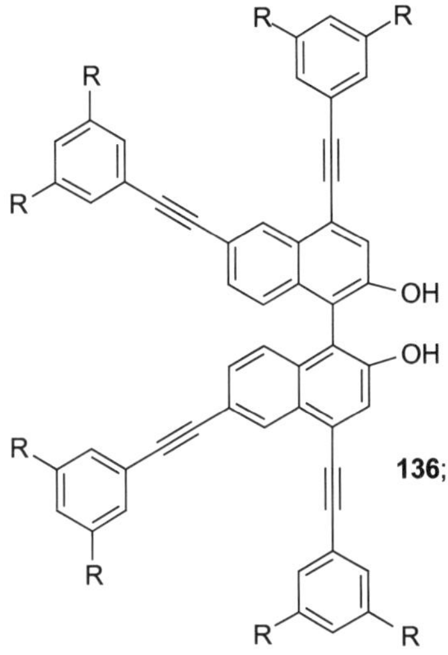

136; R =

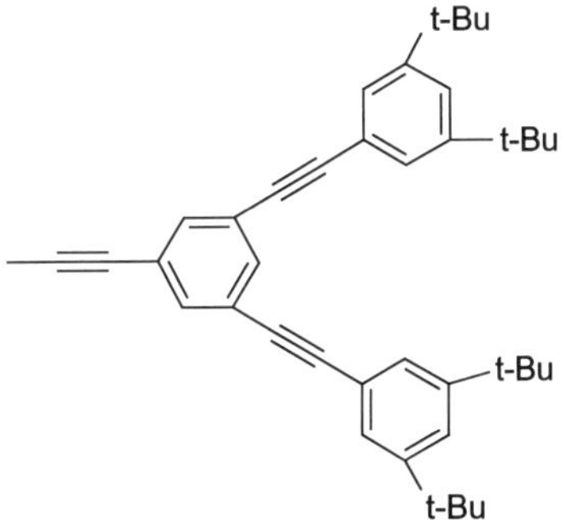

The ends of an aptamer are derivatized with fluorophore **137** and quencher **138**.[303] When cocaine is applied to the aptamer, it folds to form a three-way junction so that **137** and **138** become neighbours. EET then causes quenching of the emission (FE = 0.6).

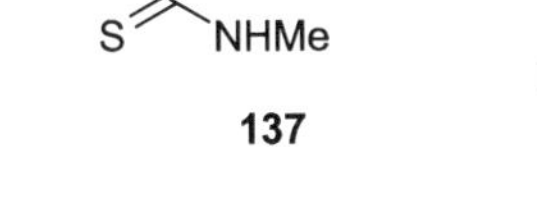

5.4.2.4 Polarity Input

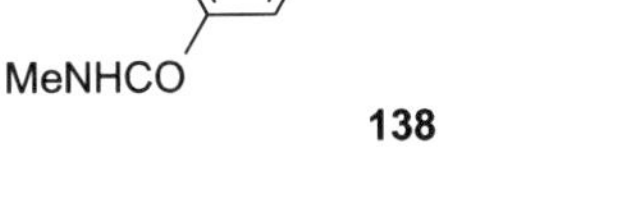

An early case to show well-characterized fluorescence switching 'on' with decreasing environmental polarity (upon engulfment in bovine serum albumin) is Daniel and Weber's **139**.[180] Following Kosower's extensive studies,[304] it is clear that the aniline donor and the naphthalene acceptor gives **139** a TICT excited state[106] which is stabilized in polar media. As mentioned in section

5.2.2.1.2, **12**[38] also contained the amino–NBD fluorophore whose emission intensity (and wavelength) shows similar environmental sensitivity.[305]

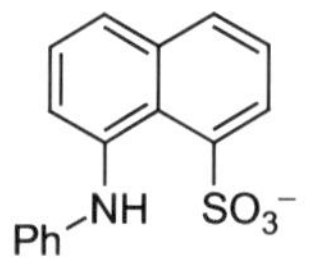

139

Fluorescent PET sensors for polarity,[39,306–309] such as **140**,[306] **141**[39] or **142**,[308] are based on an amine functionality. Their weakness of sensitivity to trace acids is repaired in cases like **143**,[310] where 9-cyanoanthracene is the fluorophore and a dioxyphenyl unit is the electron donor. In all these cases, fluorescence switching from 'on' to 'off' with increasing polarity is due to the stabilization of the radical ion pair formed after PET (Figure 5.10).

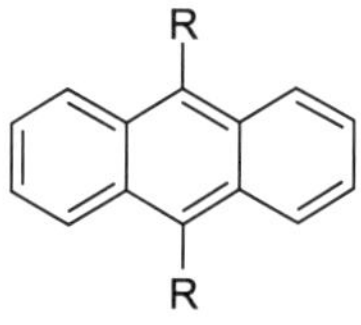

140; R = CH$_2$NHCH$_2$Ph

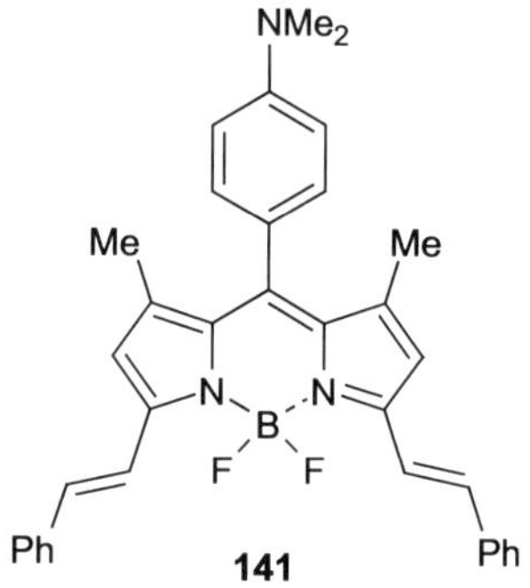

141

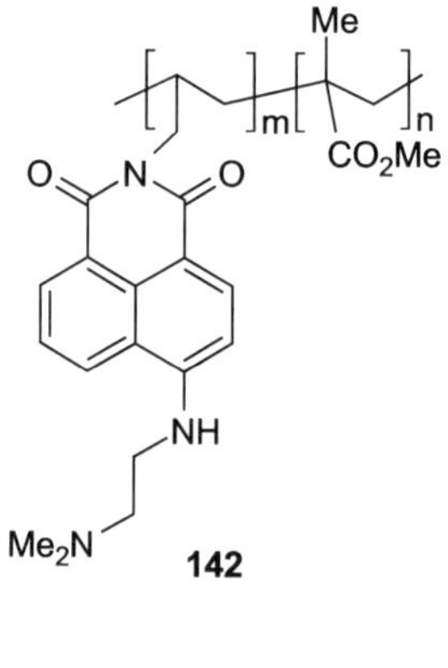

142

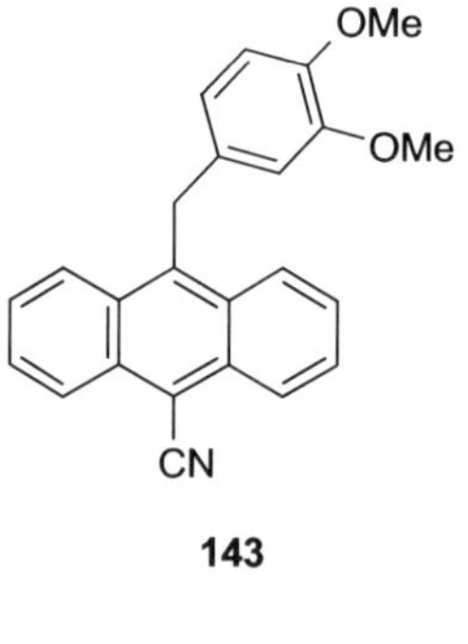

143

5.4.3 Temperature Input

The Arrhenius relationship between rate constants and temperature is well known to chemistry undergraduates.[311] When such a process is made to compete with fluorescence, it is clear that the observable emission intensity can decrease strongly with increasing temperature in suitable instances, for example **144**.[312] The process of interest in **144** is a photodissociation of the N1–C5 bond,

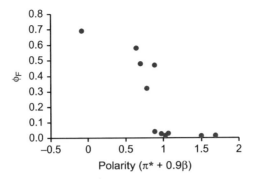

Figure 5.10 Fluorescence quantum yield of **143** as a function of solvent polarity, as measured by $(\pi^* + 0.9\beta)$. The π^* and β are Taft dipolarity and hydrogen bond basicity respectively.
Redrawn from D. C. Magri, J. F. Callan, A. P. de Silva, D. B. Fox, N. D. McClenaghan and K. R. A. S. Sandanayake, *J. Fluoresc.*, 2005, **15**, 769 with permission from Springer publishers.

which is largely reversible owing to intramolecular radical recombination within the solvent cage. Since the temperature is smoothly signalled by the fluorescence intensity, **144** can serve as a luminescent thermometer.[193,194] Molecule **144** is also a NOT logic gate with temperature input and fluorescence output as given in the truth table (Table 5.4).

Thermally driven polymeric NOT gates are also known. These are versions of the YES gates **82**[195] from section 5.2.3 where the fluorophore is altered so as to be emissive in water environments. A suitable fluorophore would be **80**[189] (featured in section 5.2.2.8). Polymer **145**[189] demonstrates this line of reasoning.

5.4.4 Light Dose Input

Rotaxanes are some of the most sophisticated structures to exhibit fluorescence output according to NOT logic in response to a light dose input.[313] Tian has some important examples.[314] Leigh's **146.147** is illustrated here.[315] Fast PET occurs from the anthryl fluorophore to the pyridinium unit, provided that their separation is minimized by locating the macrocycle on the glycylglycine unit.

Table 5.4 Truth table for **144**.

Input Temperature	Output Fluorescence[a]
low (15 °C)	high (0.56)
high (60 °C)	low (0.24)

[a]Quantum yields in hexane, λ_{exc} 380 nm, λ_{em} 436 nm

Initially, the macrocycle is positioned on the *E*-fumaric acid diamide unit by strong hydrogen bonding, but irradiation at 312 nm isomerizes the latter to the *Z*-form so that the macrocycle is no longer stable there. The input-induced FE value is 0.01.

Turning to 'simpler' structures, we see that the fluorescence emission from the phosphorus(V) porphyrin unit of **148** can be modulated with a light dose input (345 nm) directed at the azobenzene moieties.[316,317] Azobenzene *E–Z* photoisomerization then takes place, though EET from the azobenzene to the phosphorus(V) porphyrin will be competitive. Excitation of the porphyrin unit results in the quenching of fluorescence output at 565 nm due to PET to the porphyrin from the azobenzenes. However the emission from *E*-**148** is twice as high as that from *Z*-**148**, probably because of the shorter centre-to-centre distance for PET in the latter. So we see light dose-driven NOT logic action if we start with *E*-**148**. Thermal relaxation of *Z*-**148** in the dark recovers *E*-**148**.

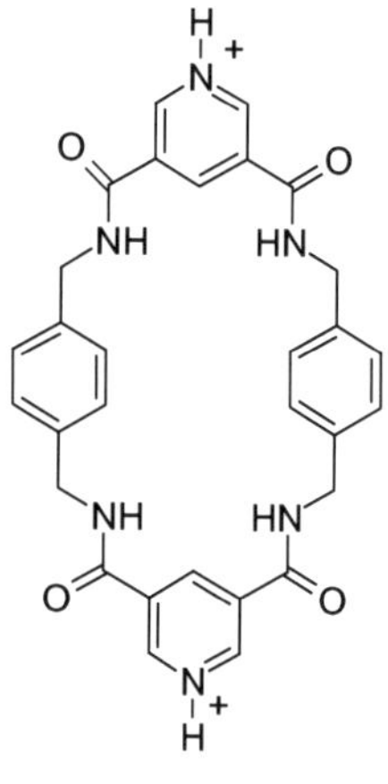

146

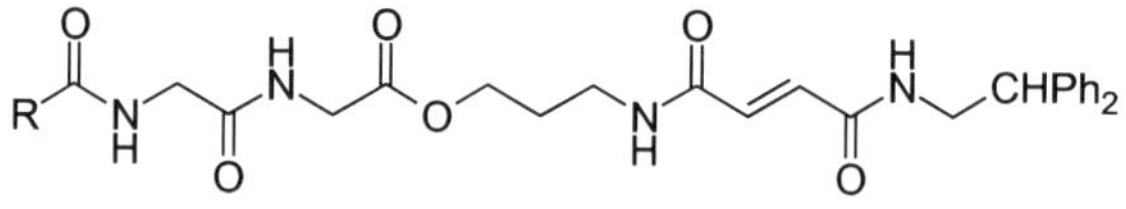

147; R = 9-anthryl

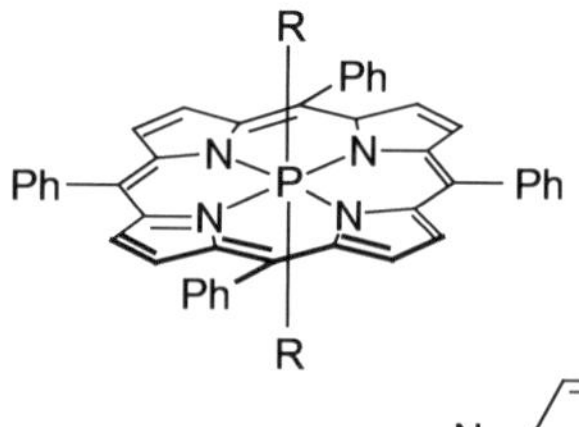

148; R = O—⟨aryl⟩—N=N—⟨aryl⟩—NO$_2$

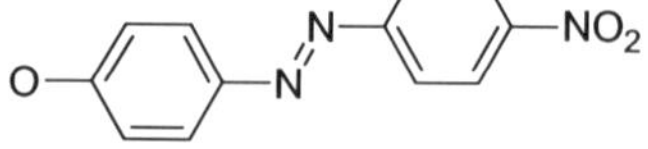

5.5 Irreversible NOT

As mentioned in section 5.3, irreversible interactions are also useful in molecular logic studies, especially when no help is forthcoming from supramolecular science. A selection of these cases is considered below.

5.5.1 Anion Input

Porphyrin **149** carrying 2′,6′-di-*t*-butylphenol units at the four *meso* positions offers an interesting case of fluorescence switching 'off' with input of F^-.[318] The facts that (a) the pale yellow solution (at 10^{-5} M) also becomes strongly blue, (b) the 'off' state persists even when F^- is washed away and (c) air is required for this transformation hint at the mechanism. The 'off' state is oxoporphyrinogen **150**. Reset can be arranged by adding ascorbic acid which reduces **150** back to **149**. Compound **149** can also be viewed as a non-volatile memory (see section 11.2 concerning memories or R–S latches).[204,319] F^- also serves as input to subphthalocyanine **151** to displace irreversibly the axial phenolate moiety. Tian's group find a slow loss of absorbance at 565 nm and emission at 580 nm (excited at 510 nm) in this case (Figure 5.11).[320]

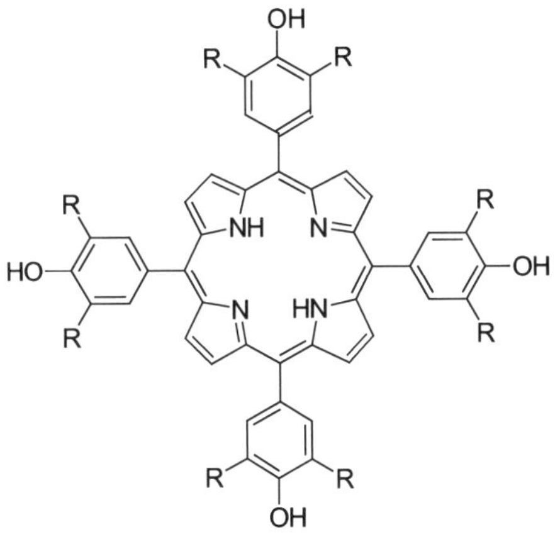

149; R = t-Bu

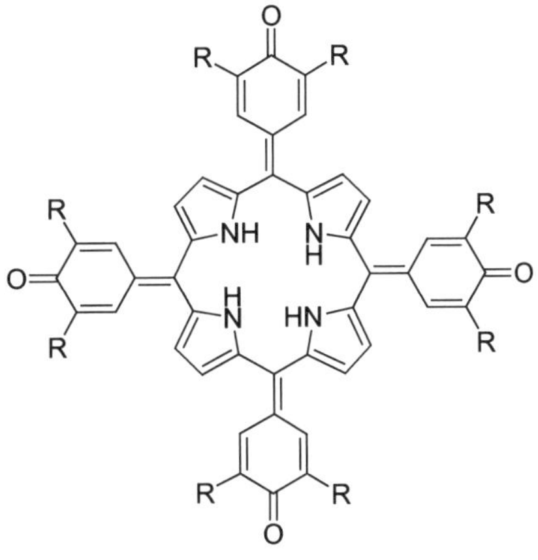

150; R = t-Bu

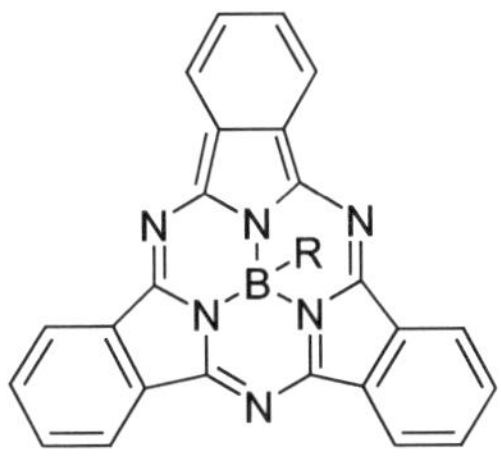

151; R = O—⟨⟩—CHO

5.5.2 Oligonucleotide Input

Though anionic, oligonucleotides possess sufficient rich structural attributes to deserve a section of their own. Stojanovic's approach with oligonucleotides has

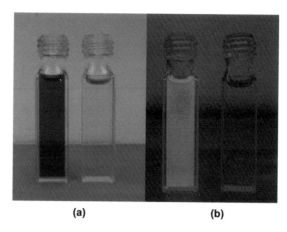

Figure 5.11 (a) Colour under ambient light and (b) fluorescence under 365 nm excitation of **151** in THF without (left) and with F$^-$ (right).
Reproduced from S. Xu, K. C. Chen and H. Tian, *J. Mater. Chem.*, 2005, **15**, 2676 with permission from the Royal Society of Chemistry.

been featured under YES logic in section 5.3.1.3. Suitable modification of that deoxyribozyme to **152** sets the scene for a NOT logic operation.[244] In a sense, this would be a logic reconfiguration, which could have been discussed in Chapter 6 or 13. The oligonucleotide substrate is the same, **103**, as that used for YES logic. When excited at 480 nm, fluorescein emission at 525 nm is weak owing to quenching by EET to the tetramethylrhodamine acceptor. However, feeding of substrate **103** to the deoxyribozyme **152** causes the fluorescein and tetramethylrhodamine units to separate so that fluorescein emission at 525 nm becomes strong ('high' output). Now, a new oligonucleotide **153** (which will be our input) can be designed to hybridize with the deoxyribozyme **152** by stretching it out so that the catalytic stem–loop-like motif is lost. Thus, when input **153** is present (input level 'high'), the substrate remains as it is and the fluorescein emission stays weak (output level 'low'), *i.e.* NOT logic. The fact that observation of the fluorescence acceptor (tetramethylrhodamine) instead would give YES logic corresponds to a wavelength reconfigurable case and will be mentioned in Chapter 6. Kolpashchikov has a fast-acting variant based on DNA four-way junctions.[321] Here, a stem–loop with a fluorophore and EET-based quencher at its terminals plays a starring role.

5.5.3 Protein Input

As found with many lanthanide complexes carrying sensitizer units,[92] Nagano's **154**.Tb^{3+} emits strongly following excitation of the amidocarbostyril chromophore.[91] Leucine aminopeptidase selectively cleaves off the leucine terminal of **154**.Tb^{3+}. The aniline unit so exposed launches a PET process[89,90] to the

Single Input–Single Output Systems

amidocarbostyril chromophore. This prevents the formation of its triplet excited state, which is needed for efficient Tb^{3+} emission *via* EET.

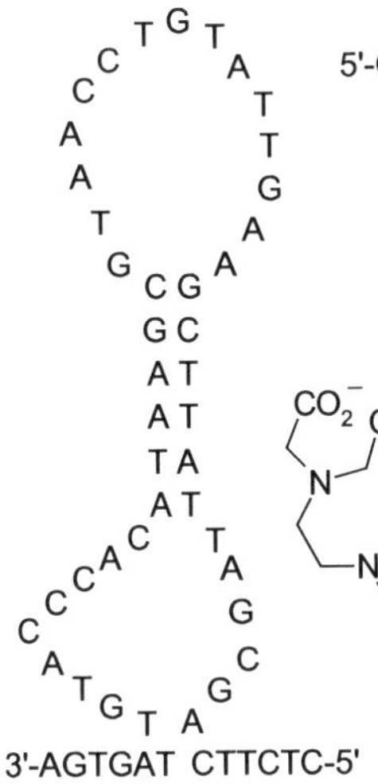

[Structure 152: DNA hairpin with loop sequences, paired with 5'-CATTGGACATAACTT-3' (153); and structure 154 showing a chelator-linked amidocarbostyril chromophore with i-But group]

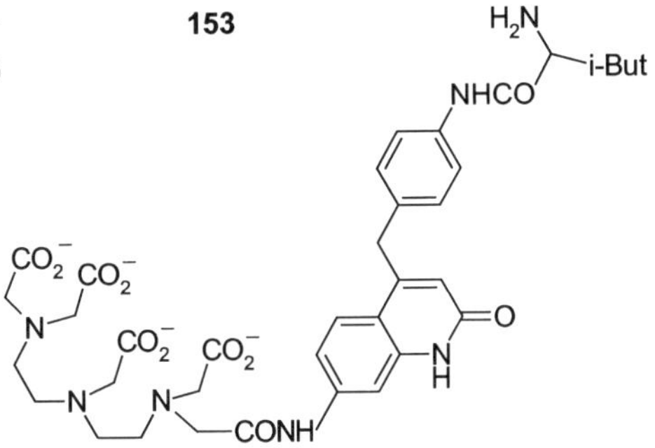

5.6 PASS 1

PASS 1 logic corresponds to a situation where the output stays at a 'high' level whatever the value of the input. In the electronic case, a live output unconnected to the input line would achieve this result. Common electronic logic gates – NOT and OR – can be wired together to give PASS 1 logic as shown in Figure 3.2. For molecular cases using a chemical input and fluorescence output, we can turn to a simple fluorophore devoid of receptors. Among many examples would be **155**, which would give a 'high' fluorescence signal (when appropriately excited) whether or not, say, H^+ is present. If an absorbance output is required, monitoring at the absorption band maximum wavelength is sufficient. If a transmittance output is desired, observation well away from absorption bands would be necessary. Such wavelength-dependent observations will be discussed in Chapter 13. Of course, we can avoid using a chromophore altogether to get the same result.

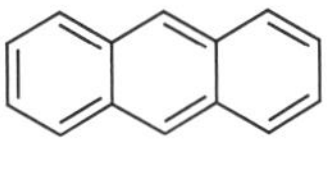

[Structure 155: anthracene]

5.7 PASS 0

PASS 0 is the opposite instance (*c.f.* section 5.6), where the output remains 'low' whatever the input situation. In the electronic case, all this requires is an earthed output unconnected to the input. The same outcome can be arranged by employing common electronic gates – NOT and AND gates – as shown in Figure 3.2. A molecular implementation with chemical input and fluorescence output would be any molecule without a fluorophore. Even observation at a wavelength remote from the emission band of a fluorophore, say **155**, would suffice. The absorbance output would correspond to PASS 0 logic if the monitoring is conducted outside the absorption bands. If transmittance of light is employed as output, observation at a wavelength near the absorption band maximum of a simple chromophore without receptors would do the trick.

References

1. J. Jortner and M. A. Ratner (ed.), *Molecular Electronics*, Blackwell, Oxford, 1997.
2. H. Haick and D. Cahen, *Progr. Surf. Sci.*, 2008, **83**, 217.
3. J. L. He, B. Chen, A. K. Flatt, J. J. Stephenson, C. D. Doyle and J. M. Tour, *Nature Mater.*, 2006, **5**, 63.
4. H. W. Song, Y. S. Kim, Y. H. Jang, H. J. Jeong, M. A. Reed and T. H. Lee, *Nature*, 2009, **462**, 1039.
5. J. Kushmerick, *Nature*, 2009, **462**, 994.
6. D. K. James and J. M. Tour, in *Encyclopedia of Nanoscience and Nanotechnology*, ed. J. A. Schwarz, C. Contescu and K. Putyera, Dekker, New York, 2004, p. 2177.
7. L. A. Bumm, J. J. Arnold, M. T. Cygan, T. D. Dunbar, T. P. Burgin, L. Jones, D. L. Allara, J. M. Tour and P. S. Weiss, *Science*, 1996, **271**, 1705.
8. E. B. Sandell, *Colorimetric Determination of Traces of Metals*, Interscience, London, 3rd edn, 1959.
9. E. Bishop (ed.), *Indicators*, Pergamon, Oxford, 1972.
10. A. Fernández-Gutiérrez and A. Muñoz de la Peña, in *Molecular Luminescence Spectroscopy. Methods and Applications. Part 1*, ed. S. G. Schulman, Wiley, New York, 1985, p. 371.
11. R. Y. Tsien, *Am. J. Physiol.*, 1992, **263**, C723.
12. A. W. Czarnik (ed.), *Fluorescent Chemosensors for Ion and Molecule Recognition*, ACS Books, Washington DC, 1993.
13. A. P. de Silva, H. Q. N. Gunaratne, T. Gunnlaugsson, A. J. M. Huxley, C. P. McCoy, J. T. Rademacher and T. E. Rice, *Chem. Rev.*, 1997, **97**, 1515.
14. A. W. Czarnik and J. -P. Desvergne (ed.), *Chemosensors for Ion and Molecule Recognition*, Kluwer, Dordrecht, 1997.
15. K. Rurack, *Spectrochim. Acta A. Mol. Biomol. Spectrosc.*, 2001, **57**, 2161.
16. L. Prodi, *New J. Chem.*, 2005, **29**, 20.
17. J. F. Callan, A. P. de Silva and D. C. Magri, *Tetrahedron*, 2005, **61**, 8551.
18. D. W. Domaille, E. L. Que and C. J. Chang, *Nature Chem. Biol.*, 2008, **4**, 168.
19. E. L. Que, D. W. Domaille and C. J. Chang, *Chem. Rev.*, 2008, **108**, 1517.

20. O. Trapp, *Angew. Chem. Int. Ed.*, 2008, **47**, 8158.
21. A. P. Demchenko, *Introduction to Fluorescence Sensing*, Springer, New York, 2009.
22. A. P. de Silva, T. S. Moody and G. D. Wright, *Analyst*, 2009, **134**, 2385.
23. X. H. Qian, Y. Xiao, Y. F. Xu, X. F. Guo, J. H. Qian and W. P. Zhu, *Chem. Commun.*, 2010, 6418.
24. J. Y. Han and K. Burgess, *Chem. Rev.*, 2010, **110**, 2709.
25. B. L. Feringa and W. R. Browne (ed.), *Molecular Switches*, Wiley-VCH, Weinheim, 2011.
26. A. P. de Silva, *Isr. J. Chem.*, 2011, **51**, 16.
27. *The Molecular Probes Handbook 11th Ed.*, www.invitrogen.com/site/us/en/home/References/Molecular-Probes-The-Handbook.html
28. W. S. Han, H. Y. Lee, S. H. Jung, S. J. Lee and J. H. Jung, *Chem. Soc. Rev.*, 2009, **38**, 1904.
29. G. P. Gorbenko, N. O. Mchedlev-Petrossyan and T. A. Chernaya, *J. Chem. Soc. Faraday Trans.*, 1998, **94**, 2117.
30. A. P. de Silva, H. Q. N. Gunaratne and T. Gunnlaugsson, *Chem. Commun.*, 1996, 1967.
31. A. P. de Silva and H. Q. N. Gunaratne, *J. Chem. Soc. Chem. Commun.*, 1990, 186.
32. C. Saudan, V. Balzani, P. Ceroni, M. Gorka, M. Maestri, V. Vicinelli and F. Vogtle, *Tetrahedron*, 2003, **59**, 3845.
33. G .S. Cox, N. J. Turro, N. C. Yang and M. J. Chen, *J. Am. Chem. Soc.*, 1984, **106**, 422.
34. D. J. Cram, R. A. Carmack and R. C. Helgeson, *J. Am. Chem. Soc.*, 1988, **110**, 571.
35. J. P. Geue, N. J. Head, D. L. Ward and S. F. Lincoln, *Dalton Trans.*, 2003, 521.
36. T. Gunnlaugsson, B. Bichell and C. Nolan, *Tetrahedron*, 2004, **60**, 5799.
37. A. P. de Silva and K. R. A. S. Sandanayake, *J. Chem. Soc. Chem. Commun.*, 1989, 1183.
38. T. Oe, M. Morita and T. Toyo'oka, *Anal. Sci.*, 1999, **15**, 1021.
39. K. Rurack, M. Kollmannsberger, U. Resch-Genger, W. Rettig and J. Daub, *Chem. Phys. Lett.*, 2000, **329**, 363.
40. S. Kenmoku, Y. Urano, K. Kanda, H. Kojima, K. Kikuchi and T. Nagano, *Tetrahedron*, 2004, **60**, 11067.
41. S. Ast, H. Müller, R. Flehr, T. Klamroth, B. Walz and H. J. Holdt, *Chem. Commun.*, 2011, **47**, 4685.
42. P. A. Panchenko, Y. V. Fedorov, V. P. Perevalov, G. Jonusauskas and O. A. Fedorova, *J. Phys. Chem. A*, 2010, **114**, 4118.
43. A. Minta, J. P. Y. Kao and R. Y. Tsien, *J. Biol. Chem.*, 1989, **264**, 8171.
44. A. P. de Silva, H. Q. N. Gunaratne, A. T. M. Kane and G. E. M. Maguire, *Chem. Lett.*, 1995, 125.
45. B. Ozmen and E. U. Akkaya, *Tetrahedron Lett.*, 2001, **41**, 9185.
46. S. Gaillard, A. Yakovlev, C. Luccardini, M. Oheim, A. Feltz and J.-M. Mallet, *Org. Lett.*, 2007, **9**, 2629.

47. T. Gunnlaugsson, T. C. Lee and R. Parkesh, *Org. Biomol. Chem.*, 2003, **1**, 3265.
48. T. Gunnlaugsson, T. C. Lee and R. Parkesh, *Tetrahedron*, 2004, **60**, 11239.
49. C. J. Chang, E. M. Nolan, J. Jaworski, S. C. Burdette, M. Shang and S. J. Lippard, *Chem. Biol.*, 2004, **11**, 203.
50. Z. C. Xu, K. H. Baek, H. N. Kim, J. N. Cui, X. H. Qian, D. R. Spring, I. J. Shin and J. Yoon, *J. Am. Chem. Soc.*, 2010, **132**, 601.
51. A. El Majzoub, C. Cadiou, I. Déchamps-Olivier, B. Tinant and F. Chuburu, *Inorg. Chem.*, 2011, **50**, 4029.
52. Y. Wang, X. Y. Hu, L. Wang, Z. B. Shang, J. B. Chao and W. J. Jin, *Sensors Actuators B*, 2011, **156**, 126.
53. H. Sakamoto, J. Ishikawa, S. Nakao and H. Wada, *Chem. Commun.*, 2000, 2395.
54. L. Prodi, C. Bargossi, M. Montalti, N. Zaccheroni, N. Su, J. S. Bradshaw, R. M. Izatt and P. B. Savage, *J. Am. Chem. Soc.*, 2000, **122**, 6769.
55. E. M. Nolan and S. J. Lippard, *J. Am. Chem. Soc.*, 2003, **125**, 14270.
56. X. F. Guo, X. H. Qian and L. H. Jia, *J. Am. Chem. Soc.*, 2004, **126**, 2272.
57. J. V. Mello and N. S. Finney, *J. Am. Chem. Soc.*, 2005, **127**, 10124.
58. M. Verma, A. F. Chaudhry, M. T. Morgan and C. J. Fahrni, *Org. Biomol. Chem.*, 2010, **8**, 363.
59. A. F. Chaudhry, M. Verma, M. T. Morgan, M. M. Henary, N. Siegel, J. M. Hales, J. W. Perry and C. J. Fahrni, *J. Am. Chem. Soc.*, 2010, **132**, 737.
60. P. Li, X. Duan, Z. Z. Chen, Y. Liu, T. Xie, L. B. Fang, X. R. Li, M. Yin and B. Tang, *Chem. Commun.*, 2011, **47**, 7755.
61. H. X. Wang, D. L. Wang, Q. Wang, X. Y. Li and C. A. Schalley, *Org. Biomol. Chem.*, 2010, **8**, 1017.
62. J. L. Bricks, A. Kovalchuk, C. Trieflinger, M. Nofz, M. Buschel, A. I. Tolmachev, J. Daub and K. Rurack, *J. Am. Chem. Soc.*, 2005, **127**, 13522.
63. T. Schwarze, H. Müller, C. Dosche, T. Klamroth, W. Mickler, A. Kelling, H. G. Löhmannsröben, P. Saalfrank and H. J. Holdt, *Angew. Chem. Int. Ed.*, 2007, **46**, 1671.
64. T. D. James, K. R. A. S. Sandanayake and S. Shinkai, *Nature*, 1995, **374**, 345.
65. S. Brasselet and W. E. Moerner, *Single Mol.*, 2000, **1**, 17.
66. L. Zang, R. Liu, M. W. Holman, K. T. Nguyen and D. M. Adams, *J. Am. Chem. Soc.*, 2002, **124**, 10640.
67. R. A. Bissell, E. Calle, A. P. de Silva, S. A. de Silva, H. Q. N. Gunaratne, J.-L. Habib-Jiwan, S. L. A. Peiris, R. A. D. D. Rupasinghe, T. K. S. D. Samarasinghe, K. R. A. S. Sandanayake and J.-P. Soumillion, *J. Chem. Soc. Perkin Trans.*, 1992, **2**, 1559.
68. A. P. de Silva and R. A. D. D. Rupasinghe, *J. Chem. Soc. Chem. Commun.*, 1985, 1669.
69. W. F. Jager, T. S. Hammink, O. van den Berg and F. C. Grozema, *J. Org. Chem.*, 2010, **75**, 2169.
70. G. Nishimura, Y. Shiraishi and T. Hirai, *Chem. Commun.*, 2005, 5313.

71. A. P. de Silva, S. S. K. de Silva, N. C. W. Goonesekera, H. Q. N. Gunaratne, P. L. M. Lynch, K. R. Nesbitt, S. T. Patuwathavithana and N. L. D. S. Ramyalal, *J. Am. Chem. Soc.*, 2007, **129**, 3050.
72. L. J. Shen, X. Y. Lu, H. Tian and W. H. Zhu, *Macromolecules*, 2011, **44**, 5612.
73. R. J. Wandell, A. H. Younes and L. Zhu, *New J. Chem.*, 2010, **34**, 2176.
74. C. J. Fahrni, L. Yang and D. G. VanDerveer, *J. Am. Chem. Soc.*, 2003, **125**, 3799.
75. J. Cody, S. Mandal, L. Yang and C. J. Fahrni, *J. Am. Chem. Soc.*, 2008, **130**, 13023.
76. F. Pragst and E. Weber, *J. Prakt. Chem.*, 1976, **318**, 51.
77. H. Tian, J. Gan, K. Chen, Q. L. Song and X. Y. Hou, *J. Mater. Chem.*, 2002, **12**, 1262.
78. A. P. de Silva, H. Q. N. Gunaratne, J.-L. Habib-Jiwan, C. P. McCoy, T. E. Rice and J.-P. Soumillion, *Angew. Chem. Int. Ed. Engl.*, 1995, **34**, 1728.
79. A. P. de Silva and T. E. Rice, *Chem. Commun.*, 1999, 163.
80. J. Gan, K. Chen, C. P. Chang and H. Tian, *Dyes Pigm.*, 2003, **57**, 21.
81. A. P. de Silva and S. A. de Silva, *J. Chem. Soc. Chem. Commun.*, 1986, 1709.
82. P. Kele, K. Nagy and A. Kotschy, *Angew. Chem. Int. Ed.*, 2006, **45**, 2565.
83. H. A. Klok and M. Möller, *Makromol. Chem. Phys.*, 1996, **197**, 1395.
84. H. Sakamoto, T. Anase, H. Osuga and K. Kimura, *React. Func. Polym.*, 2011, **71**, 569.
85. K. R. Gee, A. Rukavishnikov and A. Rothe, *Combinatorial Chem. High Throughput Screening*, 2003, **6**, 363.
86. R. A. Bissell, A. P. de Silva, H. Q. N. Gunaratne, P. L. M. Lynch, G. E. M. Maguire and K. R. A. S. Sandanayake, *Chem. Soc. Rev.*, 1992, **21**, 187.
87. S. Y. Hsueh, C. C. Lai and S. H. Chiu, *Chem. Eur. J.*, 2010, **16**, 2997.
88. B. Clapham and A. J. Sutherland, *Chem. Commun.*, 2003, 84.
89. A. P. de Silva, H. Q. N. Gunaratne and T. E. Rice, *Angew. Chem. Int. Ed. Engl.*, 1996, **35**, 2116.
90. A. P. de Silva, H. Q. N. Gunaratne, T. E. Rice and S. Stewart, *Chem. Commun.*, 1997, 1891.
91. T. Terai, K. Kikuchi, S. Y. Iwasawa, T. Kawabe, Y. Hirata, Y. Urano and T. Nagano, *J. Am. Chem. Soc.*, 2006, **128**, 6938.
92. J. C. G. Bunzli, *Chem. Rev.*, 2010, **110**, 2729.
93. T. Gunnlaugsson and J. P. Leonard, *Chem. Commun.*, 2004, 782.
94. T. Gunnlaugsson and J. P. Leonard, *Chem. Commun.*, 2005, 3114.
95. A. Thibon and V. Pierre, *J. Am. Chem. Soc.*, 2009, **131**, 434.
96. M. J. Minch and S. S. Shah, *J. Org. Chem.*, 1979, **44**, 3252.
97. D. Georgescauld, J. P. Desmasez, R. Lapouyade, A. Babeau, H. Richard and M. A. Winnik, *Photochem. Photobiol.*, 1980, **31**, 539.
98. M. D. P. de Costa, A. P. de Silva and S. T. Pathirana, *Canad. J. Chem.*, 1987, **69**, 1416.

99. K. Kalyanasundaram, *Photochemistry in Microheterogeneous Systems*, Academic, Orlando, 1987.
100. S. Ercelen, A. S. Klymchenko and A. P. Demchenko, *Anal. Chim. Acta*, 2002, **464**, 273.
101. M. Baruah, W. Qin, C. Flors, J. Hofkens, R. A. L. Vallée, D. Beljonne, M. Van der Auweraer, W. M. De Borggraeve and N. Boens, *J. Phys. Chem. A*, 2006, **110**, 5998.
102. S. Uchiyama, K. Iwai and A. P. de Silva, *Angew. Chem. Int. Ed.*, 2008, **47**, 4667.
103. Y. Ando, Y. Homma, Y. Hiruta, D. Citterio and K. Suzuki, *Dyes Pigm.*, 2009, **83**, 198.
104. C. Xu, W. S. Liu and W. W. Qin, *J. Phys. Chem. A*, 2011, **115**, 4288.
105. T. Morozumi, T. Anada and H. Nakamura, *J. Phys. Chem. B*, 2001, **105**, 2923.
106. W. Rettig, *Top. Curr. Chem.*, 1994, **169**, 253.
107. N. S. Bazilevskaya and A. S. Cherkasov, *Opt. Spectrosc.*, 1965, **18**, 30.
108. T. C. Werner and D. M. Hercules, *J. Phys. Chem.*, 1969, **2**, 225.
109. T. Morozumi, H. Hiraga and H. Nakamura, *Chem. Lett.*, 2003, **32**, 146.
110. J. Kawakami, H. Kimura, M. Nagaki, H. Kitahara and S. Ito, *J. Photochem. Photobiol. A: Chem.*, 2004, **161**, 141.
111. M. Montalti, L. Prodi and N. Zaccheroni, *J. Fluoresc.*, 2000, **10**, 71.
112. V. G. Young, H. L. Quiring and A. G. Sykes, *J. Am. Chem. Soc.*, 1997, **119**, 12477.
113. I. Leray, F. O'Reilly, J.-L. Habib-Jiwan, J.-P. Soumillion and B. Valeur, *Chem. Commun.*, 1999, 795.
114. C. M. Browne, G. Ferguson, M. A. McKervey, D. L. Mulholland and M. Parvez, *J. Am. Chem. Soc.*, 1985, **107**, 2703.
115. T. Hirano, K. Kikuchi, Y. Urano, T. Higuchi and T. Nagano, *Angew. Chem. Int. Ed.*, 2000, **39**, 1052.
116. E. M. Nolan and S. J. Lippard, *Acc. Chem. Res.*, 2009, **42**, 193.
117. E. Tamanini, A. Katewa, L. M. Sedger, M. H. Todd and M. Watkinson, *Inorg. Chem.*, 2009, **48**, 319.
118. Z. C. Xu, J. Yoon and D. R. Spring, *Chem. Soc. Rev.*, 2010, **39**, 1996.
119. Y. H. Lau, P. J. Rutledge, M. Watkinson and M. H. Todd, *Chem. Soc. Rev.*, 2011, **40**, 2848.
120. S. A. de Silva, A. Zavaleta, D. E. Baron, O. Allam, E. V. Isidor, N. Kashimura and J. M. Percarpio, *Tetrahedron Lett.*, 1997, **38**, 2237.
121. G. K. Walkup, S. C. Burdette, S. J. Lippard and R. Y. Tsien, *J. Am. Chem. Soc.*, 2000, **122**, 5644.
122. S. C. Burdette, G. K. Walkup, B. Spingler, R. Y. Tsien and S. J. Lippard, *J. Am. Chem. Soc.*, 2001, **123**, 7831.
123. A. P. de Silva, *J. Comput. Theor. Nanosci.*, 2011, **8**, 409.
124. R. Y. Tsien, *Chem. Eng. News*, 1994, **72**(29), 34.
125. C. C. Woodroofe and S. J. Lippard, *J. Am. Chem. Soc.*, 2003, **125**, 11458.

126. S. R. Adams, A. Harootunian, Y. J. Buechler, S. S. Taylor and R. Y. Tsien, *Nature*, 1991, **349**, 694.
127. C. C. Cain and R. F. Murphy, *J. Cell. Biol.*, 1988, **106**, 269.
128. A. P. de Silva, H. Q. N. Gunaratne, T. Gunnlaugsson and P. L. M. Lynch, *New J. Chem*, 1996, **20**, 871.
129. M. Licchelli, A. O. Biroli, A. Poggi, D. Sacchi, C. Sangermani and M. Zenna, *Dalton Trans.*, 2003, 4537.
130. K. Ashikaga, S. Ito, M. Yamamoto and Y. Nishijima, *J. Am. Chem. Soc.*, 1988, **110**, 198.
131. T. C. Barros, P. B. Filho, V. G. Toscano and M. J. Politi, *J. Photochem. Photobiol. A: Chem.*, 1995, **89**, 141.
132. T. Cheng, Y. F. Xu, S. Zhang, W. P. Zhu, X. H. Qian and L. Duan, *J. Am. Chem. Soc.*, 2008, **130**, 16160.
133. B. K. Paul, S. Mahanta, R. B. Singh and N. Guchhait, *Spectrochimica Acta Part A*, 2011, **79**, 197.
134. Z. C. Xu, S. J. Han, C. M. Lee, J. Yoon and D. R. Spring, *Chem. Commun.*, 2010, **46**, 1679.
135. X. Michalet, F. F. Pinaud, L. A. Bentolila, J. M. Tsay, S. Doose, J. J. Li, G. Sundaresan, A. M. Wu, S. S. Gambhir and S. Weiss, *Science*, 2005, **307**, 538.
136. M. J. Ruedas-Rama and E. A. H. Hall, *Anal. Chem.*, 2008, **80**, 8260.
137. E. M. Nolan and S. J. Lippard, *Chem. Rev.*, 2008, **108**, 3443.
138. K. Rurack, M. Kollmannsberger, U. Resch-Genger and J. Daub, *J. Am. Chem. Soc.*, 2000, **122**, 968.
139. S. Yoon, A. E. Albers, A. P. Wong and C. J. Chang, *J. Am. Chem. Soc.*, 2005, **127**, 16030.
140. R. T. Bronson, M. Montalti, L. Prodi, N. Zaccheroni, R. D. Lamb, N. K. Dally, R. M. Izatt, J. S. Bradshaw and P. B. Savage, *Tetrahedron*, 2004, **60**, 11139.
141. Y. Koide, Y. Urano, K. Hanaoka, T. Terai and T. Nagano, *ACS Chem. Biol.*, 2011, **6**, 600.
142. T. Schwarze, W. Mickler, C. Dosche, R. Flehr, T. Klamroth, H. G. Löhmannsröben, P. Saalfrank and H. J. Holdt, *Chem. Eur. J.*, 2010, **16**, 1819.
143. T. Schwarze, C. Dosche, R. Flehr, T. Klamroth, H. G. Löhmannsröben, P. Saalfrank, E. Cleve, H. J. Buschmann and H. J. Holdt, *Chem. Commun.*, 2010, **46**, 2034.
144. F. M. Harold, *The Vital Force: A Study of Bioenergetics*, Freeman, New York, 1986.
145. Y. Nakahara, T. Kida, Y. Nakatsuji and M. Akashi, *Chem. Commun.*, 2004, 224.
146. S. Bhattacharya and A. Gulyani, *Chem. Commun.*, 2003, 1158.
147. J. H. Fendler, *Membrane Mimetic Chemistry*, Wiley, New York, 1982.
148. R. A. Bissell, A. J. Bryan, A. P. de Silva and C. P. McCoy, *J. Chem. Soc. Chem. Commun.*, 1994, 405.
149. R. Martínez-Máñez and F. Sancenón, *Chem. Rev.*, 2003, **103**, 4419.

150. T. Gunnlaugsson, M. Glynn, G. M. Tocci, P. E. Kruger and F. M. Pfeffer, *Coord. Chem. Rev.*, 2006, **250**, 3094.
151. R. M. Duke, E. B. Veale, F. M. Pfeffer, P. E. Kruger and T. Gunnlaugsson, *Chem. Soc. Rev.*, 2010, **39**, 3936.
152. M. E. Moragues, R. Martínez-Máñez and F. Sancenón, *Chem. Soc. Rev.*, 2011, **40**, 2593.
153. Y. Zhou, Z. Xu and J. Yoon, *Chem. Soc. Rev.*, 2011, **40**, 2222.
154. F. Oton, A. Tárraga, M. D. Velasco, A. Espinosa and P. Molina, *Chem. Commun.*, 2004, 1658.
155. M. Alfonso, A. Espinosa, A. Tárraga and P. Molina, *Org. Lett.*, 2011, **13**, 2078.
156. Y. Kubo, M. Tsukahara, S. Ishihara and S. Tokita, *Chem. Commun.*, 2000, 653.
157. G. Hennrich, H. Sonnenschein and U. Resch-Genger, *Tetrahedron Lett.*, 2001, **42**, 2805.
158. L. Fabbrizzi, A. Leone and A. Taglietti, *Angew. Chem. Int. Ed.*, 2001, **40**, 3066.
159. R. Prohens, G. Martorell, P. Ballester and A. Costa, *Chem. Commun.*, 2001, 1456.
160. J.-M. Lehn, *Supramolecular Chemistry*, VCH, Weinheim, 1995.
161. C. Siering, S. Grimme and S. R. Waldvogel, *Chem. Eur. J.*, 2005, **11**, 1877.
162. C. Siering, H. Kerschbaumer, M. Nieger and S. R. Waldvogel, *Org. Lett.*, 2006, **8**, 1471.
163. E. V. Anslyn, *J. Org. Chem.*, 2007, **72**, 687.
164. F. B. Xu, L. H. Weng, L. J. Sun and Z. Z. Zhang, *Organometallics*, 2000, **19**, 2658.
165. F. Fages, J.-P. Desvergne, H. Bouas-Laurent, P. Marsau, J.-M. Lehn, F. Kotzyba-Hibert, A.-M. Albrecht-Gary and M. Al-Joubbeh, *J. Am. Chem. Soc.*, 1989, **111**, 8672.
166. R. W. Wagner, T. E. Johnson and J. S. Lindsey, *J. Am. Chem. Soc.*, 1996, **118**, 11166.
167. J. Otsuki, A. Yasuda and T. Takido, *Chem. Commun.*, 2003, 608.
168. P. P. Neelakandan, M. Hariharan and D. Ramaiah, *J. Am. Chem. Soc.*, 2006, **128**, 11334.
169. D. Ramaiah, P. P. Neelakandan, A. K. Nair and R. R. Avirah, *Chem. Soc. Rev.*, 2010, **39**, 4158.
170. L. Pu, *Chem. Rev.*, 2004, **104**, 1687.
171. H. L. Liu, H. P. Zhu, X. L. Hou and L. Pu, *Org. Lett.*, 2010, **12**, 4172.
172. J. Lin, Z. B. Li, H. C. Zhang and L. Pu, *Tetrahedron Lett.*, 2003, **45**, 103.
173. D. B. Cordes, S. Gamsey and B. Singaram, *Angew. Chem. Int. Ed.*, 2006, **45**, 3829.
174. D. L. Robertson and G. F. Joyce, *Nature*, 1990, **344**, 467.
175. A. D. Ellington and J. W. Szostak, *Nature*, 1990, **346**, 818.
176. C. Tuerk and L. Gold, *Science*, 1990, **249**, 505.
177. B. A. Sparano and K. Koide, *J. Am. Chem. Soc.*, 2005, **127**, 14954.

178. T. J. Drake and W. Tan, *Appl. Spectrosc.*, 2004, **58**, 269A.
179. S. Tyagi and F. R. Kramer, *Nature Biotechnol.*, 1996, **14**, 303.
180. E. Daniel and G. Weber, *Biochemistry*, 1966, **5**, 1893.
181. D. Whitten, L. Chen, R. Jones, T. Bergstedt, P. Heeger and D. McBranch, in *Optical Sensors and Switches*, ed. V. Ramamurthy and K. S. Schanze, Dekker, New York, 2001, p. 189.
182. B. Kratochvil and D. A. Zatko, *Anal. Chem.*, 1964, **36**, 527.
183. V. Goulle, A. Harriman and J.-M. Lehn, *J. Chem. Soc. Chem. Commun.*, 1993, 1034.
184. G. R. Deviprasad, B. Keshavan and F. D'Souza, *J. Chem. Soc. Perkin Trans*, 1998, **1**, 3133.
185. P. Yan, M. W. Holman, P. Robustelli, A. Chowdhury, F. I. Ishak and D. M. Adams, *J. Phys. Chem. B*, 2005, **109**, 130.
186. E. W. Miller, S. X. Bian and C. J. Chang, *J. Am. Chem. Soc.*, 2007, **129**, 3458.
187. R. L. Zhang, Z. L. Wang, Y. S. Wu, H. B. Fu and J. N. Yao, *Org. Lett.*, 2008, **10**, 3065.
188. D. P. Kennedy, C. M. Kormos and S. C. Burdette, *J. Am. Chem. Soc.*, 2009, **131**, 8578.
189. S. Uchiyama, K. Takehira, T. Yoshihara, S. Tobita and T. Ohwada, *Org. Lett.*, 2006, **8**, 5869.
190. K. Muthramu and V. Ramamurthy, *J. Photochem.*, 1984, **26**, 57.
191. Y. Ooyama, M. Sumomogi, T. Nagano, K. Kushimoto, K. Komaguchi, I. Imae and Y. Harima, *Org. Biomol. Chem.*, 2011, **9**, 1314.
192. Y. Ooyama, A. Matsugasako, K. Oka, T. Nagano, M. Sumomogi, K. Komaguchi, I. Imae and Y. Harima, *Chem. Commun.*, 2011, **47**, 4448.
193. N. Chandrasekharan and L.A. Kelly, *Rev. Fluoresc.*, 2004, **1**, 21.
194. S. Uchiyama, A.P. de Silva and K. Iwai, *J. Chem. Educ.*, 2006, **83**, 720.
195. S. Uchiyama, Y. Matsumura, A. P. de Silva and K. Iwai, *Anal. Chem.*, 2003, **75**, 5926.
196. C. Gota, S. Uchiyama, T. Yoshihara, S. Tobita and T. Ohwada, *J. Phys. Chem. B*, 2008, **112**, 2829.
197. S. Uchiyama, Y. Matsumura, A. P. de Silva and K. Iwai, *Anal. Chem.*, 2004, **76**, 1793.
198. M. Engeser, L. Fabbrizzi, M. Licchelli and D. Sacchi, *Chem. Commun.*, 1999, 1191.
199. Z. Wang, D. Q. Zhang and D. B. Zhu, *Tetrahedron Lett.*, 2005, **46**, 4609.
200. T. Saika, T. Iyoda, K. Honda and T. Shimidzu, *J. Chem. Soc. Chem. Commun.*, 1992, 591.
201. G. M. Tsivgoulis and J.-M. Lehn, *Adv. Mater.*, 1997, **9**, 627.
202. D. Gosztola, M. P. Niemczyk and M. R. Wasielewski, *J. Am. Chem. Soc.*, 1998, **120**, 5118.
203. F. M. Raymo and S. Giordani, *Proc. Natl. Acad. Sci. USA*, 2002, **99**, 4941.
204. M. Irie, *Chem. Rev.*, 2000, **100**, 1685.

205. H. Durr and H. Bouas-Laurent (ed.), *Photochromism. Molecules and Systems*, Elsevier, Amsterdam, 1990.
206. A. Zweig, *Pure Appl. Chem.*, 1973, **33**, 389.
207. A. J. Myles and N. R. Branda, *J. Am. Chem. Soc.*, 2001, **123**, 177.
208. R. T. Hayes, M. R. Wasielewski and D. Gosztola, *J. Am. Chem. Soc.*, 2000, **122**, 5563.
209. T. A. Moore, D. Gust, P. Mathis, J. C. Mialocq, C. Chachaty, R. V. Bensasson, E. J. Land, D. Doizi, P. A. Liddell, W. R. Lehman, G. A. Nemeth and A. L. Moore, *Nature*, 1984, **307**, 630.
210. A. P. de Silva, H. Q. N. Gunaratne and T. Gunnlaugsson, *Tetrahedron Lett.*, 1998, **39**, 5077.
211. M. Y. Chae and A. W. Czarnik, *J. Am. Chem. Soc.*, 1992, **114**, 9704.
212. J. Yoon, N. E. Ohler, D. H. Vance, W. D. Aumiller and A. W. Czarnik, *Tetrahedron Lett.*, 1997, **38**, 3845.
213. G. Hennrich, H. Sonnenschein and U. Resch-Genger, *J. Am. Chem. Soc.*, 1999, **121**, 5073.
214. M. Sandor, F. Geistmann and M. Schuster, *Anal. Chim. Acta*, 1999, **388**, 19.
215. G. X. Zhang, D. Q. Zhang, S. W. Yin, X. D. Yang, Z. G. Shuai and D. B. Zhu, *Chem. Commun.*, 2005, 2161.
216. T. Toyo'oka and K. Imai, *Anal. Chem.*, 1984, **56**, 2461.
217. T. Toyo'oka, T. Suzuki, Y. Saito, S. Uzu and K. Imai, *Analyst*, 1989, **114**, 413.
218. S. Uchiyama, K. Takehira, S. Kohtani, K. Imai, R. Nakagaki, S. Tobita and T. Santa, *Org. Biomol. Chem.*, 2003, **1**, 1067.
219. S. Uzu, S. Kanda, K. Nakashima and K. Imai, *Analyst*, 1990, **115**, 1477.
220. P. Herman, Z. Murtaza and J. R. Lakowicz, *Anal. Biochem.*, 1999, **272**, 87.
221. C. J. Purnell and R. F. Walker, *Analyst*, 1985, **110**, 893.
222. T. Santa, D. Matsumura, C. Huang, C. Kitada and K. Imai, *Biomed. Chromatogr.*, 2002, **16**, 523.
223. S. Uchiyama, T. Santa and K. Imai, *Analyst*, 2000, **125**, 1839.
224. S. Tal, H. Sahman, Y. Abraham, M. Botoshansky and Y. Eichen, *Chem. Eur. J.*, 2006, **12**, 4858.
225. J. J. Lee and B. D. Smith, *Chem. Commun.*, 2009, 1962.
226. M. Onoda, S. Uchiyama, A. Endo, H. Tokuyama, T. Santa and K. Imai, *Org. Lett.*, 2003, **5**, 1459.
227. M. Onoda, S. Uchiyama, T. Santa and K. Imai, *Anal. Chem.*, 2002, **74**, 4089.
228. J. K. Weltman, R. P. Szaro, A. R. Frackelston, R. M. Dowben, J. R. Bunting and R. E. Cathou, *J. Biol. Chem.*, 1973, **218**, 3173.
229. H. J. Verhey, C. H. W. Bekker, J. W. Verhoeven and J. W. Hofstraat, *New J. Chem.*, 1996, **20**, 809.
230. T. Matsumoto, Y. Urano, T. Shoda, H. Kojima and T. Nagano, *Org. Lett.*, 2007, **9**, 3375.
231. T. J. Dale and J. Rebek, *J. Am. Chem. Soc.*, 2006, **128**, 4500.

232. S. Bencic-Nagale, T. Sternfeld and D. R. Walt, *J. Am. Chem. Soc.*, 2006, **128**, 5041.
233. T. J. Dale and J. Rebek, *Angew. Chem. Int. Ed.*, 2009, **48**, 7850.
234. F. Tanaka, N. Mase and C. F. Barbas, *J. Am. Chem. Soc.*, 2004, **126**, 3692.
235. M. Montalti, A. Credi, L. Prodi and M. T. Gandolfi, *Handbook of Photochemistry*, CRC Press, Boca Raton, 3rd edn, 2006.
236. H. Siegerman, in *Techniques of Electroorganic Synthesis. Part II*, ed. N. L.Weinberg, Wiley, New York, 1975.
237. A. P. de Silva, S. A. de Silva, A. S. Dissanayake and K. R. A. S. Sandanayake, *J. Chem. Soc. Chem. Commun.*, 1989, 1054.
238. E. Sasaki, H. Kojima, H. Nishimatsu, Y. Urano, K. Kikuchi, Y. Hirata and T. Nagano, *J. Am. Chem. Soc.*, 2005, **127**, 3684.
239. H. Kojima and T. Nagano, *Adv. Mater.*, 2000, **12**, 763.
240. M. J. Plater, I. Greig, M. H. Helfrich and S. H. Ralston, *J. Chem. Soc. Perkin Trans.*, 2001, **1**, 2553.
241. D. Q. Zhang, S. Y. Zhao and H. X. Liu, *Progr. Chem.*, 2008, **20**, 1396.
242. F. D'Souza, G. R. Deviprasad and Y. Y. Hsieh, *Chem. Commun.*, 1997, 533.
243. L. M. Adleman, *Science*, 1994, **266**, 1021.
244. M. N. Stojanovic, T. E. Mitchell and D. Stefanovic, *J. Am. Chem. Soc.*, 2002, **124**, 3555.
245. A. P. de Silva, H. Q. N. Gunaratne and C. P. McCoy, *Nature*, 1993, **364**, 42.
246. R. R. Breaker and G. F. Joyce, *Chem. Biol.*, 1995, **2**, 655.
247. X. Chen, Y. F. Wang, Q. Liu, Z. Z. Zhang, C. H. Fan and L. He, *Angew. Chem. Int. Ed.*, 2006, **45**, 1759.
248. N. Carmi, L. A. Shultz and R. R. Breaker, *Chem. Biol.*, 1996, **3**, 1039.
249. J. Elbaz, O. Lioubashevski, F. Wang, F. Remacle, R. D. Levine and I. Willner, *Nature Nanotechnol.*, 2010, **5**, 417.
250. H. J. Issaq, *Adv. Protein Chem.*, 2003, **65**, 249.
251. B. K. Wetzl, S. M. Yarmoluk, D. B. Craig and O. S. Wolfbeis, *Angew. Chem. Int. Ed.*, 2004, **43**, 5400.
252. R. A. John, in *Enzyme Assays*, ed. R. Eisenthal and M. J. Danson, Oxford University Press, Oxford, 2nd edn, 2002, p. 49.
253. V. Sharma, R. S. Agnes and D. S. Lawrence, *J. Am. Chem. Soc.*, 2007, **129**, 2742.
254. E. D. Matayoshi, G. T. Wang, G. A. Krafft and J. W. Erickson, *Science*, 1990, **247**, 954.
255. J. Grimshaw and A. P. de Silva, *Chem. Soc. Rev.*, 1981, **10**, 181.
256. V. Derycke, R. Martel, J. Appenzeller and P. Avouris, *Nano Lett.*, 2001, **1**, 453.
257. Z. H. Chen, J. Appenzeller, Y. M. Lin, J. Sippel-Oakley, A. G. Rinzler, J. Y. Tang, S. J. Wind, P. M. Solomon and P. Avouris, *Science*, 2006, **311**, 1735.

258. P. Avouris, *Nature Nanotechnol.*, 2007, **2**, 605.
259. E. C. P. Smits, S. G. J. Mathijssen, P. A. van Hal, S. Setayesh, T. C. T. Geuns, K. A. H. A. Mutsaers, E. Cantatore, H. J. Wondergem, O. Werzer, R. Resel, M. Kemerink, S. Kirchmeyer, A. M. Muzafarov, S. A. Ponomarenko, B. de Boer, P. W. M. Blom and D. M. de Leeuw, *Nature*, 2008, **455**, 956.
260. A. P. de Silva, H. Q. N. Gunaratne and P. L. M. Lynch, *J. Chem. Soc. Perkin Trans.*, 1995, **2**, 685.
261. N. R. Cha, S. Y. Moon and S. K. Chang, *Tetrahedron Lett.*, 2003, **44**, 8265.
262. L. S. Bark and A. Rixon, *Analyst*, 1970, **95**, 786.
263. E. U. Akkaya, M. E. Huston and A. W. Czarnik, *J. Am. Chem. Soc.*, 1990, **112**, 3590.
264. S. Y. Moon, N. R. Cha, Y. H. Kim and S. K. Chang, *J. Org. Chem.*, 2004, **69**, 181.
265. G. Klein, D. Kaufmann, S. Schurch and J.-L. Reymond, *Chem. Commun.*, 2001, 561.
266. Y. Zheng, J. Orbulescu, X. Ji, F. M. Andreopoulos, S. M. Pham and R. M. Leblanc, *J. Am. Chem. Soc.*, 2003, **125**, 2680.
267. S. B. Jedner, R. James, R. N. Perutz and A. K. Duhme-Klair, *J. Chem. Soc. Dalton Trans.*, 2001, 2327.
268. T. Gunnlaugsson, A. P. Davis and M. Glynn, *Chem. Commun.*, 2001, 2556.
269. T. Gunnlaugsson, A. P. Davis, G. M. Hussey, J. Tierney and M. Glynn, *Org. Biomol. Chem.*, 2004, **2**, 1856.
270. C. M. G. dos Santos, T. McCabe and T. Gunnlaugsson, *Tetrahedron Lett.*, 2007, **48**, 3135.
271. M. A. McKervey and D. L. Mulholland, *J. Chem. Soc. Chem. Commun.*, 1977, 438.
272. T. Gareis, C. Huber, O. S. Wolfbeis and J. Daub, *Chem. Commun.*, 1997, 1717.
273. L. Fabbrizzi, M. Licchelli and P. Pallavicini, *Acc. Chem. Res.*, 1999, **32**, 846.
274. A. W. Czarnik, *Acc. Chem. Res.*, 1994, **27**, 302.
275. P. Grandini, F. Mancin, P. Tecilla, P. Scrimin and U. Tonellato, *Angew. Chem. Int. Engl.*, 1999, **38**, 3061.
276. Y. Díaz-Fernández, A. Pérez-Gramatges, V. Amendola, F. Foti, C. Mangano, P. Pallavacini and S. Patroni, *Chem. Commun.*, 2004, 1650.
277. E. Brasola, F. Mancin, E. Rampazzo and U. Tonellato, *Chem. Commun.*, 2003, 3026.
278. R. Grigg, J. M. Holmes, S. K. Jones and W. D. J. A. Norbert, *J. Chem. Soc. Chem. Commun.*, 1994, 185.
279. E. B. Veale, G. M. Tocci, F. M. Pfeffer, P. E. Kruger and T. Gunnlaugsson, *Org. Biomol. Chem.*, 2009, **7**, 3447.
280. W. X. Liu and Y. B. Jiang, *Org. Biomol. Chem.*, 2007, **5**, 1771.
281. S. S. Sun and A. J. Lees, *Chem. Commun.*, 2000, 1687.

282. J. Michl and V. Bonačić-Koutecký, *Electronic Aspects of Organic Photochemistry*, Wiley, New York, 1990.
283. M. Klessinger and J. Michl, *Excited States and Photochemistry of Organic Molecules*, VCH, New York, 1995.
284. M. D. Pratt and P. D. Beer, *Tetrahedron*, 2004, **60**, 11227.
285. J. Y. Kwon, N. J. Singh, H. N. Kim, K. S. Kim and J. Yoon, *J. Am. Chem. Soc.*, 2004, **126**, 8892.
286. J. F. Callan, A. P. de Silva and N. D. McClenaghan, *Chem. Commun.*, 2004, 2048.
287. J. B. LePecq and C. A. Paoletti, *J. Mol. Biol.*, 1967, **27**, 87.
288. W. C. Tse and D. L. Boger, *Acc. Chem. Res.*, 2004, **37**, 61.
289. D. L. Boger, B. E. Fink and M. P. Hedrick, *J. Am. Chem. Soc.*, 2000, **122**, 6382.
290. F. D'Souza, *J. Am. Chem. Soc.*, 1996, **118**, 923.
291. B. Ferrer, G. Rogez, A. Credi, R. Ballardini, M. T. Gandolfi, V. Balzani, Y. Liu, H. R. Tseng and J. F. Stoddart, *Proc. Natl. Acad. USA*, 2006, **103**, 18411.
292. J. Yoon and A. W. Czarnik, *Bioorg. Med. Chem.*, 1993, **1**, 267.
293. A. W. Czarnik, *ACS Symp. Ser.*, 1994, **561**, 314.
294. J. L. Geng, P. Liu, B. H. Liu, G. J. Guan, Z. P. Zhang and M. Y. Han, *Chem. Eur. J*, 2010, **16**, 3720.
295. W. Kloppfer, *Adv. Photochem.*, 1977, **10**, 311.
296. G. Springsteen and B. Wang, *Chem. Commun.*, 2001, 1608.
297. D. H. Kim and M. S. Han, *Bioorg. Med. Chem. Lett.*, 2003, **13**, 2453.
298. E. Kimura and T. Koike, *Chem. Soc. Rev.*, 1998, **27**, 179.
299. B. K. Kaletas, R. M. Williams, B. Konig and L. De Cola, *Chem. Commun.*, 2002, 776.
300. I. Hamachi, T. Nagase and S. Shinkai, *J. Am. Chem. Soc.*, 2000, **122**, 12065.
301. L. Prodi, F. Bolletta, M. Montalti, N. Zaccheroni, P. Huszthy, E. Samu and B. Vermes, *New J. Chem.*, 2000, **24**, 781.
302. V. J. Pugh, Q. Hu and L. Pu, *Angew. Chem. Int. Ed.*, 2000, **39**, 3638.
303. M. N. Stojanovic, P. de Prada and D. W. Landry, *J. Am. Chem. Soc.*, 2001, **123**, 4928.
304. E. M. Kosower, *Acc. Chem. Res.*, 1982, **15**, 259.
305. R. A. Kenner and A. A. Aboderin, *Biochemistry*, 1971, **10**, 4433.
306. R. A. Bissell, A. P. de Silva, W. T. M. L. Fernando, S. T. Patuwathavithana and T. K. S. D. Samarasinghe, *Tetrahedron Lett.*, 1991, **32**, 425.
307. H. F. Ji, R. Dabestani, G. M. Brown and R. L. Hettich, *Photochem. Photobiol.*, 1999, **69**, 513.
308. I. Grabchev, J.-M. Chovelon and X. H. Qian, *J. Photochem. Photobiol. A: Chem.*, 2003, **158**, 37.
309. M. J. Hall, L. T. Allen and D. F. O'Shea, *Org. Biomol. Chem.*, 2006, **4**, 776.
310. D. C. Magri, J. F. Callan, A. P. de Silva, D. B. Fox, N. D. McClenaghan and K. R. A. S. Sandanayake, *J. Fluoresc.*, 2005, **15**, 769.

311. F. Daniels and R. A. Alberty, *Physical Chemistry*, Wiley, New York, 3rd edn, 1967.
312. A. P. de Silva, H. Q. N. Gunaratne, K. R. Jayasekera, S. O'Callaghan and K. R. A. S. Sandanayake, *Chem. Lett.*, 1995, 123.
313. H. Tian, J. Gan, K. Chen, Q. L. Song and X. Y. Hou, *J. Mater. Chem.*, 2002, **12**, 1262.
314. D. H. Qu, Q. C. Wang, J. Ren and H. Tian, *Org. Lett.*, 2004, **6**, 2085.
315. E. M. Perez, D. T. F. Dryden, D. A. Leigh, G. Teobaldi and F. Zerbetto, *J. Am. Chem. Soc.*, 2004, **126**, 12210.
316. D. R. Reddy and B. G. Maiya, *Chem. Commun.*, 2001, 117.
317. D. R. Reddy and B. G. Maiya, *J. Phys. Chem. A.*, 2003, **107**, 6326.
318. A. Shundo, J. P. Hill and K. Ariga, *Chem. Eur. J.*, 2009, **15**, 2486.
319. F. Pina, M. J. Melo, M. Maestri, R. Ballardini and V. Balzani, *J. Am. Chem. Soc.*, 1997, **119**, 5556.
320. S. Xu, K. C. Chen and H. Tian, *J. Mater. Chem.*, 2005, **15**, 2676.
321. A. Lake, S. Shang and D. M. Kolpashchikov, *Angew. Chem. Int. Ed.*, 2010, **49**, 4459.

CHAPTER 6
Reconfigurable Single Input–Single Output Systems

6.1 Introduction

Up to now, we have discussed molecules each of which was designed to perform a given logic operation. Much of semiconductor-based computing hardware is also based on hard-wired logic circuits of a given configuration, with flexibility of the overall computation being introduced through software. A notable exception is found in erasable programmable read-only memories (EPROM).[1] A more recent innovation has been the advent of field-programmable logic arrays (FPGA).[2] These have the ability to change the configuration of a logic circuit upon command. This naturally allows the creation of multiple logic circuits each optimized to carry out different tasks. Molecular logic gates can be reconfigured too. One reason is that, unlike electrons in semiconductors, chemical species and light are blessed with diversity. Many types of chemicals and many colours of light which are all distinguishable one from another, can serve as input/outputs.[3] Another reason is that molecules can be interrogated in many ways. The following sections will be classified according to the reconfiguring variable. A similar organization will be employed when the double-input versions are treated in Chapter 8, where more cases will be featured. Switching between positive and negative logic conventions will naturally lead to different logic types but we will largely stick with positive logic in this book.

6.2 Nature of Inputs

A nice example of how different inputs produce different logic behaviours is Schmittel and Lin's **1**.[4] It clearly demonstrates YES logic action when the input

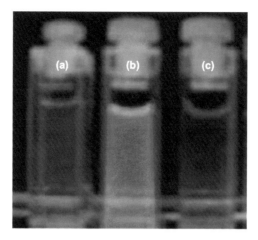

Figure 6.1 Luminescence, excited at 365 nm, of **1** (a) alone, (b) with Pb^{2+} and (c) with Cu^{2+}.
Reprinted from M. Schmittel and H. W. Lin, *Angew. Chem. Int. Ed.*, 2007, **46**, 893, with permission from Wiley-VCH.

is Pb^{2+}, by showing a substantial luminescence enhancement (Figure 6.1). The steric crowding within **1** leads to twisting of the amine units so that they begin to lose conjugation with the rest of the π-system. The Pb^{2+}-induced luminescence enhancement can then arise by a suppression of a TICT excited state[5,6] (see section 4.2). However the use of open-shell Cu^{2+} as input introduces luminescence quenching pathways (EET/PET) so that NOT logic emerges instead. It is important that **1** shows a medium-level luminescence which can be manipulated upwards or downwards with Pb^{2+} and Cu^{2+} respectively. If not, the case would be best developed as a double-input logic gate (Chapter 7).

A brief consideration of the influence of input level is warranted at this point. For instance, 'off–on–off' switching systems[7,8] could be discussed here because they involve a 'low', 'medium' and 'high' input level. Selecting either 'high' or 'medium' H^+ levels would give H^+-driven PASS 0 or H^+-driven YES logic. However, these are deferred to section 12.2, concerning ternary logic.

6.3 Output Observation Technique

Examples where the logic type depends on the observation technique arise sometimes from the field of multi-mode sensing.[9–11] The irreversible response of **2**[12] towards Cu^{2+} and Hg^{2+} serves as an illustration: uv–visible spectral monitoring elicits a strong absorbance signal at 556 nm due to the production of the rhodamine B chromophore only from Cu^{2+} input. This YES logic response is to be contrasted with the approximate PASS 0 logic behaviour shown by Hg^{2+}. However, the tables are turned when the fluorescence output of the rhodamine B-based product is the target, with only Hg^{2+} delivering a significant signal. The reason for the PASS 0 logic behaviour of Cu^{2+} in this case is the strong fluorescence quenching caused by the open-shell ion.

2

Schmittel and Lin's **1**[4] (discussed in section 6.2) goes further by using four output channels provided by luminescence,[13–15] uv-visible absorption,[16] voltammetry[17] and electrochemiluminescence.[18] For instance, Pb^{2+} induces no change in absorbance of the MLCT absorption band, gives a 1.7-fold enhancement in MLCT emission and leaves electrochemiluminescence intensity unchanged. These outputs correspond to PASS 0, YES and PASS 0, respectively. A 180 mV anodic shift in cyclic voltammetry is also observed. The anodic current output in the cyclic voltammogram (Figure 6.2) at a potential of 0.6 V (vs. Fc), which can serve as the output, drops from 1.6 to 0.3 μA. So we have NOT logic here. Martínez-Máñez's **3**[19] is an earlier case of this type which employs the first three of these channels. In this case, Pb^{2+} induces a blue shift in absorption, an enhancement in fluorescence and an anodic shift in cyclic voltammetry. Structures like **3** have TICT excited states.[5,6] Similar patterns can be seen in Zn^{2+}-induced phenomena of **4**.[20] While the studies of **3** and related compounds[21] involve multiple targets, they are not presented as sets to the molecular device for logic evaluation.

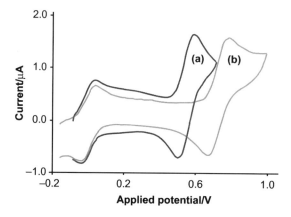

Figure 6.2 Cyclic voltammograms of **1** (a) alone and (b) with 0.2 M Pb^{2+} in acetonitrile solution with 0.1 M Bu_4PF_6, versus Ag wire. The wave at 0.0 V is the ferrocene internal standard.
Adapted from M. Schmittel and H. W. Lin, *Angew. Chem. Int. Ed.*, 2007, **46**, 893 (supplementary information) with permission from Wiley-VCH.

6.4 Observation Wavelength

Given that the light intensity at several wavelengths can be observed simultaneously without much trouble, we have a rare variable for logic devices which reminds us of quantum phenomena. This aspect will be discussed in Chapter 13. Some of the examples of the single-input type which are to be found there could have featured here. Just for a taste, we briefly mention Stojanovic and Stefanovich's deoxyribozyme work discussed in section 5.5.2 where the observation of green fluorescence from a fluorescein unit corresponded to NOT logic whereas the tetramethylrhodamine-based orange emission showed YES logic instead.[22]

References

1. C. Maxfield, *From Bebop to Boolean Boogie*, Newnes, Oxford, 2009.
2. S. Hauck and A. DeHon (ed.), *Reconfigurable Computing*, Elsevier, Burlington, MA, 2008.

3. D. Margulies, G. Melman and A. Shanzer, *Nature Mater.*, 2005, **4**, 768.
4. M. Schmittel and H. W. Lin, *Angew. Chem. Int. Ed.*, 2007, **46**, 893.
5. W. Rettig, *Top. Curr. Chem.*, 1994, **169**, 253.
6. M. M. Martin, P. Plaza, Y. H. Meyer, F. Badaoui, J. Bourson, J. P. Lefebvre and B. Valeur, *J. Phys. Chem.*, 1996, **100**, 6879.
7. A. P. de Silva, H. Q. N. Gunaratne and C. P. McCoy, *Chem. Commun.*, 1996, 2399.
8. S. A. de Silva, A. Zavaleta, D. E. Baron, O. Allam, E. Isidor, N. Kashimura and J. M. Percarpio, *Tetrahedron Lett.*, 1997, **38**, 2237.
9. T. Suzuki, Y. Ishigaki, T. Iwai, H. Kawai, K. Fujiwara, H. Ikeda, Y. Kano and K. Mizuno, *Chem. Eur. J.*, 2009, **15**, 9434.
10. S. H. Kim, H. S. Choi, J. H. Kim, S. J. Lee, D. T. Quang and J. S. Kim, *Org. Lett.*, 2010, **12**, 560.
11. H. Zhang, X. X. Kou, Q. O. Zhang, D. H. Qu and H. Tian, *Org. Biomol. Chem.*, 2011, **9**, 4051.
12. L. J. Tang, F. F. Li, M. H. Liu and R. Nandhakumar, *Spectrochim. Acta Part A*, 2011, **78**, 1168.
13. A. P. de Silva, H. Q. N. Gunaratne, T. Gunnlaugsson, A. J. M. Huxley, C. P. McCoy, J. T. Rademacher and T. E. Rice, *Chem. Rev.*, 1997, **97**, 1515.
14. B. Valeur and M. N. Berberan-Santos, *Molecular Fluorescence*, Wiley-VCH, Weinheim, 2nd edn, 2012.
15. J. R. Lakowicz, *Principles of Fluorescence Spectroscopy*, Springer, New York, 3rd edn, 2006.
16. E. B. Sandell, *Colorimetric Determination of Traces of Metals*, Interscience, London, 3rd edn, 1959.
17. P. D. Beer and P. A. Gale, *Angew. Chem. Int. Ed.*, 2001, **40**, 487.
18. M. M. Richter, *Chem. Rev.*, 2004, **104**, 3003.
19. D. Jiménez, R. Martínez-Máñez, F. Sancenón and J. Soto, *Tetrahedron Lett.*, 2004, **45**, 1257.
20. A. C. Benniston, A. Harriman, D. J. Lawrie, A. Mayeux, K. Rafferty and O. D. Russell, *Dalton Trans.*, 2003, 4762.
21. T. Ábalos, D. Jiménez, R. Martínez-Máñez, F. Sancenón, J. Soto, A. M. Costero, M. Parra and S. Gil, *Dalton Trans.*, 2010, **39**, 3449.
22. M. N. Stojanovic, T. E. Mitchell and D. Stefanovic, *J. Am. Chem. Soc.*, 2002, **124**, 3555.

CHAPTER 7
Double Input–Single Output Systems

7.1 Introduction

The two previous chapters have illustrated the Boolean nature of single-input, single-output molecular devices, but the popular view of logic in a computing context requires the demonstration of double-input systems. Each double-input, single-output Boolean logic type will be considered separately, as far as practicable. The nature of the device, inputs and outputs will be other means of organization when required. The order of logic types considered here differs from that described in Chapter 3 but is used for ease of organization of material.

7.2 AND

AND is the logic type that many people recognize, perhaps because its human analogy embraces those universal values of cooperation and unity. Cultures old and new possess lines like 'United we stand, divided we fall'[1] and '(We are) better together'.[2] AND logic is contained in the word 'synergy' which is popular in business circles. Any one 'high' input by itself is powerless to produce the output, but the output comes alive when both 'high' inputs are applied at the same time. The general relationship between the inputs and output can be illustrated with an output–input response surface,[3–7] which is an expansion of the stimulus–response curve discussed in section 2.5.

Because of the large amount of material available, we will organize it primarily according to whether the inputs can be distinguished by the device or not. Any physical connection (or not) between the inputs will be another primary organizational tool.

Monographs in Supramolecular Chemistry No. 12
Molecular Logic-based Computation
By A Prasanna de Silva
© The Royal Society of Chemistry 2013
Published by the Royal Society of Chemistry, www.rsc.org

7.2.1 Distinguishable and Separate Inputs

Distinguishability of inputs in electronic contexts is arranged by sending the signals along separate wires. Where chemical inputs to molecular devices are concerned, the 'ports' on the latter will choose the appropriate input species from a mixture in solution if the receptors employed are selective enough. Thankfully, chemistry is at a stage of development sufficient to offer a range of receptors which selectively bind various input species.

7.2.1.1 Cation Inputs

The first molecular logic gate **1** was constructed using the PET-based 'lumophore–spacer$_1$–receptor$_1$–spacer$_2$–receptor$_2$' model.[8] Figure 7.1 outlines the general operation of this model, where guests G$_1$ and G$_2$ are captured reversibly by receptor$_1$ and receptor$_2$ respectively. If binding is irreversible, the system still permits single-use computations. A PET process to the lumophore can originate at each receptor. Each PET process is stopped by binding of the guest to the appropriate receptor. Luminescence is released only when both PET processes stop. The device is powered by the absorbed exciting light. The frontier orbital energy diagram for the situation in Figure 7.1a is shown in Figure 7.2. As indicated in section 4.5, PET applies to fluorescence and other forms of luminescence. The general case of luminescence is given in Figures 7.1 and 7.2.

1; R = n-Pr

We see that a fluorescence signal is released by the anthracene unit of **1** only when both chemical inputs H$^+$ and Na$^+$ are present in a sufficient concentration, thus satisfying the AND logic truth table (Table 7.1). The anthracene unit would normally fluoresce blue when it is exposed to ultraviolet light. In system **1**, however, the fluorescence is quenched because of a faster process, PET,[9-11] in which an electron is transferred to the anthracene unit from either the amine or dialkoxybenzene unit. The amine can act as a receptor for H$^+$ and the benzo-15-crown-5 ether can capture Na$^+$. If one of these two receptor sites is occupied, PET will still occur from the other, and no fluorescence should be observed (Figures 7.1b and 7.1c). If, however, both receptor sites are filled, *i.e.* H$^+$ and Na$^+$ ions are present, both PET paths are prevented (because the electrons are now tied up with binding ions) and strong fluorescence should result (Figure 7.1d).

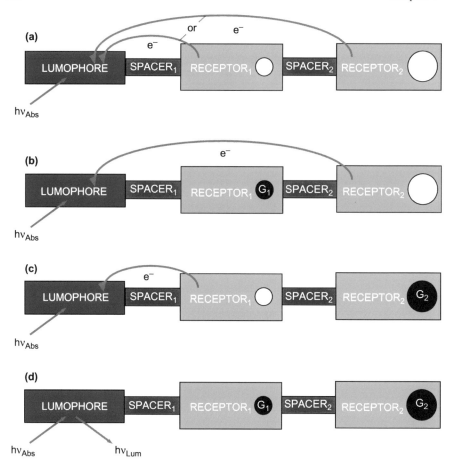

Figure 7.1 G_1, G_2-driven AND logic arising from the 'lumophore–spacer$_1$–receptor$_1$–spacer$_2$–receptor$_2$' system which produces luminescence as output. When two PET paths are shown, it is understood that only the faster one would occur.

When the experiment is conducted on **1**, both inputs are required to turn 'on' the fluorescence of the anthracene unit above a reasonable threshold. With 10^{-3} M H$^+$ added fluorescence increases by a factor of only 1.7, with Na$^+$ (10^{-2} M) the enhancement is only 1.1, but with both there is a fluorescence enhancement of 6 due to the blocking of both of the PET channels (Figure 7.3). This was the first instance where molecules were endowed with computational capabilities in the primary literature.[8] It is still remarkable that a small molecule manages to marshal two inputs (from free-swimming chemical species), an ultraviolet light power supply and a blue fluorescence output while providing the internal data processing required. Quantitative data concerning **1** is summarized in Table 7.2 along with

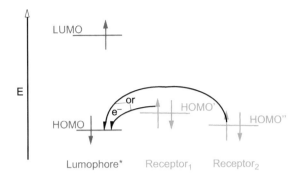

Figure 7.2 Approximate frontier orbital energy diagram corresponding to the case in Figure 7.1a. Diagrams corresponding to the cases in Figure 7.1b–d can be derived by lowering the HOMO energy of the appropriate receptor below that of the lumophore. The electron occupation in the excited lumophore is shown.

Table 7.1 Truth table for **1**.[a]

$Input_1$ H^+	$Input_2$ Na^+	Output Fluorescence[b]
none	none	low (0.012)
none	high (10^{-2} M)	low (0.013)
high (10^{-3} M)	none	low (0.020)
high (10^{-3} M)	high (10^{-2} M)	high (0.068)

[a]10^{-5} M in methanol:2-propanol (1:1, v/v).
[b]Quantum yields, λ_{exc} 387 nm, λ_{em} 446 nm.

those for several other AND gates. Of course, AND gates are common in modern semiconductor computers[12–14] and are well-known to mathematicians.[15,16]

The first example of intrinsically molecular numeracy (see section 9.2) contains an H^+, Ca^{2+}-driven AND gate **2**[17] which essentially exchanges Tsien's famous Ca^{2+} receptor[32] for Pedersen's even more famous benzocrown ether[33] within **1** while preserving the PET design.

By rearranging the format to a 'receptor$_1$–spacer$_1$–lumophore–spacer$_2$–receptor$_2$' (with smaller separation distances for the PET pathways) from the original 'lumophore–spacer$_1$–receptor$_1$–spacer$_2$–receptor$_2$'[8] there is a strong increase in the Na^+, H^+-induced fluorescence enhancement (FE) factor to 25.[18] Compound **3** uses Na^+ and H^+ inputs again being captured by a benzo-15-crown-5 ether and tertiary amine respectively. The choice of these receptors permits rapid PET to quench emission so that the 'off' states are virtually non-fluorescent under the operational conditions.

Table 7.2 Collected data for some molecular AND logic gates.

Device[a]	Input$_1$[b](hi,lo)[c]	Input$_2$[b](hi,lo)[c]	Output[d](hi,lo)[e]	Power[f]	Characteristics[g]
1^8 in MeOH: 2-PrOH	$H^+(10^{-3},0)$	$Na^+(10^{-2},0)$	$Flu_{446}(6.0,1.7)$	$h\nu_{387}$ $\log\varepsilon$ 4.0	$\log\beta_{H+}$ 4.5 $\log\beta_{Na+}$ 2.6
2^{17} in H_2O	$H^+(10^{-6},10^{-9.5})$	$Ca^{2+}(10^{-2.3},0)$	$Flu_{419}(25,2.5)$	$h\nu_{369}$ $\log\varepsilon$ 4.0	$\log\beta_{H+}$ 7.8 $\log\beta_{Ca2+}$ 6.0
3^{18} in MeOH	$H^+(10^{-3},0)$	$Na^+(10^{-2},0)$	$Flu_{428}(92,4)$	$h\nu_{377}$ $\log\varepsilon$ 4.1	$\log\beta_{Na+}$ 2.7
4^{19} in TADS[h]	$H^+(10^{-3},10^{-11})$	$Na^+(10^{-0.4},0)$	$Flu_{435}(18,2.4)$	$h\nu_{378}$ $\log\varepsilon$ 4.0	$\log\beta_{H+}$ 8.6 $\log\beta_{Na+}$ 1.9
5^{20} in MeOH	$H^+(10^{-1.8},0)$	$Cs^+(10^{-1},0)$	$Flu_{429}(4.0,1.5)$	$h\nu_{379}$ $\log\varepsilon$ 4.1	$\log\beta_{Cs+}$ 2.6
6^{21} in MeOH:H_2O	$H^+(10^{-2},10^{-12})$	$Na^+(0.6,0)$	$Flu_{432}(58,16)$	$h\nu_{368}$	$\log\beta_{H+}$ 7.8 $\log\beta_{Na+}$ −0.3
7^{22} in THF	$Na^+(10^{-5},0)$	$Zn^{2+}(10^{-5},0)$	$Flu_{430}(103,2)$	$h\nu_{363}$	
8^{23} in THF:H_2O	$Na^+(10^{-4.7},0)$	$K^+(10^{-4.7},0)$	$Flu_{435}(5.8,1.0)$	$h\nu_{270}$	$\log\beta_{Na+}$ 3.3
$9.10.11^{24}$ in 12^i	$H^+(10^{-8},10^{-12})$	$Ca^{2+}(0.2,0)$	$Flu_{620}(10,4)$	$h\nu_{450}$	$\log\beta_{H+}$ 10.3 $\log\beta_{Ca2+}$ 1.5
$13.14.QD^{25}$ in MeCN:H_2O	$H^+(10^{-8.7},10^{-6.2})$	$Na^+(10^{-3},0)$	$Flu_{560}(9,1.7)$	$h\nu_{415}$	
15^{26} in MeOH	$K^+(10^{-2},0)$	$F^-(10^{-2},0)$	$Flu_{397}(2.0,1.1)$	$h\nu_{343}$	$\log\beta_{K+}$ 4.0 $\log\beta_{F-}$ 2.5
$16.17.18^{27}$ in DCB[k]	$K^+(10^{-4.4},0)$	$Im^j(10^{-4.4},0)$	$Flu_{606}(17,3)$	$h\nu_{430}$	
19^{28} in CH_2Cl_2	$Na^+(10^{-3},0)$	$C_{60}(10^{-4.7},0)$	$Flu_{420}(4.5,1.7)$	$h\nu_{370}$	
20^{29} in MeCN	$H^+(10^{-2},0)$	$V_{app}^l(0.9,0)$	$Flu_{413}(15,3)$	$h\nu_{367}$ $\log\varepsilon$ 4.0	E^m 0.74
21^{30} in MeCN	$H^+(−, 0)$	$V_{app}^n(−0.4,0)$	Abs_{452}	$h\nu_{452}$ $\log\varepsilon$ 2.6	E^o −0.4
22^{31} in H_2O	t-BuOH(0.3,0)	$G\beta CD^p(10^{-3},0)$	$Pho_{493}(6250,1)$	$h\nu_{313}$	$\log\beta_{CD}$ 2.9

[a]The device is taken as a given molecular structure in a given medium.
[b]Chemical concentration unless noted otherwise.
[c]High (hi) and low (lo) input values. All concentrations are in M units.
[d]Output is given as the intensity of fluorescence or phosphorescence (or the absorbance) at a given wavelength. All wavelengths are in nm units.
[e]High and low output values. The low output value is the highest of the expected 'low' output values observed under the three different input conditions. The lowest of the expected 'low' output values observed is scaled to unity.
[f]Power supply is light at a given wavelength. The extinction coefficient (ε) of the device at that wavelength is given in units of $M^{-1}cm^{-1}$.
[g]Characteristics of device operation regarding inputs include chemical binding constants (β) in units of M^{-1}. Characteristics of device operation regarding outputs include emission quantum yields (ϕ_{em}).
[h]Tetramethylammoniumdodecylsulfate, 2×10^{-2} M in H_2O.
[i]Triton X-100 (**12**) in water.
[j]Imidazole.
[k]o-Dichlorobenzene
[l]Applied potential in volts, versus Ag.
[m]Standard potential in volts, versus Ag.
[n]Applied potential in volts, versus sce.
[o]Standard potential in volts, versus sce.
[p]Glucosyl-β-cyclodextrin.

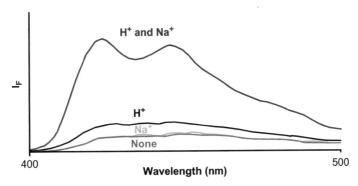

Figure 7.3 Fluorescence emission spectra of AND logic gate **1** ($\lambda_{ex} = 387\,\text{nm}$) when exposed to $10^{-3}\,\text{M}\,\text{H}^+$, $10^{-2}\,\text{M}\,\text{Na}^+$, both of them or neither of them. Redrawn from A. P. de Silva, H. Q. N. Gunaratne and C. P. McCoy, *Nature*, 1993, **364**, 42 with permission from Nature Publishing Group.

A simple lengthening of one of the ethyl groups on the amine in $\mathbf{3}^{18}$ to an octyl chain (and shortening of the other ethyl to a methyl) produces $\mathbf{4}^{19}$ which allows the operation of a molecular logic gate within a well-defined nanospace for the first time. Earlier work on sensing in aqueous detergent micelles[34] teaches us that **4** is a 'receptor$_1$–spacer$_1$–lumophore–spacer$_2$–receptor$_2$' system with a targeting/anchoring unit attached. It is clear that **4** is incorporated into the micelle while still allowing the analytes (H^+ and Na^+) to bind to their receptors and while preserving its fluorescent switching behaviour. The 3 nm radius of the micelle[35] shows that molecular computational elements can clearly operate in very tiny spaces.

An interesting feature of $\mathbf{5}$,[20] which distinguishes it from previous H^+, Na^+-driven fluorescent AND cases,[8,18] is that it also targets H^+, Cs^+-driven AND behaviour. Its 'receptor$_1$–spacer$_1$–fluorophore–spacer$_2$–receptor$_2$–spacer$_2$–fluorophore–spacer$_1$–receptor$_1$' format allows folded structures to come into play where one large cation can be sandwiched between two crown ether units. When the amine is bound to H^+, the binding of a Cs^+ to the benzocrown receptors switches 'on' fluorescence since all three possible PET channels are blocked.

Just like compound **4**[19] was derived from the early AND gate **3**[18] by adding a micelle-anchoring unit, case **6** is obtained by attaching a precursor of **3** to an amino-terminated polymer bead.[21] Though it would have fitted naturally in section 7.2.7, **6** is mentioned here owing to this logical connection to **3** and **4**. AND logic gates like **6** serve as identification tags for small objects and this is part of the molecular computational identification (MCID) technique discussed in Chapter 14.[21]

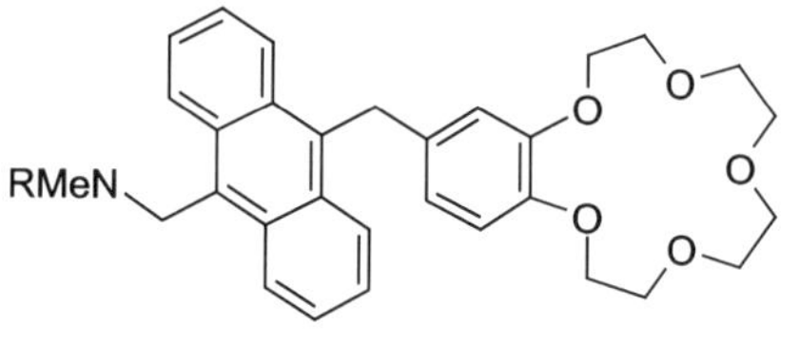

Features of the basic structure of **3**[18] can be discerned within Bag and Bharadwaj's **7**[22], which is a nice variation employing Zn^{2+} as one input alongside Na^+ as the other one. Many other open-shell transition metal ions can be substituted for Zn^{2+} as the input without losing efficacy, in spite of the notoriety of open-shell metal ions as luminescence quenchers. Care needs to be taken when open-shell metal ions cause fluorescence enhancements.[36,37]

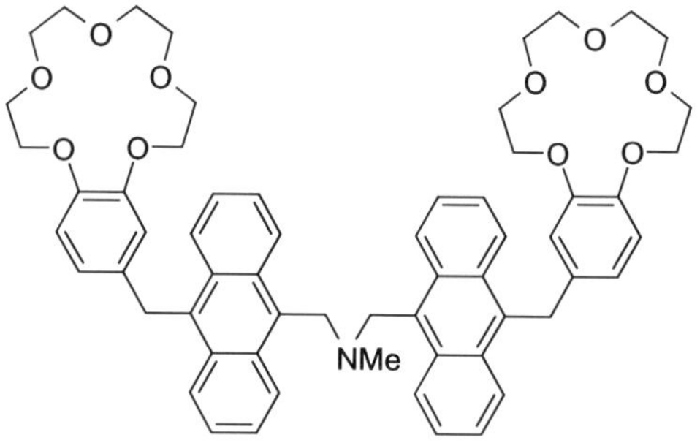

Classical fluorescent sensors for Mg^{2+} involve excited state intramolecular proton transfer (ESIPT) of phenolic imines,[38] as does a new case due to Callan.[39] Callan's team now interpret the ion-induced fluorescence behaviour of tripodal compound **8**[23] containing similar motifs as Na^+ binding to two of the pods in the enol form to give an emission at 355 nm, but the remaining pod in its keto form then binds to K^+ to produce a strong emission at 445 nm. K^+ on its own can only bind to the enol form and produce a weak emission at 355 nm. Na^+, K^+-driven AND logic is therefore visible at

445 nm. Further studies to clarify this very interesting behaviour would be welcome.

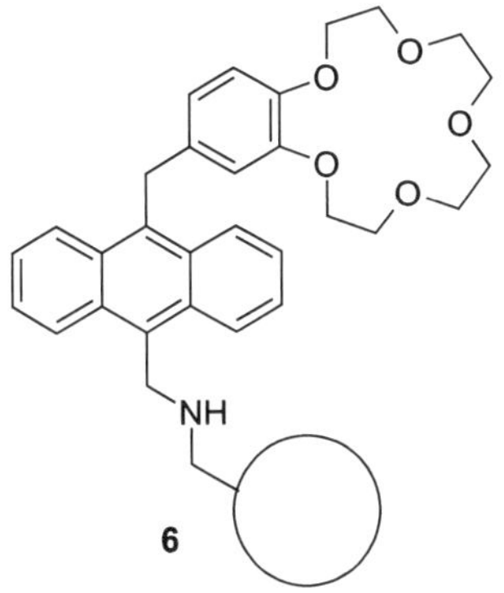

Many classical analytical reagents of a 'lumophore–receptor' construction also show features of AND logic. The importance of pH control is stressed during assays of metal ions, for instance. Therefore, alkaline conditions (H^+ input 'low') produce no signal (output 'low') even if the metal ion was present (metal input 'high') since the hydroxide ion competes successfully for it. Upon adjustment of pH to optimal near-neutral conditions (H^+ input 'high'), the presence of metal ion (metal input 'high') produces an analytically useful fluorescence signal (output 'high').[40]

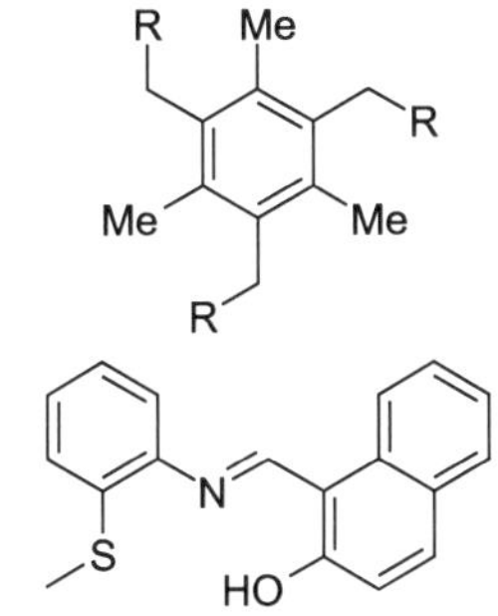

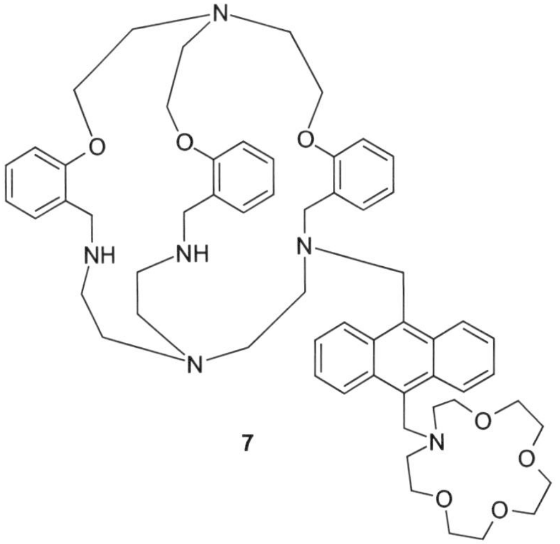

An interesting instance of self-assembly is seen when Schneider uses a double end-functionalized diphenylmethane derivative **23** and Zn^{2+} to produce a cationic cyclophane.[41,42] Fluorophore **24** is highly emissive in non-polar environments, but is hardly fluorescent in water due to a twisted internal charge transfer (TICT) excited state.[43,44] So **24** switches 'on' its fluorescence upon inclusion in the cyclophane which provides some isolation from water due to the aromatic walls provided by the diphenylmethane unit. Such behaviour has precedence in experiments with **24** complexed with classical cyclophanes.[45,46] Thus, the fluorescence output of **24** is controlled by the two inputs of **23** and Zn^{2+}. A related case from de Shayes has Ca^{2+} and a cousin of **23** (though not cationic in this case) driving the fluorescence switching 'on' of **24**.[47]

Self-assembly of systems composed of many more components, *e.g.* detergent micelles[24] are also useful in this regard. Aqueous micelles made from **12** contain hydrophobic proton receptor **9** and the hydrophobic Ca^{2+} receptor **10**. The hydrophobic lumophore **11**, which has a long-lived excited state,[48] is also included in the micelle so that it can be quenched quite efficiently by **9** and **10** by PET even though the components are not covalently linked. While it is true that intramolecular PET within systems like **1**[8] are faster, the pseudointramolecular system created within the micelle of about 3 nm radius[35] limits the separation distances between the components **9**, **10** and **11** sufficiently to allow significant PET rates. Addition of H^+ and Ca^{2+} at sufficient concentration allows the arrest of both PET pathways so that the luminescence of **11** to be enhanced by a factor of 2.4. Even though the luminescence enhancement factor is not large, the avoidance of substantial synthesis is a clear advantage. Such self-assembled systems will allow laboratories without synthesis capabilities to participate in new research concerning molecular logic and computation. Pallavicini's prior work on self-assembled fluorescent PET systems[49] needs to be noted here.

Self-assembly is also crucial when the receptor$_1$ **13** (targeting H^+) is exchanged via its thiol group onto the surface of CdSe/ZnS core-shell quantum dot fluorophores;[50] 20% of this receptor is converted to the receptor$_2$ **14** (targeting Na^+) via imine formation. This system due to Callan *et al.* is an important extension of PET-based 'receptor$_1$–spacer$_1$–lumophore–spacer$_2$–receptor$_2$' cases like **3**[18] into the nanomaterials domain.[25]

7.2.1.2 Cation and Anion Inputs

Systems that directly detect ion pairs are rare, even though ion pairs play an important role within both chemistry and biology.[51,52] The PET-based 'receptor$_1$–spacer$_1$–fluorophore–spacer$_2$–receptor$_2$' motif allows the use of fluorescence emission as a means of signalling the presence of the ion pair according to AND logic. Though ion pair detectors are rare,[53] **15** neatly targets K^+ and F^- together. James' team[26] use a boronic acid moiety to complex F^- which facilitates the complexation of K^+ by the benzocrown ether due to electrostatic attraction. The binding of each receptor to its target cuts off a PET process so that strong fluorescence is produced when the K^+F^- ion pair is encountered.

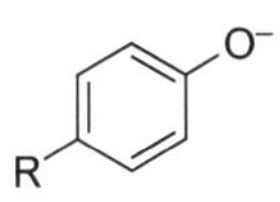

9; R = t-Bu

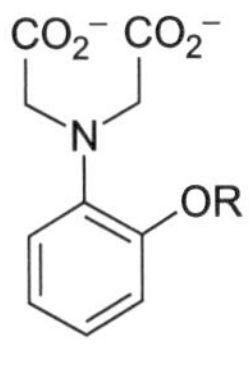

10; R = n-octyl

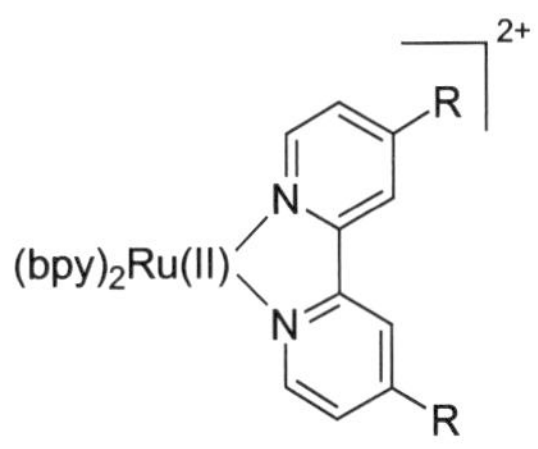

11; R = n-nonyl

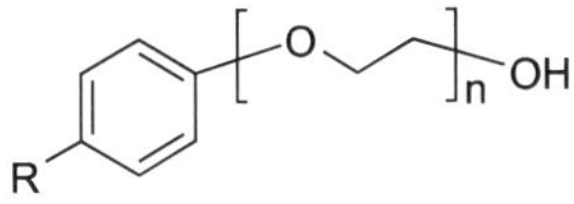

12; R = 1,1,3,3-tetramethylbutyl

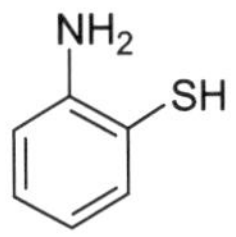

13

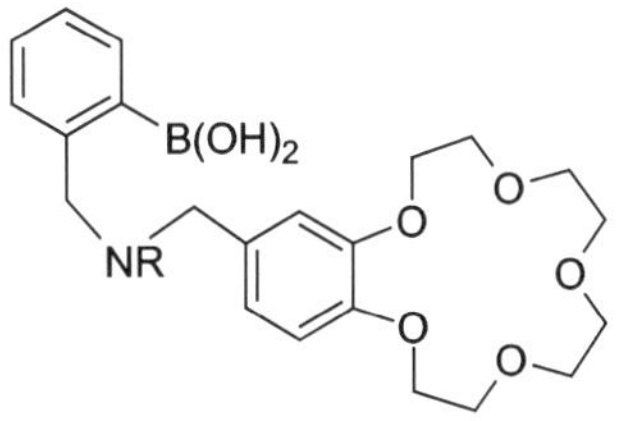

15; R = 1-pyrenylmethyl

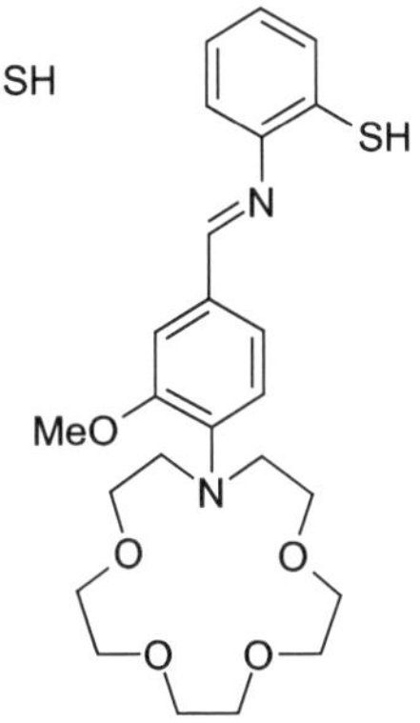

14

It might be surprising to find that even the simple 'lumophore–spacer–receptor' system can give rise to AND gate behaviour. An old case **25** due to Czarnik[54] can be viewed, in its free base form, as a H^+, PO_4^{3-}-driven AND gate. His group adapt a partially protonated polyamine receptor (as shown within **25**) at neutral pH to serve as a receptor to capture a HPO_4^{2-} anion. Only when the hydroxyl group is coordinated to the secondary amine does fluorescence occur. H^+ transfers intramolecularly to the secondary amine from the hydroxyl group. This suppresses PET from the benzylic amine to the anthracene. All the prior protonations do not influence the fluorescence output themselves but only prepare the system for this final protonation of the secondary amine to cause the light emission. This case can now be viewed as having H^+ and PO_4^{3-} as inputs. Fluorescence switching 'on' can be seen only in the presence of both inputs, *i.e.* AND logic. From a mechanistic viewpoint, HPO_4^{2-} binding only occurs after binding H^+ to the polyamine and to PO_4^{3-}. Nevertheless, mass action-type equilibria ensure that the system will show regular AND logic behaviour. Simpler assignment of input roles to H^+ is possible in the case of the polyquaternary salt **26**[55] since only the benzylic amine is available as a receptor and it does not protonate at neutral pH. However, PO_4^{3-} is subject to protonation under such conditions. This approach to anion recognition can be seen to build upon an earlier example of cooperative binding due to Lehn,[56] which can be interpreted as H[+], ATP[−]-driven AND behaviour. Tàrraga and Molina demonstrate Pb^{2+}, H$_2$PO$_4$[−]-driven AND logic in the fluorescence output at 422 nm of the simpler structure **27**.[57] Moro and Mohr's **28** uses Zn^{2+} and ATP inputs.[58] PET keeps the system silent until Zn^{2+} is bound, but in so doing, the 4-NH is deprotonated which opens a new radiationless deactivation path. This deprotonation is stopped only when ATP is bound at the vacancy in the Zn^{2+} coordination sphere. When the output is taken as the binding event itself (monitored by NMR shifts), Beer's isophthalamide podand-strapped calix[4]diquinone and Sessler's calix[4]arene-strapped calix[4]pyrrole show Na[+], Cl[−]-driven[59] and Cs[+], F[−]-driven[60] AND logic respectively. In both these instances, the components of the ion pair do not show significant receptor-binding when alone.

7.2.1.3 Cation and Neutral Inputs

Maligaspe and D'souza[27] use the crown-appended porphyrinZn(II) **16** complexed to C$_{60}$-appended imidazole **17** via the Zn(II) centre and bound to C$_{60}$-attached ammonium ion **18** via the crown as the starting state of the device. The fluorescence of the porphyrinZn(II) unit is heavily quenched due to PET from it to the C$_{60}$ moieties. Now if this ternary complex **16.17.18** is treated with imidazole as input$_1$, **17** is displaced thus ridding the porphyrinZn(II) unit of one of the PET pathways which ruined its fluorescence. However, the PET pathway involving **18** is still around.[61] Similarly, treatment of **16.17.18** with K[+] as input$_2$ displaces **18** but fluorescence is not recovered owing to the continued presence of **17** close to **16**.[62] It is only when input$_1$ and input$_2$ are both simultaneously present that **16** is freed from the shackles to **17** and to **18**. Strong red fluorescence is the upshot (Figure 7.4). C$_{60}$ features again in Zhang and Zhu's observation that it and Na[+] produce a fluorescence enhancement in **19**,[28] though the mechanism of action is less clear. On the other hand, Zhang and Zhu's **29**,[63] shows clearly understandable AND logic action in its fluorescence since H[+] is needed to block PET arising from the nitrogen lone pair and since cysteine is required to reduce the nitroxide radical and block PET to the radical centre.

Double Input–Single Output Systems

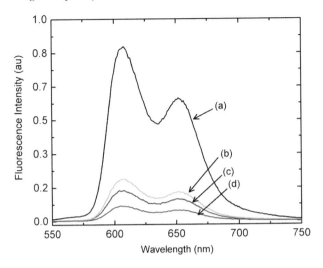

Figure 7.4 Fluorescence spectra ($\lambda_{ex} = 430$ nm) of **16** in 1,2-dichlorobenzene, (a) alone, (b) with **17**, (c) with **18** and (d) with **17** and **18**.
Reprinted from E. Maligaspe and F. D'souza, *Org. Lett.*, 2010, **12**, 624 with permission from the American Chemical Society.

19; R = 9-anthryl

20; R = 9-anthrylmethyl

22

21

23; R = NMe$_2^+$(CH$_2$)$_6$NH(CH$_2$)$_2$NH$_2$

24

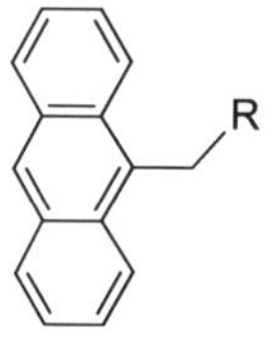

25; R = NH(CH$_2$)$_3$NH$^+$[(CH$_2$)$_3$NH$_3^+$]$_2$

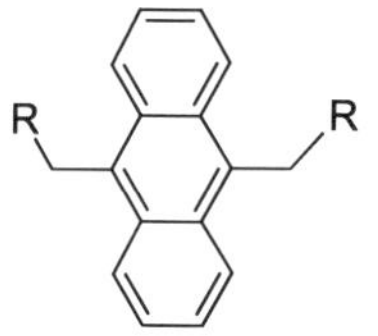

26; R = NH(CH$_2$)$_3$N(Me)$_2^+$(CH$_2$)$_3$NMe$_3^+$
46; R = N(CH$_3$)CH$_2$CH$_2$N(CH$_3$)$_2$
47; R = N(CH$_2$CH$_2$)$_2$O
48; R = Ph-4-N(CH$_2$CO$_2^-$)$_2$
51; R = CH$_2$N(CH$_2$CH$_2$)$_2$O
103; R = NH(C=S)NHPh

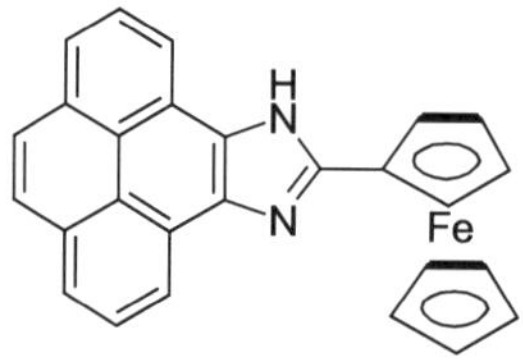

27

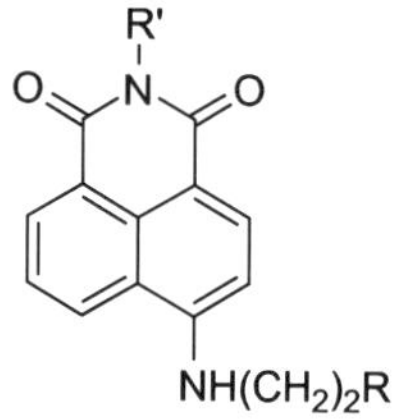

28; R = bis(2-pyridylmethyl)amino, R' = (CH$_2$)$_5$CO$_2$H

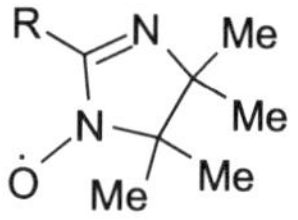

29; R = 1-pyrenyl

7.2.1.4 Cation and Biomolecule Inputs

Even large biomolecules can be employed as inputs. Ca^{2+}-dependent protein kinase Cα (PKCα) phosphorylates serine hydroxyls in suitable peptides. By optimizing a rationally designed lead compound, Chen, Ye and Lawrence[64] find that peptide **30** undergoes a Ca^{2+}, PKCα-induced FE of 3.6. The non-peptidic portion of **30** is inspired by Tsien's powerful fluorescent PET sensors for intracellular Ca^{2+} such as **31**[65,66] (which are noted in Table 5.1). We see how, upon PKCα-induced phosphorylation of the serine hydroxyl of **30**, a Ca^{2+}-binding pocket not dissimilar to that of **31** is produced. Importantly, this causes a conformational change which decouples the iminodiacetate unit from the neighbouring benzene ring. The electron-rich aniline unit effectively disappears so that PET from the latter to the difluorofluorescein fluorophore is halted and a 'high' fluorescence output is generated.[67] Peptide **30** is a compromise between the optimum Ca^{2+}-binding pocket (which favours a small separation between the serine and the iminodiacetate) and the optimum fit of the peptide into the PKCα pocket (which favours a large separation between the serine and the difluorofluorescein–phenyliminodiacetate unit). Interestingly, rather favourable Michaelis–Menten kinetic parameters V_{max} (8.5×10^{-6} M min^{-1} mg^{-1}) and K_M (2×10^{-5} M) are found as well.

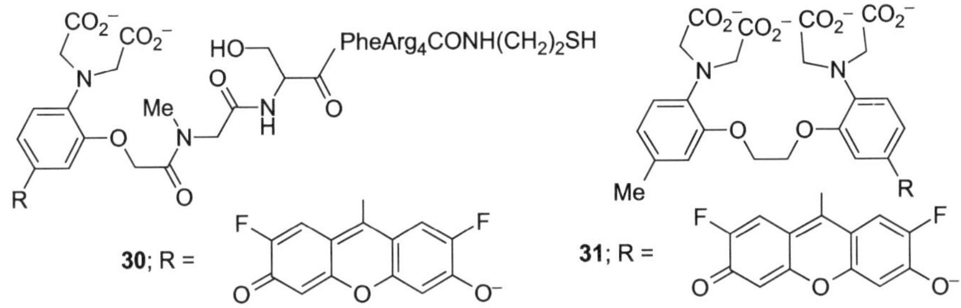

30; R =

31; R =

A smaller biomolecule, a bisphosphoundecapeptide **32**, is employed in our next example[68] alongside Zn^{2+} as the other input. This peptide is related to the C-terminal domain (CTD) of RNA polymerase II and is bound by the group IV WW domain of the Pin1 protein.[69] A mutant of the WW domain has a cysteine near the active site whose thiol group is reacted with the maleimide unit of **33** to produce the logic device. The FE of 1.6 ($\lambda_{exc} = 340$ nm, $\lambda_{em} = 440$ nm) which is achieved when both inputs are supplied at a 'high' level, is significant since several variants of the device and of the biomolecular input do not produce anything even close. The fluorescence enhancement occurs by restricting the rotation about the alkene bond during the excited state of the stilbazole unit. Similar fluorescence enhancement by guest-induced rigidification of alkene bonds are known.[70] The experiments are actually conducted with the Zn^{2+} complex of the device in the knowledge that the Zn^{2+}–bispicolylamine unit is a strong binder for phosphate groups.[71] The protein-based binding site and the Zn^{2+}-bispicolylamine unit act cooperatively to give a higher binding constant (1.2×10^6 M^{-1}) than the mutant WW domain alone (1.4×10^5 M^{-1}).

HO$_2$CThrSerProSerTyrpSerProThrpSerProSerNH$_2$

32

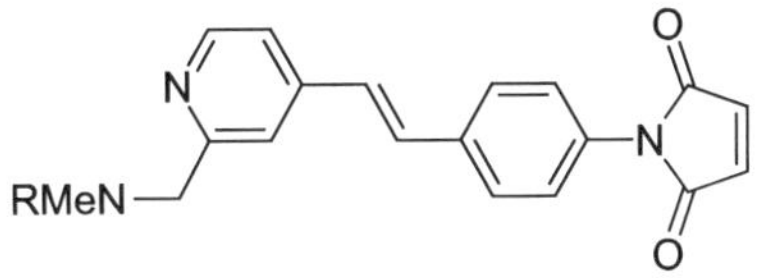

33; R = 2-pyridylmethyl

Another case of induced rigidification is best discussed here, in spite of it using anion and biomolecule inputs. Malachite green (**34**) is a non-fluorescent dye which becomes fluorescent only when rigidly bound into an RNA-based aptamer pocket which is organized only when flavin mononucleotide anion (**35**) is bound into a nearby pocket.[72] The aptamer and **35** can be considered the inputs. Malachite green would then be the device, with excitation at 610 nm as the power supply and emission at 645 nm as the output. Notably, **35** is transparent to 610 nm light. Potentially, cells can be genetically engineered to express the particular aptamer so that important cell components like **35** can be tracked by diffusing **34** into the cell.[73] Schneider's fluorescent AND logic device **24** (section 7.2.1.1) is strikingly similar in being activated only when a pocket is organized within diphenylmethane **23** by the binding of Zn^{2+}.[41,42]

35; R = $CH_2(CHOH)_3CH_2OPO_3^{2-}$
84; R = $CH_2(CHOH)_3CH_2OH$

7.2.1.5 Cation and Redox Inputs

Magri's **20**[29] is an AND gate with a fluorescence output controlled by oxidation and H^+ inputs. Like the amine, the tetrathiafulvalene (TTF) unit serves as a PET donor to the anthracene fluorophore. The PET process from the amine can be stopped as usual by protonation. The TTF unit needs to be selectively oxidized to its dication before photoinduced electron transfers stop. Even then, a PET process from the anthracene to the TTF dication is apparently prevented only by the Marcus inverted region.[29] The redox reactions can be induced exclusively by chemical reagents, like Fe^{3+}, so that the AND behaviour occurs on a molecular-scale. Cleaner and quantitative versions of these redox reactions are best achieved via voltammetry with a bulk metal electrode, however.

A previous example of this type, Stoddart's and Balzani's catenane **21**,[30] uses the redox input as a reduction of both bipyridinium units to their radical cation form (−0.44 V *vs.* sce). A bipyridinium unit exerts a stronger attraction to the polyether ring than the secondary ammonium, which in turn dominates over the mono-reduced bipyridinium moiety. The latter can hold onto the polyether ring only when competing against a secondary amine. The polyether ring's presence on the ammonium unit is taken as output 'high'. Output 'low' corresponds to the polyether ring living on a bipyridinium-based moiety. Movement of the polyether away from the bipyridinium-based unit only happens when **21** is simultaneously protonated and reduced.

7.2.1.6 Neutral Inputs

When an unusual strategy is employed, AND logic can arise from a structure as simple as **22**. When derivatized with heavy atoms and embedded in low temperature glasses, naphthalenes are strongly phosphorescent.[74] However, such phosphorescence usually cannot survive in aerated fluid solutions at room temperature. β-Cyclodextrin encapsulation can help in overcoming this problem,[75] but O_2 is still capable of quenching phosphorescence. Nocera's group encapsulates **22** within the extended cylinder of glucosyl β-cyclodextrin and caps the complex with *t*-butanol in order to cut off the access of O_2 to the phosphor.[31] *t*-Butanol and glucosyl β-cyclodextrin are the inputs to **22** which gives phosphorescence output. Related cases are available in a later review.[76]

Optical outputs are useful for easy communication between molecules and people. Electrical outputs, though usually emerging from larger devices, are also common. Nevertheless other forms of output can diversify and enrich the field. Such a case is thrown up by Fujita.[77] A host with a large hydrophobic cavity self-assembles from six (diaminoethane)Pd^{2+} and four exo-tridentate **36** ligands.[78] For instance, the curved aliphatic **37** and the flat aromatic **38** will, if present together, enter the host in water, as evidenced by strong absorbance at 437 nm. Optimal packing of the void volume of the host occurs in this way alone. The stoichiometry is 1:1:1 for the host.**37.38**, as found by ^{1}H NMR spectroscopic analysis. Further, nuclear Overhauser enhancement correlations are seen for protons of **37** and **38**, proving they are neighbours. The real output is the complexation event itself, even though one of its manifestations employs an optical channel.

36; R = 4-pyridyl

7.2.2 Indistinguishable and Separate Inputs

The previous paragraphs considered sets of examples where the case for molecular AND logic can be made easily because the inputs, outputs and even the power supply (which is provided by the incident light in absorbance- or emission-based experiments) are distinguishable from each other so that the analogy with individually wired electronic situations is there for all to see. However, chemistry is rich enough to throw up several other general situations which also can be recognized as involving AND logic.

Let us consider the luminescent PET system 'receptor–spacer–lumophore–spacer–receptor' where the terminal receptors are identical. This would be a particular type of homobireceptor[79] system quite common in supramolecular chemistry. The particularity is that emission would only emerge when each receptor is occupied by its target, *i.e.* when two copies of the target species have arrived at two sites in the device and been bound. In other words, this is two-input AND logic. As discussed previously under the heterobireceptor[79] cases such as **3**,[18] a PET process originates from each receptor when it is unoccupied by its target and the situation is no different in the present case. While transparent from a mechanistic or molecular-level viewpoint, the practical implementation is less clear-cut in this situation, involving what has been called 'degenerate' inputs.[80] This is because many experiments will not allow simple stoichiometric titration of the input species owing to incomplete binding arising from finite binding constants. An additional complication is that some input species such as H^+ will also be generated by the solvent such as water. Both these situations can be minimized by working in relatively inert (poorly competitive) media and by dosing-in equivalents of the input species. However, many investigations tend to be done in water because of its obvious biological relevance. Then we have to look to finer arguments to support the logic interpretation of experiments.

In the case of an ionic input, the luminescence switch 'on' for the 'receptor–spacer–lumophore–spacer–receptor' system is expected to occur at higher target concentrations than for the corresponding 'lumophore–spacer–receptor' because the binding of the second input species will be electrostatically repelled by the first input species already present at the other receptor close by.[81] For instance, the H^+-induced fluorescence switching 'on' of **39** corresponds to a pK_a value of 5.9 whereas the corresponding value of the monoreceptor version **40** is 7.2.[82] The difference of 1.3 pH units can be attributed to the repulsion of two protonated amines across the anthracene-9,10-dimethyl skeleton.

39; R = $CH_2N(CH_2CH_2OH)_2$
40; R = H

A higher emission enhancement factor, when compared with the corresponding monoreceptor, can also be expected because the homobireceptor case has a statistical advantage for launching the PET process.[81] A 6-fold higher fluorescence enhancement factor is measured for **39** *c.f.* **40**.[82] A 4-fold higher PET rate constant is also found.[82] Additionally, cases involving H^+ can show PET acceleration in the monoprotonated case due to electrostatic attraction of the incoming electron from the unbound receptor. The latter point is illustrated by the decrease of luminescence as **41** is monoprotonated as the pH value goes from 3.8 to 1.8.[83] The PET rate is poor in unprotonated **41** due to a near-neutral driving force whereas the rate is substantial for the structurally related **42** which has a more electron-rich arylamino receptor. We can conclude that

39,[82] **42**[83] and **43**[84–86] are H$^+$, H$^+$-driven AND gates based on PET mechanisms. It is noteworthy that fluorescence from an organic $\pi\pi^*$ excited state[87] and luminescence from a metal complex with a MLCT (metal to ligand charge transfer) excited state[87] are the outputs from **39** and **42** respectively.

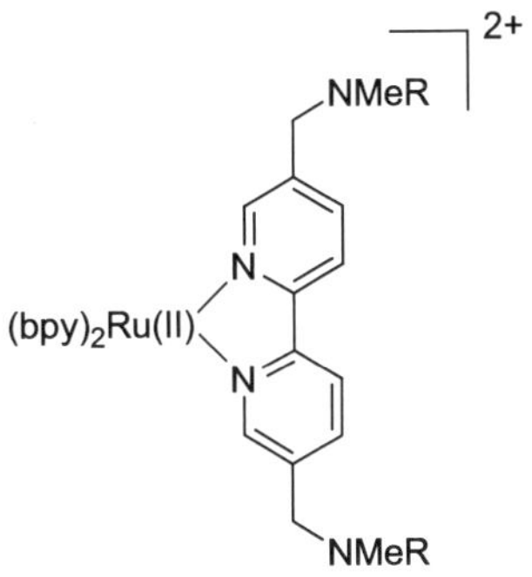

41; R = Ph
42; R = Ar

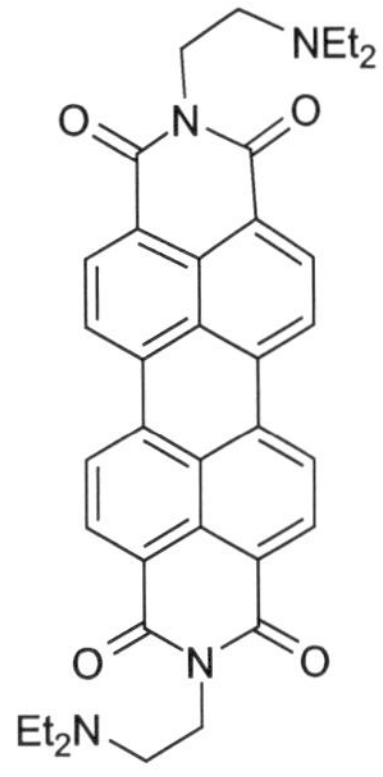

We collect a body of examples which have not been examined in a logical context. Luminescence output of the f–f type originating in lanthanide ions can be seen in Tb^{3+}.**44**[88] and Eu^{3+}.**45**[89] respectively. These gates are H$^+$, H$^+$- and K$^+$, K$^+$-driven respectively. A series of Zn^{2+}, Zn^{2+}-driven cases **46**,[90] **47**[91,92] **48**[93] (structures given alongside **26**) and **49**[94] all contain tricyclic lumophores except for **50**.[95] Two more cases with tricyclic lumophores are H$^+$, H$^+$-driven **51**[96] (structure given alongside **26**) and Li$^+$, Li$^+$-driven **52**[97] where the macrocyclic receptors in the latter (and the aprotic solvent) are responsible for the alkali cation sensitivity.

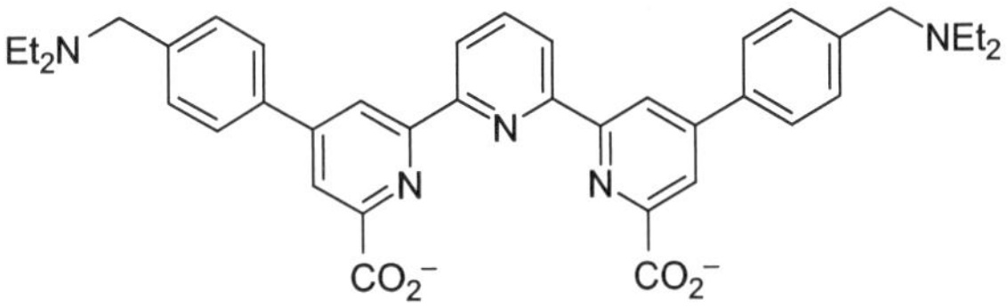

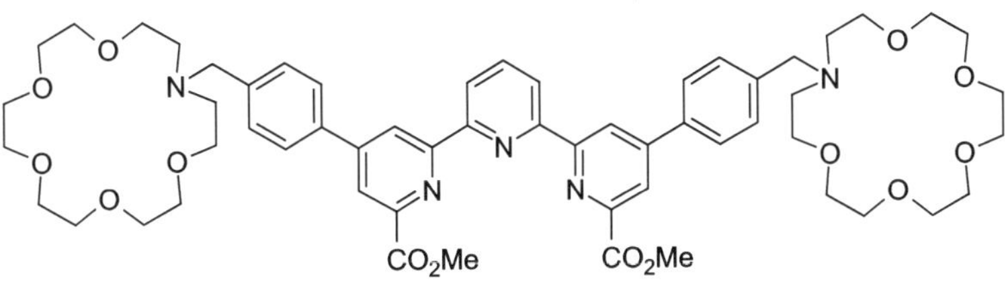

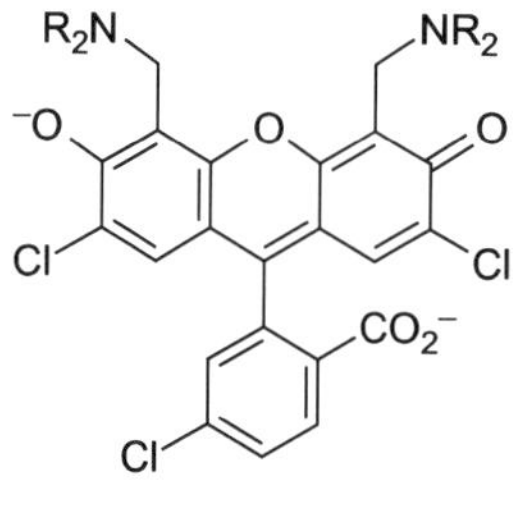

49; R = 2-pyridylmethyl

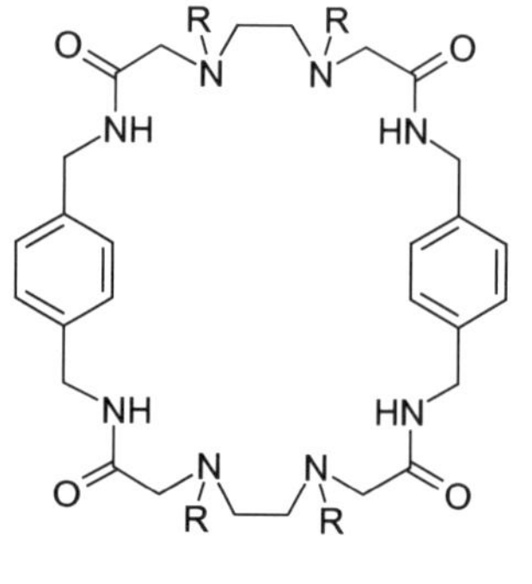

50; R = CH$_2$CO$_2^-$

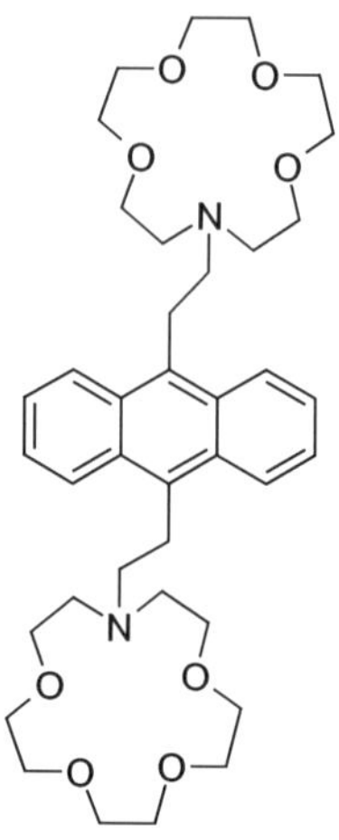

52

Homoditopic systems without PET mechanisms can also occasionally produce AND logic action. Li and Schmehl's example **53**[98] and a relative[99] depend upon rigidification to prevent double bond torsion and therefore release fluorescence. The two crown receptors accept a Cs$^+$ each but the larger-than-optimal size of Cs$^+$ allows another copy of **53** to stack on top which locks everything in place. Such locking only occurs after two Cs$^+$ are bound.

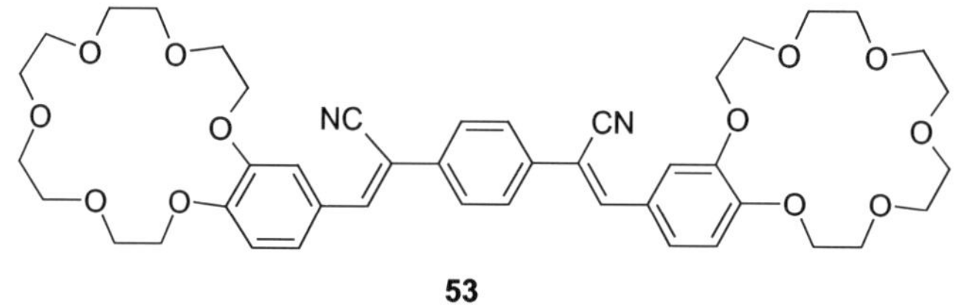

53

7.2.3 Distinguishable and Connected Inputs

Input chemical species can be connected before presentation to a molecular device but the idea of connected inputs in a semiconductor device context would have to mean that the two input lines are physically joined. These chemical inputs are functional groups, which could have existed on separated molecules, but are now deliberately coupled within a new multifunctional compound. Such coupling produces a chelate effect where the advantages of higher effective concentration and reduced entropy loss (upon binding) are seen.[100,101] A molecular device, with the appropriate geometric disposition of 'ports', can accept the input array presented by the multifunctional compound at lower concentrations than possible with separate input species. If the geometry is inappropriate, the input array (at the lower concentrations) cannot be accepted point-for-point by the device and hence a 'low' output is returned by the AND gate. Thus, the nature of the connector in the input array plays a critical role.

Imagine glucosammonium (**54**) being split into glucose and ammonium units which can then be separately targeted by an aminomethylphenylboronic acid[102] and an azacrown ether[103] respectively. Indeed, PET has been arrested with these, or similar, bindings to release fluorescence emission.[102,104] For instance, sugar 1,3-diol binding to aminomethylphenylboronic acid to produce the boronate ester leads to a stronger B–N bond which stops PET. Then, a molecular device like **55**[105,106] with the two receptors at the appropriate distance of separation could target glucosamine but the pH value needs to be chosen so that glucosamine is protonated but not the nitrogen centre in the azacrown ether. Cooper and James obtain a good fluorescence enhancement with **55** in response to glucosamine at physiological pH. It is mostly insensitive to simple ammonium ions and to glucose at similar concentrations,[105,106] owing to the chelate effect exerted by the glucosammonium ion.[100,101] The two PET processes present in **55** are arrested by the pair of functional groups binding to their 'correct' receptors. Thus the heterobireceptor system **55** is a glucosammonium-driven AND gate.[105,106] A related case due to James[107] will be held back until section 14.3.1.

A previous heterobireceptor system **56**[108] has only one PET process since the guanidinium unit is not sufficiently electroactive. Therefore, proper AND action cannot be expected even though the bifunctional input array, γ-aminobutyric acid (GABA), is bound adequately. Indeed the selectivity of fluorescence detection of GABA, as compared to its component functional groups, is not good. Wang's heterobireceptor system **57**[109] uses elements from **55** and **56** to target glucarate and again pays the price of the poor electroactivity of the guanidinium unit with a reduced selectivity of detection. Nevertheless, **55–57** are the vanguard of luminescent sensors empowered with AND logic to detect small multifunctional molecules selectively, many of which are found within cell signalling pathways.

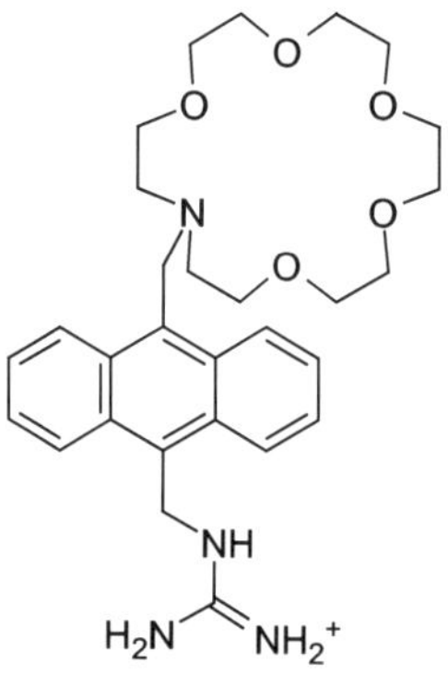

7.2.4 Indistinguishable and Connected Inputs

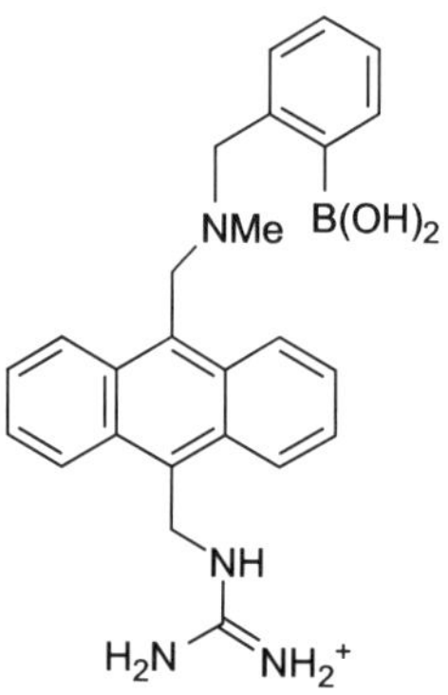

We now look at the one remaining category suggested by the two organizational parameters. Again, the increased detectability of the input array compared to the individual units will be an indication of AND logic behaviour of the device, though such information is available only occasionally.

Improved selectivity for alkanediammonium ions (log $\beta = 5.5$ for 1,4-butanediammonium) compared with the monoammonium counterpart (log $\beta = 2.9$ for ethylammonium) is seen with our **58** in chloroform–methanol solution.[110] Monoammonium ions at the same concentration of 10^{-5} M (of the optimally detectable diammonium) are not detected because only one guest binds to one receptor and the other receptor will not receive the guest owing to electrostatic repulsion. The second receptor still activates a PET process and keeps the fluorescence switched 'off'. At much higher concentrations, the second monoammonium ion binds to switch the fluorescence 'on'. Additionally, 1,4-butanediammonium ions win quite decisively in the battle among various alkanediammonium species, as regards detectability by **58** with its rather rigid backbone.

1,7-Heptanediammonium ions are detected best by **59**[111] compared with other alkanediammonium species, due to the rigid backbone. The monomer emission quantum yield of the anthracene fluorophore increased from 0.03 to 0.32 during this detection, which can be seen as 1,7-heptanediammonium-driven AND action. The longer-wavelength excimer emission seen in guest-free **59**, due to the face-to face orientation of the fluorophores, drops off concurrently, corresponding to NAND logic. Wavelength-reconfigurable logic (section 8.7.1) receives another nice example here, besides arising from a non-PET mechanism.

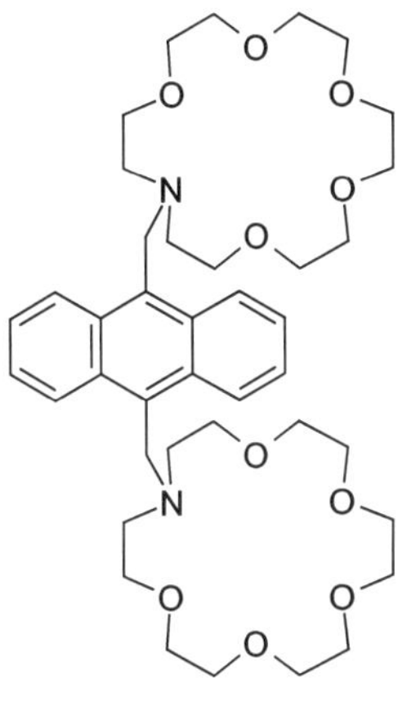

58 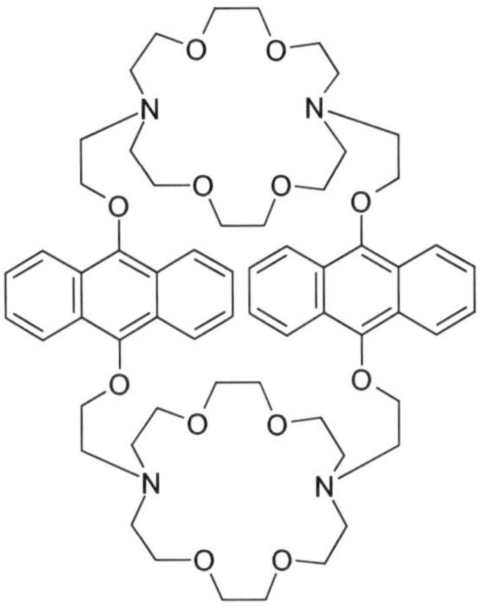   **59**

Continuing with the rigid backbone theme, ureidopyridinium moieties attached *via* methylene groups to the 9,10-positions of the rigid anthracene fluorophore in **60** provide urea receptor units for α,ω-dicarboxylates such as 1,4-phenylenediacetate.[112] The pyridinium units engage in PET from the anthracene fluorophore to weaken fluorescence in DMSO, though a significant exciplex emission component is also seen. 1,4-Phenylenediacetate binding reduces the positive charge density of the pyridinium units and causes a FE value of 3.0 by suppressing PET. Acetate achieves the same FE value but at much higher concentrations, because two equivalents of the anion need to arrive nearly independently (against electrostatic repulsion) at the receptors in order to block both PET processes. There is no intramolecular advantage, as seen in the binding of the second carboxylate of 1,4-phenylenediacetate. Thus **60** shows improved selectivity for sensing 1,4-phenylenediacetate owing to its exploitation of two receptors. The two phosphate faces of pyrophosphate can be similarly engaged.[113]

Owing to the rigid backbone again, Shinkai's **61**[114] is a selective glucose detector compared to monoreceptor versions.[115] Both of these are fluorescent PET systems. Each receptor of **61** binds a diol unit within glucose. Actually the binding diols in the pair are inequivalent, *i.e.* distinguishable to an experimentalist by using various techniques, but the homobireceptor-based device does not distinguish these diols. A pinnacle of this line of research is the fluorimetric discrimination between *D*- or *L*-glucose by enantiomers of **62**.[116]

Each enantiomer arising from **62** also has a second life in fluorimetric detection of tartaric acid enantiomers.[107] While both tartaric acid enantiomers produce nicely different amounts of fluorescence enhancement at pH 8.3, shifting to pH 5.6 gives fluorescence enhancement with one enantiomer and quenching with the other.

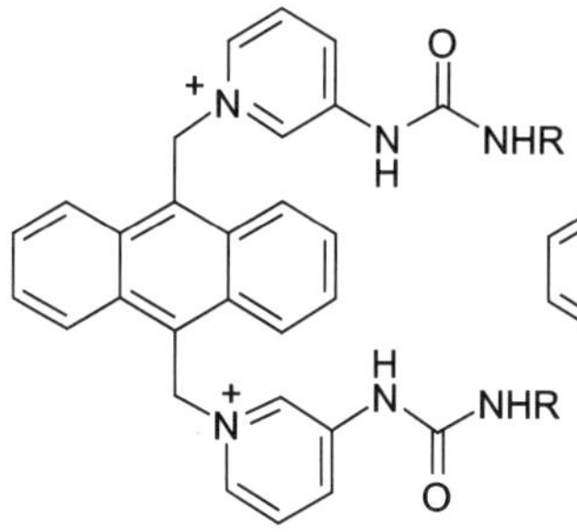

60; R = n-Pr 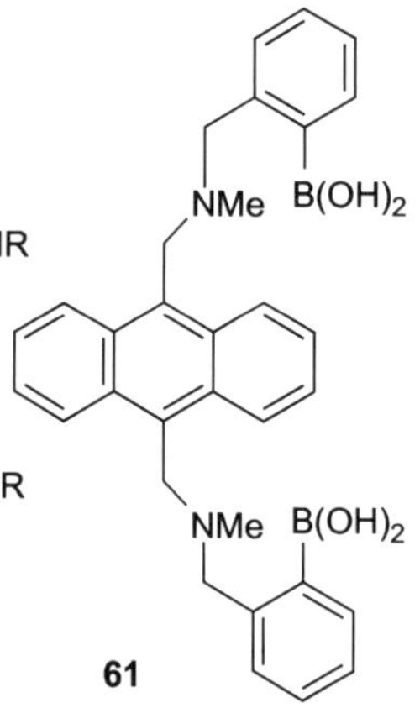   **61** 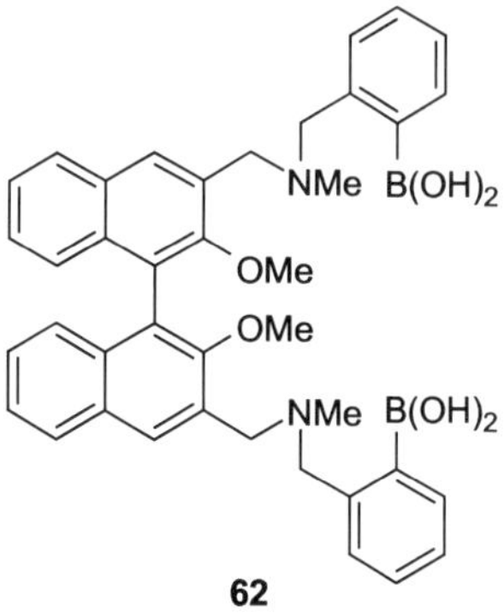   **62**

Pyrophosphate serves as a double input of phosphate faces to the Zn^{2+} centres of two molecules of **63**.Zn^{2+} so that the pyrene fluorophores are able to produce an excimer emission as the output.[117] Though the selectivity for pyrophosphate *vs.* ATP is not great, Hong's team has improved this situation with a different design involving a longer synthetic route[118] and this core receptor has been put to work by Smith.[119] If each Zn^{2+} is also considered as an input, this work would belong in Chapters 10 and 14 where related cases are on show. A case conceptually related to **63**.Zn^{2+} due to Prodi[120] was mentioned in section 4.7.

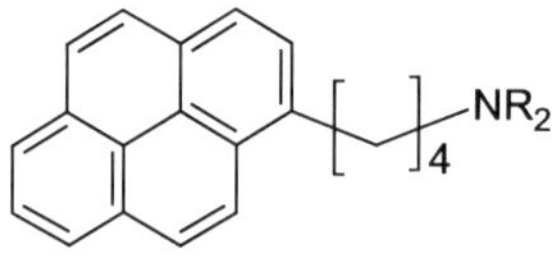

63; R = 2-pyridylmethyl

7.2.5 Light Dose Input(s)

Sensitized molecular photochemistry[121] necessarily requires a sensitizer and a dose of light as the two inputs. Then and only then does the output of the sensitized photoproduct arise. The substrate would be the device. Reactions using photogenerated catalysts[122] would also come under this general category. One example out of many photosensitized reactions would be phosphite **64**, which rearranges to phosphonate **65** under irradiation with light with wavelength > 300 nm if benzophenone is present to act as a triplet sensitizer.[123]

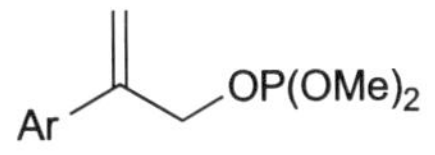

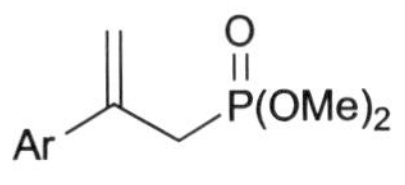

64: Ar-C(=CH₂)-CH₂-OP(OMe)₂

65: Ar-C(=CH₂)-CH₂-P(=O)(OMe)₂

Light and ions can serve as mixed inputs for AND gates. In the majority of cases, the light dose must be large enough to cause a measurable degree of photochemical change. If light absorption is taken as the output, such systems translate to photochromics,[124] which are chemically responsive[125–140] owing to the availability of an internal receptor or active site. Photochromics have the property of being switched from their colourless state to the coloured state by ultraviolet (uv) irradiation, and being returned to the colourless state by visible irradiation. Unlike, say, ion-induced fluorescence switching, photochromics latch in their colourless or coloured states for a period of time after the input stimulus has ceased. Eventually they return thermally to the thermodynamically stable state. Because of this memory effect, photochromics will feature in Chapter 11, which concerns history-dependent systems. For the purpose of combinational logic experiments, the memory effects will be removed by returning the system to the same starting state each time.

Among these receptor-equipped photochromics is Inouye's **66**.[133] Unlike classical photochromic spiropyrans,[141] **66** produces little coloration upon uv irradiation. Strong coloration is seen if Li$^+$ is present during irradiation, *i.e.* AND logic with Li$^+$ and ultraviolet light inputs along with visible absorption as output. Inouye's **66**[133] shares many conceptual similarities with Diederich's **67**.[135] Inouye also published previous examples without logic interpretations.[132] Other dual-mode transducers with electrochemical stimulation and light irradiation as inputs are also available.[142–146] Related systems where the metal ion can be held by the phenolate oxygen develop the coloured ring-opened form either by metal ion or by uv irradiation. These are OR gates and can be compared with the examples in section 7.3.

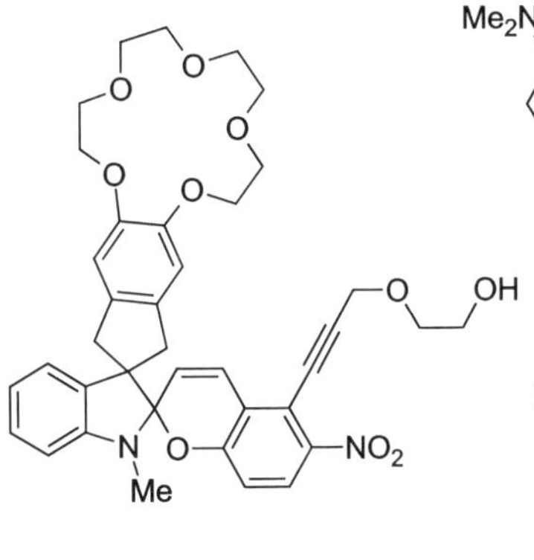

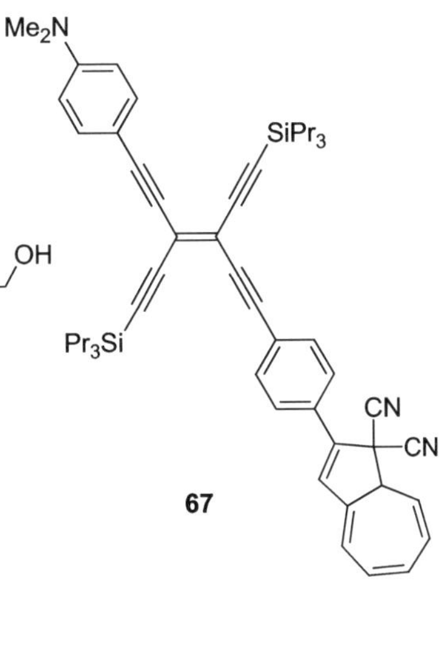

66

67

A different structural type can be seen in **68**, due to Pina, Balzani and colleagues, where a 365 nm ultraviolet light dose allows isomerization of *E*-**68** to its *Z*-form. 10^{-1} M H$^+$ then cyclizes the latter to the product **69** which fluoresces at 515 nm.[147–149] Therefore both high [H$^+$] (pH 1) and radiation dose are necessary to produce **69**. Compounds such as **68** have the added interest of acting as photochromic memories capable of locking and erasing besides the usual write–read operations.

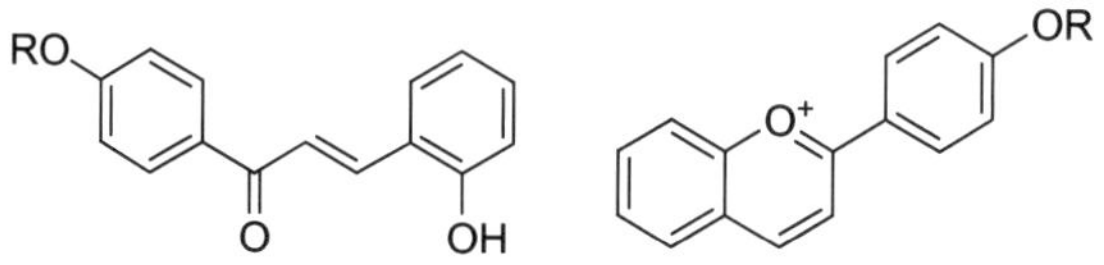

68; R = H
138; R = Me

69; R = H
139; R = Me

If we return to spiropyran photochromics, we can examine a case where chemical reception emerges only from the merocyanine form. To set the stage, let us recall that amidines are basic enough to deprotonate alcohols to a significant extent. Also, CO_2 can be used to neutralize alkoxide in a reversible manner.[150] Zwitterionic merocyanine **70**, which arises from the photoinduced ring opening of **71**, can be deprotonated and cyclized to **72** in this way.[140] The pale yellow-coloured **72** is no longer photochromic. Given that uv light and CO_2 can be applied as inputs to this system while monitoring the absorbance output at various wavelengths, there is potential for logic operations. Of course, more complex logic operations can be produced if visible light is also considered as an input, not only for **71** but also for **66**,[133] **67**[135] and **68**[147,149] (see section 10.9).

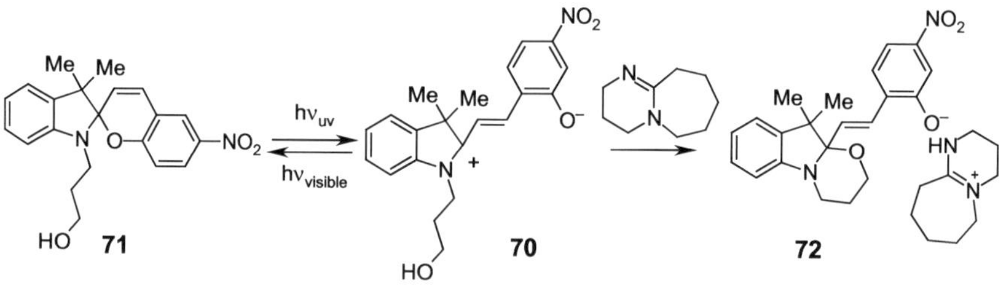

Other examples of all-optical AND logic are available,[151–154] even though an old claim of this kind in the conference literature apparently has not crossed into the peer-reviewed primary literature.[155] We focus on **73** and **74**, due to the teams of Wasielewski and Levine respectively, which use different methods. The tetra-chromophore system **73** is held together by *m*-substituted benzene rings. The 4-amino-1,8-naphthalimide is initially pumped at 420 nm to cause PET from it to the 1,4:5,8-naphthalenediimide. The other chromophores become involved only if the naphthalenediimide radical anion is pumped at 480 nm. Now the extra electron within the naphthalenediimide is passed to the 1,8-naphthalimide and then on to the 1,2:4,5-benzenediimide. So the absorption signature of the benzenediimide radical anion at 720 nm (the output) is only observed if the two femtosecond laser pulses at 420 nm (input$_1$) and at 480 nm (input$_2$) are applied sequentially (2 ns separation). It is notable, however, that conventional AND gates require simultaneous, and not sequential, application of inputs (a 2 ns delay would matter in devices running near gigaHertz rates). Nevertheless **73** is a rather fast gate because it resets in 25 ns.

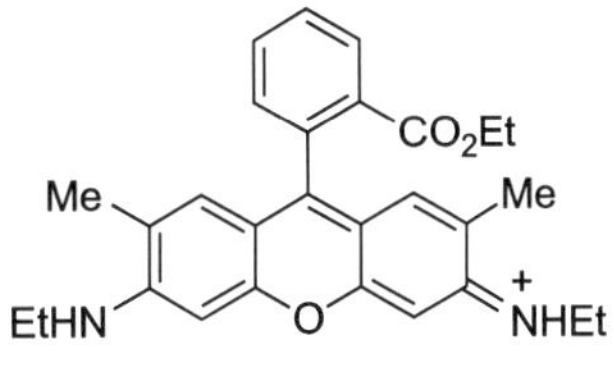

74

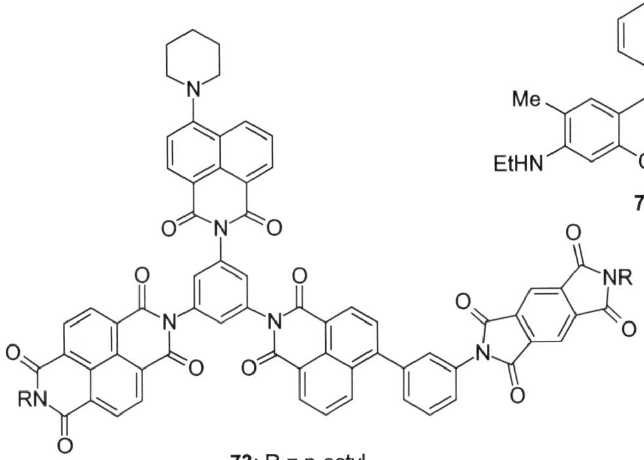

73; R = n-octyl

Wasielewski's previous publication on **75**[156,157] was the forerunner of the discussion concerning **73**.[151] The donor–acceptor molecule **75**[156,157] was a fast molecular switch with potential logic capabilities. The two terminal porphyrin donor moieties were independently capable of reducing the central perylene tetracarboxydiimide acceptor moiety via PET. When only one porphyrin was excited by a femtosecond laser pulse, the absorption due to the perylene tetracarboxydiimide radical anion was seen. If both porphyrins are excited simultaneously by a higher intensity laser pulse, two PET processes to the central acceptor can occur, giving a dianion, with a different absorption band.

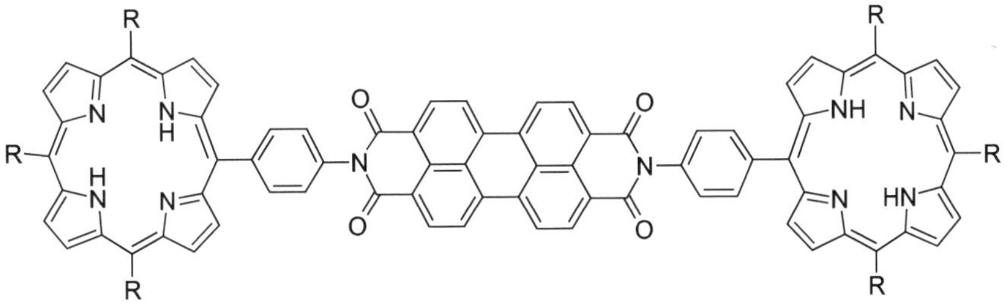

75; R = n-pentyl

It was imagined that **75** could perform the AND logic operation if two light beams of different wavelengths were used as inputs, and if the dianion absorption was taken as the output. This would be possible for **75** because excitation of the perylene tetracarboxydiimide moiety would induce the first PET process, and the second input light dose should have a wavelength matching the porphyrin absorption. The speed of such light-driven molecular switches and gates would be testimony to the lightness of their only 'moving part' – the electron. It can also be imagined how two laser pulses of sub-threshold intensity at the same wavelength can be combined to create the dianion absorption, *i.e.* AND logic. Observation of the radical anion absorption instead would then correspond to XOR logic (see section 7.7.1). This would be a case of logic reconfiguring caused by the observation wavelength (Chapter 8). Further, having AND and XOR gates available would make half-adder action (Chapter 9) possible.

Light intensities can be regarded as the inputs in a minority of cases which rely exclusively on easily detected fluorescence phenomena (so that significant doses of light are not required). Levine noticed that two-photon processes in general could be candidates for two-input AND logic, especially when the output of the process is a detectable fluorescence.[152] This has additional utility, as discussed later, if the fluorescence is from an upper excited state. A fundamental photochemical generalisation, the Vavilov–Kasha rule, stands in the way of this requirement. Thankfully the 'rule' makes quite a few exceptions.[158] The common laser dye **74** is a suitable example where a relatively low intensity laser beam at 532 nm (input$_1$) would allow one-photon absorption to the lowest excited state. Now another laser beam of similar intensity (and colour) (input$_2$) is directed onto the sample such that the combined intensity enables two-photon absorption which populates the next higher excited state. This obviously requires judicious choice of the beam intensities. Now a weak but detectable fluorescence is seen at 430 nm. AND and all 15 other double-input, single-output logic gates are described by Zhang *et al.* with bacteriorhodopsin as the active medium, two laser beams (blue and yellow) as inputs and a chosen transmittance as the output.[159] Another laser beam is required for biasing purposes. The rich photochemical cycle of bacteriorhodopsin is the source of this versatility.

Now we feature a carotene–porphyrin–fullerene[160] triad[161] due to the Gust, Moore and Moore trio. Laser excitation of the prophyrin unit (at 575 nm) rapidly causes PET to the fullerene and a subsequent electron transfer from the carotene to the porphyrin radical cation. When monitored at 980 nm, this charge-separated biradical state lives in a glass at 77 K for 1.3 µs. However, if the experiment is repeated in a magnetic field of 20 mT, the lifetime is extended to 2.0 µs. In essentially field-free conditions, all three triplet levels of the biradical couple with the singlet state but an applied magnetic field breaks the degeneracy of the triplet levels so that only the mid-energy level remains energetically capable of coupling with the singlet. If the laser light dose and the magnetic field are the inputs and the transient absorbance at a suitable delay time (*ca.* 1 µs) is the output, AND logic emerges. For instance, if the magnetic field were missing, the absorbance would decay too fast for a 'high' value to be registered at the chosen delay.

We finish this section by turning to complex chemical systems.[162] Ashkenasy's team consider a self-replicating peptide[163] where the replication depends on dimerization of templates. This can be prevented by attaching a bulky but photoremovable group. The replication rate then depends on the availability of a light dose and also on the template concentration.[164] We need to note here that the word 'system' is often used in this book to refer to small assemblages such as 'fluorophore–spacer–receptor' systems. Though small, these also show clear emergent properties, such as switchable emission signals. However, larger chemical networks bring out much higher degrees of complexity. Some of these libraries are computed to show build-up of a particular library member when two effector compounds are provided as inputs.[165]

7.2.6 Biopolymeric AND Gates

Until now, we have employed a classification based on input type. However, now is the time to depart from this classification in order to accommodate an important category of logic devices. Molecular logic is not only found within small molecules. Many native enzymes and enzyme cascades can be considered for the same purpose, theoretically[166–168] or experimentally.[6,169–171] Importantly, this can be done without any need for molecular synthesis in-house because the enzymes and their natural substrates can be bought in. This is clearly an advantage for laboratories without synthesis expertise.[172] Enzyme-based logic was available from Conrad and Zauner earlier but the output signals turned out to be small, and they were dependent on the threshold chosen (see section 8.5).[6]

Enzymes also bring the celebrity of biomolecules into the field of molecular logic, as well as allowing cascading of several enzymes to permit physical integration of individual molecular logic gates. One example of this is the chaining of glucose dehydrogenase (GDH) and horseradish peroxidase (HRP) with their common cofactor, the $NADH/NAD^+$ redox couple (Figure 7.5).[173] GDH processes glucose with the aid of the cofactor NAD^+ into gluconic acid (and NADH). HRP similarly processes H_2O_2 with the aid of the cofactor NADH into H_2O (and NAD^+). Therefore the output gluconic acid is produced

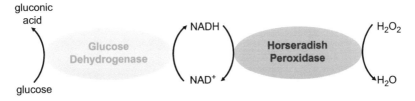

Figure 7.5 Glucose dehydrogenase and horseradish peroxidase cascade which leads to glucose, H_2O_2-driven AND logic with gluconic acid output.

from this enzyme set only if glucose and H_2O_2 are both present as inputs. If H_2O_2 is absent, glucose cannot be processed by GDH in a steady-state fashion because the starting NAD^+ concentration is deliberately held at zero as part of the operating specification of the device. Also, NADH needs H_2O_2 in order to produce NAD^+ from HRP. Of course, if glucose is absent, there can be no gluconic acid. This is glucose, H_2O_2-driven AND logic with gluconic acid output.

Cytochrome c in its native form displays a nice case of AND logic in Konermann's hands.[174] Urea and H^+ inputs in water get together to denature the protein reversibly so that an amino acid (tryptophan) fluorophore avoids quenching by a haem porphyrin which is located at the protein active site. The concentrations of urea (4.8 M) and H^+ (1.3×10^{-3} M) are chosen to be low enough so that neither can act as a denaturant on its own.

The approach by Sivan et al.[175] uses a synthetic modification of the enzyme α-chymotrypsin as it hydrolyses **76** to the coloured phenolate product. The integration of the reactive group of **76** with the chromophore suggests an ICT mechanism to the switching action. A lysine unit on the gateway to the active site of the enzyme is modified by derivatization with **77** to form an amide. Irradiation at 334 nm converts the E-form of the derivatized **77** into its Z-version which blocks the gateway to the enzyme active site. Thus, the substrate **76** and the inhibitor **78** (proflavine, a classical drug) are rendered unemployed, since the enzyme is disabled. Irradiation at 420 nm restores the E-form of **77** amide, opens the gateway for **76** and **78** to reach the active site and, therefore, if the inhibitor is reduced to a non-inhibitory version, **76** can be processed (output 'high'). Reduction of **78** is achieved photochemically (436 nm) in the absence of air. The corresponding oxidation back to **78** can be accomplished by simple aeration and, so, the reduction of **78** and the 420 nm radiation to give the E-**77** amide are the two inputs (in their 'high' states) of the AND gate. Some cross talk is unavoidable at such close wavelengths of photochemical irradiation.

76; R = n-pentyl   **77**   **78**

If biomolecules are celebrities, then DNA is in the A-list. Oligonucleotide-based logic gates are therefore key entrants into the field of molecular logic.[176,177] They also allow the use of oligonucleotides as both inputs and outputs for physically integrating gates in a vectorial stream,[178] and will therefore be discussed in section 9.2. Ribonucleotide-based analogues are even more generally applicable.[179] Pleasingly, some aptazyme-based cases do not require labelling of the principal molecular device.[180] However, this approach remains distinct from the situation in semiconductor electronics where electrons constrained in wires perform the integrator role, permitting the close co-existence of millions of gates.[181]

Gate **79**[182] is an example of a DNA-based 2-input AND logic function where a fluorescence output results from two sequential molecular-scale events. The logic gate **79** is modified at the 3'-terminus with a fluorescein moiety to provide the fluorescence output signal. The two inputs are **80**, a complementary oligonucleotide, and **81**, which binds in the minor groove of double-stranded DNA. When the duplex forms between **79** and **80** the minor groove binding site for **81** is created. The proximity of the fluorescein unit of **79** to **81** in the 1 : 1 complex results in electronic energy transfer (EET)[11,183] from **81** to fluorescein and a strong fluorescence emission from the latter at 520 nm if **81** is excited at 350 nm. For **79** alone, 350 nm excitation is wasted because the fluorescein unit absorbs poorly at this wavelength. So no significant 520 nm emission is seen. Neither of the inputs alone causes a significant change in the fluorescence as expected of an AND logic function. Other biopolymeric AND logic gates are known as parts of more complex systems,[184] and therefore will be reserved for discussion at several points within Chapter 10, *e.g.* Figure 10.5.

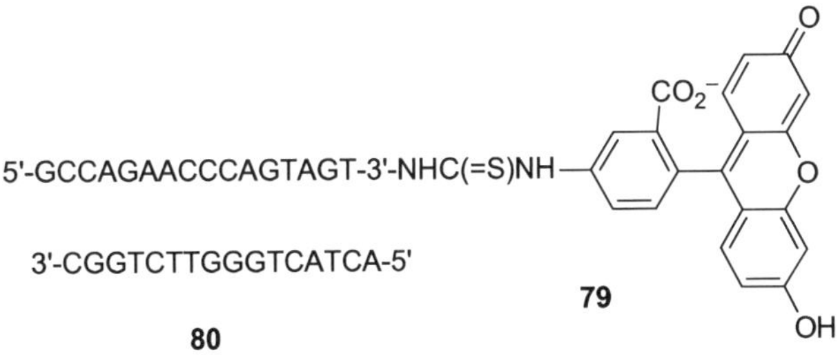

5'-GCCAGAACCCAGTAGT-3'-NHC(=S)NH—

3'-CGGTCTTGGGTCATCA-5'

79

80

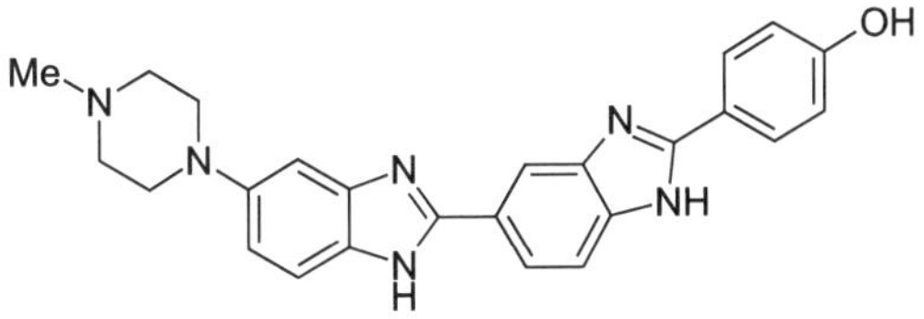

81

7.2.7 AND Gates using Molecule-based Materials

We continue with a classification based on the device rather than the input. This is done in order to accommodate yet another important category of logic devices – those based on materials of one kind or another. Metals, glassy carbon, organic polymer films and gels, inorganic polymer particles are some of the examples featured here. The material is intrinsic to the observation technique in some cases.

Conventional semiconductor logic gates such as AND are produced by wired arrays of simpler switches such as diodes or (in newer systems) transistors. Molecular electronic implementations can therefore do the same with a good chance for success. Heath, Stoddart and colleagues use a monolayer of **82** sandwiched between metal and metal oxide layers and outer metallic contacts to produce diode behaviour,[185,186] several examples of which are known.[187,188] These diodes can now be wired conventionally to give logic behaviour, *e.g.* AND (see Chapter 3, Figure 3.4).

82; R = —CH$_2$—N$^+$⟨⟩—⟨⟩N$^+$—

In a similar vein, two nanosheet diodes built from polymer Langmuir–Blodgett films are operated at different wavelengths to produce an AND logic gate.[189] One nanosheet is constructed with a phenanthrene- and an electron-acceptor-based (dinitrobenzene) acrylate–acrylamide copolymer film and the other with an anthracene- and an electron-donor-based (dimethylamine) acrylate–acrylamide copolymer film. The input signals are light doses possessing excitation wavelengths for the two chromophores, phenanthrene and anthracene, each of which can be selectively excited. When the phenanthrene layer is excited at 300 nm, PET occurs to the dinitrobenzene-containing film, and a photocurrent of 70 pA, similar to the dark current, is recorded. When the anthracene layer is excited at 380 nm, ET occurs from the dimethylamino layer to the anthracene film, again with a low photocurrent of 70 pA. However, when both chromophore-containing polymers are excited simultaneously, charge transport occurs from the phenanthrene to the dinitrobenzene to the dimethylamino to the anthracene layer with a high photocurrent of 190 pA. Thus, AND logic is demonstrated with two optical inputs and an electrical output. We note that the first experimental molecular AND gates[8] also used two PET processes, but with a single acceptor which was responsible for the fluorescence output. The two inputs were also chemical, rather than optical. In concluding this paragraph, it is worth noting that all electrical or electrochemical experiments concerning AND logic would belong in this section because they need bulk metal electrodes.

To emphasize this point, we discuss a cyclic voltammetry experiment on glucose-6-phosphate dehydrogenase, immobilized on multi-walled carbon nanotubes deposited on a glassy carbon electrode.[3] The enzyme is needed to catalyse the redox reaction between the two inputs, glucose-6-phosphate and NAD$^+$. One of the products, NADH, is detected as a current in the cyclic voltammogram.

Schneider's and Hamachi's groups interpret gel transitions in terms of AND logic.[190,191] The former case employs a soft polymer material based on the motif **83** which, depending on the protonation state, leaves it open to the gamut of molecular interactions. Application of 0.03 M phosphate (at pH 7) produces a reversible 5% linear expansion whereas augmentation of phosphate with

0.1 M AMP makes the expansion jump to 20%. The corresponding volume expansion change, which matters in drug delivery applications for instance, is much larger. Gawel and Stokke achieve a similar response by providing additional DNA-based crosslinks within an acrylamide/bisacrylamide hydrogel.[192] The DNA-based crosslinks are chosen to hybridize across two sections. Both these sections can be disrupted by the addition of two separate single-strand oligonucleotides (which serve as the inputs) which begin with a suitable toe-hold and follow up with strand displacement. The gel swells because of the decrease in the cross link density and also because of the accumulation of the input oligonucleotides which draw in more water.

Me—⟩—CONHn-Decyl
Me—⟩—CONH—CH₂CH₂—NH—CH₂CH₂—NH₂
83

Nandi's sol–gel case based on aqueous methylcellulose depends on its gelation above 50 °C to create hydrophobic regions.[193] At neutral pH, the fluorescence of riboflavin **84** (structure given alongside **35**) dispersed in this medium is 85-fold stronger in the gel state owing to its location in the hydrophobic regions where hydration-caused Born–Oppenheimer holes[194] are less likely. As observed previously in water,[195] emission quenching also occurs under alkaline conditions in both sol and gel owing to deprotonation of the imide. The fluorescence output thus fits H^+, temperature-driven AND logic. Some hysteresis is seen during thermal cycling of the device due to slow kinetics associated with gel reversal.

Stoddart, Zink and their groups joined forces to build mesoporous SiO_2 nanoparticles[196] that release a fluorescent cargo through their pores only when simultaneously interrogated by two inputs in the form of OH^- and a light dose at 448 nm. The pores are blocked by cucurbit[6]uril stoppers attracted to ammonium stalks. Application of OH^- opens the pores by removing this attraction but the cargo hardly diffuses out at all. Irradiation causes rapid *E–Z* isomerization of the azobenzene units followed by similarly rapid thermal *Z–E* isomerization. The impelling action so-created pushes the cargo out (Figure 7.6).

7.3 OR

Previous sections on AND logic showed our reliance on receptors binding their guests selectively so that each guest could serve as the correct input to each port of the molecular device. Now we head to the opposite extreme. Molecular OR logic might therefore seem an antithesis because most chemical research aims for selective reactions, except for notable exceptions, such as the non-selective

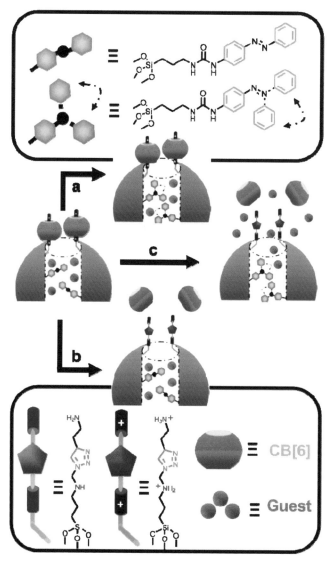

Figure 7.6 Stoddart and Zink's mesoporous SiO$_2$ nanoparticles, (a) excited with 448 nm light, (b) after addition of OH$^-$, (c) after simultaneous excitation with 448 nm light and addition of OH$^-$.
Reprinted from S. Angelos, Y. W. Yang, N. M. Khashab, J. F. Stoddart and J. I. Zink, *J. Am. Chem. Soc.*, 2009, **131**, 11344 with permission from the American Chemical Society.

reactions sought during photoaffinity labelling of enzymes[197–199] and tagging of polymer beads for combinatorial chemistry.[200,201] Indeed, well-behaved OR logic suggests perfectly unselective output production by different inputs. Unselective behaviour of a given receptor towards, say, two different guests allows us to approach OR logic devices containing a single receptor alone, *i.e.* double-input logic devices with a single port.

While old examples of molecular OR logic can be located, *e.g.* in **85**,[202] the first deliberately designed gate of this type was **86**.[203] 'Fluorophore–spacer–receptor' PET system **86** has an aromatic amino acid receptor[204,205] which acts as an electron donor towards the excited diarylpyrazoline fluorophore, resulting in negligible emission. This receptor is sufficiently unselective to Ca^{2+} or Mg^{2+} so that binding of either ion, supplied at high enough concentration, blocks the electron-rich sites of the receptor and arrests PET. Each ion produces essentially identical extents of switching 'on' (Table 7.3). This similarity is due to essentially identical conformational changes produced upon complexation. Each ion-bound state effectively decouples the amine substituent from the oxybenzene unit so that PET is similarly suppressed. This also means that the charge density difference between the two cations is of secondary importance in these conformationally switchable systems. It is also notable that a single-receptor system is sufficient in this case to achieve a two-input logic gate. Quantitative data for **86** and a few other cases are summarized in Table 7.4.

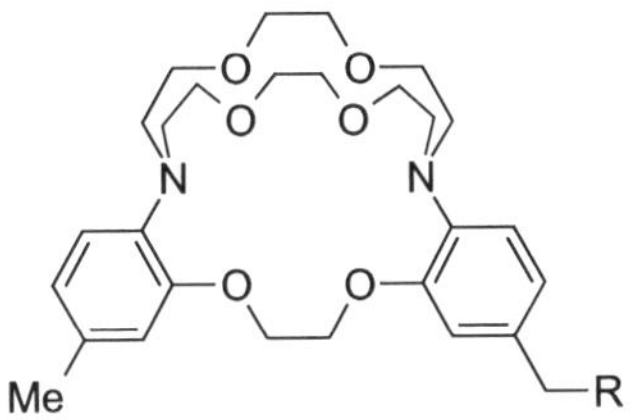

85; R = 9-anthryl 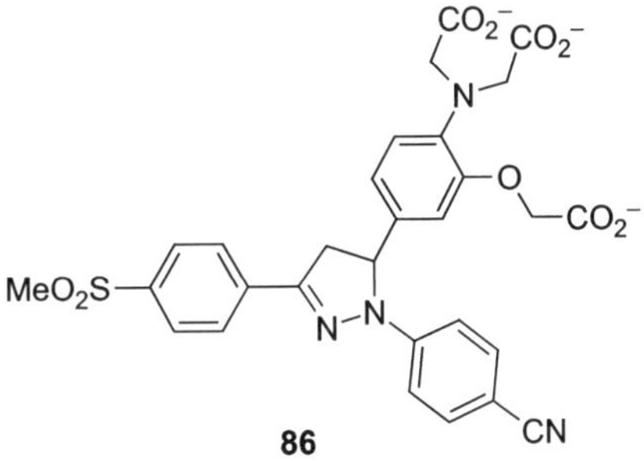   **86**

The case **85** mentioned above,[202] gives almost equal fluorescence enhancements when sufficient quantities of K^+ or Rb^+ are added, probably for a similar reason as for **86**. However, the relative rigidity of the cryptand moiety exacts a price. Only when neither input species was present was the fluorescence seen to be switched 'off' due to fast PET. Cases like **87**,[104] of the same PET

Table 7.3 Truth table for **86**.[a]

$Input_1$ Ca^{2+}	$Input_2$ Mg^{2+}	Output Fluorescence[b]
none	none	low (0.0042)
none	high (0.5 M)	high (0.24)
high (10^{-3} M)	none	high (0.28)
high (10^{-3} M)	high (0.5 M)	high (0.28)

[a] 10^{-5} M in water at pH 7.3.
[b] Quantum yields, λ_{exc} 389 nm, λ_{em} 490 nm.

Table 7.4 Collected data for a few molecular OR logic gates.

Device[a]	Input$_1$[b](hi,lo)[c]	Input$_2$[b](hi,lo)[c]	Output[d](hi,lo)[e]	Power[f]	Characteristics[g]
85[202] in MeOH	K$^+$($10^{-1.3}$,0)	Rb$^+$($10^{-1.3}$,0)	Flu$_{413}$(11,9)	hv$_{368}$ logε 4.0	
86[203] in H$_2$O	Mg^{2+}(0.5,0)	Ca^{2+}(10^{-3},0)	Flu$_{490}$(67,57)	hv$_{389}$ logε 4.5	logβ$_{Mg2+}$ 2.8 logβ$_{Ca2+}$ 4.9
87[104] in MeOH	Na$^+$(10^{-2},0)	K$^+$(10^{-2},0)	Flu$_{414}$(47,18)	hv$_{368}$ logε 4.0	logβ$_{Na+}$ 3.1 logβ$_{K+}$ 4.4
88[204] in MeCN	Na$^+$(10^{-2},0)	Hg^{2+}(10^{-5},0)	φ[h](1.9,1.6)	hv$_{369}$	

[a]The device is taken as a given molecular structure in a given medium.
[b]Chemical concentration unless noted otherwise.
[c]High (hi) and low (lo) input values. All concentrations are in M units.
[d]Output is given as the intensity of fluorescence at a given wavelength, except for one case (see later). All wavelengths are in nm units.
[e]High output values when the 'low' output value observed is scaled to unity. The first number (hi) is the highest and the second number (lo) is the lowest of the 'high' output values observed under the three different input conditions.
[f]Power supply is light at a given wavelength. The extinction coefficient (ε) of the device at that wavelength is given in units of M^{-1}cm^{-1}.
[g]Characteristics of device operation regarding inputs include chemical binding constants (β) in units of M^{-1}. Characteristics of device operation regarding outputs include emission quantum yields (φ$_{em}$).
[h]Quantum yield of photodimerization of anthracene units.

design but without a conformational switch as in **86**, produce less unselective FE values due to the charge density effect (besides others). Other instances of this kind are known.[206,207] The OR logic effect is gradually degraded as we go from **86** to **85** to **87**.

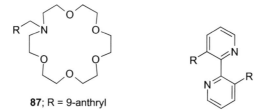

87; R = 9-anthryl

88; R = (CH$_2$OCH$_2$)$_2$CH$_2$O-9-anthryl

Singh and Kumar's **89**[208] is a relatively small molecule, and a pair of atomic ions (Zn^{2+} and Cd^{2+}) achieves a similar output. OR logic is seen here because these post-transition metal ions bind to the quinoline lone electron pairs in **89** and destabilize the normally low-lying n–π* excited states. Such states are weak to moderate emitters.[11,209,210] The lowest excited state now becomes π–π* (ICT) in character and these emit efficiently. So the fluorescence output is switched 'on' with either metal ion. Actually, the chances are that **89** can be pared down to leave two or even one quinoline unit and still maintain the OR logic action, albeit with higher input levels.

Fallis and Aldridge's decamethylferrocene **90** has an electron-withdrawing dimesitylboranyl substituent which leads to a reduction potential of –0.18 V (*vs.* Fc^+/Fc) in acetonitrile.[211] Upon encountering the Lewis base CN^- (input$_1$), the substituent changes to an electron-releasing cyanoborate so that the reduction potential changes to –0.69 V. Tetrazolium violet **91** has an intermediate reduction potential so that a mixture of **90** and **91** in the presence of CN^- leads to the reduced form of **91** whose deep violet colour (or the absorbance at 500 nm) is a convenient output. The tetrazolium dyes are well known as sensors of reducing environments in cells.[212] F^- (input$_2$) can also be used as an input to give the same final result because of the similar anion-induced change in potential.

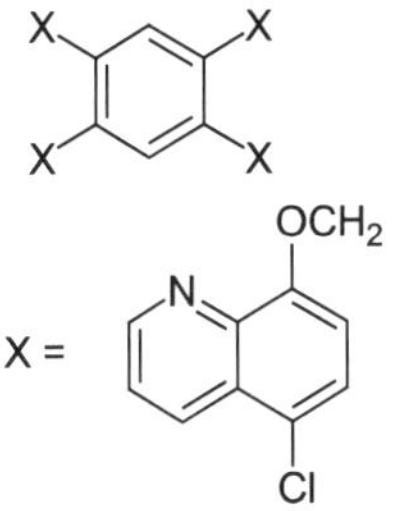

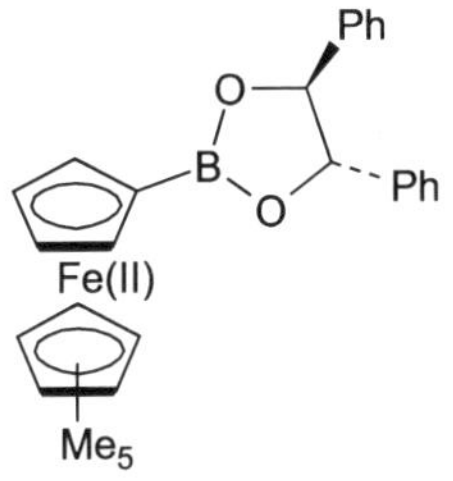

Ultraviolet absorption and NMR spectroscopy serve to indicate the intramolecular photodimerization of two anthracene units within **88**.[204] Desvergne and Tucker find that the quantum yield of this reaction is enhanced by Na^+, Hg^{2+} or both. Hg^{2+} rotates the aryl–aryl bond of the 2,2'-bipyridine unit so that the two nitrogens are ready for chelation. Rather similarly, Na^+ pulls the podands together for chelation. Photodimerization efficiency is the beneficiary in either case because the two anthracene units are brought closer together.

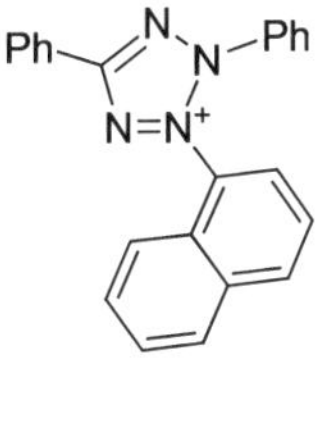

Fujita's host[77] (discussed under AND logic in section 7.2.1.6) also can be reconfigured to display OR logic, by using different inputs. OR logical recognition and binding of two molecular inputs can be arranged with this host, if they are small aromatic species. For instance, **92** will enter the host in a 2:1 stoichiometry (**92**.host). The same applies for **93**. However, if **92** and **93** are both presented to the host, a new 1:1:1 complex of **92.93**.host is produced. In other words, the host will form a complex of some kind whether one or the other or both of the guests are offered to it. This is **92, 93**-driven OR logic where the complexation event (as determined by NMR peak shifts) is the output.

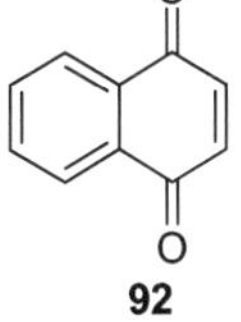

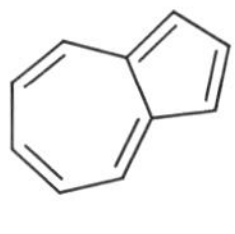

Though achieving OR logic with oligonucleotides is harder than producing YES, NOT and AND gates,[213] newer refinements allow this too.[179,214] Miyoshi et al.[215] introduce the special behaviour of telomeres to the field. Telomeres are found at the termini of chromosomes and have interesting tertiary structures under defined conditions. For instance, d(G_4T_4)$_3G_4$ forms a quadruplex in the presence of K^+, and the complementary d(C_4A_4)$_3C_4$ folds into a structure called an i-motif in acidic medium (pH 5). A duplex forms between these two complementary strands only at low K^+ and low H^+ (pH 8), when there are no competing structures being produced. This can be developed into an easily observable logic device by labelling the G-rich strand with fluorophore **94** and the C-rich strand with fluorescence quencher **95**. So the emission of **94** at 536 nm is quenched in the duplex via the EET mechanism due to the **95** close by, under the conditions of low K^+ and low H^+. All the other three input conditions give >28-fold higher emission. So we have K^+, H^+-driven OR logic with 536 nm output. However, the slow folding kinetics is noted by the authors as a weakness. Ma, Leung and their team also exploit these ideas to produce the same gate controlled by the same inputs, but without strand labelling.[216] They depend on the binding of triphenylmethane dyes to the G-quadruplex, as well as to the i-motif, so that the dye becomes emissive by rigidification.

It was only a matter of time before the major technique in oligonucleotide research – the polymerase chain reaction (PCR) method of amplifying DNA[217] – was coopted into molecular logic research. Nojima et al.[218] note that the basic PCR method adds two short primers which bind to the 3′-ends of the separated strands of the starting DNA piece. These provide the anchors for the polymerase to build up from. Thus, basic PCR is intrinsically a case of AND logic when the amplified product DNA is taken as the output. OR logic is achieved by Nojima et al.[218] by premixing one of the two primers and by replacing the other one with two separate sequences. These two sequences are the inputs. As long as at least one of the latter is present, the polymerase can build up the strand from there so that it can go into the amplification cycle. Convenient fluorescence readout is arranged by including the gene for green fluorescent protein in the starting DNA double strand and by working up for appropriate transcription and translation.

7.3.1 OR Gates using Molecule-based Materials

What oligonucletides can do in terms of OR logic, polypeptide-based enzymes can do too, even though an electrode is involved. Consider Katz and Pita's device composed of a mixture of ethyl butyrate, glucose and urea.[219] Application of the enzymes esterase and glucose oxidase as inputs, either alone or together, will produce acidification (ΔpH = -1.2 in 60 min.) as an output in a weakly buffered solution. The high selectivity allows esterase to target ethyl butyrate (to produce butyric acid) and for glucose oxidase to attack glucose (to generate gluconic acid) independently. Resetting of the system is achieved by applying urease to target urea to produce ammonia which neutralizes the acid. Furthermore, the acidification output can be converted into a cell current by using an electrode modified with poly(4-vinylpyridine) brushes which is permeable to redox-active $[Fe(CN)_6]^{3-}$ only when the polymer is protonated and extended. Such a situation can be exploited to yield logically controlled fuel cells.[220] The acidification output can also be viewed through Faradaic impedance spectroscopy because the extended polymer has a high electron transfer conductance whereas the shrunken polymer does not. The acidification output of the parent case is convertible into a fluorescence signal from a pH sensor attached to a glass microparticle.[221] Visualization is achieved via fluorescence microscopy.

OR logic gates receive a molecular electronic implementation with a conventional wiring of molecular diodes (see Chapter 3, Figure 3.5) in the hands of Collier et al.[185,186] A somewhat related case involves crossed microfluidic channels with underlying electrodes so that electrochemiluminescence (ECL) can be generated.[222] Channels 1 and 2 carry $Ru(NH_3)_6^{3+}$ whereas channel 3 carries a mixture of $Ru(bpy)_3^{2+}$ and tripropylamine. The latter mixture is famous[223] because the Ru(II) complex can be electro-oxidized to the Ru(III) state which is then reduced by the amine to produce the excited state of $Ru(bpy)_3^{2+}$. Electric current passes and ECL arises only when $Ru(bpy)_3^{2+}$ is oxidized and $Ru(NH_3)_6^{3+}$ is reduced (diode behaviour). Similar electrochemical diode actions are known.[224] ECL occurs when voltage inputs (1.9 V vs. Ag and of the correct polarity) are applied across channels 1 & 3 or channels 2 & 3 or both. As seen in the previous paragraph, copies of these diode assemblies can be linked to produce other logic gates, e.g. NAND.

Like Katz' cyclic voltammetry AND logic experiment on enzymes immobilized on multi-walled carbon nanotubes deposited on a glassy carbon electrode,[3] Kim, Oh and co-workers produce glucose, sucrose-driven OR logic with immobilized glucose oxidase and invertase.[225] The current output is produced by oxidation of glucose, whether it is supplied directly or produced from sucrose by invertase catalysis.

Chemical waves arising from the Belousov–Zhabotinskii (B-Z) reaction can be constrained to occur in limited spaces where a catalyst resides. Synchronous waves (output 'high') arise in a OR logical fashion on a gel containing the B-Z reactants when input waves propagate along a network of connected compartments with a particular geometry.[226] A degree of logic reconfiguring can be arranged by changing the geometry (see Chapters 6 and 8). We note that a molecular phenomenon lies at the heart of these experiments.

Matsui and Miyashita's nanosheet devices[189] (section 7.2.7) also include OR logic.[227]

7.4 NOR

The difficulties of integrating molecular logic gates have been discussed in section 2.2. Although NOR logic is as legitimate as any of the other 15 double-input, single-output logic gates, common computer literature represents it as a particular integration of NOT and OR gates (as its name and its electronic symbol also suggest). Such physical integration requires additional molecule–molecule linking that throws up even more obstacles to molecular implementation. On the other hand, functional integration of NOT and OR logic operations would be sufficient to achieve the same objective. From a more general viewpoint, the input–output pattern of a single molecular device operating via a set of mechanisms is then recognized as fitting the truth table of a multi-gate array. This idea can be extended to the design of new switches that integrate a number of logic functions within a single molecular structure[228,229] so that gates with more complex logic would emerge.

In a fluorescent photoionic context we can argue as follows. NOT logic represents switching 'off' of fluorescence when an ionic species arrives. Two-input OR logic represents unselective switching 'on' of fluorescence when either of two ionic species arrives. So NOR logic corresponds to switching 'off' of fluorescence when either of two ionic species arrives. For example, using the 'fluorophore–spacer–receptor' motif consisting of an anthracene fluorophore, a methylene spacer and a 2,2'-bipyridyl receptor within **96** to bind either H^+ or Zn^{2+} it is possible to achieve NOR logic.[230] When the 2,2'-bipyridyl unit complexes either H^+ or Zn^{2+}, it becomes increasingly reducible due to its cationic nature and increased planarity.[231,232] This reducibility allows a PET process to occur from the anthracene fluorophore to the 2,2'-bipyridyl receptor in order to quench the fluorescence emission (Table 7.5). The required non-selectivity of the input-induced luminescence response is not as stringent as with OR gates provided that the quenching is efficient enough. There is a related case controlled by redox-active metal ions, *e.g.* Cu^{2+} and Fe^{3+}, so that fluorescence is equally quenched.[233] Quantitative data for **96** and a few other cases are summarized in Table 7.6.

Table 7.5 Truth table for **96**.[a]

$Input_1$ H^+	$Input_2$ Zn^{2+}	Output Fluorescence[b]
low (10^{-7} M)	none	high (1.0)
low (10^{-7} M)	high (10^{-3} M)	low (0.13)
high (10^{-2} M)	none	low (0.13)
high (10^{-2} M)	high (10^{-3} M)	low (0.12)

[a]10^{-5} M in methanol : water (1 : 1, v/v).
[b]Relative quantum yields, λ_{exc} 368 nm, λ_{em} 405, 425, 440 nm.

Table 7.6 Collected data for a few molecular NOR logic gates.

Device[a]	Input$_1$[b] (hi,lo)[c]	Input$_2$[b] (hi,lo)[c]	Output[d] (hi,lo)[e]	Power[f]	Characteristics[g]
96[230] in MeOH:H$_2$O	H$^+$(10^{-2},10^{-7})	Zn^{2+}(10^{-3},0)	Flu$_{425}$(8.3,1.1)	hv$_{368}$ logε 4.0	logβ$_{H+}$ 4.2, logβ$_{Zn2+}$ 5.2
97[230] in MeCN	H$^+$(10^{-3},0)	Hg^{2+}($10^{-2.8}$,0)	Flu$_{440}$(10,1.1)	hv$_{374}$ logε 4.3	logβ$_{H+}$ 4.8[h], logβ$_{Hg2+}$ 3.7
98[234] in CH$_2$Cl$_2$	K$^+$($10^{-4.3}$,0)	F$^-$(–,0)	Flu$_{450}$(18,3)	hv$_{350}$ logε 4.9[j]	logβ$_{K+}$ 7.4[i], logβ$_{F-}$ 4.5
99[235] in MeCN	H$^+$(10^{-4},0)	Zn^{2+}(10^{-4},0)	Flu$_{516}$(500,1.5)	hv$_{485}$ logε 5.1[k]	logβ$_{Zn2+}$ 7.0
100[236] in EtOH:H$_2$O	H$^+$(10^{-4},10^{-7})	101($10^{-2.8}$,0)	Flu$_{409}$(4.0,2.7)	hv$_{321}$	logβ$_{H+}$ 5.0

[a]The device is taken as a given molecular structure in a given medium.
[b]Chemical concentration unless noted otherwise.
[c]High (hi) and low (lo) input values. All concentrations are in M units.
[d]Output is given as the intensity of fluorescence (or the absorbance) at a given wavelength. All wavelengths are in nm units.
[e]High and low output values. The low output value is the highest of the expected 'low' output values observed under the three different input conditions. The lowest of the expected 'low' output values observed is scaled to unity.
[f]Power supply is light at a given wavelength. The extinction coefficient (ε) of the device at that wavelength is given in units of M^{-1} cm^{-1}.
[g]Characteristics of device operation regarding inputs include chemical binding constants (β) in units of M^{-1}. Characteristics of device operation regarding outputs include emission quantum yields (φ).
[h]Measured in MeOH : H$_2$O.
[i]Measured in CH$_2$Cl$_2$: MeOH.
[j]At 342 nm.
[k]Measured at 498 nm.

The pyrene-based fluorescence of **102**[237] was demonstrated by Fages *et al.* to be switched 'off' by Zn^{2+}. H$^+$ gives a similar result, thereby making **102** a NOR logic gate like **96**. Several other cases of this type are known.[235,236,238]

96; R = 9-anthrylCH$_2$CH$_2$, R' = Me

97; R = 2-pyridyl

98; X = On-Pr, Y = —≡—C$_6$H$_4$—BMesityl$_2$

99; R = (BODIPY core: Me groups with N,B,N, F, F)

100

101

102; R = 1-pyrenylmethyl

Crown-appended porphyrinZn(II) **16**[27] is nicely fluorescent, as expected for Zn^{2+} complexes of porphyrins.[239] Application of C_{60}-appended imidazole **17** as input$_1$ binds it to the Zn(II) centre and sets off a PET process from the porphyrinZn(II) unit to the C_{60} moiety which switches the emission 'off'.[62] Similarly, treatment of **16** with C_{60}-attached ammonium ion **18** as input$_2$ binds the latter to the crown and causes a PET process again.[61] The fluorescence is quenched as a result. Addition of both inputs forms ternary complex **16.17.18** which is lumbered with two possible PET processes and no fluorescence is seen again. It is notable that Maligaspe and D'souza employed **16**, **17** and **18** differently to achieve AND logic (see section 7.2.1.3). A rather similar NOR logic action is seen in Gunnlaugsson's **103**[240,241] (structure given alongside **26**) with its two thiourea receptors which target alkanedicarboxylates (or pyrophosphate) to cause a PET process as each receptor engages a carboxylate (or phosphate) functionality. In this case, the NOR logic action is controlled by two inputs which are indistinguishable and connected.

It is also possible to develop NOR logic gates using a spatially overlapping 'fluorophore–receptor' ICT system. Pyrazoline **97**[230] fits the bill if the two ionic species are H^+ and Hg^{2+}. H^+ kills the fluorescence of this internal charge transfer (ICT) fluorophore, a 1,3-di(2'-pyridyl) pyrazoline, by transferring electronic excitation into vibrational quanta of N⋯H–O bonds.[194] The latter are formed when H_3O^+ docks into the bay region with its three nitrogen lone electron pairs. Hg^{2+} quenches the fluorescence of **97** in a different, but equally efficient, way. Hg^{2+} binds to the 2,2':6',6''-terpyridyl look-alike and the excited state of the complex takes on a ligand-to-metal charge transfer (LMCT) character verging on a PET process. The LMCT state arises from the facile one-electron reducibility of Hg^{2+}. He and Yam also build an ICT system, but with an electron-donating calixcrown on one terminus and an electron-accepting triarylborane on the other.[234] These two receptors within **98** are served by K^+ and F^- inputs, respectively. Blocking either receptor reduces the degree of ICT character of the excited state so that the initial blue emission is quenched considerably.

Chiu's team[242] present a macrocycle **104** based on a crown ether for K^+ binding, and a pair of pyridinium units for inciting charge transfer (CT) interactions with electron-rich π-systems in the Mulliken sense.[243] As a complement, they present clip **105** with a pair of electron-rich tetrathiafulvalene

Double Input–Single Output Systems

units held parallel to each other at a separation which is ideal to receive a pyridinium unit as a sandwich filling. So macrocycle **104** and clip **105** complex together quite nicely, and give rise to a strong absorbance at 533 nm to announce the CT interaction. Upon adding K^+ or NH_4^+, the crown ether of **104** grabs the newcomer and divorces from the clip **105**. Thus, the absorbance at 533 nm is largely lost. The threading of the macrocycle by the clip can be recovered by mopping up the K^+ with [2,2,2]cryptand or by deprotonating the NH_4^+ with $(C_2H_5)_3N$. Then the absorbance at 533 nm is mostly recovered. So this is a K^+, NH_4^+-driven NOR gate when we take the absorbance at 533 nm as the output. Interestingly, the excision of the carbonyl groups of **104** destroys its K^+ (or NH_4^+)-binding ability and the entire logic experiment collapses.

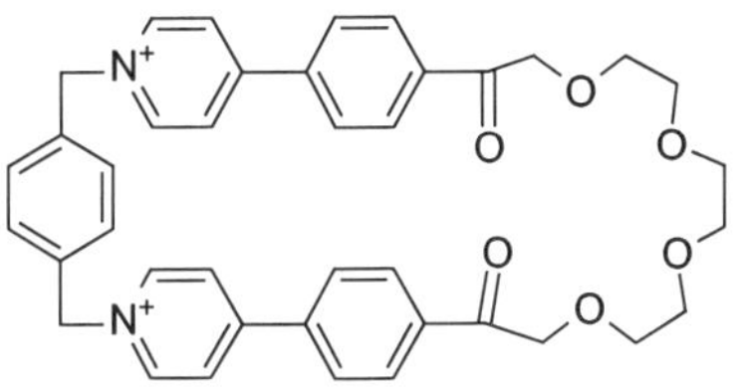

104

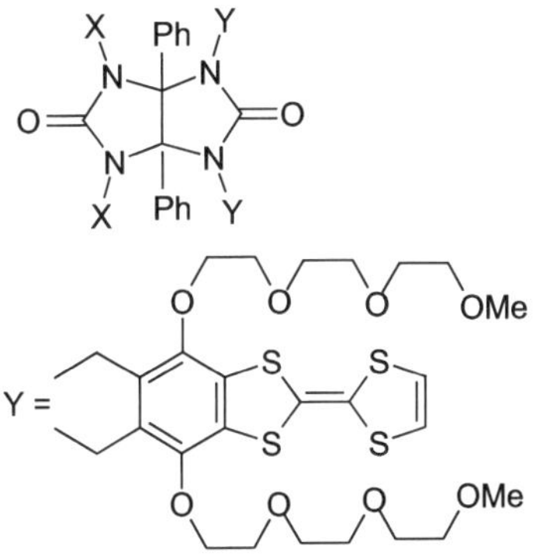

105; X, Y =

A materials-based case can be seen in the common dye **106**[244] which is normally monoprotonated in aqueous media and is fluorescent. This is switched 'off' upon diprotonation at high acidities as the ICT nature is lessened. Compound **106** is held in a polyanionic Nafion® film on an ITO (indium tin oxide) electrode. Upon reduction, monoprotonated **106** switches 'off' its fluorescence because the exciting wavelength is no longer absorbed. If monoprotonated **106** is considered as the starting state of the device, we have switching 'off' of the fluorescence output if a reduction or protonation input is supplied, *i.e.* reduction, H^+-driven NOR logic behaviour. This case can be compared and contrasted with the oxidation, H^+-driven AND logical fluorescence output seen on a molecular scale by Magri.[29] A voltage, Cu^{2+}-driven NOR gate with luminescence output also emerges from a bipyridylRu(II) anchored on nanocrystalline TiO_2 as part of a photoelectrochemical cell.[245] These cells will be discussed further in section 7.7 on XOR logic.

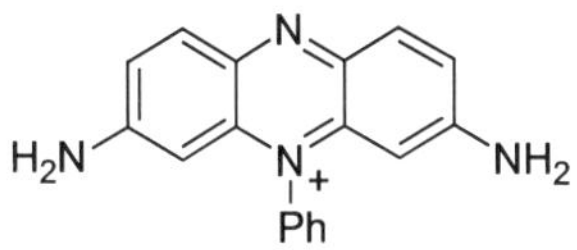

106

7.5 NAND

NAND logic can be discussed in much the same way as NOR logic was considered previously.

Iwata and Tanaka discussed **107**[246] in terms of AND logic but now it is seen to be the first molecular NAND gate. Also, this work is an early example of quantum chemical calculations supporting a PET mechanism. The fluorescence of the heteroaromatic unit is quenched only when Ca^{2+} and SCN^- ions are simultaneously present at suitably high concentrations. Since the Ca^{2+} is large in comparison with the crown ether cavity, the ester carbonyl group probably coordinates, as in the lariat ethers.[103] The large residual charge is counteracted by the binding of a SCN^- to the Ca^{2+} in an apical fashion. Only when the electron rich SCN^- is held near the fluorophore in this way does the fluorescence-quenching PET process kick in. Way back in the 1980s, Wolfbeis described the quenching of fluorescence of dibenzo-18-crown-6 ether **108** in the presence of K^+ and I^-,[247] which is now interpretable as NAND logic. It is likely that the oxidizable I^- is ion-paired to the crown-bound K^+ in much the same way as in the system involving **107**, Ca^{2+} and SCN^-.

107

108

NAND logical exploitation of tandem complexation is also seen in Garcia-Espana and Pina's **109**[248] which is brightly emissive at pH 6. Addition of protons to take it to pH 2 causes no change in fluorescence. Addition of ATP at pH 6 also maintains the status quo. However the addition of ATP and protons to produce pH 2 causes strong quenching due to the complexation of the multiprotonated forms of **109** and ATP involving π–π and CT interactions. The small, but significant, fluorescence changes in Lehn's[49] experiment with aminoacridinium compound **110**, H^+ and ATP is similar and from nearly a decade before. Lu's team reported a similar result concerning an anthrylmethylarginine derivative[249] nearly a decade after. Similarly, Fabbrizzi's **111**[250] first receives Zn^{2+} into the tetraamine cavity and then an anionic nitrobenzoate guest is held by coordination of the carboxylate to the free apical position of the Zn^{2+}. Provided Zn^{2+} is present as the binding site, the electron deficient nitrobenzoate engages in PET with the anthracene fluorophore of **111** to switch the fluorescence 'off'.

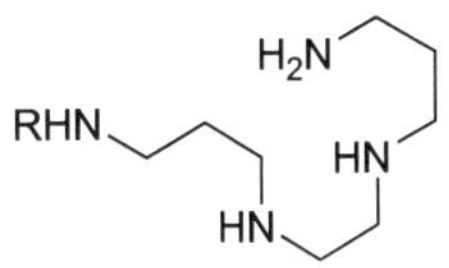

109; R = 9-anthrylmethyl

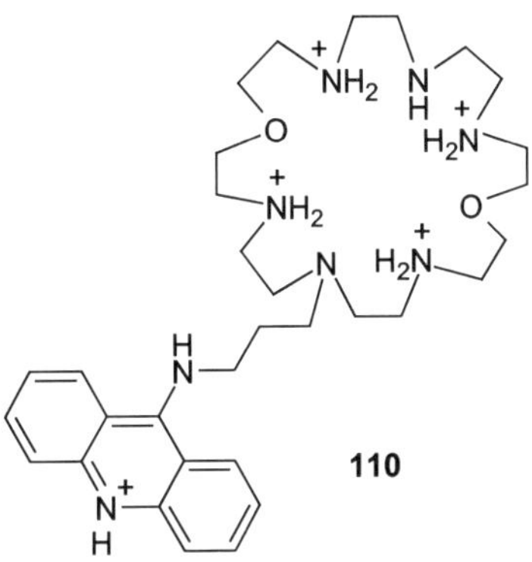

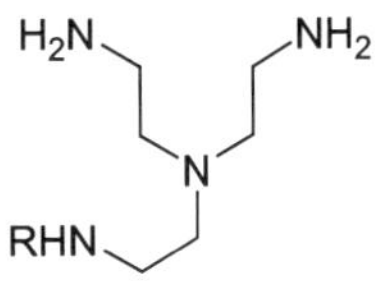

111; R = 9-anthrylmethyl

110

Though the emission of most lanthanide complexes is not quenched by O_2, the delayed emission quantum yield of **112**[251] drops below, say, 0.01 only when H^+ and O_2 are simultaneously present. Protonation of the phenanthridine side-chain of **112** causes its triplet excited state to decrease until it energetically approaches the Tb(III) $^5D_4$ excited state. This leads to equilibriation of these two excited states and sharing of their properties. Thus the metal-centred state displays the O_2 sensitivity usually only found in organic triplets. The data in Table 7.7 indicate that the three logical 'high' states vary significantly in their quantum yields so that the output threshold needs careful selection for NAND logic to be expressed.

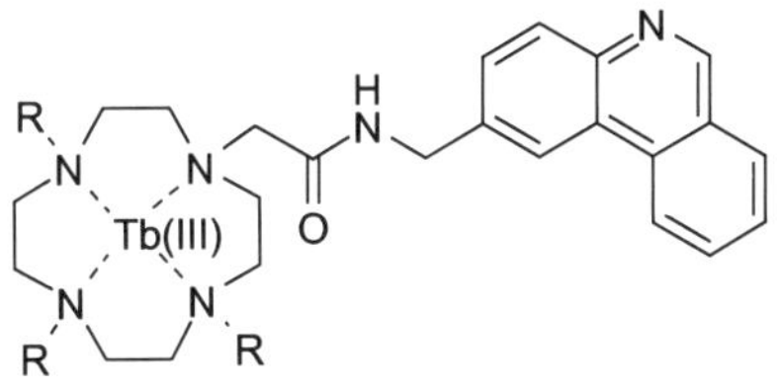

112; R = $CH_2P(Me)O_2^-$

Table 7.7 Truth table for **112**.[a]

Input$_1$ H^+	Input$_2$ O_2	Output Luminescence[b]
low (10^{-9} M)	none	high (0.12)
low (10^{-9} M)	high (160 mmHg)	high (0.025)
high (10^{-3} M)	none	high (0.046)
high (10^{-3} M)	high (160 mmHg)	low (0.0009)

[a] 10^{-5} M in water.
[b] Quantum yields, λ_{exc} 304 nm, λ_{em} 548 nm.

Zhang and Zhu's group[252] develop a combination of ion-sensitive photochromic switches[125–140] with other photochemical phenomena by addressing excimer formation. The output is the intensity of the intramolecular excimer band of **113**. The spiropyran photochrome[141] ring opens to the merocyanine on irradiation with uv light, and can then complex with Zn^{2+}. EET occurs from the excimer to the metal complex, owing to good spectral matching of the excimer emission to acceptor absorption. A significant decrease in the excimer emission is seen. This effect is not seen when either one of the inputs (uv light and Zn^{2+}) is absent, which corresponds to NAND logic. However this is more complicated because there is a third input, visible light, which converts the complex back into the initial spiropyran, but the output excimer state is not affected by visible irradiation. We also note that a simpler fluorophore emitting similar wavelengths (around 478 nm) should have achieved the same end result,[138] even though a broader band and a longer excited-state lifetime are available from **113**. NAND logic for an absorbance output can be discerned in the work of Rurack.[253]

When an experiment requires no new syntheses, as in the case of Akkaya and Baytekin's study of the fluorescent DNA reagent **114**,[254] it deserves particular appreciation. Reagent **114** even binds to a single mononucleotide pair such as the one involving adenine and thymine bases (**115** and **116** respectively). In fact it is likely that **114** stabilizes the Watson–Crick base pair *via* two amidine–phosphate interactions as well as some degree of aromatic π–π stacking. Nevertheless, the experiment requires substantial dilution of water with dimethylsulfoxide in order to be successful. Then a significant drop in the intensity of the fluorescence spectrum is seen at 455 nm. Such hypochromic effects are common when dyes intercalate into double-stranded DNA,[255] but a shift of the emission peak is equally important in causing the observed effect in this case. Akkaya and Baytekin believe that a deprotonation of an amidinium unit within **114** may play a role. No such intensity drop is found if the mononucleotides are applied singly. This case is interesting because Watson–Crick base pairing is used for the first time for small-molecule logic design. It also provides a nice bridge between molecular logic concepts and DNA computation[256,257] which has been exploited by Ghadiri[182] for instance. Data from several of these cases are summarized in Table 7.8.

$RCH_2O(CH_2)_{10}OCH_2R$

113; R = 1-pyrenyl

114

Table 7.8 Collected data for a few molecular NAND logic gates.

Device[a]	Input$_1$[b] (hi,lo)[c]	Input$_2$[b] (hi,lo)[c]	Output[d] (hi,lo)[e]	Power[f]	Characteristics[g]
112[251] in H$_2$O	H$^+$(10^{-3}, 10^{-9})	O$_2$(160^h,0)	Lum$_{548}$(133,28)	hν_{304} logε 3.5[j]	logβ_{H+} 5.7[i]
107[246] in MeCN	Ca^{2+}($10^{-3.7}$,0)	SCN$^-$($10^{-3.7}$,0)	Flu$_{487}$(9,7)	hν_{290} logε 4.5	logβ_{Ca2+} 4.7
109[248] in H$_2$O	H$^+$(10^{-2}, 10^{-6})	ATP(10^{-2},0)	Flu$_{419}$(4.0,3.6)	hν_-	logβ_{ATP} 5.8[k]
114[254] in DMSO	**115**($10^{-3.4}$,0)	**116**($10^{-3.4}$,0)	Flu$_{455}$(2.1,1.9)	hν_{345}	

[a]The device is taken as a given molecular structure in a given medium.
[b]Chemical concentration unless noted otherwise.
[c]High (hi) and low (lo) input values. All concentrations are in M units, except where noted otherwise.
[d]Output is given as the intensity of fluorescence or luminescence at a given wavelength. All wavelengths are in nm units.
[e]High output values when the 'low' output value observed is scaled to unity. The first number (hi) is the highest and the second number (lo) is the lowest of the 'high' output values observed under the three different input conditions.
[f]Power supply is light at a given wavelength. The extinction coefficient (ε) of the device at that wavelength is given in units of M^{-1} cm^{-1}.
[g]Characteristics of device operation regarding inputs include chemical binding constants (β) in units of M^{-1}. Characteristics of device operation regarding outputs include emission quantum yields (ϕ_{em}).
[h]mmHg.
[i]Inverse Stern–Volmer quenching constant with O$_2$ = 58 mmHg.
[j]At 375 nm for monoprotonated **112**.
[k]At pH 2.

115   **116**

The DNA-based AND gate **79** is converted to a NAND gate[182] by using **117** instead of **81** and switching the excitation wavelength to 490 nm so that the fluorescein unit is directly excited. In the absence of any inputs, the excitation of **79** results in strong fluorescence emission at 520 nm, *i.e.* logic output 'high'. Addition of either input **80** or **117** alone does not alter the emission. However, when both inputs are present, the duplex **79.80** is formed and **117** intercalates therein. The proximity of **117** to the fluorescein moiety again allows EET but this time from fluorescein to **117**. Thus, the 520 nm emission is switched 'off'. This fits nicely with NAND logic. Thus, the AND gate has been reconfigured into a NAND gate by choosing a different input species (see Chapter 8). Given that NAND logic is represented in electronics as an AND gate feeding into a NOT gate via a connecting wire, this example also illustrates the idea of functional integration.[230] Instead of building separate molecular gates and trying to join them physically to pass the output of one as the input of the other, molecular behaviour of sufficient complexity can be chosen so that the net input–output pattern emulates the truth table of the gate required.

117

Turning to protein-related devices, electronic energy transfer (EET) is the reason for the quenching of tryptophan fluorescence output by **118**.[258] Increasing temperature also causes fluorescence quenching, as seen for other systems. The tryptophan residues in bovine serum albumin behave similarly. Overall, we have **118**, temperature-driven NAND logic if the fluorescence quantum yield threshold is chosen carefully. Fluorescence lifetime also can serve as an output in some of these cases. Gentili developed the important thesis that the fluorescence response to these inputs is weak and non-sigmoidal (Figure 7.7), as seen by Conrad and Zauner for instance,[6] and therefore best interpreted in terms of fuzzy logic.[7] Peptide-based networks due to Ashkenasy were discussed in section 7.2.5 under light-driven AND logic, where a peptide dimerizes and serves as a template for replication.[164] Arranging for a fluorophore and a quencher to approach each other during the templation process leads to fluorescence quenching via EET. Then NAND logic emerges when the output is taken as the fluorescence emission.[259] Many other gates are thought to be achievable via these networks.[260]

118

Enzyme cascades can also be coopted to build NAND gates. While all-electronic NAND gates can be used to create any desired logic array,[13,14] most molecular NAND gates cannot make this claim owing to the difficulty of serial gate integration (Chapter 10 is full of examples of serial integration, however). Katz begins to break this deadlock by assembling sets of enzymes so that an AND gate feeds into a NOT gate with the result that NAND logic arises, rather reminiscent of semiconductor systems.[261] The principles involved here have been outlined in section 7.2.6 where other examples from the Katz–Willner stable were featured.

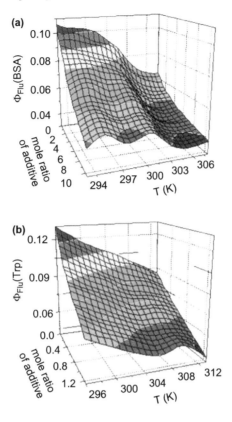

Figure 7.7 Fluorescence quantum yield of (a) BSA and (b) tryptophan as a function of additive **118** and of temperature.
Reprinted from P. L. Gentili, *J. Phys. Chem. A*, 2008, **112**, 11992 with permission from the American Chemical Society.

7.6 INHIBIT

INHIBIT logic is another type, which can be viewed as an integration of NOT and AND operations though in a different connectivity from that seen with NAND logic. The NOT operation is only applied to one input. For instance, $input_2$ is a disabling input which when 'high' kills the output irrespective of the state of $input_1$. In other words, INHIBIT logic has one input holding a veto over the other. This INHIBIT($input_2$) gate has non-commutative behaviour, *i.e.* the two inputs cannot be interchanged if the gate action is to be maintained. As noted in Chapter 3, there are two closely related gates, INHIBIT($input_2$) and INHIBIT($input_1$) within the Boolean set of double-input, single-output devices. However, different chemical species would maintain their distinct properties so that a disabling input would remain so irrespective of whether it is labelled $input_1$ or $input_2$. We will take pains in this section to point out the disabling input by stating it in a phrase or by noting it within parentheses after the gate name.

Table 7.9 Truth table for **119**.[a]

Input$_1$ H$^+$	Input$_2$ O$_2$	Output Luminescence[b]
low (10^{-11} M)	low ($<10^{-6}$ M)	low (2)
low (10^{-11} M)	high ($10^{-3.6}$ M)	low (2)
high ($10^{-2.9}$ M)	low ($<10^{-6}$ M)	high (86)
high ($10^{-2.9}$ M)	high ($10^{-3.6}$ M)	low (3)

[a]10^{-5} M in water.
[b]Relative intensities, λ_{exc} 330 nm, λ_{em} 548 nm.

Though three-input INHIBIT logic was known previously,[230] an early two-input case comes from the work of Gunnlaugsson, Parker and colleagues which is based on **119**.[262,263] Input$_1$ is H$^+$ which binds to the quinoline nitrogen to bring the absorption band of the latter into the region of the exciting wavelength (330 nm). Strong atomic line-emission is observed by sensitization of the terbium centre by the quinolinium triplet, but only if O$_2$ is barred (Table 7.9). O$_2$ is the inhibitory input$_2$ which quenches the long-lived emission from the terbium ion, because the terbium excited state equilibriates with the pumping triplet of the quinolinium group. Specifically, **119** would be an H$^+$, O$_2$-driven INHIBIT(O$_2$) gate.

119; R = CH$_2$P(Me)O$_2^-$

Example **120**[264] has its fluorescence switched 'on' by H$^+$ but only as long as K$^+$ is absent. The 1,3-dialternate calixcrown binds K$^+$ via ion–dipole and cation–π interactions.[265] The bound K$^+$ electrostatically ejects H$^+$ from the anilinium moiety (if it had been previously protonated), but is apparently incapable of sufficient direct interaction with the aniline nitrogen lone pair and so PET is re-established from the aniline to the anthracene. A somewhat related case, **121**, of H$^+$, Cu^{2+}-driven INHIBIT(Cu^{2+}) is also known.[266] More quantitative data concerning these and a few more examples can be found in Table 7.10.

Though **122** was a fluorescent PET sensor in a previous life,[271] it is now a O$_2$, Eu^{3+}-driven INHIBIT(O$_2$) logic gate where f–f luminescence is the output.[267] The emission originates from **122**.Eu^{3+}. The rather uncompetitive CH$_3$CN solvent allows the amino group as well as the carbonyl groups to dominate the solvation sphere of Eu^{3+} in the 1:1 complex. Such intimate contact between **122** and Eu^{3+} allows the following scenario: efficient optical excitation of the naphthalimide unit, inter-system crossing to its lowest-energy triplet excited state, and EET to the lowest-energy excited state $^5D_0$ of Eu^{3+}. The fact

Table 7.10 Collected data for a few molecular INHIBIT logic gates.

Device[a]	Input$_1$[b](hi,lo)[c]	Input$_2$[b](hi,lo)[c]	Output[d](hi,lo)[e]	Power[f]	Characteristics[g]
120[264] in MeCN:H$_2$O	H$^+$(10^{-2},–)	K$^+$(10$^{-2.6}$,0)	Flu$_{420}$(–,–)	hν$_{369}$	logβ_{H+} 5.1 logβ_{K+} 5.1
121[266] in H$_2$O	H$^+$(10^{-4}, 10^{-8})	Cu^{2+}(10$^{-4.4}$,0)	Flu$_{416}$(50,5)	hν$_{368}$	logβ_{H+} 7.1,5.0 logβ_{Cu2+} 14.4
122[267] in MeCN	Eu^{3+}(10$^{-4.4}$,0)	O$_2$(–[h],–[i])	Lum$_{614}$(140,<2)	hν$_{332}$	logβ_{Eu3+} 6.3
123[268] in CHCl$_3$	H$^+$(10^{-1},0)	K$^+$(10^{-4},0)	Flu$_{530}$(40,3)	hν$_{400}$	
124 + 125[269] in CHCl$_3$	Cl$^-$(10$^{-3.6}$,0)	Et$_4$N$^+$ (10$^{-3.4}$,0)	Abs$_{2000}$(5,1.2)	hν$_{2000}$	
126[270] in H$_2$O	T(35,10)	H$^+$(10^{-5},10^{-9})	Flu$_{530}$(10,1.1)	hν$_{444}$	

[a]The device is taken as a given molecular structure in a given medium.
[b]Chemical concentration unless noted otherwise. Input$_2$ is the disabling input.
[c]High (hi) and low (lo) input values. All concentrations are in M units.
[d]Output is given as the intensity of fluorescence or luminescence (or the absorbance) at a given wavelength. All wavelengths are in nm units.
[e]High and low output values. The low output value is the highest of the expected 'low' output values observed under the three different input conditions. The lowest of the expected 'low' output values observed is scaled to unity.
[f]Power supply is light at a given wavelength.
[g]Characteristics of device operation regarding inputs include chemical binding constants (β) in units of M^{-1}. Characteristics of device operation regarding outputs include emission quantum yields (ϕ).
[h]Air-equilibrated.
[i]De-oxygenated.

that triplet–triplet absorption of **122** is observed in the presence of Eu^{3+} is good evidence for this order of events. Additionally, Eu^{3+} blocks PET from the amine to the naphthalimide. However, f–f emission (ϕ_{Lum} = 0.1) only emerges in the absence of O$_2$, since the latter quenches the naphthalimde triplet state which is formed by thermally activated back-EET from the ^{5}D$_0$ state. A similar situation, but for Tb^{3+}, was seen at the start of this section.[263] Lanthanide-based systems such as these are interesting complements to fluorescent cases owing to their unique optical features including long emission lifetimes, which have been exploited for switching.[88,89,272,273]

Compound **123**[268] is special because PET produces an observable charge transfer (CT) emission when the N-oxide attaches onto H$^+$.[274,275] In most cases, PET causes only the loss of the characteristic emission of the fluorophore from a locally excited (LE) state. However, in the present instance, the PET is arrested by attaching K$^+$ to the benzocrown ether and so the CT emission subsides as well. K$^+$ is therefore the disabling input of this INHIBIT gate. The fact that **123**'s emission can be observed from the CT or LE states, or even at intermediate wavelengths, is notable and that it can receive H$^+$ or Zn^{2+} at the N-oxide oxygen centre, besides receiving K$^+$ or Ba^{2+} at the crown, is an added distinction. Because of this flexibility in inputs and outputs this case will also be mentioned in sections 10.8 and 10.10. From another viewpoint, **123** combines PET with ICT processes (because the N-oxide is integrated into the fluorophore). Such combined switching mechanisms are rare, too.[37]

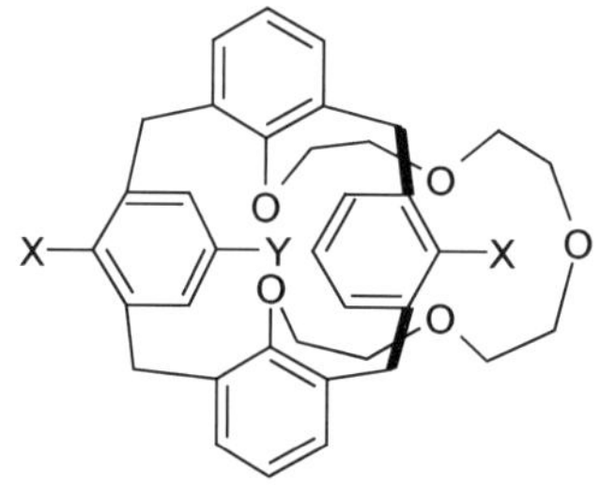

120; X = OEt, Y = 9-anthrylmethylamino

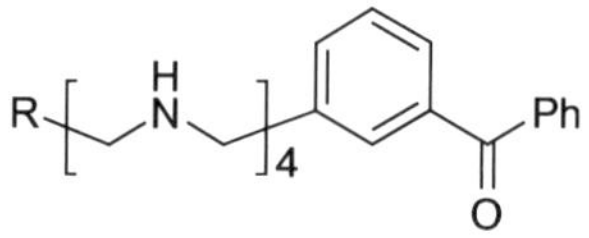

121, R = 9-anthryl

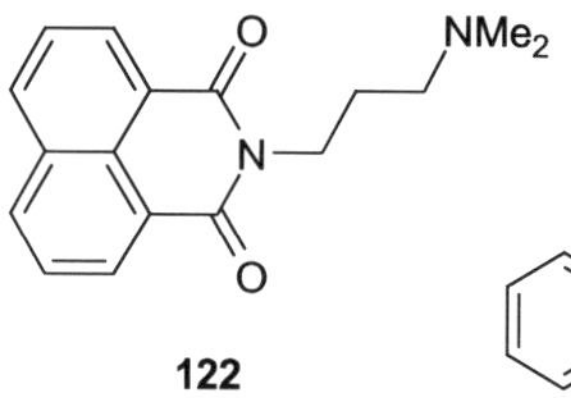

122

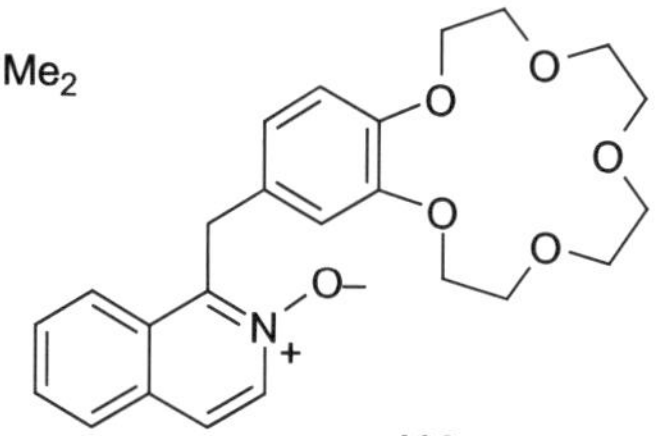

123

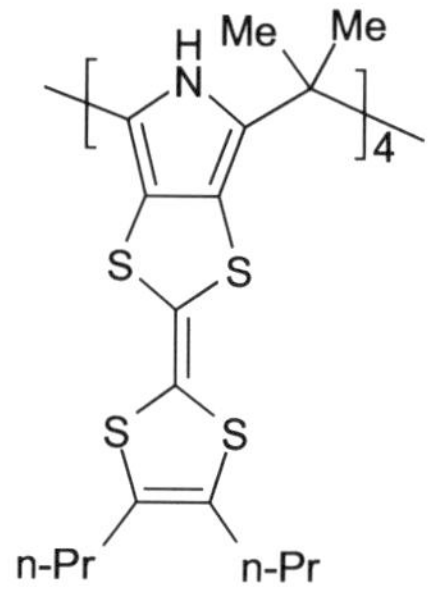

124

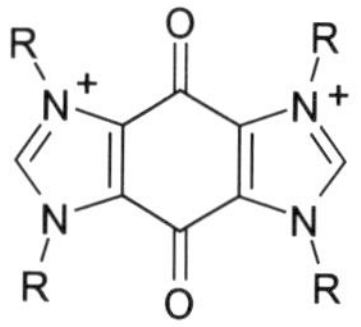

125; R = Mesityl

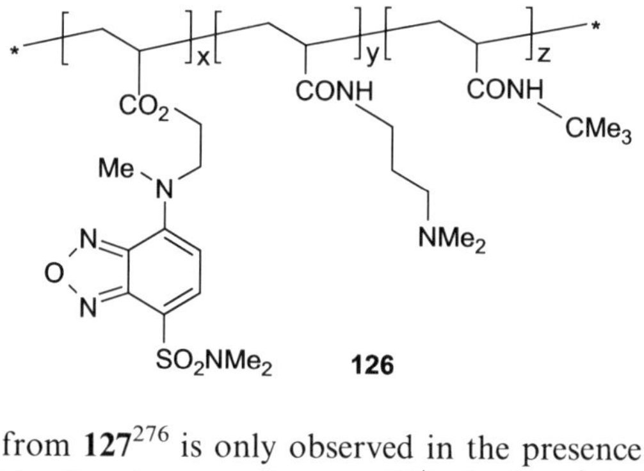

126

Strong fluorescence output from **127**[276] is only observed in the presence of Zn^{2+} (or Cd^{2+} or Pd^{2+}) and in the absence of excess H^+. Such H^+ levels displace Zn^{2+} from the bis(picolyl)amine receptor and the pyridinium groups so-formed encourage PET from the fluorophore. In the absence of this problem, Zn^{2+} blocks PET from the tertiary amine to the fluorophore thus producing emission. So this is H^+, Zn^{2+}-driven INHIBIT(H^+) logic, with high levels of H^+ causing low output in all situations. Another example of this exact type depends on Zn^{2+}-binding to an enol **128**[277] to produce a species absorbing at 355 nm. The facts that **128** is a reversibly photochemically produced enol and an additional input (OH^-) is applied do not affect the point being made.

Double Input–Single Output Systems

127; R = bis(2-pyridylmethyl)amino

128

In a grand coalition, the teams of Kadish, Fukuzumi, Bielawski and Sessler[269] produce a unique output for any logic gate type – an electron paramagnetic resonance (EPR) signal. Though this possibility was mentioned by Zhang and Zhu,[63] no experimental demonstration has been available until now. Calixpyrrole **124** usually exists in the 1,3-dialternate conformation, but the binding of Cl$^-$ to the four N–H groups changes it to the cone version. Two of these cones can form a capsule around electron-poor **125**. Owing to the electron-richness of (**124**.Cl$^-$)$_2$, a thermal electron transfer occurs. The radical ion pair so produced announces itself by a well-resolved EPR spectrum as well as by a near-IR absorption signature (Figure 7.8). The latter is expected for related radical ion species.[278] However, the capsular guest **125** can be replaced by Et$_4$N$^+$ so that the radical ion pair signatures disappear. This is Cl$^-$, Et$_4$N$^+$-driven INHIBIT(Et$_4$N$^+$) logic.

A series of INHIBIT gates created as components of half-subtractor systems will be discussed in the corresponding section (section 9.3).[279–282] These use cation and anion input pairs which neutralize each other. Those that do not undergo mutual annihilation are also known, *e.g.* K$^+$ and F$^-$.[283] Some cases of this type, *e.g.* Na$^+$ and Cl$^-$, show small changes of the output which need careful thresholding.[284]

A rare two-photon absorption output is demonstrated with the aid of **129** by Samoc, Humphreys and their colleagues.[285] A dithienylethene photochromic[286] is outfitted with alkynylRu(II) centres. The latter can not only be oxidized to the Ru(III) state reversibly, but they can also be reversibly converted to the vinylidene complex by protonation. Naturally, **129** can also be reversibly photocyclized but memory effects must be avoided by resetting to the original state in order to demonstrate combinatorial logic, as mentioned previously. The availability of multiple stable states, each with its characteristic non-linear optical behaviour, creates several possibilities of reconfigurable logic (Chapter 8) but we just note electrochemical oxidation and light dose as suitable inputs to result in INHIBIT(voltage) logic in the two-photon absorption output monitored at 1150 nm. Monitoring at 1550 nm is less distinctive.

Now we consider a series of cases based on soluble polymers. The case of **126**[270] employs poly(acrylamide) chain folding phenomena[287] to create a molecular logic operation, specifically H$^+$, T-driven INHIBIT(H$^+$) logic. Aqueous solutions of linear polymers or copolymers with balanced

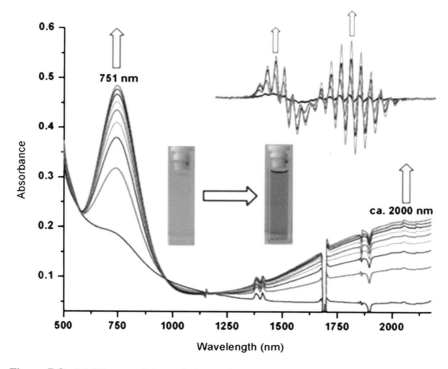

Figure 7.8 Visible–near infra red absorption spectra and EPR spectra (inset) for an equimolar mixture of **124** and **125**, when treated with increasing amounts of Cl$^-$.
Reprinted from J. S. Park, E. Karnas, K. Ohkubo, P. Chen, K. M. Kadish, S. Fukuzumi, C. W. Bielawski, T. W. Hudnall, V. M. Lynch and J. L. Sessler, *Science*, 2010, **329**, 1324 with permission from the American Association for the Advancement of Science.

hydrophobicity and hydrophilicity are quite sensitive to temperature. High temperatures cause the chains to contract into a globular form. In fact, this contraction can occur quite sharply. The current example possesses the hydrophobic *t*-butyl and the hydrophilic dimethylaminopropyl units to achieve this delicate balance. The third co-monomer (which is only present at 0.1%) contains the fluorophore, which is a benzofurazan system with electron-donor amino and electron-acceptor sulfonamide units. This fluorophore's emission is quenched when exposed to water, which is the case when the copolymer chain is in the extended form in cold water. The situation changes in hot water as the chain curls up to hide the fluorophore within its folds. Then fluorescence is revived. However this switching 'on' of fluorescence can be prevented by supplying an H$^+$ input. Protonation of the dimethylaminopropyl unit hugely enhances its hydrophilicity to leave the copolymer chain extended at all temperatures examined. The mutual repulsion of protonated dimethylaminopropyl units must also contribute to this chain extension. Thus the fluorophore is never hidden from water and the emission is never switched 'on' in acid medium.

129; R = —≡—⟨C₆H₄⟩—≡—Ru(II)(Cl)(X)₂

X = Ph₂P(CH₂)₂PPh₂

Pasparakis, Alexander and their team also achieve H^+, T-driven INHIBIT(H^+) logic using a polymer related to **126** by replacing the dimethylaminopropyl units with glucosyloxyethyl moieties. Catechol derivative **132**, shown in section 5.4.2.3, is non-fluorescent until it is bound to phenylboronic acid. Both these compounds are added to the polymer solution.[288] Fluorescence emerges only when two conditions are met. One, the glucose units are hidden from water and prevented from binding to the phenylboronic acid by curling the polymer up at high temperature. Two, the H^+ level is low enough to give good binding of phenylboronic acid to the catechol derivative.

Yet another case based on a soluble anionic poly(phenylfluorene) derivative shows lipase, β-galactosidase-driven INHIBIT(β-galactosidase) logic.[289] The anionic polymeric fluorophore is associated with a galactose double-derivatized with nitrophenyl ether and cationic fatty acid ester units. No fluorescence is initially seen because of this close association of the fluorophore with the nitroarene, which leads to PET. The lipase input cleaves off the cationic fatty acid so that the neutral galactose fragment escapes the clutches of the fluorophore. On the other hand, the β-galactosidase input produces the strongly light-absorbing nitrophenolate ICT chromophore so that fluorescence is ruined again.

7.6.1 INHIBIT Gates using Molecule-based Materials

When DNA oligonucleotides are used as inputs, outputs and devices, there is the possibility that cascades can be arranged so that serial logic gate integration is achieved. Ghadiri's group[290] adapt the Winfree's strand displacement strategy[291–294] so that the logic device is immobilized on polymer beads. The output oligonucleotide is released into solution, which can be filtered and applied as input to the next on-bead logic gate. This allows the isolation of the gate from downstream input oligonucleotides until needed, which avoids unexpected hybridizations. Let us examine the implementation of INHIBIT logic according to this approach (Figure 7.9). Displacement of one oligonucleotide of a duplex with another can be done by ensuring an overhang or a toehold in one strand of the duplex. Then a new, and longer, oligonucleotide can exploit the toehold by hybridizing there. Further encroachment is possible if the new oligonucleotide contains a contiguous run which can take over the role of the shorter strand. A carefully chosen pair of oligonucleotides, N-N_T and N_T, are inputs where the latter is the disabler. The gate itself is a complementary strand of N-N_T

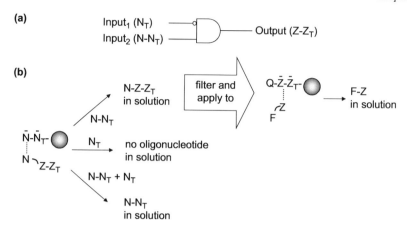

Figure 7.9 (a) INHIBIT logic using oligonucleotide N-N_T as input$_1$ and oligonucleotide N_T as input$_2$. The device is based upon hybridization with oligonucleotide run N and its toehold N_T. (b) Hybridization is shown with a red dotted line. The top row alone produces free unquenched fluorescent oligonucleotide in solution when the resulting solution from the first gate is filtered and applied to the second gate. The bottom row operates correctly only if N_T is applied before N-N_T.

immobilized on a polymer bead, which is hybridized to the oligonucleotide N-Z-Z_T to leave a dangling run of Z-Z_T. Strand N-Z-Z_T will be displaced only if input$_2$ (N-N_T) is applied alone. For correct operation, the application of 'high' levels of both input$_1$ (N_T) and input$_2$ (N-N_T) together needs to done so that N_T is applied first. Then the complementary strand of N-N_T on-bead is hybridized separately to N (from N-Z-Z_T) and to N_T (from input$_1$) and will not suffer strand displacement from N-N_T (input$_2$) due to insufficient driving force.

When the output from the device is taken to be the Z-Z_T run, we see that the above description, given the caveats, fits INHIBIT logic. However, it is advantageous to produce a light signal as the final outcome. So the output of the INHIBIT gate (actually N-Z-Z_T) is fed as input into a separate on-bead YES gate to give a fluorescence output which reports the presence of the Z-Z_T run to us. The YES gate has a complementary strand of Z-Z_T with an attached fluorescence quencher (operating via EET) which is hybridized to the short strand Z attached to a fluorophore such as cyanine dye. The on-bead YES gate is therefore poorly fluorescent as it stands. The supply of input N-Z-Z_T displaces the fluorophore-appended Z strand into solution. Since the latter is now free from the clutches of the quencher, a 'high' fluorescence signal is generated. The ease of serial integration is seen even within this simple implementation, though Ghadiri's group[290] discuss more complex cases.

As discussed under AND logic (section 7.2.7), geometric distortion of gel slabs can be an easily observable logic output. Here, we note Asoh and Akashi's example[295] where *N*-isopropylacrylamide (which forms thermoresponsive polymers as seen in section 7.6) and acrylic acid (which forms pH-responsive polymers) are co-polymerized in a clever way. The monomers at room temperature and neutral pH are subject to electrophoresis so that the acrylate anions accumulate near the anode before photopolymerization. The gel slab is therefore essentially a bilayer composed of the two types of polymers. Lowering the H$^+$ concentration to 10^{-12} M ('low' input$_1$) and raising the temperature to 40 °C ('high' input$_2$) produced extensive slab curvature ('high' output) as the poly(acrylate) layer expands and the poly(N-isopropylacrylamide) shrinks.

Bulk membrane opt(r)odes, composed of PVC for example, illustrate INHIBIT logic in a natural way[296] because the reversible protonation of a lipophilic acid–base indicator (**130**) is opposed by the arrival of Na$^+$ transported by a suitable lipophilic receptor. The cation exchange is facilitated by an anion of the tetraphenylborate family. If the absorbance of **130.**H$^+$ is monitored at 660 nm, H$^+$, Na$^+$-driven INHIBIT(Na$^+$) is found. As is typical of ICT chromophores such as **130**, the logic type reconfigures to IMPLICATION(Na$^+$ ⇒ H$^+$) if the absorbance of free base **130** is monitored at 435 nm instead. So this example could have featured in Chapter 8 too.

130; R = n-heptadecyl

7.7 XOR

Though mathematicians give the XOR gate equal weight with another 15 2-input gates,[16] XOR logic has received more than the usual amount of attention because it is an essential part of semiconductor numeracy (see Chapter 9). The first XOR logic operation was produced by Balzani's and Stoddart's teams.[297] Self-assembled pseudorotaxane **131.132** is non-emissive owing to the PET-type CT processes. Either H$^+$ or n-Bu$_3$N can dissociate the pseudorotaxane components by binding with **131** or **132** respectively. The protonation of the crown ether oxygens of **131** succeeds because of the poorly solvating CH$_2$Cl$_2$: CH$_3$CN medium. Overall, bright fluorescence is observed from **131** or protonated **131**, both of which happen to emit in the same wavelength range. Clearly, the addition of 1 : 1 H$^+$ and n-Bu$_3$N, a case of acid–base neutralization, gives no change from the original non-emissive situation. Thus the four rows of the XOR truth table (Table 7.11) are reproduced. Acid–base neutralization has also been adopted in more recent research on arithmetic systems (see Chapter 9).[279–282] Data on one of these cases are summarized in Table 7.12. However, more general approaches employing non-interfering inputs would be welcome so that larger logic arrays can be built in an incremental fashion. This can be done by reaching out to ICT systems as seen in the next paragraph.

Table 7.11 Truth table for **131.132**.[a]

Input$_1$ n-Bu$_3$N	Input$_2$ CF$_3$SO$_3$H	Output Fluorescence[b]
none	none	low (<1)
none	high (6×10^{-4} M)	high (88)
high (6×10^{-4} M)	none	high (94)
high (6×10^{-4} M)	high (6×10^{-4} M)	low (10)

[a] 3×10^{-5} M **131** and 2×10^{-5} M **132** in dichloromethane:acetonitrile (9:1, v/v).
[b] Relative intensities, λ_{exc} 277 nm, λ_{em} 343 nm.

Table 7.12 Collected data for a few molecular XOR logic gates.

Device[a]	Input$_1$[b] (hi,lo)[c]	Input$_2$[b] (hi,lo)[c]	Output[d] (hi,lo)[e]	Power[f]	Characteristics[g]
131.132[297] in CH$_2$Cl$_2$:MeCN	Bu$_3$N($10^{-3.2}$,0)	H$^+$($10^{-3.2}$,0)	Flu$_{343}$(88,10)[h]	hv$_{277}$	
133[279] in DMF	BuO$^-$ (10^{-1},0)	H$^+$(10^{-1},0)	%T$_{417}$(160,1)	hv$_{417}$ logε 5.4	
134[17] in H$_2$O	H$^+$(10^{-6},$10^{-9.5}$)	Ca^{2+}($10^{-2.3}$,0)	%T$_{390}$(4.1,1.5)	hv$_{390}$ logε 4.1	logβ_{H+} 7.0 logβ_{Ca2+} 5.9
135.136[298] in CH$_2$Cl$_2$:MeCN	**137**(10^{-2},0)	H$^+$(10^{-2},0)	Flu$_{335}$(2.4,1.3)	hv$_{270}$ logε 5.0	logβ_{H+} 8.0,7.1

[a] The device is taken as a given molecular structure in a given medium.
[b] Chemical concentration unless noted otherwise.
[c] High (hi) and low (lo) input values. All concentrations are in M units.
[d] Output is given as the intensity of fluorescence (or the transmittance) at a given wavelength. All wavelengths are in nm units.
[e] High and low output values. The high output value is the lower of the expected 'high' output values observed under the two different input conditions. The low output value is the higher of the expected 'low' output values observed under the other two different input conditions. The lower of the expected 'low' output values observed is scaled to unity.
[f] Power supply is light at a given wavelength. The extinction coefficient (ε) of the device at that wavelength is given in units of M^{-1}cm^{-1}.
[g] Characteristics of device operation regarding inputs include chemical binding constants (β) in units of M^{-1}. Characteristics of device operation regarding outputs include emission quantum yields (ϕ).
[h] Lower limits.

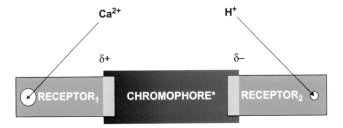

Figure 7.10 Schematic representation of ICT excited state of **134**. Note that the chromophore is strongly coupled to the receptors.

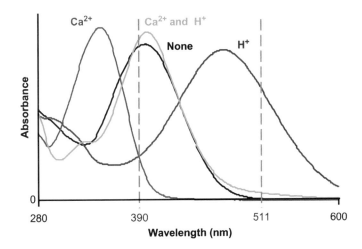

Figure 7.11 Ultraviolet–visible absorption spectra of **134** in the presence/absence of H^+ and Ca^{2+} inputs. The transmittance is monitored at the specific wavelength of 390 nm in order to extract XOR logic behaviour. Similar monitoring at 511 nm produces IMPLICATION($H^+ \Rightarrow Ca^{2+}$) logic. Adapted from A. P. de Silva and N. D. McClenaghan, *J. Am. Chem. Soc.*, 2000, **122**, 3965 with permission from the American Chemical Society.

Our photoionic approach to XOR logic uses Ca^{2+} and H^+ inputs applied to **134**[17,299] while transmittance is monitored at a wavelength close to the ion-free absorption band maximum. The chromophore within **134** is of the ICT type[11] and has the two ion receptors attached at the opposite ends, *i.e.* a 'receptor$_1$–chromophore–receptor$_2$' system (Figure 7.10). So the excited state dipole arising from the charge transfer (from the aniline nitrogen to the quinoline) will suffer opposite energetic influences when each ion arrives. The binding of Ca^{2+} by the Tsien receptor[32] incorporating the aniline destabilizes the ICT state to produce a blue-shift in the uv–vis absorption spectrum, whereas the quinoline nitrogen receptor/H^+ combination does the opposite. The monitored transmittance at 390 nm rises in each case because the absorption band moves away from the monitoring wavelength (Figure 7.11). These two ionic influences virtually cancel each other when they are applied simultaneously. So the monitored transmittance falls again to the original ion-free level. A uv–vis spectrophotometer directly measures the light transmitted though the sample cuvet so that no external mathematical manipulations need be done (say, to convert measured transmittance into an absorbance) to extract an output from the molecular logic device **134**.

Returning to acid–base neutralization, a 1 : 1 mixture of dendrimer **135** and metal complex **136** in $CH_3CN:CH_2Cl_2$ (1 : 1) is used as the molecular device.[298] Compound **135** has photoactive naphthalenes at the periphery and an ion-binding cyclam core. Irradiation at 280 nm largely excites the naphthalene, but its emission (335 nm) can be lost by EET to **136**, appearing as emission of the latter at 630 nm. Such EET only occurs if the cyclam is diprotonated so that it binds to two cyanide units of the dianionic **136**. This is our starting state for the device. Addition of more acid directly protonates **136** and releases it from the grip of **135**. Then the 335 nm intensity increases considerably. Addition of a tertiary amine **137** forms neutral **135** which also falls away from **136**. Again, the 335 nm intensity rises. Addition of equivalent amounts of acid and **137** causes no nett effect due to neutralization, and hence the starting state is maintained. So we have XOR logic, where a ‘high’ output is registered only when just one or other input is ‘high’.

133

134

135; R = 2-naphthylmethyl

137

$[(bpy)Ru(II)(CN)_4]^{2-}$

136

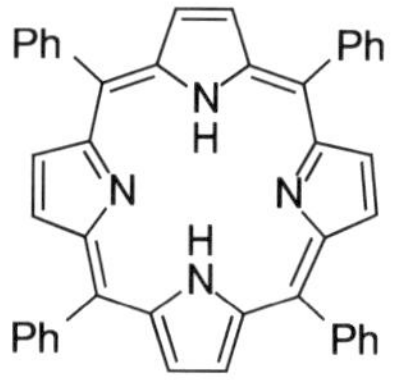

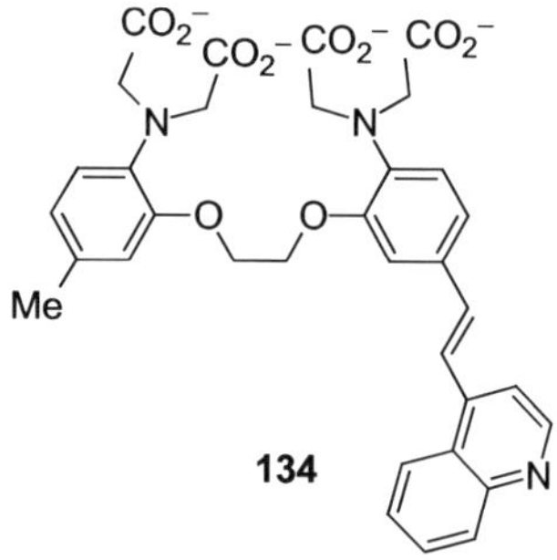

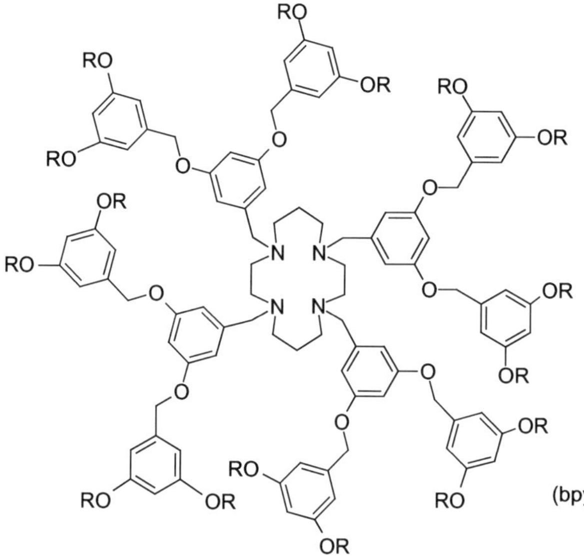

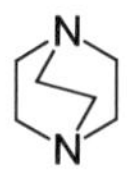

Let us now turn to biomolecular XOR gates. For instance, the enzyme malate dehydrogenase converts malate to oxalacetate. NAD^+ is taken to NADH concurrently so that the absorbance at 339 nm allows the reaction rate to be monitored. The influence of Ca^{2+} and Mg^{2+} on this rate is interpreted by Zauner and Conrad as XOR logic action.[6] Each of these dications individually accelerates the reaction at low concentrations but high ionic strength has a retarding effect. Thus 0.03 M of these dications causes significant absorbance increases but there is no nett absorbance change when both inputs are applied together.

Other enzymes are co-opted into cascades in Willner's and Katz's laboratories so that XOR gates are achieved,[169,300–302] among many others. Some of the chemistry involved has been discussed in section 7.2.6 and more discussion of related cases will be presented in Chapter 10.

Oligonucleotide-based enzymes are also good sources of XOR logic gates.[177] Because of their general expansion into more complex logic systems,[303] we will defer their discussion to section 9.2. Oligonucleotides themselves can be persuaded to self-assemble into two-dimensional arrays which can produce XOR logic too.[304]

7.7.1 Light Dose Input(s)

A fluorescent case controlled by optical inputs,[305] **138** (structure given alongside **68**), is a close structural and functional analogue of **68**. One novel twist introduced by Pina, Balzani and colleagues is that $[Co(CN)_6]^{3-}$ co-stars alongside **138** at pH 3.6. While **138** rapidly leads to a fluorescent cyclized product **139** (structure given alongside **69**) under the influence of light and H^+, $[Co(CN)_6]^{3-}$ slowly hydrolyses and ties up H^+ in the form of HCN. So when suitably high photon numbers have been absorbed, the fluorescent cyclized product is turned back to **138** in the alkalinized solution. The second novel twist is the design of an experiment with two isoenergetic light flashes (of 355 nm) such that either flash alone produces a large amount of fluorescent product **139** whereas both together leads to the alkalinized solution where the amount of **139** is very low. The latter situation of low fluorescence almost corresponds to the pre-irradiative condition, *i.e.* XOR logic applies.

Case **74**, discussed in section 7.2.5 with respect to AND logic, also contains an XOR logic gate within.[152] Whereas the fluorescence output at 430 nm from the second excited state produced AND logic from two 532 nm inputs, the emission output at 571 nm from the lowest excited state corresponds to XOR logic. The latter is naturally zero when both inputs are absent. Provision of either relatively low intensity laser beam at 532 nm leads, equally naturally, to strong fluorescence at 571 nm. Combination of both input beams gives sufficient intensity to kick-start two-photon absorption to populate the second excited state at the expense of the lowest. So the 571 nm emission is less than might have been expected (on the basis of one-photon processes exclusively). Overall, this means that **74** on its own can serve as a half-adder (Section 9.2).

The 571 nm emission of **74**, or the output from the XOR gate component, can be used to feed a second half-adder **93** whose lowest excited state is suitably lower than that of **74**. Electronic energy transfer (EET) is the link. The additional input (Input$_3$) required to run **93** is a third relatively low-intensity laser beam at 694 nm. Similar to those seen with **74**, **93**'s lowest and higher excited state emission at 746 and 375 nm respectively produce XOR and AND logic respectively. The authors even speculate that a full adder (section 10.21.2) can be approached by putting the outputs of the XOR and AND gates through an appropriate one-photon broad-band absorber.

7.7.2 XOR Gates using Molecule-based Materials

Nanocrystalline TiO$_2$ electrodes with adsorbed [Fe^{II}(CN)$_6$]$^{4-}$ species[306] or acetato ruthenium clusters[307] in suitable photoelectrochemical cells produce photocurrents whose sign depends on the redox state of the complexed species and whether the TiO$_2$ substrate or the complexed species is being photo-selected. It is then possible to arrange XOR logic by mutual cancellation of photocurrents arising from two light beams with different excitation wavelengths. For this scheme to be successful, the output has to be chosen as the modulus of the photocurrent, *i.e.* a prior computation has to be applied by the human operator. Molecular-scale XOR gates previously used mutual annihilation of acid and base.[297] Given that the applied potential can be varied at the same time as the excitation wavelength in the case of [Fe^{II}(CN)$_6$]$^{4-}$ species,[306] the former can be employed as a means of reversibly reconfiguring the logic pattern (Chapter 8).

A similar photoelectrochemical situation is available in a prior case,[308,309] except that a gold substrate is required to build mixed self-assembled monolayers. One of the adsorbates is a sulfide-based phthalocyanine which acts as a PET donor to dioxygen or a bipyridinium compound in acetonitrile solution so as to produce a cathodic current. The other sulfide-based adsorbate is a tris (bipyridyl)Ru(II) spaced from a bipyridinium unit which constitutes a PET system on its own. However, the Ru(III) state so-produced obtains an electron from a trialkylamine in solution and produces an anodic current.

The nanosheet approach[189] discussed under AND logic can be adapted to include XOR logic as well.[227,310]

Now we move away from photoelectrochemistry to electrochemistry, but the issue of human-applied prior computation raises its head again. A gold electrode is derivatized with an aptamer strand which folds up when presented with a cocaine input so that an electroactive methylene blue (MB) terminal approaches the metal surface. The same electrode also carries a DNA stem–loop motif with an MB terminal which is held close to the metal surface. When a complementary DNA strand is offered as input, the stem–loop is opened out and rigidified so that the MB unit is held at a distance. The output signal is defined by the authors as the modulus of the difference in voltammetric current as compared to the starting state, which is the prior computation. If we follow this definition, cocaine, DNA-driven XOR logic emerges.[311] A case of XOR behaviour controlled by $Fe(CN)_6^{2-}$ and $IrCl_6^{2-}$ also fits here.[312] A gold electrode modified by hexadecanethiol only conducts an electric current through ferrocenethiol intermediaries. Importantly, the electrochemical potential gradient between the latter and the redox species in solution controls the direction of the current, *i.e.* rectification is seen.

7.8 XNOR

The first example of a molecular-scale XNOR logic gate, which combines NOT and XOR operations, involves pseudorotaxane **140.141**[313] and its electrochromic behaviour. This complex displays a charge transfer (CT) absorption band at 830 nm because of the electron richness of **140** and electron deficiency of **141** (first row of truth table 7.13). The 'high' state of $input_1$ is a reduction potential, large enough to reduce the component **140** within the complex **140.141**. Upon reduction, the CT interaction between the two components is sufficiently weakened that dethreading occurs with concurrent loss of the absorption band. The 'high' state of $input_2$ is an oxidation potential (0.5 V) sufficient to oxidize component **140** within **140.141**. Again a dethreading process occurs with the loss of the CT absorption. The fourth row of the truth table (Table 7.13) can be imagined as the simultaneous application of oxidation and reduction potentials resulting in a mutual annihilation, *i.e.* insufficient potential for a redox process. Then no perturbation of the complex **140.141** occurs. Thus the CT absorption is seen again, as in the first row of the truth table. An annihilation between acid and base inputs is involved in a case from Akkaya's laboratory[281] which will be featured under half-subtractors (section 9.3). A mutual 'annihilation', in terms of chemical effect, can be arranged with Fe^{3+} and F^- inputs, when they are applied to a calixcrown derivative outfitted with pyrenamide fluorophores. Fluorescence output is quenched when Fe^{3+} alone is present owing to PET/EET. F^- achieves the same result by deprotonating the pyrenamide in the water-poor solvent employed.[314] Equivalents of H^+ in CH_2Cl_2–MeOH solution are the inputs to Schmittel's **142**[315] where the non-protonated and di-protonated forms are highly emissive but the monoprotonated form is not. It needs to be noted that each phenthroline moiety chelates a proton[231] or not in the range of acidities encountered during these experiments.

Table 7.13 Truth table for **140.141**.[a]

$Input_1$ Potential[b]	$Input_2$ Potential[b]	Output Extinction coeff[c]
low (0 V)	low (0 V)	high (5×10^3)
low (0 V)	high (0.5 V)	low (0)
high (–0.3 V)	low (0 V)	low (0)
high (–0.3 V)	high (0.5 V)	high (5×10^3)

[a] 5×10^{-4} M **140** and 5×10^{-4} M **141** in acetonitrile.
[b] *vs.* sce.
[c] in $M^{-1} cm^{-1}$; λ_{abs} 830 nm.

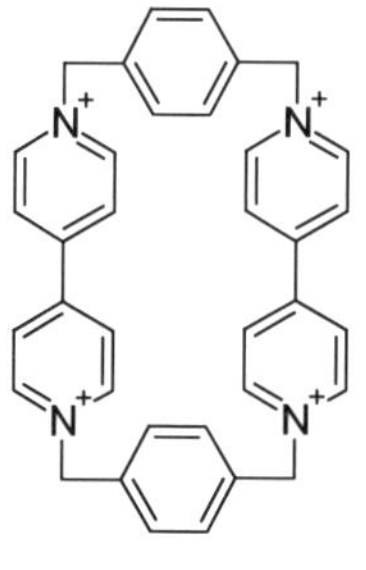

141

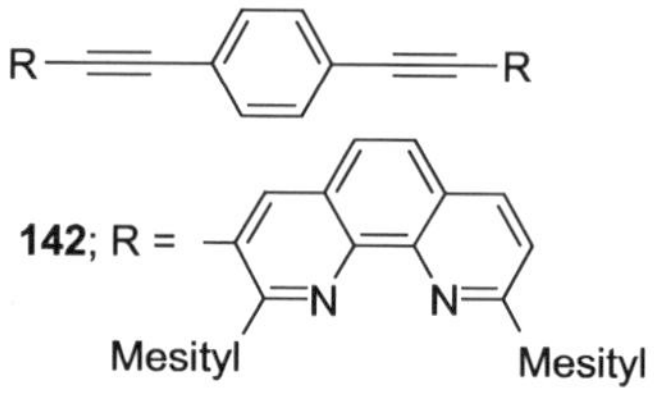

140

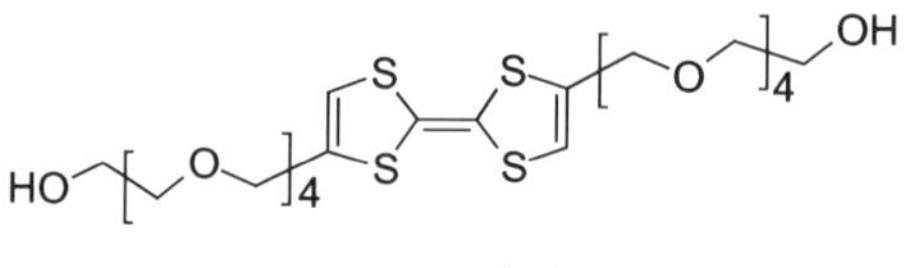

142; R = Mesityl / Mesityl (with phenanthroline)

XNOR logic also emerges from a biofuel cell in Dong's hands,[316] though this is best deferred to Chapter 8 owing to its reconfigurability.

7.9 IMPLICATION

As we saw under INHIBIT logic (section 7.6), the arbitrary nature of assigning chemical agents, say, to input$_1$ or input$_2$ becomes important when non-commutative molecular gates are being considered. IMPLICATION and reverse IMPLICATION are also of this type, represented by input$_1 \Rightarrow$ input$_2$ and input$_1 \Leftarrow$ input$_2$ respectively (Chapter 3). Therefore, simple interchange of the designation of the two inputs would change an IMPLICATION gate to its reverse counterpart. We will exclusively use the IMPLICATION description while carefully noting the input relationship within parentheses following the gate name.

IMPLICATION(Na$^+ \Rightarrow$ Ag$^+$) logic is seen in the fluorescence output of Rurack and Daub's **143**.[317] The aza-15-crown-5 unit selectively binds Na$^+$ to negate its electron rich nitrogen atom so that a PET process arises from the phenylazadithiaoxa-12-crown-4 motif to the extended borondipyrromethene fluorophore (with its ICT excited state). Thus the initially strong fluorescence of **143** is quenched once Na$^+$ arrives. Presentation of Ag$^+$ will allow its selective binding to the phenylazadithiaoxa-12-crown-4 receptor so that the PET process is stopped and fluorescence is revived (Table 7.14). Quantitative data on this and a few other cases are collected in Table 7.15. Ag$^+$ is featured again in the IMPLICATION(Ag$^+ \Rightarrow$ Cysteine) gate[318] based on a guanine quadruplex

Table 7.14 Truth table for **143**.[a]

Input$_1$ Na$^+$	Input$_2$ Ag$^+$	Output Fluorescence[b]
none	none	high (0.18)
none	high (5 × 10^{-4} M)	high (0.21)
high (2 × 10^{-2} M)	none	low (0.04)
high (2 × 10^{-2} M)	high (5 × 10^{-4} M)	high (0.19)

[a] 2 × 10^{-6} M in acetonitrile.
[b] Quantum yields, λ_{exc} 572–604 nm, λ_{em} 703–717 nm.

Table 7.15 Collected data for a few molecular IMPLICATION(input$_1$ ⇒ input$_2$) logic gates.

Device[a]	Input$_1^b$ (hi,lo)[c]	Input$_2^b$ (hi,lo)[c]	Output^d (hi,lo)[e]	Power[f]	Characteristics[g]
143[317] in MeCN	Na$^+$($10^{-1.7}$,0)	Ag$^+$($10^{-3.3}$,0)	Flu$_{710}^h$(5.3,4.5)	$h\nu_{588}^h$ logε 4.3[i]	
134[299] in H$_2$O	H$^+$(10^{-6},$10^{-9.5}$)	Ca^{2+}($10^{-2.3}$,0)	%T$_{511}$(4.0,3.7)	$h\nu_{390}$ logε 4.1	logβ_{H+} 7.0 logβ_{Ca2+} 5.9
144[319] in MeCN	Cu^{2+}(10^{-3},0)	NO$_3^-$(10^{-3},0)	Flu$_{490}$(6.5,5.0)	$h\nu_{375}$	logβ_{Cu2+} 4.4
145[320] in EtOH	Zn^{2+}($10^{-3.5}$,0)	P$_2$O$_7^{2-}$($10^{-3.6}$,0)	Flu$_{604}$(8.7,5.0)	$h\nu_{466}$	

[a]The device is taken as a given molecular structure in a given medium.
[b]Chemical concentration unless noted otherwise.
[c]High (hi) and low (lo) input values. All concentrations are in M units.
[d]Output is given as the intensity of fluorescence (or the absorbance) at a given wavelength. All wavelengths are in nm units.
[e]High output values when the 'low' output value observed is scaled to unity. The first number (hi) is the highest and the second number (lo) is the lowest of the 'high' output values observed under the three different input conditions.
[f]Power supply is light at a given wavelength. The extinction coefficient (ε) of the device at that wavelength is given in units of M^{-1} cm^{-1}.
[g]Characteristics of device operation regarding inputs include chemical binding constants (β) in units of M^{-1}. Characteristics of device operation regarding outputs include emission quantum yields (ϕ_{em}).
[h]Average value.
[i]Average value at 265 nm.

which binds triphenylmethane dyes to render them fluorescent. However, Ag$^+$ disrupts the G-quadruplex so that the dye is displaced to leave it non-emissive. Addition of cysteine competes for Ag$^+$ to allow the fluorescent dye–G-quadruplex to reform.

However, the first example of IMPLICATION logic was an aspect of **134**,[299] which was discussed in some detail under XOR logic (section 7.7). The broadness of **134**'s absorption spectra coupled with the relative complexity of the ion-induced shifts permits the emergence of additional logic types at other observation wavelengths. IMPLICATION(H$^+$ ⇒ Ca^{2+}) logic shows up when the transmittance at 511 nm is observed (Figure 7.11).

3-Aminocarbazole **144** carries a pyridylmethyl side-chain[319] so that it displays a charge transfer emission (somewhat similar to that seen by Perez-Inestrosa[268]) at 430 nm. Cu^{2+} quenches this emission owing to the usual reasons of EET and PET. NO$_3^-$ has no effect on this emission. When the ion pair is presented together, a significant new emission is seen at 520 nm. The perturbed CT emission arises from Cu^{2+} being bound to the pyridyl nitrogen of **144**, while NO$_3^-$ hydrogen-bonds to the amino group and is geometrically able to interact with the Cu^{2+} centre as well. If the emission intensity at the intermediate wavelength of 490 nm is monitored as the output, IMPLICATION(Cu^{2+} ⇒ NO$_3^-$) logic is the result. A case interpretable as IMPLICATION(Zn^{2+} ⇒ P$_2$O$_7^{2-}$) logic arises when the fluorescence of **145** at pH 7.4 and at 604 nm is found to fall drastically only in the presence of Zn^{2+}.[320]

The Zn^{2+}complex of **145** emits at 630 nm, though with a reduced quantum yield. $P_2O_7^{2-}$ wrests the Zn^{2+} from **145** to recover the 604 nm emission.

143

144; R = 3-pyridylmethyl

145

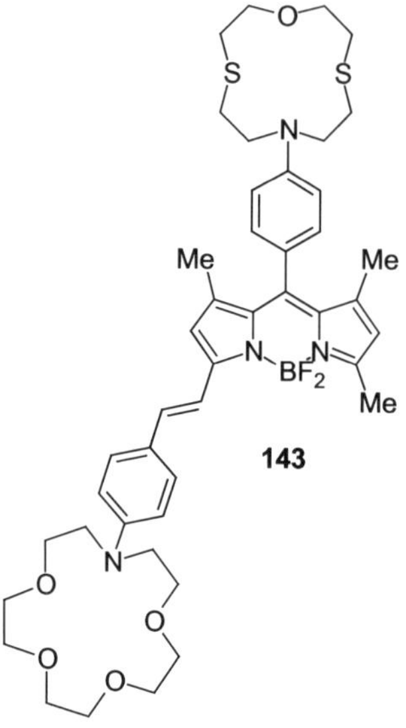

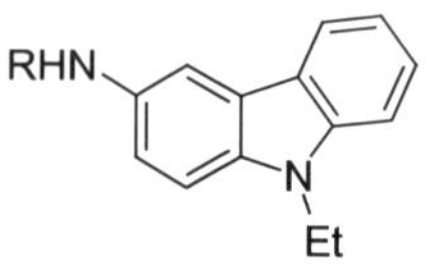

Another case of this general type allows low levels of **146** in common beverages to be measured with the aid of **147** provided that Cu^{2+} is present,[321] *i.e.* IMPLICATION($Cu^{2+} \Rightarrow$ **146**) logic. Binding of Cu^{2+} to the 1,10-phenanthroline unit quenches emission of the fluorophore via PET and perhaps EET. Binding of **146** to the Cu^{2+} centre bound in **147** enhances the emission. While the authors consider a **146**-induced change in the redox potential of the metal and hence a change in the PET rate, another possibility is that **146** pulls the Cu^{2+} centre away from the phenanthroline lumophore to attenuate the quenching powers of the metal. In a way, this would be opposite to the situation of bringing a Cu^{2+} centre close to a fluorophore as discussed by Reymond.[322] Measurement of **146** can also be accomplished by the strong luminescence that arises when **146** combines with the drug tetracycline and Eu^{3+} to form a ternary complex.[323]

146

147

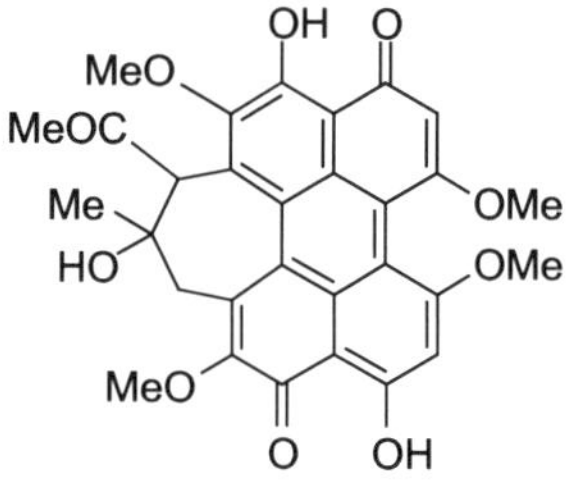

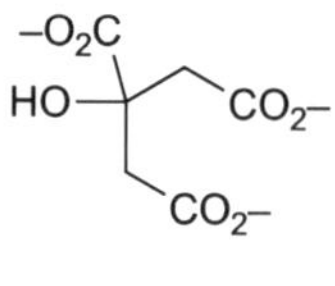

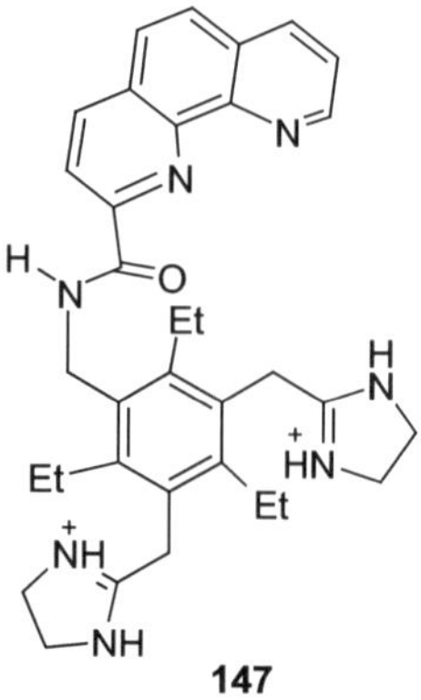

Double Input–Single Output Systems

The rotaxane **148.149**[324] has an anthracene fluorophore as a stopper on the axle composed of peptide and alkane regions. In its polymeric version, the wheel carrying pyridine units could be shunted close to the anthracene in non-polar solvents such as dichloromethane, owing to hydrogen-bonding between the axle peptides and the ring amides. If H^+ is made available, the fluorescence is quenched by PET to the neighbouring pyridinium units. So, fluorescence output becomes 'low' only when polar solvents such as dimethylsulfoxide are absent. IMPLICATION($H^+ \Rightarrow$ dimethylsulfoxide) applies.

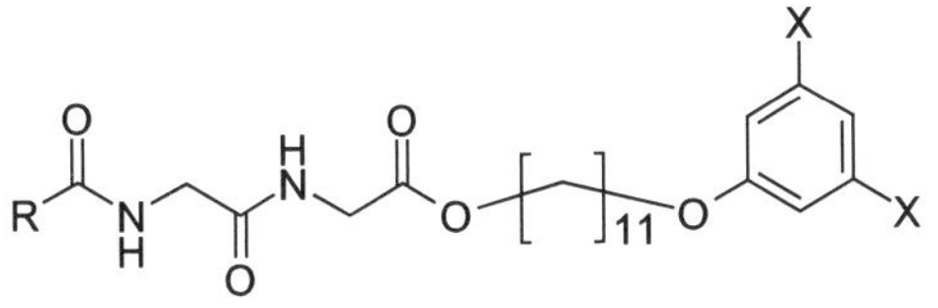

148; R = 9-anthryl, X = t-but

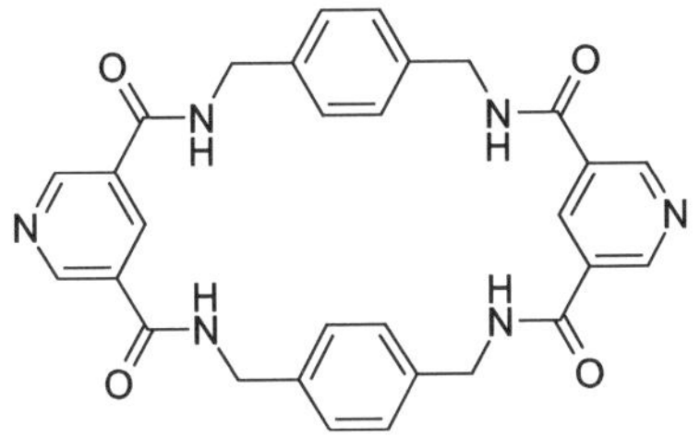

149

We have already encountered INHIBIT logic gates based on the temperature-induced chain folding of poly(acrylamide).[270] Such chain folding leads to turbidity caused by polymer insolubility. The critical solution temperature of the polymer outfitted with a photochromic azobenzene[325] under defined conditions is 20.5 °C. When the polymer is irradiated at 334 nm, the critical temperature shifts to 23.7 °C. If the temperature input is taken as 'low' when <20.5 °C and 'high' when 20.5–23.7 °C, the polymer becomes insoluble only when it is unirradiated and when the temperature input is 'high'. This becomes IMPLICATION(temperature $\Rightarrow$ light dose) logic if solubility is taken as the output.

7.10 TRANSFER

Being another non-commutative gate, TRANSFER needs to be referred to either one of its two inputs. Whether a given input species is labelled input$_1$ or input$_2$ is arbitrary however. Here is an illustration. Though discussed above under NAND logic, **114**[254] also demonstrates TRANSFER logic when the output is observed at 412 nm. Since this is the earliest example of logic reconfiguring by changing the observation wavelength it will be discussed properly in Chapter 8. TRANSFER(input$_1$) logic can be achieved operationally by a YES gate acting on input$_1$ alone whereas input$_2$ is not involved with the gate at all. Such cases can be arranged with any of the single-input YES logic gates discussed in Chapter 5. For instance, **40** with fluorescence output[326] whose amine receptor accepts input$_1$ (H^+) and ignores input$_2$ (say, Na^+) would serve the current purpose of TRANSFER(H^+) logic.

7.11 NOT TRANSFER

Though not reported as such, any of the single-input NOT logic gates discussed in Chapter 5 would serve as a NOT TRANSFER gate, provided that the second input chosen is one that does not interact with the gate. So, **150**[327] with fluorescence output would switch 'off' when H^+ is supplied as the input but would not respond to Na^+. Under these conditions, we have NOT TRANSFER(H^+) logic. The statements concerning non-commutative behaviour, mentioned in the previous section, apply here too. Rurack and Daub's **143**[317] is a double-input device (discussed in section 7.9) which displays NOT TRANSFER(Ag^+) logic as far as the absorbance at 265 nm is concerned.

Following Raymo and Giordani's[328] use of non-luminescent dyes to show how the output of one molecular switch could be fed as input into another via the intermediacy of protons, Guo et al.[329] demonstrate a case involving fluorescence. Compound **151** is a simple PET pH sensor[326] whose fluorescence is switched 'on' when protonated. As noted before, the spiropyran photochromic **152** converts into the merocyanine **153** via uv irradiation. Memory aspects are put aside until Chapter 11. The increased basicity of **153** leads to the acceptance of a proton from **151**.H^+, thus switching 'off' fluorescence. Irradiation with visible light converts the cation **153**.H^+ back into the spiropyran **152** and releases the proton, allowing it to join **151** again and switch fluorescence back 'on'. Competitive light absorption issues appear to be minor. Ultraviolet and visible light doses are input$_1$ and input$_2$ respectively. The light doses are chosen so that influence of the uv light is dominant. Under these conditions, the fluorescence output of the system of **152** and **151**.H^+ behaves according to NOT TRANSFER(ultraviolet light dose) logic.

The action of **152** relates to photogenerated acids and bases,[330] though many of these are irreversible. The combination of photogenerated acids and pH sensors is also of interest for imaging on polymer films[331] and in sunburn detection.[332]

When the temperature, light dose-driven IMPLICATION gate based on a poly(acrylamide) carrying a photochromic azobenzene[325] is changed to a fulgimide version, the critical temperature does not shift upon irradiation. The polymer solubility becomes 'low' when the temperature input is 'high' irrespective of the light dose input, and corresponds to NOT TRANSFER(temperature) logic.

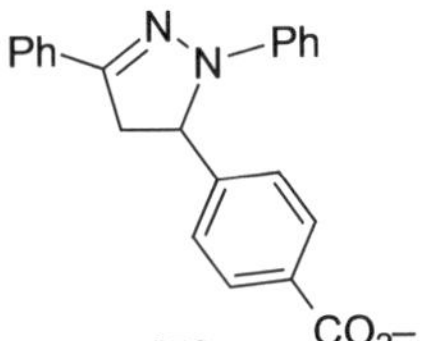

[Structures 150, 151, 152, 153]

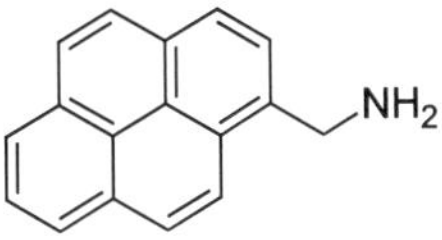

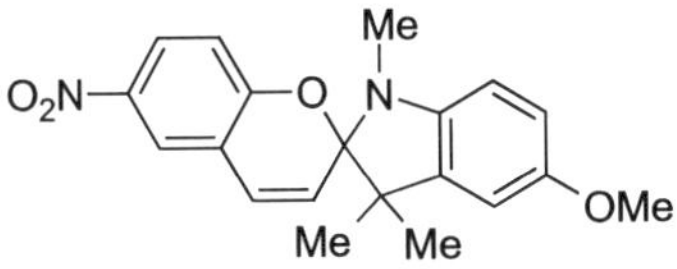

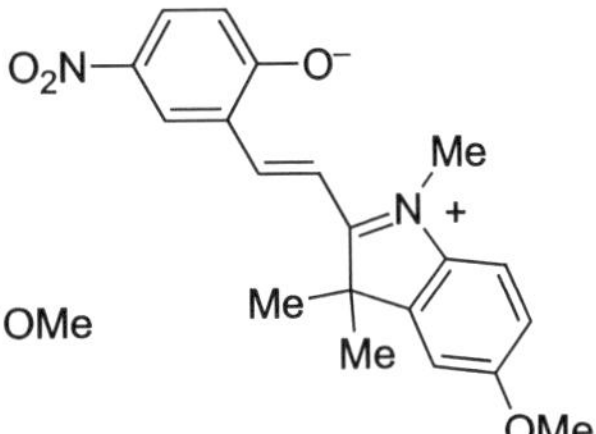

7.12 PASS 0 and PASS 1

Previous sections have considered each of the 16 double-input, single-output logic gates that are possible for binary data, except for PASS 0 and PASS 1. We saw in Chapter 5 that single-input, single-output PASS 1 and PASS 0 are easily demonstrated, *e.g.* with fluorophores and non-fluorophores respectively, in the case of simple ion inputs and fluorescence output. A very similar situation applies now, in that a pair of simple ion inputs will have no interaction with a fluorophore so that fluorescence will be maintained at a high level whatever the ion input status, *i.e.* double-input-driven PASS 1 with a fluorescence output. Furthermore, the corresponding PASS 0 arises when two simple ion inputs interrogate a solution containing a non-fluorophore or no fluorophore at all. PASS 0 and PASS 1 gates, whether of the double-input or single-input variety, are useful as tags for molecular computational identification of small objects (to be discussed in Chapter 14).[21]

PASS 1 gates have been workhorses in biology laboratories as stains (whether fluorescent or not) for many years.

This completes the set of 16 gates that this chapter set out to catalogue. However, it is important to note that this completion is only a beginning of the attempt to generalize rather simple molecular logic phenomena from as many viewpoints as possible.

References

1. J. Jacobs, *Aesops's Fables*, P. F. Collier & Son, New York, 1909, p. 14.
2. J. Johnson, *Brushfire Records*, Haleiwa, Hawaii, 2006.
3. V. Privman, V. Pedrosa, D. Melnikov, G. Strack, M. Pita, M. Simonian and E. Katz, *Biosens. Bioelectron.*, 2009, **25**, 695.

4. V. Privman, M. A. Arugula, J. Halamek, M. Pita and E. Katz, *J. Phys. Chem. B*, 2009, **113**, 5301.
5. D. Melnikov, G. Strack, M. Pita, V. Privman and E. Katz, *J. Phys. Chem. B*, 2009, **113**, 10472.
6. K. P. Zauner and M. Conrad, *Biotechnol. Prog.*, 2001, **17**, 553.
7. P. L. Gentili, *J. Phys. Chem. A*, 2008, **112**, 11992.
8. A. P. de Silva, H. Q. N. Gunaratne and C. P. McCoy, *Nature*, 1993, **364**, 42.
9. R. A. Bissell, A. P. de Silva, H. Q. N. Gunaratne, P. L. M. Lynch, G. E. M. Maguire, C. P. McCoy and K. R. A. S. Sandanayake, *Top. Curr. Chem.*, 1993, **168**, 223.
10. A. J. Bryan, A. P. de Silva, S. A. de Silva, R. A. D. D. Rupasinghe and K. R. A. S. Sandanayake, *Biosensors*, 1989, **4**, 169.
11. A. P. de Silva, H. Q. N. Gunaratne, T. Gunnlaugsson, A. J. M. Huxley, C. P. McCoy, J. T. Rademacher and T. E. Rice, *Chem. Rev.*, 1997, **97**, 1515.
12. J. Millman and A. Grabel, *Microelectronics*, McGraw-Hill, New York, 2nd edn, 1988.
13. A. P. Malvino and J. A. Brown, *Digital Computer Electronics*, Glencoe, New York, 3rd edn, 1993.
14. A. L. Sedra and K. C. Smith, *Microelectronic Circuits*, Oxford University press, Oxford, 5th edn, 2003.
15. G. Boole, *An Investigation of the Laws of Thought*, Dover, New York, 1958.
16. M. Ben-Ari, *Mathematical Logic for Computer Science*, Prentice-Hall, Hemel Hempstead, 1993.
17. A. P. de Silva and N. D. McClenaghan, *J. Am. Chem. Soc.*, 2000, **122**, 3965.
18. A. P. de Silva, H. Q. N. Gunaratne and C. P. McCoy, *J. Am. Chem. Soc.*, 1997, **119**, 7891.
19. S. Uchiyama, G.D. McClean, K. Iwai and A.P. de Silva, *J. Am. Chem. Soc.*, 2005, **127**, 8920.
20. D. C. Magri, G. D. Coen, R. L. Boyd and A. P. de Silva, *Inorg. Chim. Acta.*, 2006, **568**, 156.
21. A. P. de Silva, M. R. James, B. O. F. McKinney, P. A. Pears and S. M. Weir, *Nature Mater.*, 2006, **5**, 787.
22. B. Bag and P. K. Bharadwaj, *Chem. Commun.*, 2005, 5133.
23. N. Kaur, N. Singh, D. Cairns and J. F. Callan, *Org. Lett.*, 2009, **11**, 2229.
24. A. P. de Silva, C. M. Dobbin, T. P. Vance and B. Wannalerse, *Chem. Commun.*, 2009, 1386.
25. N. Kaur, N. Singh, B. McCaughan and J. F. Callan, *Sensors Actuators B*, 2010, **144**, 88.
26. S. J. M. Koskela, T. M. Fyles and T. D. James, *Chem. Commun.*, 2005, 945.
27. E. Maligaspe and F. D'souza, *Org. Lett.*, 2010, **12**, 624.

28. G. X. Zhang, D. Q. Zhang, Y. C. Zhou and D. B. Zhu, *J. Org. Chem.*, 2006, **71**, 3970.
29. D. C. Magri, *New J. Chem.*, 2009, **33**, 457.
30. P. R. Ashton, V. Baldoni, V. Balzani, A. Credi, H. D. A. Hoffmann, M.-V. Martínez-Díaz, F. M. Raymo, J. F. Stoddart and M. Venturi, *Chem. Eur. J*, 2001, **7**, 3482.
31. A. Ponce, P. A. Wong, J. J. Way and D. G. Nocera, *J. Phys. Chem.*, 1993, **93**, 11137.
32. R. Y. Tsien, *Biochemistry*, 1980, **19**, 2396.
33. C. J. Pedersen, *Science*, 1988, **241**, 536.
34. R. A. Bissell, A. J. Bryan, A. P. de Silva and C. P. McCoy, *J. Chem. Soc. Chem. Commun.*, 1994, 405.
35. K. Sumaru, H. Matsuoka, H. Yamaoka and G. D. Wignall, *Phys. Rev. E*, 1996, **53**, 1744.
36. B. Ramachandram and A. Samanta, *J. Phys. Chem. A*, 1998, **102**, 10579.
37. J. F. Callan, A. P. de Silva, J. Ferguson, A. J. M. Huxley and A. M. O'Brien, *Tetrahedron*, 2004, **60**, 11125.
38. V. T. Lieu and C. A. Handy, *Anal. Lett.*, 1974, **7**, 267.
39. N. Singh, N. Kaur, R. C. Mulrooney and J. F. Callan, *Tetrahedron Lett.*, 2008, **49**, 6690.
40. A. Fernández-Gutiérrez and A. Muñoz de la Peña, in *Molecular Luminescence Spectroscopy. Methods and Applications. Part 1*, ed. S. G. Schulman, Wiley, New York, 1985, p. 371.
41. H.-J. Schneider and D. Ruf, *Angew. Chem. Int. Ed. Engl.*, 1990, **29**, 1159.
42. R. Baldes and H.-J. Schneider, *Angew. Chem. Int. Ed. Engl.*, 1995, **34**, 321.
43. G. Weber and D. J. R. Laurence, *Biochem. J.*, 1954, **31**, 56.
44. E. M. Kosower, *Acc. Chem. Res.*, 1982, **15**, 266.
45. K. Odashima, A. Itai, Y. Iitaka and K. Koga, *J. Am. Chem. Soc.*, 1980, **102**, 2504.
46. F. Diederich, *Cyclophanes*, Royal Society of Chemistry, Cambridge, 1991.
47. K. L. Cole, M. A. Farran and K. DeShayes, *Tetrahedron Lett.*, 1992, **33**, 599.
48. D. Dominguez-Gutiérrez, G. De Paoli, A. Guerrero-Martínez, G. Ginocchietti, D. Ebeling, E. Eiser, L. De Cola and C. J. Elsevier, *J. Mater. Chem.*, 2008, **14**, 2248.
49. Y. Díaz-Fernández, F. Foti, C. Mangano, P. Pallavicini, S. Patroni, A. Pérez-Gramatges and S. Rodriguez-Calvo, *Chem. Eur. J.*, 2006, **12**, 921.
50. X. Michalet, F. F. Pinaud, L. A. Bentolila, J. M. Tsay, S. Doose, J. J. Li, G. Sundaresan, A. M. Wu, S. S. Gambhir and S. Weiss, *Science*, 2005, **307**, 538.
51. J. H. Fendler, *Membrane Mimetic Chemistry*, Wiley, New York, 1982.
52. H. L. Goering, E. G. Briody and J. F. Levy, *J. Am. Chem. Soc.*, 1963, **85**, 1257.
53. A. P. de Silva, G. D. McClean and S. Pagliari, *Chem. Commun.*, 2003, 2010.

54. M. E. Huston, E. U. Akkaya and A. W. Czarnik, *J. Am. Chem. Soc.*, 1989, **111**, 8735.
55. S. Y. Hong and A. W. Czarnik, *J. Am. Chem. Soc.*, 1993, **115**, 3330.
56. M. W. Hosseini, A. J. Blacker and J.-M. Lehn, *J. Am. Chem. Soc.*, 1990, **112**, 3896.
57. M. Alfonso, A. Espinosa, A. Tàrraga and P. Molina, *Org. Lett.*, 2011, **13**, 2078.
58. A. J. Moro, P. J. Cywinski, S. Korsten and G. J. Mohr, *Chem. Commun.*, 2010, **46**, 1085.
59. M. D. Lankshear, A. R. Cowley and P. D. Beer, *Chem. Commun.*, 2006, 612.
60. S. K. Kim, J. L. Sessler, D. E. Gross, C. H. Lee, J. S. Kim, V. M. Lynch, L. H. Delmau and B. P. Hay, *J. Am. Chem. Soc.*, 2010, **132**, 5827.
61. F. D'souza, R. Chitta, S. Gadde, M. E. Zandler, A. L. McCarty, A. S. D. Sandanayaka, Y. Araki and O. Ito, *Chem. Eur. J.*, 2005, **11**, 4416.
62. F. D'souza, G. R. Deviprasad, M. E. Zandler, V. T. Hoang, A. Klykov, M. VanStipdonk, A. Perera, M. E. El-Khouly, M. Fujitsuka and O. Ito, *J. Phys. Chem. A*, 2002, **106**, 3243.
63. H. M. Wang, D. Q. Zhang, X. F. Guo, L. Y. Zhu, Z. G. Shuai and D. B. Zhu, *Chem. Commun.*, 2004, 670.
64. C. A. Chen, R. H. Ye and D. S. Lawrence, *J. Am. Chem. Soc.*, 2002, **124**, 3840.
65. G. Grynkiewicz, M. Poenie and R. Y. Tsien, *J. Biol. Chem.*, 1985, **260**, 3440.
66. A. Minta, J. P. Y. Kao and R. Y. Tsien, *J. Biol. Chem.*, 1989, **264**, 8171.
67. A. P. de Silva, H. Q. N. Gunaratne, A. T. M. Kane and G. E. M. Maguire, *Chem. Lett.*, 1995, 125.
68. T. Anai, E. Nakata, Y. Koshi, A. Ojida and I. Hamachi, *J. Am. Chem. Soc.*, 2007, **129**, 6232.
69. M. A. Verdecia, M. E. Bowman, K. P. Lu, T. Hunter and J. P. Neel, *Nature Struct. Biol.*, 2000, **7**, 639.
70. K. R. A. S. Sandanayake, K. Nakashima and S. Shinkai, *J. Chem. Soc. Chem. Commun.*, 1994, 1621.
71. A. Ojida, M. Mito-oka, K. Sada and I. Hamachi, *J. Am. Chem. Soc.*, 2004, **126**, 2454.
72. M. N. Stojanovic and D. M. Kolpashchikov, *J. Am. Chem. Soc.*, 2004, **126**, 9266.
73. M. Famulok, *Nature*, 2004, **430**, 976.
74. D. S. McClure, *J. Chem. Phys.*, 1949, **17**, 905.
75. J. D. Bolt and N. J. Turro, *Photochem. Photobiol.*, 1982, **35**, 305.
76. Y. L. Peng, Y. T. Wang, Y. Wang and W. J. Jin, *J. Photochem. Photobiol. A: Chem.*, 2005, **173**, 301.
77. M. Yoshizawa, M. Tamura and M. Fujita, *J. Am. Chem. Soc.*, 2004, **126**, 6846.
78. M. Fujita, *Chem. Soc. Rev.*, 1998, **27**, 417.
79. J.-M. Lehn, *Supramolecular Chemistry*, VCH, Weinheim, 1995.

80. D. Margulies, G. Melman and A. Shanzer, *J. Am. Chem. Soc.*, 2006, **128**, 4865.
81. A. P. de Silva, T. P. Vance, M. E. S. West and G. D. Wright, *Org. Biomol. Chem.*, 2008, **6**, 2468.
82. R. A. Bissell, E. Calle, A. P. de Silva, S. A. de Silva, H. Q. N. Gunaratne, J. L. Habib-Jiwan, S. L. A. Peiris, R. A. D. D. Rupasinghe, T. K. S. D. Samarasinghe, K. R. A. S. Sandanayake and J.-P. Soumillion, *J. Chem. Soc. Perkin Trans.*, 1992, **2**, 1559.
83. R. Grigg and W. D. J. A. Norbert, *J. Chem. Soc. Chem. Commun.*, 1992, 1298.
84. L. M. Daffy, A. P. de Silva, H. Q. N. Gunaratne, C. Huber, P. L. M. Lynch, T. Werner and O. S. Wolfbeis, *Chem. Eur. J.*, 1998, **4**, 1810.
85. L. Huang and S. W. Tam-Chang, *J. Fluoresc.*, 2011, **21**, 213.
86. N. I. Georgiev, A. R. Sakr and V. B. Bojinov, *Dyes Pigm.*, 2011, **91**, 332.
87. J. R. Lakowicz, *Principles of Fluorescence Spectroscopy*, Springer, New York, 3rd edn, 2006.
88. A. P. de Silva, H. Q. N. Gunaratne and T. E. Rice, *Angew. Chem. Int. Ed. Engl.*, 1996, **35**, 2116.
89. A. P. de Silva, H. Q. N. Gunaratne, T. E. Rice and S. Stewart, *Chem. Commun.*, 1997, 1891.
90. M. E. Huston, K. W. Haider and A. W. Czarnik, *J. Am. Chem. Soc.*, 1988, **110**, 4460.
91. K. Kubo and A. Mori, *Chem. Lett.*, 2003, **32**, 926.
92. K. Kubo and A. Mori, *J. Mater. Chem.*, 2005, **15**, 2902.
93. T. Gunnlaugsson, T. C. Lee and R. Parkesh, *Tetrahedron*, 2004, **60**, 11239.
94. C. J. Chang, E. M. Nolan, J. Jaworski, S. C. Burdette, M. Shang and S. J. Lippard, *Chem. Biol.*, 2004, **11**, 203.
95. M. B. Inoue, E. Medrano, M. Inoue, A. Raitsimring and Q. Fernando, *Inorg. Chem.*, 1997, **36**, 2335.
96. J. P. Geue, N. J. Head, D. L Ward and S. F. Lincoln, *Aust. J. Chem.*, 2003, **56**, 301.
97. J. P. Geue, N. J. Head, D. L. Ward and S. F. Lincoln, *Dalton Trans.*, 2003, 521.
98. W. S. Xia, R. H. Schmehl and C. J. Li, *Chem. Commun.*, 2000, 695.
99. W. S. Xia, R. H. Schmehl and C. J. Li, *J. Am. Chem. Soc.*, 1999, **121**, 5599.
100. E. V. Anslyn and D. A. Dougherty, *Modern Physical Organic Chemistry*, University Science Books, Mill Valley, CA, 2006.
101. R. H. Crabtree (ed.), *Encyclopedia of Inorganic Chemistry*, Wiley, New York, 2006.
102. A. Ikeda and S. Shinkai, *Chem. Rev.*, 1997, **97**, 1713.
103. G. W. Gokel, *Crown Ethers and Cryptands*, Royal Society of Chemistry, Cambridge, 1991.
104. A. P. de Silva and S. A. de Silva, *J. Chem. Soc. Chem. Commun.*, 1986, 1709.

105. C. R. Cooper and T. D. James, *Chem. Commun.*, 1997, 1419.
106. C. R. Cooper and T. D. James, *J. Chem. Soc. Perkin Trans.*, 2000, **1**, 963.
107. J. Z. Zhao, T. M. Fyles and T. D. James, *Angew. Chem. Int. Ed.*, 2004, **43**, 3461.
108. A. P. de Silva, H. Q. N. Gunaratne, C. McVeigh, G. E. M. Maguire, P. R. S. Maxwell and E. O'Hanlon, *Chem. Commun.*, 1996, 2191.
109. W. Q. Yang, J. Yan, H. Fang and B. H. Wang, *Chem. Commun.*, 2003, 792.
110. A. P. de Silva and K. R. A. S. Sandanayake, *Angew. Chem. Int. Ed. Engl.*, 1990, **29**, 1173.
111. F. Fages, J.-P. Desvergne, K. Kampke, H. Bouas-Laurent, J.-M. Lehn, J.-P. Konopelski, P. Marsau and Y. Barrans, *J. Chem. Soc. Chem. Commun.*, 1990, 655.
112. K. Ghosh, G. Masanta and A. P. Chattopadhyay, *Tetrahedron Lett.*, 2007, **48**, 6129.
113. A. W. Czarnik, *ACS Symp. Ser.*, 1994, **561**, 314.
114. T. D. James, K. R. A. S. Sandanayake and S. Shinkai, *Angew. Chem. Int. Ed. Engl.*, 1994, **33**, 2207.
115. T. D. James, K. R. A. S. Sandanayake and S. Shinkai, *J. Chem. Soc. Chem. Commun.*, 1994, 477.
116. T. D. James, K. R. A. S. Sandanayake and S. Shinkai, *Nature*, 1995, **374**, 345.
117. H. K. Cho, D. H. Lee and J. I. Hong, *Chem. Commun.*, 2005, 1690.
118. D. H. Lee, S. Y. Kim and J. I. Hong, *Angew. Chem. Int. Ed.*, 2004, **43**, 4777.
119. R. G. Hanshaw, S. M. Hilkert, H. Jiang and B. D. Smith, *Tetrahedron Lett.*, 2004, **45**, 8721.
120. L. Prodi, R. Ballardini, M. T. Gandolfi and R. Roverai, *J. Photochem. Photobiol. A Chem.*, 2000, **136**, 49.
121. N. J. Turro, V. Ramamurthy and J. C. Scaiano, *Modern Molecular Photochemistry of Organic Molecules*, University Science Books, Mill Valley, CA, 2010.
122. R. S. Stoll and S. Hecht, *Angew. Chem. Int. Ed.*, 2010, **49**, 5054.
123. D. Shukla, C. Lu, N. P. Schepp, W. G. Bentrude and L. J. Johnston, *J. Org. Chem.*, 2000, **65**, 6167.
124. H. Dürr and H. Bouas-Laurent(ed.), *Photochromism. Molecules and Systems*, Elsevier, Amsterdam, 1990.
125. J. Philips, A. Mueller and F. Przystal, *J. Am. Chem. Soc.*, 1965, **87**, 4020.
126. L. Taylor, J. Nicholson and R. Davis, *Tetrahedron Lett.*, 1967, 1585.
127. T. Tamaki and K. Ichimura, *J. Chem. Soc. Chem. Commun.*, 1989, 1477.
128. J. D. Winkler, K. Deshayes and B. Shao, *J. Am. Chem. Soc.*, 1989, **111**, 769.
129. M. Inouye, M. Ueno, T. Kitao and K. Tsuchiya, *J. Am. Chem. Soc.*, 1990, **112**, 8977.
130. K. Kimura, T. Yamashita and M. Yokoyama, *J. Chem. Soc. Perkin Trans.*, 1992, **2**, 613.

131. M. Inouye, K. Kim and T. Kitao, *J. Am. Chem. Soc.*, 1992, **114**, 778.
132. M. Inouye, Y. Noguchi and K. Isagawa, *Angew. Chem. Int. Ed.*, 1994, **33**, 1163.
133. M. Inouye, K. Akamatsu and H. Nakazumi, *J. Am. Chem. Soc.*, 1997, **119**, 9160.
134. J. T. C. Wojtyk, P. M. Kazmaier and E. Buncel, *Chem. Commun.*, 1998, 1703.
135. L. Gobbi, P. Seiler and F. Diederich, *Angew. Chem. Int. Ed.*, 1999, **38**, 674.
136. F. M. Raymo and S. Giordani, *J. Am. Chem. Soc.*, 2001, **123**, 4651.
137. F. M. Raymo and S. Giordani, *Org. Lett.*, 2001, **3**, 1833.
138. F. M. Raymo and S. Giordani, *J. Am. Chem. Soc.*, 2002, **124**, 2004.
139. S. Giordani, M. A. Cejas and F. M. Raymo, *Tetrahedron*, 2004, **60**, 10973.
140. T. A. Darwish, R. A. Evans, M. James, N. Malic, G. Triani and T. L. Hanley, *J. Am. Chem. Soc.*, 2010, **132**, 10748.
141. Y. Hirshberg, *J. Am. Chem. Soc.*, 1956, **78**, 2304.
142. J. Daub, J. Salbeck, T. Knochel, C. Fischer, H. Kunkely and K. M. Rapp, *Angew. Chem. Int. Ed. Engl.*, 1989, **28**, 1494.
143. J. Achatz, C. Fischer, J. Salbeck and J. Daub, *J. Chem. Soc. Chem. Commun.*, 1991, 504.
144. A. K. Newell and J. H. P. Utley, *J. Chem. Soc. Chem. Commun.*, 1992, 800.
145. T. Saika, T. Iyoda, K. Honda and T. Shimidzu, *J. Chem. Soc. Perkin Trans.*, 1993, **2**, 1181.
146. S. H. Kawai, S. L. Gilat and J.-M. Lehn, *J. Chem. Soc. Chem. Commun.*, 1994, 1011.
147. F. Pina, M. Maestri and V. Balzani, *Chem. Commun.*, 1999, 107.
148. F. Pina, M. J. Melo, M. Maestri, R. Ballardini and V. Balzani, *J. Am. Chem. Soc.*, 1997, **119**, 5556.
149. F. Pina, A. Roque, M. J. Melo, M. Maestri, L. Belladelli and V. Balzani, *Chem. Eur. J.*, 1998, **4**, 1184.
150. P. G. Jessop, D. J. Heldebrant, X. W. Li, C. A. Eckert and C. L. Liotta, *Nature*, 2005, **436**, 1102.
151. A. S. Lukas, P. J. Bushard and M. R. Wasielewski, *J. Am. Chem. Soc.*, 2001, **123**, 2440.
152. F. Remacle, S. Speiser and R. D. Levine, *J. Phys. Chem. B*, 2001, **105**, 5589.
153. D. Gust, T. A. Moore and A. L. Moore, *Chem. Commun.*, 2006, 1169.
154. D. Kuciauskas, P. A. Liddell, A. L. Moore, T. A. Moore and D. Gust, *J. Am. Chem. Soc.*, 1998, **120**, 10880.
155. R. R. Birge, in *Nanotechnology: Research and Perspectives*, ed. B. C. Crandall and J. Lewis, MIT Press, Cambridge MA, 1992, p. 149.
156. M. R. Wasielewski, M. P. O'Neil, D. Gosztola, M. P. Niemczyk and W. A. Svec, *Pure Appl. Chem.*, 1992, **64**, 1319.
157. M. P. O'Neil, M. P. Niemczyk, W. A. Svec, D. Gosztola, G. L. Gaines III and M. R. Wasielewski, *Science*, 1992, **257**, 63.

158. N. J. Turro, V. Ramamurthy, W. Cherry and W. Farneth, *Chem. Rev.*, 1978, **78**, 125.
159. T. H. Zhang, C. P. Zhang, G. H. Fu, Y. D. Li, L. Q. Gu, G. Y. Zhang, Q. W. Song, B. Parsons and R. R. Birge, *Opt. Eng.*, 2000, **39**, 527.
160. D. Kuciauskas, P. A. Liddell, A. L. Moore, T. A. Moore and D. Gust, *J. Am. Chem. Soc.*, 1998, **120**, 10880.
161. T. A. Moore, D. Gust, P. Mathis, J. C. Mialocq, C. Chachaty, R. V. Bensasson, E. J. Land, D. Doizi, P. A. Liddell, W. R. Lehman, G. A. Nemeth and A. L. Moore, *Nature*, 1984, **307**, 630.
162. J. H. R. Reek and S. Otto (ed.), *Dynamic Combinatorial Chemistry*, Wiley-VCH, Weinheim, 2010.
163. G. Ashkenasy and M. R. Ghadiri, *J. Am. Chem. Soc.*, 2004, **126**, 11140.
164. Z. Dadon, M. Samiappan, E. Y. Safranchik and G. Ashkenasy, *Chem. Eur. J.*, 2010, **16**, 12096.
165. P. T. Corbett, J. K. M. Sanders and S. Otto, *Angew. Chem. Int. Ed.*, 2007, **46**, 8858.
166. M. Sugita, *J. Theor. Biol.*, 1961, **1**, 415.
167. A. Arkin and J. Ross, *Biophys. J.*, 1994, **67**, 560.
168. D. Bray, *Nature*, 2002, **376**, 307.
169. R. Baron, O. Lioubashevski, E. Katz, T. Niazov and I. Willner, *Angew. Chem. Int. Ed.*, 2006, **45**, 1572.
170. A. Doron, M. Portnoy, M. Lion-Dagan, E. Katz and I. Willner, *J. Am. Chem. Soc.*, 1996, **118**, 8937.
171. E. Katz and V. Privman, *Chem. Soc. Rev.*, 2010, **39**, 1835.
172. T. K. Tam, E. Katz and M. Pita, *Sensors Actuators*, 2009, **140**, 1.
173. R. Baron, O. Lioubashevski, E. Katz, T. Niazov and I. Willner, *Org. Biomol. Chem.*, 2006, **4**, 989.
174. A. S. Deonarine, S. M. Clark and L. Konermann, *Future Gener. Comp. Sys*, 2003, **19**, 87.
175. S. Sivan, S. Tuchman and N. Lotan, *BioSystems*, 2003, **70**, 21.
176. J. Macdonald, D. Stefanovic and M. N. Stojanovic, *Sci. Am.*, 2008, **299**, 84.
177. M. N. Stojanovic, T. E. Mitchell and D. Stefanovic, *J. Am. Chem. Soc.*, 2002, **124**, 3555.
178. M. N. Stojanovic, S. Semova, D. Kolpashchikov, J. Macdonald, C. Morgan and D. Stefanovic, *J. Am. Chem. Soc.*, 2005, **127**, 6914.
179. R. Penchovsky and R. R. Breaker, *Nature Biotechnol.*, 2005, **23**, 1424.
180. A. Ogawa and M. Maeda, *Chem. Commun.*, 2009, 4666.
181. J. S. Kilby, *ChemPhysChem*, 2001, **2**, 483.
182. A. Saghatelian, N. H. Volcker, K. M. Guckian and M. R. Ghadiri, *J. Am. Chem. Soc.*, 2003, **125**, 346.
183. B. Valeur and M. N. Berberan-Santos, *Molecular Fluorescence*, Wiley-VCH, Weinheim, 2nd edn, 2012.
184. A. Okamoto, K. Tanaka and I. Saito, *J. Am. Chem. Soc.*, 2004, **126**, 9458.
185. C. P. Collier, E. W. Wong, M. Belohradsky, F. M. Raymo, J. F. Stoddart, P. J. Kuekes, R. S. Williams and J. R. Heath, *Science*, 1999, **285**, 391.

186. C. P. Collier, G. Mattersteig, E. W. Wong, Y. Luo, K. Beverly, J. Sampaio, F. M. Raymo, J. F. Stoddart, P. J. Kuekes, R. S. Williams and J. R. Heath, *Science*, 2000, **289**, 1172.
187. R. M. Metzger, *Acc. Chem. Res.*, 1999, **32**, 950.
188. A. C. Brady, B. Hodder, A. S. Martin, J. R. Sambles, C. P. Ewels, R. Jones, P. R. Briddon, A. M. Musa, C. A. F. Panetta and D. L. Mattern, *J. Mater. Chem.*, 1999, **9**, 2271.
189. J. Matsui, M. Mitsuishi, A. Aoki and T. Miyashita, *Angew. Chem. Int. Ed.*, 2003, **42**, 2272.
190. H.-J. Schneider, T. J. Liu, N. Lomadze and B. Palm, *Adv. Mater.*, 2004, **16**, 613.
191. H. Komatsu, S. Matsumoto, S. I. Tamaru, K. Kaneko, M. Ikeda and I. Hamachi, *J. Am. Chem. Soc.*, 2009, **131**, 5580.
192. K. Gawel and B. T. Stokke, *Soft Matter*, 2011, **7**, 4615.
193. A. Saha, S. Manna and A. K. Nandi, *Soft Matter*, 2009, **5**, 3992.
194. M. D. P. De Costa, A. P. de Silva and S. T. Pathirana, *Canad. J. Chem.*, 1987, **65**, 1416.
195. R. F. Chen, *Biochemical Fluorescence: Concepts*, Dekker, New York, 1975.
196. S. Angelos, Y. W. Yang, N. M. Khashab, J. F. Stoddart and J. I. Zink, *J. Am. Chem. Soc.*, 2009, **131**, 11344.
197. H. Bayley and J. R. Knowles, in *Affinity Labelling*, ed. W. B. Jakoby and M. Wilchek, Academic, Orlando, 1977, p. 69.
198. T. Nagase, S. Shinkai and I. Hamachi, *Chem. Commun.*, 2001, 229.
199. H. X. Wang, E. Nakata and I. Hamachi, *ChemBioChem*, 2009, **10**, 2560.
200. M. H. J. Ohlmeyer, R. N. Swanson, L. W. Dillard, J. C. Reader, G. Asouline, R. Kobayashi, M. Wigler and W. C. Still, *Proc. Natl. Acad. Sci. USA*, 1993, **90**, 10922.
201. H. P. Nestler, P. A. Bartlett and W. C. Still, *J. Org. Chem.*, 1994, **59**, 4723.
202. A. P. de Silva, H. Q. N. Gunaratne and K. R. A. S. Sandanayake, *Tetrahedron Lett.*, 1990, **31**, 5193.
203. A. P. de Silva, H. Q. N. Gunaratne and G. E. M. Maguire, *J. Chem. Soc. Chem. Commun.*, 1994, 1213.
204. G. McSkimming, J. H. R. Tucker, H. Bouas-Laurent and J.-P. Desvergne, *Angew. Chem. Int. Ed.*, 2000, **39**, 2167.
205. L. A. Levy, E. Murphy, B. Raju and R. E. London, *Biochemistry*, 1988, **27**, 4041.
206. P. Ghosh, P. K. Bharadwaj, S. Mandal and S. Ghosh, *J. Am. Chem. Soc.*, 1996, **118**, 1553.
207. P. Ghosh, P. K. Bharadwaj, J. Roy and S. Ghosh, *J. Am. Chem. Soc.*, 1997, **119**, 11903.
208. P. Singh and S. Kumar, *New J. Chem.*, 2006, **30**, 1553.
209. M. Kadarkaraisamy and A. G. Sykes, *Inorg. Chem.*, 2006, **45**, 779.
210. L. Prodi, F. Bolletta, N. Zaccheroni, C. I. F. Watt and N. J. Mooney, *Chem. Eur. J.*, 1998, **4**, 1090.
211. A. E. J. Broomsgrove, D. A. Addy, C. Bresner, I. A. Fallis, A. L. Thompson and S. Aldridge, *Chem. Eur. J.*, 2008, **14**, 7525.

212. T. Mosmann, *J. Immunol. Meth.*, 1983, **65**, 55.
213. A. A. Margolin and M. N. Stojanovic, *Nature Biotechnol.*, 2005, **23**, 1374.
214. M. N. Stojanovic, D. B. Nikic and D. Stefanovic, *J. Serb. Chem. Soc.*, 2003, **68**, 321.
215. D. Miyoshi, M. Inoue and N. Sugimoto, *Angew. Chem. Int. Ed.*, 2006, **45**, 7716.
216. D. L. Ma, M. H. T. Kwan, D. S. H. Chan, P. Lee, H. Yang, V. P. Y. Ma, L. P. Bai, Z. H. Jiang and C. H. Leung, *Analyst*, 2011, **136**, 2692.
217. K.B. Mullis, *Angew. Chem. Int. Ed. Engl.*, 1994, **33**, 1209.
218. T. Nojima, T. Yamamoto, H. Kimura and T. Fujii, *Chem. Commun.*, 2008, 3771.
219. X. M. Wang, J. Zhou, T. K. Tam, E. Katz and M. Pita, *Bioelectrochem.*, 2009, **77**, 69.
220. L. Amir, T. K. Tam, M. Pita, M. M. Meijler, L. Alfonta and E. Katz, *J. Am. Chem. Soc.*, 2009, **131**, 826.
221. V. Bychkova, A. Shvarev, J. Zhou, M. Pita and E. Katz, *Chem. Commun.*, 2010, **46**, 94.
222. W. Zhan and R. M. Crooks, *J. Am. Chem. Soc.*, 2003, **125**, 9934.
223. A. J. Bard and L. R. Faulkner, *Electrochemical Methods: Fundamentals and Applications*, Wiley, New York, 2nd edn, 2001.
224. P. G. Pickup, W. Kutner, C. R. Leidner and R. W. Murray, *J. Am. Chem. Soc.*, 1984, **106**, 1991.
225. K. W. Kim, B. C. Kim, H. J. Lee, J. B. Kim and M. K. Oh, *Electroanalysis*, 2011, **23**, 980.
226. O. Steinbock, P. Kettunen and K. Showalter, *J. Phys. Chem.*, 1996, **100**, 18970.
227. J. Matsui, K. Abe, M. Mitsuishi, A. Aoki and T. Miyashita, *Langmuir*, 2009, **25**, 11061.
228. A. P. de Silva, *Nature Mater.*, 2005, **4**, 15.
229. H. Tian, *Angew. Chem. Int. Ed.*, 2010, **49**, 4710.
230. A. P. de Silva, I. M. Dixon, H. Q. N. Gunaratne, T. Gunnlaugsson, P. R. S. Maxwell and T. E. Rice, *J. Am. Chem. Soc.*, 1999, **121**, 1393.
231. M. Cesario, C. O. Dietrich, A. Edel, J. Guilheim, J. P. Kintzinger, C. Pascard and J.-P. Sauvage, *J. Am. Chem. Soc.*, 1986, **108**, 6250.
232. M. Yagi, T. Kaneshima, Y. Wada, K. Takemura and Y. Yokoyama, *J. Photochem. Photobiol. A: Chem.*, 1994, **84**, 27.
233. H. J. Jung, N. Singh, D. Y. Lee and D. O. Jang, *Tetrahedron Lett.*, 2009, **50**, 5555.
234. X. M. He and V. W. W. Yam, *Org. Lett.*, 2011, **13**, 2172.
235. B. Turfan and E. U. Akkaya, *Org. Lett.*, 2002, **4**, 2857.
236. Z. X. Wang, G. R. Zheng and P. Lu, *Org. Lett.*, 2005, **17**, 3669.
237. J.-E. Sohna Sohna, P. Jaumier and F. Fages, *J. Chem. Res.*, 1999, 134.
238. D. F. H. Wallach and L. T. Steck, *Anal. Chem.*, 1963, **35**, 1035.
239. M. Gouterman, in *Porphyrins*, ed. D. Dolphin, Academic Press, New York, 1978, vol. 3, p. 41.

240. T. Gunnlaugsson, A. P. Davis, J. E. O'Brien and M. Glynn, *Org. Lett.*, 2002, **4**, 2449.
241. T. Gunnlaugsson, A. P. Davis, J. E. O'Brien and M. Glynn, *Org. Biomol. Chem.*, 2005, **3**, 48.
242. P. N. Cheng, P. T. Chiang and S. H. Chiu, *Chem. Commun.*, 2005, 1285.
243. C. J. Bender, *Chem. Soc. Rev.*, 1986, **15**, 475.
244. T. Komura, G. Y. Niu, T. Yamaguchi and M. Asano, *Electrochimica Acta*, 2003, **48**, 631.
245. M. Biancardo, C. Bignozzi, D. C. Hugh and G. Redmond, *Chem. Commun.*, 2005, 3918.
246. S. Iwata and K. Tanaka, *J. Chem. Soc. Chem. Commun.*, 1995, 1491.
247. O. S. Wolfbeis and H. Offenbacher, *Monatsh. Chem.*, 1984, **115**, 647.
248. M. T. Albelda, M. A. Bernardo, E. Garcia-Espana, M. L. Godino-Salido, S. V. Luis, M. J. Melo, F. Pina and C. Soriano, *J. Chem. Soc. Perkin Trans.*, 1999, **2**, 2545.
249. G. Q. Zong, L. Xian and G. X. Lu, *Tetrahedron Lett.*, 2007, **48**, 3891.
250. G. De Santis, L. Fabbrizzi, M. Licchelli, A. Poggi and A. Taglietti, *Angew. Chem., Int. Ed. Engl.*, 1996, **35**, 202.
251. D. Parker and J. A. G. Williams, *Chem. Commun.*, 1998, 245.
252. X. F. Guo, D. Q. Zhang, T. X. Wang and D. B. Zhu, *Chem. Commun.*, 2003, 914.
253. K. Rurack, A. Koval'chuck, J. L. Bricks and J. L. Slominskii, *J. Am. Chem. Soc.*, 2001, **123**, 6205.
254. H. T. Baytekin and E. U. Akkaya, *Org. Lett.*, 2000, **2**, 1725.
255. C. V. Kumar and E. H. Asuncion, *J. Am. Chem. Soc.*, 1993, **115**, 8547.
256. L. M. Adleman, *Science*, 1994, **266**, 1021.
257. L. M. Adleman, *Sci. Am.*, 1998, **279**, 54.
258. P. L. Gentili, F. Ortica and G. Favaro, *J. Phys. Chem. B*, 2008, **112**, 16793.
259. M. Samiappan, Z. Dadon and G. Ashkenasy, *Chem. Commun.*, 2011, **47**, 710.
260. N. Wagner and G. Ashkenasy, *Chem. Eur. J.*, 2009, **15**, 1765.
261. J. Zhou, M. A. Arugula, J. Halamek, M. Pita and E. Katz, *J. Phys. Chem. B*, 2009, **113**, 16065.
262. T. Gunnlaugsson, D. A. Mac Dónaill and D. Parker, *Chem. Commun.*, 2000, 93.
263. T. Gunnlaugsson, D. A. Mac Dónaill and D. Parker, *J. Am. Chem. Soc.*, 2001, **123**, 12866.
264. J. H. Bu, Q. Y. Zheng, C. F. Chen and Z. T. Huang, *Org. Lett.*, 2004, **6**, 3301.
265. A. Casnati, A. Pochini, R. Ungaro, C. Bocchi, F. Ugozzoli, R. J. M. Egberink, H. Struijk, R. Lugtenberg, F. de Jong and D. N. Reinhoudt, *Chem. Eur. J.*, 1996, **2**, 436.
266. G. Nishimura, K. Ishizumi, Y. Shiraishi and T. Hirai, *J. Phys. Chem. B*, 2006, **110**, 21596.

267. M. de Sousa, M. Kluciar, S. Abad, M. A. Miranda, B. de Castro and U. Pischel, *Photochem. Photobiol. Sci.*, 2004, **3**, 639.
268. J. M. Montenegro, E. Perez-Inestrosa, D. Collado, Y. Vida and R. Suau, *Org. Lett.*, 2004, **6**, 2353.
269. J. S. Park, E. Karnas, K. Ohkubo, P. Chen, K. M. Kadish, S. Fukuzumi, C. W. Bielawski, T. W. Hudnall, V. M. Lynch and J. L. Sessler, *Science*, 2010, **329**, 1324.
270. S. Uchiyama, N. Kawai, A. P. de Silva and K. Iwai, *J. Am. Chem. Soc.*, 2004, **126**, 3032.
271. B. Ramachandram, G. Saroja, N. B. Sankaran and A. Samanta, *J. Phys. Chem. B*, 2000, **104**, 11824.
272. S. Pandya, J. H. Yu and D. Parker, *Dalton Trans.*, 2006, 2757.
273. T. Gunnlaugsson and J. P. Leonard, *Chem. Commun.*, 2005, 3114.
274. D. Collado, E. Perez-Inestrosa, R. Suau, J.-P. Desvergne and H. Bouas-Laurent, *Org. Lett.*, 2002, **4**, 855.
275. D. Collado, E. Perez-Inestrosa and R. Suau, *J. Org. Chem.*, 2003, **68**, 3574.
276. S. Banthia and A. Samanta, *Eur. J. Org. Chem.*, 2005, 4967.
277. X. Y. Xie, L. Liu, D. Z. Jia, J. X. Guo, D. L. Wu and X. L. Xie, *New J. Chem.*, 2009, **33**, 2232.
278. J. Fabian and R. Zahradnik, *Angew. Chem. Int. Ed. Engl.*, 1989, **28**, 677.
279. S. J. Langford and T. Yann, *J. Am. Chem. Soc.*, 2003, **125**, 11198.
280. S. J. Langford and T. Yann, *J. Am. Chem. Soc.*, 2003, **125**, 14951.
281. A. Coskun, E. Deniz and E. U. Akkaya, *Org. Lett.*, 2005, **7**, 5187.
282. D. Margulies, G. Melman and A. Shanzer, *Nature Mater.*, 2005, **4**, 768.
283. H. Miyaji, S. R. Collinson, I. Prokes and J. H. R. Tucker, *Chem. Commun.*, 2003, 64.
284. H. Miyaji, H. K. Kim, E. K. Sim, C. K. Lee, W. S. Cho, J. L. Sessler and C. H. Lee, *J. Am. Chem. Soc.*, 2005, **127**, 12510.
285. K. A. Green, M. P. Cifuentes, T. C. Corkery, M. Samoc and M. G. Humphrey, *Angew. Chem. Int. Ed.*, 2009, **48**, 7867.
286. M. Irie, *Chem. Rev.*, 2000, **100**, 1685.
287. M. Irie, *Adv. Polym. Sci.*, 1993, **110**, 50.
288. G. Pasparakis, M. Vamvakaki, N. Krasnogor and C. Alexander, *Soft Matter*, 2009, **5**, 3839.
289. Y. J. Wang, B. J. Xin, X. R. Duan, G. W. Xing and S. Wang, *Macromol. Rapid Commun.*, 2010, **31**, 1473.
290. B. M. Frezza, S. L. Cockroft and M. R. Ghadiri, *J. Am. Chem. Soc.*, 2007, **129**, 14875.
291. G. Seelig, D. Soloveichik, D. Y. Zhang and E. Winfree, *Science*, 2006, **314**, 1565.
292. W. Fontana, *Science*, 2006, **314**, 1552.
293. C. Zhang, J. Yang and J. Xu, *Langmuir*, 2010, **26**, 1416.
294. G. Seelig, B. Yurke and E. Winfree, *J. Am. Chem. Soc.*, 2006, **128**, 12211.
295. T. Asoh and M. Akashi, *Chem. Commun.*, 2009, 3548.

296. D. Sooksawat, W. Aeungmaitrepirom, W. Ngeontae and T. Tuntulani, *New J. Chem.*, 2011, **35**, 345.
297. A. Credi, V. Balzani, S. J. Langford and J. F. Stoddart, *J. Am. Chem. Soc.*, 1997, **119**, 2679.
298. G. Bergamini, C. Saudan, P. Ceroni, M. Maestri, V. Balzani, M. Gorka, S. K. Lee, J. V. Heyst and F. Vogtle, *J. Am. Chem. Soc.*, 2004, **126**, 16466.
299. A. P. de Silva and N. D. McClenaghan, *Chem. Eur. J.*, 2002, **8**, 4935.
300. R. Baron, O. Lioubashevski, E. Katz, T. Niazov and I. Willner, *J. Phys. Chem. A*, 2006, **110**, 8548.
301. V. Privman, J. Zhou, J. Halamek and E. Katz, *J. Phys. Chem. B*, 2010, **114**, 13601.
302. J. Halamek, V. Bocharova, M. A. Arugula, G. Strack, V. Privman and E. Katz, *J. Phys. Chem. B*, 2011, **115**, 9838.
303. M. N. Stojanovic and D. Stefanovic, *J. Am. Chem. Soc.*, 2003, **125**, 6673.
304. C. D. Mao, T. H. LaBean, J. H. Reif and N. C. Seeman, *Nature*, 2000, **407**, 493.
305. F. Pina, M. J. Melo, M. Maestri, P. Passaniti and V. Balzani, *J. Am. Chem. Soc.*, 2000, **122**, 4496.
306. K. Szacilowski, W. Macyk and G. Stochel, *J. Am. Chem. Soc.*, 2006, **128**, 4550.
307. L. F. O. Furtado, A. D. P. Alexiou, L. Goncalves, H. E. Toma and K. Araki, *Angew. Chem. Int. Ed.*, 2006, **45**, 3143.
308. S. Nitahara, T. Akiyama, S. Inoue and S. Yamada, *J. Phys. Chem. B*, 2005, **109**, 3944.
309. S. Nitahara, N. Terasaki, T. Akiyama and S. Yamada, *Thin Solid Films*, 2006, **499**, 354.
310. J. Matsui, M. Mitsuishi, A. Aoki and T. Miyashita, *J. Am. Chem. Soc.*, 2004, **126**, 3708.
311. F. Xia, X. L. Zuo, R. Q. Yang, R. J. White, Y. Xiao, D. Kang, X. Gong, A. A. Lubin, A. Vallee-Belisle, J. D. Yuen, B. Y. B. Hsu and K. W. Plaxco, *J. Am. Chem. Soc.*, 2010, **132**, 8557.
312. Y. Q. Liu, A. Offenhausser and D. Mayer, *Angew. Chem. Int. Ed.*, 2010, **49**, 2595.
313. M. Asakawa, P. R. Ashton, V. Balzani, A. Credi, G. Mattersteig, O. A. Matthews, M. Montalti, N. Spencer, J. F. Stoddart and M. Venturi, *Chem. Eur. J.*, 1997, **3**, 1992.
314. M. Kumar, R. Kumar and V. Bhalla, *Tetrahedron Lett.*, 2010, **51**, 5559.
315. M. Schmittel, P. Mal and A. de los Rios, *Chem. Commun.*, 2010, **46**, 2031.
316. M. Zhou, F. Wang and S. J. Dong, *Electrochim. Acta*, 2011, **56**, 4112.
317. K. Rurack, C. Trieflinger, A. Kovalchuck and J. Daub, *Chem. Eur. J.*, 2007, **13**, 8998.
318. J. H. Guo, D. M. Kong and H. X. Shen, *Biosens. Bioelectron.*, 2010, **26**, 327.

319. S. J. Dickson, A. N. Swinburne, M. J. Paterson, G. O. Lloyd, A. Beeby and J. W. Steed, *Eur. J. Inorg. Chem.*, 2009, 3879.
320. X. J. Zhao and C. Z. Huang, *Analyst*, 2010, **135**, 2853.
321. L. A. Cabell, M. D. Best, J. J. Lavigne, S. E. Schneider, D. M. Perreault, M. K. Monahan and E. V. Anslyn, *J. Chem. Soc. Perkin Trans 2*, 2001, 315.
322. G. Klein, D. Kaufmann, S. Schurch and J.-L. Reymond, *Chem Commun.*, 2001, 561.
323. Z. H. Lin, M. Wu, M. Schaeferling and O. S. Wolfbeis, *Angew. Chem. Int. Ed.*, 2004, **43**, 1735.
324. D. A. Leigh, M. A. F. Morales, E. M. Perez, J. K. Y. Wong, C. G. Saiz, A. M. Z. Slawin, A. J. Carmichael, D. M. Haddleton, A. M. Brouwer, W. J. Buma, G. W. H. Wurpel, S. Leon and F. Zerbetto, *Angew. Chem. Int. Ed.*, 2005, **44**, 3062.
325. F. D. Jochum, F. R. Forst and P. Theato, *Macromol. Rapid Commun.*, 2010, **31**, 1456.
326. A. P. de Silva and R. A. D. D. Rupasinghe, *J. Chem. Soc. Chem. Commun.*, 1985, 1669.
327. A. P. de Silva, S. A. de Silva, A. S. Dissanayake and K. R. A. S. Sandanayake, *J. Chem. Soc. Chem. Commun.*, 1989, 1054.
328. F. M. Raymo and S. Giordani, *Org. Lett.*, 2001, **3**, 3475.
329. X. F. Guo, D. Q. Zhang, Y. C. Zhou and D. B. Zhu, *Chem. Phys. Lett.*, 2003, **375**, 484.
330. M. Shirai, K. Suyama, H. Okamura and M. Tsunooka, *J. Photopolymer Sci. Technol.*, 2002, **15**, 715.
331. H. Tian, J. Gan, K. Chen, J. He, Q. L. Song and X. Y. Hou, *J. Mater. Chem.*, 2002, **12**, 1262.
332. A. Mills, K. McDiarmid, M. McFarlane and P. Grosshans, *Chem. Commun.*, 2009, 1345.

CHAPTER 8
Reconfigurable Double Input–Single Output Systems

8.1 Introduction

As discussed under reconfigurable single-input logic devices (Chapter 6), conventional semiconductor systems are reconfigured by changing the connectivity between individual logic gates. As also mentioned in Chapter 6, molecules can be interrogated in many ways which are unavailable to semiconductor electronic devices, so that new reconfiguring approaches come into being. The classification used here is similar to that employed in Chapter 6, with the addition of new variables as necessary.

8.2 Module Connectivity within Device

Electronics-based computing makes most of its logic gates by hard-wiring transistors in different ways. Early versions of this approach (based on diode–diode logic) were exemplified in Figures 3.4 and 3.5 for AND and OR logic respectively. So the physical formation or destruction of connections gives a straightforward way of reconfiguring logic functions. Therefore, molecular-based logic systems can also achieve reconfiguring by altering the gate connectivity by changing the 'wiring'. The wiring can be metallic or molecular.

A case of metallic wiring of device components for reconfiguring of molecular electronic gates is available from Collier *et al.*[1,2] As mentioned in Chapter 7, diodes can be constructed from unimolecular layers of surfactant redox-active viologens (*e.g.* structure **82** shown in section 7.2.7) held between titanium and aluminium metal terminals where the latter also has a barrier layer of Al_2O_3. Resonant tunnelling of electrons occurs only if the titanium electrode is held at a negative voltage. In fact, a positive voltage destroys the device. These diodes can be appropriately wired to produce various gates (Chapter 3).

Monographs in Supramolecular Chemistry No. 12
Molecular Logic-based Computation
By A Prasanna de Silva
© The Royal Society of Chemistry 2013
Published by the Royal Society of Chemistry, www.rsc.org

A double-input OR gate is obtained from a triple-input version by oxidatively destroying a diode in one of the input lines. Unfortunately, such reconfiguring is irreversible, but there are parallel situations in the semiconductor electronics field, *e.g.* one-time programmable read-only memories (PROM).[3]

In some situations, the 'wiring' refers to molecular pathways for electron transfer. This analogy holds best if the electron transfer is a 'through-bond' process.[4] However, even 'through-space' electron transfers are greatly facilitated by bonds which hold the electron donor and acceptor close together.[4] Since most molecular logic gate systems are constructed by covalently bonding various modules together, the formatting of these modules can be altered by following different synthetic procedures. For instance, 'fluorophore–spacer$_1$–receptor$_1$–spacer$_2$–receptor$_2$' and 'receptor$_1$–spacer$_1$–fluorophore–spacer$_2$–receptor$_2$' formats differ in this way (Figure 8.1). This is illustrated by PET systems **1** and **2**[5] which display different logic capabilities. By starting with 'fluorophore–spacer$_1$–receptor$_1$–spacer$_2$–receptor$_2$' system **1**, using a tertiary amine (to bind H^+) and Tsien's BAPTA receptor[6] (to bind Ca^{2+}), it is possible to get an H^+-driven 'off–on–off' system (see Chapter 12) which is only enabled[7] in the presence of Ca^{2+}. The Ca^{2+}-induced enabling is due to PET suppression. The basic arm of the 'off–on–off' profile arises via H^+-induced PET suppression. The acidic arm arises from carboxylic acid groups hydrogen-bonding to the π-cloud of the anthracene fluorophore in the excited state. By changing the format to

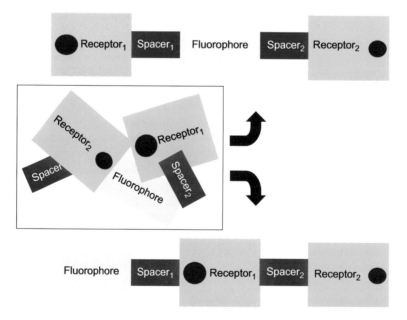

Figure 8.1 Two different ways of formatting a fluorophore, two receptors and two spacers so as to produce fluorescent PET systems with different logic behaviour.

'receptor$_1$–spacer$_1$–fluorophore–spacer$_2$–receptor$_2$', while still using the same receptors and fluorophore, no 'off–on–off' behaviour is seen for **2** because the hydrogen-bonding is geometrically impossible in this case. However, both **1** and **2** show H$^+$, Ca^{2+}-driven AND logic when the input levels are chosen appropriately.

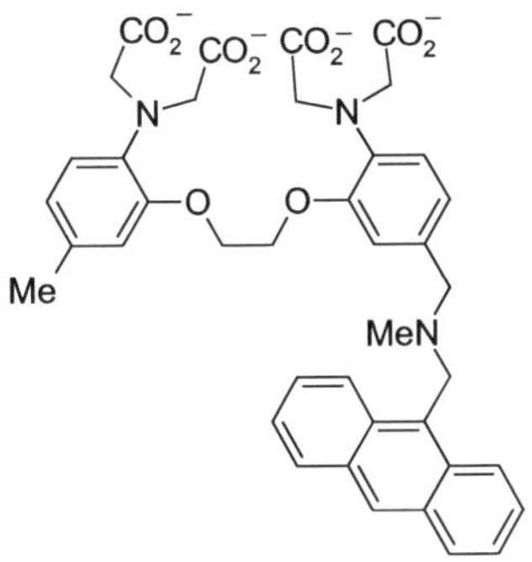

1 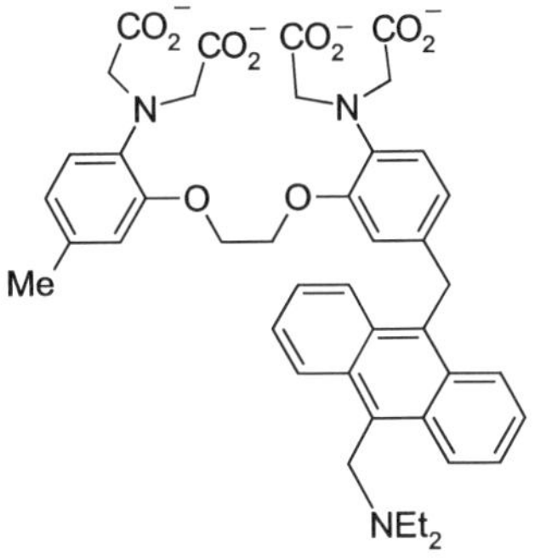   **2**

Wireless connecting of device components can be discerned in Raymo's work with cuvet arrays of photochromic solutions.[8] A single cuvet produces light dose-driven NOT logic since a dose of ultraviolet (uv) light produces a strongly coloured solution so that a probe beam of green light (563 nm) is extinguished. The output can be taken as the transmittance in the green region of the spectrum. The memory effect of the photochromic is not taken into account in this analysis (but see Chapter 11). Passing the probe beam through series or parallel arrangements of cuvets produces NAND or NOR logic gates respectively. Logic reconfiguring occurs by what are essentially simple geometric adjustments. More complex cases of this type will be taken up in Chapter 10.

8.3 Functional Group Connectivity within Input Array

The inputs can be functional groups which are connected within a more complex molecule. If this connectivity is altered, say by changing the separation distance between the functional groups, the response of the device to the same inputs will be altered, sometimes drastically. For example, the correct length of the α,ω-alkanediammonium ion to engage the receptor pair within **3** is C$_4$ which produces AND logic response, whereas the wrong length (C$_3$) gives only PASS 0 when the fluorescence output is viewed[9] – with reasonable thresholding – at an alkanediammonium concentration of 10$^{-5.3}$ M (Figure 8.2). The comparison is even starker when the C$_4$ and C$_7$ versions are examined in Figure 8.2. Many examples of this kind were collected within Chapter 7 (section 7.2.4).

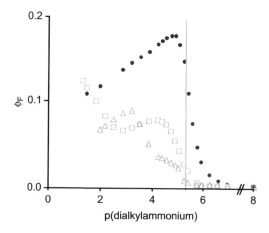

Figure 8.2 Fluorescence quantum yield of **3** as a function of concentration (in –log units) of α,ω-alkanediammonium ions. The concentration itself is in molar units. Alkane chain lengths are 3 (green squares), 4 (red circles) and 7 (blue triangles) methylene groups in the three cases. The latter two cases also show 2:1 binding occurring at higher concentrations. The concentration value of $10^{-5.3}$ M is shown by the pink line.
Redrawn from A. P. de Silva and K. R. A. S. Sandanayake, *Angew. Chem. Int. Ed. Engl.*, 1990, **29**, 1173 with permission from Wiley-VCH.

3

8.4 Functional Group Configuration within Input Array

Subtler forms of the above section are found when we employ enantiomers where the connectivity is the same, but the configuration in three-dimensional (3D) space is different. D- and L-enantiomers of glucose elicit rather different fluorescence enhancement factors from the R-enantiomer of **4**.[10] As maybe expected, the S-enantiomer of **4** shows the opposite discrimination between the glucose enantiomers. This example was discussed in section 7.2.4. Much progress has been made in enantioselective detection.[11]

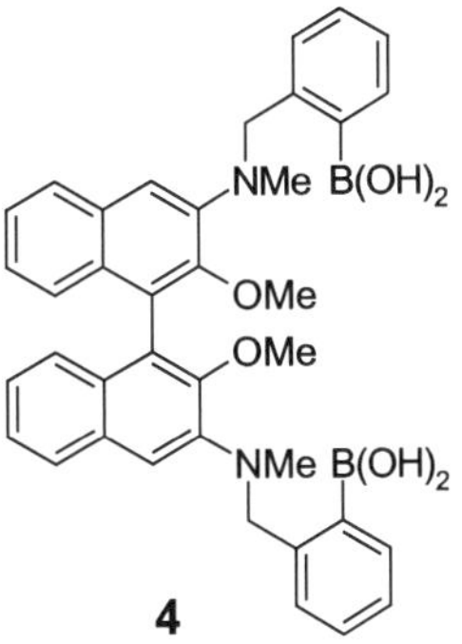

4

8.5 Nature of Inputs

Chemistry presents us with a rich variety of species that can be employed as inputs for molecular logic devices. So it should not be surprising that the application of different inputs to some devices can result in different logic responses.

For instance, Konermann's exploitation of cytochrome c as an AND gate[12] using tryptophan fluorescence as output with carefully chosen concentrations of urea and H^+ serving as the inputs was featured in section 7.2.6. By choosing the reversible denaturants 0.1 M H^+ and 0.1 M OH^- as inputs the tryptophan fluorescence emerges only when either input is present at the 'high' level. When both inputs are applied simultaneously at a 'high' level, neutralization occurs, to leave the protein at neutral pH in its native state so that the tryptophan emission is quenched by the neighbouring haem unit. So we have XOR logic arising from the replacement of urea input with OH^-.

The logic behaviour of PET-based anthrylmethylpolyamines such as **5** is also reconfigurable according to the nature/level of the ion inputs.[13] Filled-electron shell metal ions such as Zn^{2+} switch 'on' the fluorescence by blocking PET, just like H^+ does. When H^+ and Zn^{2+} are considered as inputs, OR logic is expressed. On the other hand, open-shell Cu^{2+} binds **5** the strongest of all and quenches the emission by EET/PET. So if H^+ and Cu^{2+} are considered as inputs, we have INHIBIT(Cu^{2+}) logic instead. Further, if Cu^{2+} and Zn^{2+} are considered as inputs, the descriptor becomes INHIBIT(Cu^{2+}) logic. This has also been described as NOT IMPLICATION($Zn^{2+} \Rightarrow Cu^{2+}$) logic.[13] Replacement of Zn^{2+} with Ni^{2+}, another open-shell ion, gives no fluorescence situations at all, *i.e.* Cu^{2+}, Ni^{2+}-driven PASS 0 logic prevails. The Zn^{2+}, Cu^{2+}-driven INHIBIT(Cu^{2+}) gate **6**,[14] with its fluorescence output, shows some similarity with **5**, but the similarity ends when the emission of **6** also displays Na^+, Mg^{2+}-driven AND logic. The latter is hard to pin down mechanistically at present, though the weakly coordinating solvent tetrahydrofuran might play a part.

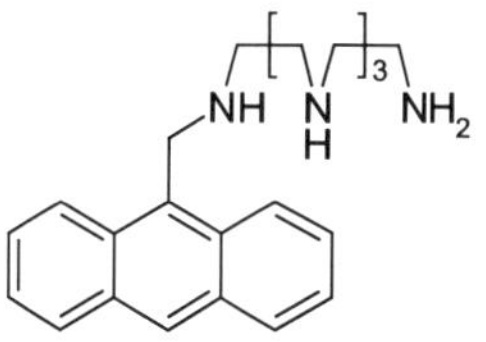

5

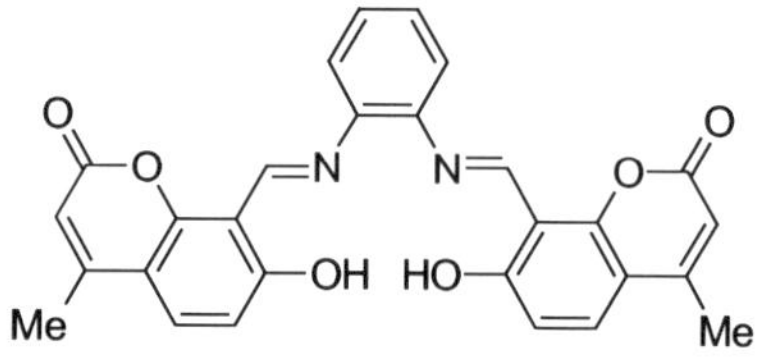

6

Fujita's experiments with a pyramidal host,[15] constructed with metal–ligand coordinate bonds, were discussed at length in sections 7.2.1.6 and 7.3 concerning AND and OR logic respectively. The AND logic arose with inputs **7** and **8** whereas OR came about with inputs **9** and **10**. Some degree of shape selectivity is involved during this logic reconfiguring.

7 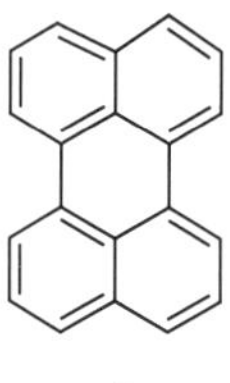 **8** 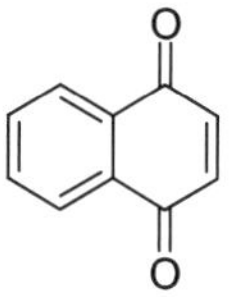 **9** 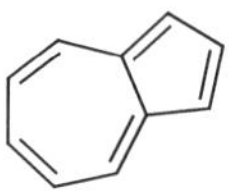 **10**

Switching between AND and NAND is arranged in Ghadiri's **11** by changing the input pair from **12** and **13** to **12** and **14**.[16] However, the detailed discussion in section 7.5 (following that in section 7.2.6) noted that a change of excitation wavelength from 350 to 490 nm is also involved.

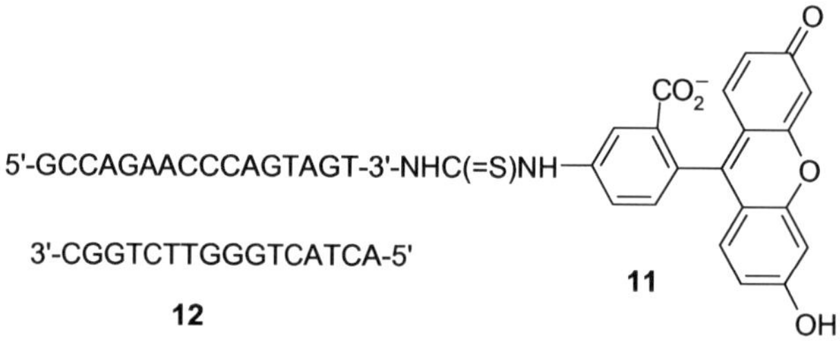

5'-GCCAGAACCCAGTAGT-3'-NHC(=S)NH—

3'-CGGTCTTGGGTCATCA-5'

12

11

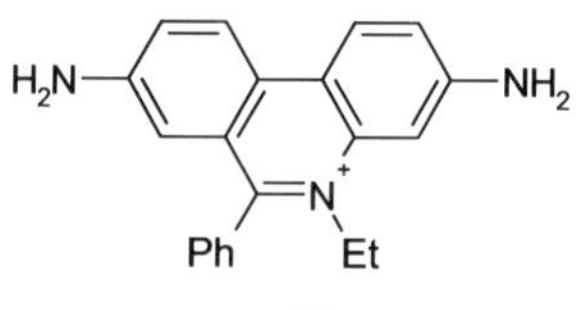

14

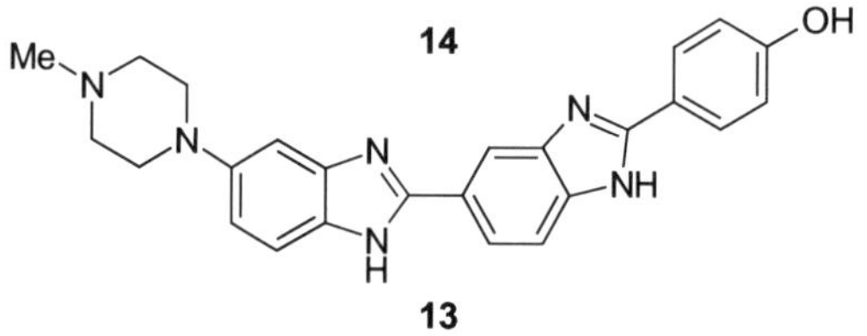

13

The spiropyran–merocyanine pair features from a photochromic viewpoint at various points in this book, but Tomizaki and Mihara develop it as a thermochromic system.[17] When attached to certain peptides and dark-adapted, the coloured merocyanine form dominates (Figure 8.3). It is also fluorescent in this situation, so that two (absorption and emission) output readouts become available. Both these outputs can be sent 'low', *i.e.* the spiropyran form comes to the fore, if the peptide is wrapped up with another polymer to create a less polar and less fluid microenvironment. Electrostatic attraction enables this wrapping so that phosphorylation of the peptide becomes a good way of control. The peptide is designed to carry two sequences which are separately targeted by two kinases (SrcN1 and PKA) which become the inputs to the peptide device. It is also tetracationic at neutral pH. Another good means of control is the charge on the second polymer, so this becomes the reconfiguring variable. If an anionic polymer such as poly(aspartate) is used, the wrapping

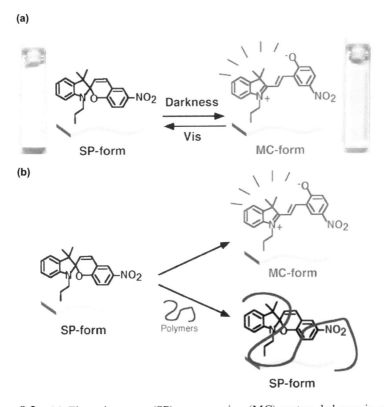

Figure 8.3 (a) The spiropyran (SP)–merocyanine (MC) system behaves in a thermochromic manner when attached to certain peptides. (b) However, this is made conditional by wrapping the peptide with other biopolymers depending on their charge and depending on the phosphorylation state of the peptide. The latter is arranged by using two kinases as inputs.
Reprinted from K. Tomizaki and H. Mihara, *J. Am. Chem. Soc.*, 2007, **129**, 8345 with permission from the American Chemical Society.

becomes less and less, as one and then both kinase inputs are applied. So the output is 'high' when both inputs are present, *i.e.* AND logic. On the other hand, if a cationic polymer such as poly(lysine) is present, the wrapping occurs even if one kinase input is made 'high'. Adding the second kinase does not make a significant difference. So the output is 'low' when either or both inputs are present, which corresponds to NOR logic. This system's versatility can be increased by specifying different threshold levels. It also contains an interesting memory aspect which is held not by the spiropyran–merocyanine pair (as seen in Chapter 11), but by the phosphorylation state of the peptide. Finally, reset is achieved by applying a phosphatase to cause dephosphorylation.

We focused on the two-photon absorption experiments of Samoc and Humphreys[18] in section 7.6 where an INHIBIT gate was driven by oxidation and light dose. The discussion in section 7.6 is not repeated here. Structure **129**, shown in Chapter 7, responds to protonation/deprotonation besides electro- and photochemical stimulation. Driving the gate, with H^+ and light dose inputs, results in IMPLICATION(H^+ ⇒ light dose) logic. The observation of two-photon absorption is done at 900 nm. When the open form of the dithienylethene unit and the vinylidene form of the Ru(II) centre coexist, there is no detectable two-photon absorption at 900 nm. Notably, this case could not be analysed as a triple input device because only six stable states are available. Eight states are needed for a full truth table analysis of a triple input–single output logic system (Chapter 10).

The final paragraphs of this section consider a subtler form of logic reconfiguring which arises by changing the level of the 'high' or 'low' input signal. However, systems amenable to this procedure can also be viewed in terms of multi-level logic. Therefore, most of these cases are discussed in Chapter 12.

The fluorescent PET system **1**[5] shows AND logic with H^+ and Ca^{2+} inputs given that the H^+ input is defined as 'high' at pH 6.7 and 'low' at pH 9.7. However, if the H^+ input is re-defined as 'high' at pH 4.5 (instead of pH 6.7), the response converts to an H^+, Ca^{2+}-driven PASS 0 gate where no input combination revives fluorescence. At such high acidities, Ca^{2+} cannot compete for its amino acid receptor because of protonation. Further mechanistic points were made in the discussion of this compound in section 8.2.

The other case mentioned in Section 8.2 (**2**)[5] switches from H^+, Ca^{2+}-driven AND to H^+, Ca^{2+}-driven TRANSFER(H^+) logic upon changing the high proton threshold value from $10^{-6.7}$ M to $10^{-4.5}$ M. The latter type of logic can be described in an operationally simpler form, *i.e.* H^+-driven YES.

The oxidation of malate by malate dehydrogenase in the presence of NAD^+ was discussed in section 7.7 as a XOR logic gate when driven by Mg^{2+} and Ca^{2+} inputs. These inputs were taken as 'high' at 0.03 M. The alteration of these 'high' levels produces OR logic instead.[19] Unfortunately, the mechanistic basis of this change is not clear because the publication is phenomenological in nature.

8.6 Output Observation Technique

There are several examples where absorption and emission spectra of a compound yield different logic behaviours.[20–25] For instance, **15** can be exploited in these two ways to produce XOR and INHIBIT logic respectively.[20,21] Transmittance and fluorescence intensity are the respective outputs. Langford and Yann[20,21] rely on the amphoteric properties of the porphyrin core, so that the inputs used are H^+ and t-BuO$^-$. We will resist further discussion because this case will be considered at length in Chapter 9. Absorption and emission spectra are also used by Gust and collaborators to achieve AND and NAND logic respectively, according to their photochromic approach (see section 7.2.5 for instance).[22] However, a third-harmonic-generating crystal is an essential part of this laser experiment.

15

Each discovery of molecular sensors which can be interrogated by several techniques can be examined for possible reconfigurable logic gate action. For instance, Schmittel's Ru(II) complex, which is interrogated by voltammetry and electrochemiluminescence besides absorption and emission spectroscopy for the purpose of metal ion sensing,[23] received attention in sections 6.2 and 6.3. Martínez-Máñez, Sancenón and their co-workers observed differing results from absorption spectroscopy, voltammetry and emission spectroscopy on **16** when faced with two inputs (Table 8.1).[24] Only Pb^{2+} causes a large absorption spectral blue shift (45 nm), whereas Pb^{2+} and Fe^{3+} (individually and together) cause substantial (>0.3 V) anodic potential shifts in the reduction wave. The absorption spectral shift depends on engagement of the amine nitrogen with Pb^{2+}. The Fe^{3+} prefers to bind the harder oxygens instead. The similar potential shifts can be understood in terms of the cation-induced stabilization of the radical anion of **16**. Also, the presence of Fe^{3+} without Pb^{2+} quenches

Table 8.1 Truth tables for observation technique–reconfigurable logic behaviour of **16**.[a]

$Input_1$ Pb^{2+}	$Input_2$ Fe^{3+}	$Output_1$ $\Delta\lambda_{Abs}$[b]	$Output_2$ ΔE[c]	$Output_3$ Fluorescence[d]
none	none	low (0)	low (0)	high (100)
none	high (5×10^{-5} M)	low (7)	high (340)	low (78)
high (5×10^{-5} M)	none	high (45)	high (360)	high (145)
high (5×10^{-5} M)	high (5×10^{-5} M)	high (45)	high (360)	high (145)

[a]In acetonitrile, concentration of **16** depends on observation technique.
[b]Starting absorption 520 nm.
[c]Anodic shift in mV. Starting wave potential –0.7 V (vs. sce).
[d]Relative intensities, λ_{exc} 520 nm, λ_{em} 610 nm.

fluorescence. Redox-active Fe^{3+} is a strong fluorescence quencher, but it cannot bind to **16** when in competition with Pb^{2+} whose binding constant is 30-fold larger. So, voltammetric output shows up as Pb^{2+}, Fe^{3+}-driven OR logic, whereas absorption spectral output is describable as TRANSFER(Pb^{2+}) and emission output comes out as IMPLICATION($Pb^{2+} \Leftarrow Fe^{3+}$).

16

8.7 Nature of Output (within a given Observation Technique)

Absorption spectroscopic measurements of different outputs are available within Willner's enzyme cascades.[26,27] The enzyme glucose dehydrogenase (GDH) uses glucose as a substrate ($input_1$) in the presence of the cofactor NAD^+ to produce gluconic acid ($output_1$). Horseradish peroxidase (HRP) takes in H_2O_2 ($input_2$) to produce H_2O with the aid of NADH. These two enzymes now run in parallel because the cofactor cycles between its two redox states (NAD^+ and NADH). If there is no NAD^+ initially, and if the system is allowed time to reach a steady state, the NADH concentration (as measured on the basis of its absorption of light) drops only on the condition that glucose is absent and H_2O_2 is present. If this absorbance change is taken as $output_2$, the coupled enzyme system manifests glucose, H_2O_2-driven INHIBIT logic.

When glucose and H_2O_2 are both present, the NAD^+ produced by HRP's processing of H_2O_2 is taken up by GDH to process glucose so that the NADH level remains almost constant. On the other hand, gluconic acid ($output_1$, measured on the basis of its absorption of light after reaction with hydroxylamine and ferric ion) accumulates only if glucose and H_2O_2 are both present *i.e.* glucose, H_2O_2-driven AND logic is seen. If H_2O_2 is not available, NADH cannot be converted to NAD^+ by HRP so that GDH cannot process glucose in turn.

8.7.1 Observation Wavelength

As was mentioned in Chapter 6, quantum possibilities emerge when several wavelengths of observation are exploited simultaneously as logic reconfiguring variables. This aspect is held back until Chapter 14. Several double-input examples are to be found there. However, a case which narrowly misses inclusion in Chapter 14 is discussed here. Although the logic output is indeed reconfigured according to the observation wavelength, each observation requires resetting of an excitation wavelength. The simultaneous use of two excitation wavelengths, though physically possible, has not been evaluated in this instance for possible chemical complications.

System **17** + **18** is due to the combined efforts of Andréasson's and Pischel's laboratories.[28] The coloured merocyanine form of a spiropyran photochromic quenches the fluorescence from the aminonaphthalimide fluorophore within **17** *via* EET. Application of a visible light dose ($\lambda > 420$ nm) as input$_1$ ring-closes the merocyanine into the colourless spiropyran form so that aminonaphthalimide emits strongly again at 512 nm ($\lambda_{exc} = 417$ nm). Arrival of a strong phosphazene base (**19**, input$_2$) does not change the above scenario. Thus the aminonaphthalimide fluorescence output obeys TRANSFER(input$_1$) logic. Contrasting NOT TRANSFER(input$_2$) behaviour is found for the fluorescence output from **18** at 416 nm ($\lambda_{exc} = 368$ nm) because it is largely immune to EET-based quenching by the merocyanine, owing to their intermolecular arrangement. On the other hand, the presence of basic input$_2$ (**19**) deprotonates **18** to activate a PET process from the amine so that the emission is killed off. This is also a demonstration of logic reversibility, a topic to be taken up in section 9.6, because each input state can be uniquely mapped to an output state.

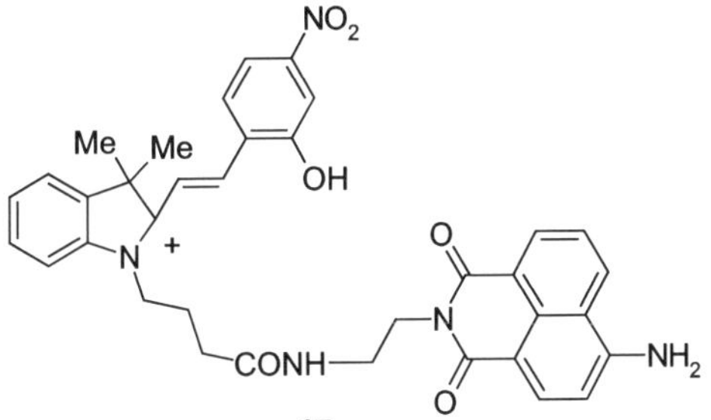

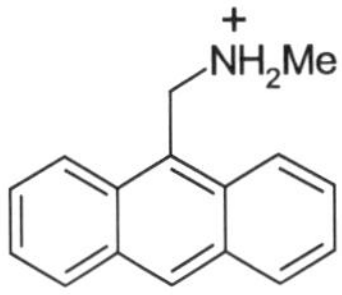

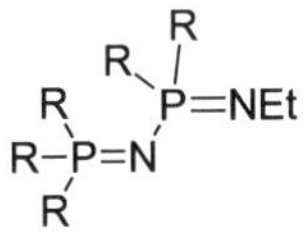

8.8 Starting State of Device

Yan uses the redox- and cation-sensitivity of **20**,[29] originating from its tetrathiafulvalene (TTF) and pyridine moieties respectively, to develop various logic types. Two of these are particularly interesting in the current context. We start with the non-fluorescent **20** (caused by a PET-type process occurring from the electron-rich TTF moiety to the rest of the structure, among other processes). The TTF part of **20** is oxidized by $NOBF_4$ (input$_1$) to its dication, and **20**$^{2+}$ is now strongly fluorescent because the PET-type process is no longer possible. Cu^{2+} (input$_2$) is chelated by the bipyridine-like component of **20**$^{2+}$, causing the fluorescence to be quenched by the redox-active nature of Cu^{2+}. $NOBF_4$, Cu^{2+}-driven INHIBIT, is the upshot, noting that Cu^{2+} is the disabling input.

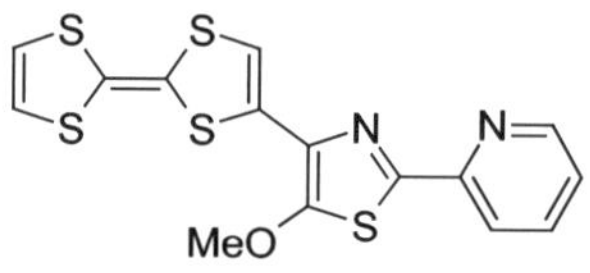

20

Now we start with pre-oxidized **20**, *i.e.* **20**$^{2+}$, and apply $NOBF_4$ and Fe^{2+} as inputs. Neither input on its own causes any change in the fluorescence, but application of both together causes the oxidation of Fe^{2+} to Fe^{3+} (by $NOBF_4$), which is a strong quencher of fluorescence after being chelated to the bipyridine-like component. So we have NAND behaviour. Though the inputs are not strictly comparable, the change of the starting state plays a role in reconfiguring the logic from INHIBIT to NAND.

Pischel, Nau and their collaborators[30] start off with **21** at pH 9, whose fluorescence output goes 'high' only when its benzimidazole unit is protonated and then encapsulated by cucubituril[7], *i.e.* AND logic. This is due to a combination of PET suppression within the 'fluorophore–spacer–receptor' system and the pK_a shift of $+2.1$ upon encapsulation. However, if the starting state is taken as **21**.H^+.cucubituril[7] at pH 7, the fluorescence decreases strongly on the addition of OH^- (to cause deprotonation and decomplexation) or of diprotonated cadaverine (to bind competitively to cucubituril[7]), *i.e.* NOR logic.

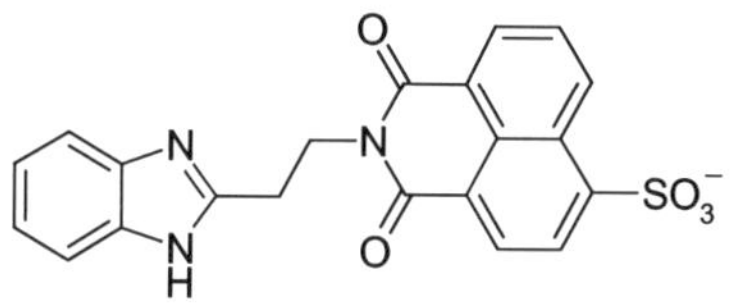

21

Dong's team constructed a glucose-fed fuel cell with two electrodes carrying immobilized enzymes.[31] The anode carries glucose oxidase and the electron relay ferrocene carboxylic acid so that a large current is passed when O_2 concentration is minimized; O_2 would compete with the relay for the electrons produced during the enzyme action. The cathode material contains bilirubin oxidase and the electron donor **22** so that current passage is best when O_2

concentration is maximized. When the electrodes are combined in the fuel cell, it is natural that the highest power arises when the O_2 level is intermediate. This level, which is close to that of air, can be obtained by balancing O_2 and N_2 delivery rates. If the starting state is the deaerated fuel cell, the power goes 'high' only when the inputs O_2 and N_2 are delivered simultaneously, *i.e.* AND logic. If the starting state is the aerated fuel cell, the power goes 'low' when either O_2 or N_2 is delivered separately. The power remains 'high' if both inputs are applied together because the condition is similar to that of aeration. Such reconfiguring is convenient because of the use of all-gas inputs, unlike most of the molecular gates described in this book. Nevertheless, the condition when neither O_2 nor N_2 is present is difficult to define.

22

Shanzer's team[32] relies on changes of starting state, among other conceptual tools, for the success of their molecular calculators, to be discussed in Chapter 10.

8.9 Applied Voltage or Redox Reagents

Molecular adsorbates in photoelectrochemical cells give rise to photocurrents which can produce XOR logic when light signals with different illumination wavelengths are used as inputs.[33-35] Applied voltage is a natural parameter to explore in such experiments. For example, Yamada's team[33,34] find that the XOR logic (section 7.7) observed at 0 V (*vs.* Ag/AgCl) changes to OR logic at −0.2 V. The inputs are light intensities at 470 and 640 nm. The cathodic current due to 470 nm absorption virtually disappears at −0.2 V so that it no longer cancels the anodic current produced by the 640 nm light. Similarly, Szacilowski's XOR gate (section 7.7.2), concerning $[Fe(II)(CN)_6]^{4-}$ on nanocrystalline TiO_2,[35] with an applied voltage of 0.25 V can be reconfigured to TRANSFER and OR logic by adjusting the voltage to 0.4 and −0.2 V respectively (Table 8.2).[36] It can be seen from Table 8.2 that the TRANSFER gate transfers $input_1$ which represents the light intensity of wavelength 400 nm. The need to take the modulus of the photocurrents manually in several cases reduces the molecular nature of the computation. It is not easy to analyse these cases by considering the applied voltage as a third input. An instance where such an analysis is possible, with a redox reagent rather than voltage,[37,38] will be discussed in section 10.15 on 2:1 multiplexers.

Table 8.2 Truth tables for voltage-reconfigurable logic behaviour of $[Fe^{II}(CN)_6]^{4-}$ on nanocrystalline TiO_2.[a]

$Input_1$ Light intensity$_{400}$	$Input_2$ Light intensity$_{460}$	$Output_1$ Current[b,c]	$Output_2$ Current[b,d]	$Output_3$ Current[b,e]
none	none	low (0.0)	low (0.0)	low (0.0)
none	high	low (0.0)	high (2.5)	high (2.5)
high	none	high (7.5)	high (2.5)	high (5.0)
high	high	high (7.5)	low (0.0)	high (9.0)

[a]In aqueous 0.1 M KNO_3.
[b]Arbitrary units.
[c]Applied voltage 0.4 V (vs. Ag/AgCl).
[d]Applied voltage 0.25 V, modulus of current is reported.
[e]Applied voltage –0.2 V, modulus of current is reported.

References

1. C. P. Collier, E. W. Wong, M. Belohradsky, F. M. Raymo, J. F. Stoddart, P. J. Kuekes, R. S. Williams and J. R. Heath, *Science*, 1999, **285**, 391.
2. C. P. Collier, G. Mattersteig, E. W. Wong, Y. Luo, K. Beverly, J. Sampaio, F. M. Raymo, J. F. Stoddart, P. J. Kuekes, R. S. Williams and J. R. Heath, *Science*, 2000, **289**, 1172.
3. C. Maxfield, *From Bebop to Boolean Boogie*, Newnes, Oxford, 2009.
4. M. N. Paddon-Row, in *Stimulating Concepts in Chemistry*, ed. F. Vogtle, J. F. Stoddart and M. Shibasaki, Wiley-VCH, Weinheim, 2000, p. 267.
5. J. F. Callan, A. P. de Silva and N. D. McClenaghan, *Chem. Commun.*, 2004, 2048.
6. R. Y. Tsien, *Biochemistry*, 1980, **19**, 2396.
7. S. A. de Silva, B. Amorelli, D. C. Isidor, K. C. Loo, K. E. Crooker and Y. E. Pena, *Chem. Commun.*, 2002, 1360.
8. F. M. Raymo and S. Giordani, *Proc. Natl. Acad. Sci. USA*, 2002, **99**, 4941.
9. A. P. de Silva and K. R. A. S. Sandanayake, *Angew. Chem. Int. Ed. Engl.*, 1990, **29**, 1173.
10. T. D. James, K. R. A. S. Sandanayake and S. Shinkai, *Nature*, 1995, **374**, 345.
11. D. Leung, S. O. Kang and E. V. Anslyn, *Chem. Soc. Rev.*, 2012, **41**, 448.
12. A. S. Deonarine, S. M. Clark and L. Konermann, *Future Gener. Comp. Sys.*, 2003, **19**, 87.
13. S. Alves, F. Pina, M. T. Albelda, E. García-España, C. Soriano and S. V. Luis, *Eur. J. Inorg. Chem.*, 2001, 405.
14. Y. Dong, J. F. Li, X. X. Jiang, F. Y. Song, Y. X. Cheng and C. J. Zhu, *Org. Lett.*, 2011, **13**, 2252.
15. M. Yoshizawa, M. Tamura and M. Fujita, *J. Am. Chem. Soc.*, 2004, **126**, 6846.
16. A. Saghatelian, N. H. Volcker, K. M. Guckian and M. R. Ghadiri, *J. Am. Chem. Soc.*, 2003, **125**, 346.
17. K. Tomizaki and H. Mihara, *J. Am. Chem. Soc.*, 2007, **129**, 8345.

18. K. A. Green, M. P. Cifuentes, T. C. Corkery, M. Samoc and M. G. Humphrey, *Angew. Chem. Int. Ed.*, 2009, **48**, 7867.
19. K. P. Zauner and M. Conrad, *Biotechnol. Prog.*, 2001, **17**, 553.
20. S. J. Langford and T. Yann, *J. Am. Chem. Soc.*, 2003, **125**, 11198.
21. S. J. Langford and T. Yann, *J. Am. Chem. Soc.*, 2003, **125**, 14951.
22. J. Andréasson, Y. Terazono, M. P. Eng, A. L. Moore, T. A. Moore and D. Gust, *Dyes Pigm.*, 2011, **89**, 284.
23. M. Schmittel and H. W. Lin, *Angew. Chem. Int. Ed.*, 2007, **46**, 893.
24. D. Jimenez, R. Martínez-Máñez, F. Sancenón, J. Soto, A. Benito and E. García-Breijo, *Eur. J. Inorg. Chem.*, 2005, 2393.
25. S. C. Wang, G. W. Men, L. Y. Zhao, Q. F. Hou and S. M. Jiang, *Sensors Actuators B*, 2010, **145**, 826.
26. R. Baron, O. Lioubashevski, E. Katz, T. Niazov and I. Willner, *Org. Biomol. Chem.*, 2006, **4**, 989.
27. R. Baron, O. Lioubashevski, E. Katz, T. Niazov and I. Willner, *Angew. Chem. Int. Ed.*, 2006, **45**, 1572.
28. P. Remon, M. Hammarson, S. M. Li, A. Kahnt, U. Pischel and J. Andréasson, *Chem. Eur. J.*, 2011, **17**, 6492.
29. C. J. Fang, Z. Zhu, W. Sun, C. H. Xu and C. H. Yan, *New J. Chem.*, 2007, **31**, 580.
30. U. Pischel, V. D. Uzunova, P. Remon and W. M. Nau, *Chem. Commun.*, 2010, **46**, 2635.
31. M. Zhou, F. Wang and S. J. Dong, *Electrochim. Acta*, 2011, **56**, 4112.
32. D. Margulies, G. Melman and A. Shanzer, *J. Am. Chem. Soc.*, 2006, **128**, 4865.
33. S. Nitahara, T. Akiyama, S. Inoue and S. Yamada, *J. Phys. Chem. B*, 2005, **109**, 3944.
34. S. Nitahara, N. Terasaki, T. Akiyama and S. Yamada, *Thin Solid Films*, 2006, **499**, 354.
35. K. Szaciłowski, W. Macyk and G. Stochel, *J. Am. Chem. Soc.*, 2006, **128**, 4550.
36. K. Szacilowski and W. Macyk, *Solid State Electron.*, 2006, **50**, 1649.
37. G. Hennrich, H. Sonnenschein and U. Resch-Genger, *J. Am. Chem. Soc.*, 1999, **121**, 5073.
38. G. Hennrich, W. Walther, U. Resch-Genger and H. Sonnenschein, *Inorg. Chem.*, 2001, **40**, 641.

CHAPTER 9
Double Input–Double Output Systems

9.1 Introduction

Molecular double input-double output logic systems allow us, among other things, to address number-handling. As mentioned in Chapter 3, this is a topic that has been close to each of us since childhood. Molecular numeracy is special because people become (and remain) numerate *via* mysterious, but molecular, processes in their brains. Furthermore, number-handling has been part of human civilization since at least 3200 BC. Figure 9.1[1] shows a document mentioning arithmetic from this time. Even older claims exist.[2] Even today, number-crunching remains the public perception of computing.

9.2 Half-adder

As discussed in section 3.4, an electronic half-adder circuit has two inputs and two output channels, which is the basis of number processing in most electronic computers. This is shown in Chapter 3 (Figure 3.6) in terms of electronic symbols. Addition needs AND logic for the carry digit and XOR for the sum digit.[3–5] In order to demonstrate the first molecular-scale version,[6,7] we selected the inputs to be Ca^{2+} and H^+, while the outputs were transmittance at 390 nm for the sum digit and fluorescence quantum yield for the carry digit.

Compound **1**[6] is an AND gate very much in the mould of the very first example of its kind,[8] but when combined in parallel with the compatible XOR gate **2**,[6,7] we have molecular-scale arithmetic for the first time. Molecular arithmetic was hampered until this point because the available AND and XOR gate molecules were not sufficiently compatible with each other to permit their parallel operation. The experimental truth table for the half-adder is shown in Table 9.1.

Monographs in Supramolecular Chemistry No. 12
Molecular Logic-based Computation
By A Prasanna de Silva
© The Royal Society of Chemistry 2013
Published by the Royal Society of Chemistry, www.rsc.org

Double Input–Double Output Systems 211

Figure 9.1 Rhind papyrus from 3200 BC, the first known document to record mathematical operations.
Image reproduced with permission from The British Museum, London.

Table 9.1 Truth table for the system $\mathbf{1+2}$.[a]

| $Input_1$ | $Input_2$ | $Output_1$ Carry Fluorescence[b] | $Output_2$ Sum Transmittance[c] |
H^+	Ca^{2+}		
0 (low, $10^{-9.5}$ M)	0 (low, $<10^{-9}$ M)	0 (low, 0.003)	0 (low, 8)
0 (low, $10^{-9.5}$ M)	1 (high, $10^{-2.3}$ M)	0 (low, 0.009)	1 (high, 40)
1 (high, 10^{-6} M)	0 (low, $<10^{-9}$ M)	0 (low, 0.005)	1 (high, 33)
1 (high, 10^{-6} M)	1 (high, $10^{-2.3}$ M)	1 (high, 0.10)	0 (low, 12)

[a]10^{-5} M each in water.
[b]Quantum yields, λ_{exc} 369 nm, λ_{em} 400, 419, 443 nm.
[c]Percentage at 390 nm, 10 cm optical path length.

1; R = 9-anthrylmethyl

2

The mechanism of action of XOR gate **2** was discussed at length in section 7.7 with the aid of Figure 7.10 so a summary will suffice here. Compound **2** is a push–pull system which has selective receptors at opposite terminals. The energy of the ICT excited state of **2** is perturbed in opposite directions when each receptor is blocked by its guest. Therefore, the absorption blue-shifts with Ca^{2+} and red-shifts with H^+. When both guests are present, the shifts cancel and the *status quo* is regained. Observation of the transmittance at 390 nm (see Figure 7.11) now gives the XOR truth table (Table 9.1).[6]

A related, but unimolecular, example from Qian's laboratory, **3**,[9] uses fluorescence and transmittance at an appropriate wavelength to achieve AND and XOR logic respectively. Instead of H^+ and Ca^{2+} inputs, **3** uses two doses of 5.2×10^{-3} M SDS (sodium dodecyl sulfate). Given that the critical micelle concentration of SDS is 8×10^{-3} M, it is only the accumulation of the two SDS doses that makes the micelle phase available for **3**.

3

The system **1** and **2** achieved a socially relevant outcome and was also notable because the parallelism was attained by deliberate mixing of gates. As shown in Chapter 3, the four rows of Table 9.1 show binary addition of 0 and 0 to give 00, 0 and 1 to give 01, 1 and 0 to give 01, 1 and 1 to give 10. In the more common decimal numbering system, these operations become the kindergarten classics: $0+0=0$; $0+1=1$; $1+0=1$ and $1+1=2$. This also establishes the ascending hierarchy of numbers 0, 1 and 2 from a molecular perspective. So we have intrinsic molecular arithmetic for the first time outside of people's brains. Suddenly, a little bit of chemistry concerned with esoteric molecules **1** and **2** takes on a wider societal significance. It is worth mentioning at this point that molecular abacuses represent another, but non-Boolean, approach to solving this problem.[10]

A Zn^{2+}, Hg^{2+}-driven unimolecular half-adder operating exclusively with absorbance outputs can be found in Akkaya's **4**.[11] Metal ion-free **4** has its absorption maximum at 697 nm, which is shifted to 673 nm with Zn^{2+} or to 674 nm with Hg^{2+}. Application of both metal ions shifts the band all the way to 630 nm (Figure 9.2). All these effects are due to decoupling of the nitrogen electron pairs from the rest of the push–pull π-electron system by binding to the metal ion. Absorbance measured at 623 and 663 nm, with the application of suitable thresholds, shows AND and XOR logic behaviour respectively. Zhu and Zhang's team[12] report similar observation of absorbance at 350 and 435 nm towards the same end during the stepwise one-electron oxidation of tetrathiafulvene (**5**) to take it to the radical cation and then to the dication. So these cases of logic reconfiguring could also have belonged in Chapters 8 and 13.

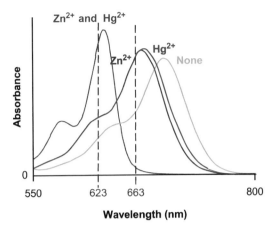

Figure 9.2 Ultraviolet–visible absorption spectra of **4** in the presence/absence of Hg^{2+} and Zn^{2+} inputs. The absorbance is monitored at the specific wavelengths of 623 and 663 nm in order to extract AND and XOR logic behaviour respectively.
Adapted from O. A. Bozdemir, R. Guliyev, O. Buyukcakir, S. Selcuk, S. Kolemen, G. Gulseren, T. Nalbantoglu, H. Boyaci and E. U. Akkaya, *J. Am. Chem. Soc.*, 2010, **132**, 8029 with permission from the American Chemical Society.

Conversely, Zhu's **6**[13] is a Zn^{2+}, Cd^{2+}-driven unimolecular half-adder operating exclusively with fluorescence outputs. Weak fluorescence is seen in **6**, due to PET from the bis(pyridylmethyl)amine receptor, which changes to a strong emission centred at 390 nm when an equivalent of Zn^{2+} or Cd^{2+} binds to the receptor. When both inputs are applied, the weaker receptor site of the bipyridyl unit is also occupied. A strong fluorescence peak at 449 nm arises as a result of the planarization of the bipyridyl unit. AND logic is therefore seen when emission is monitored at 449 nm and XOR logic is seen for the fluorescence signal monitored at 390 nm – another case of logic reconfiguring by observation at different wavelengths. Yan[14] describes a similar all-fluorescent case driven by two H^+ doses, but the need to apply opposite logic conventions to obtain AND and XOR logic is a weakness.

Accumulation of two H^+ doses as inputs also produces XOR and AND logic from the fluorescence outputs of **7** and **8** respectively.[15] Park, Yoon and their collaborators developed these cases within a microfluidic system. The aminomethylfluorescein derivative **7** is fluorescent only at intermediate pH values. Alkaline solutions permit PET from the amine units to the fluorescein.[16] On the other hand, the rhodamine derivative **8** is fluorescent only in acid.

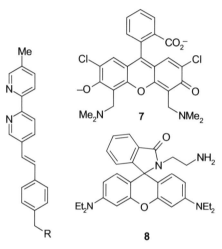

6; R = bis(2-pyridylmethyl)amino

As discussed in sections 7.2.5 and 7.7.1 under AND and XOR logic respectively,[17,18] Remacle et al.[17] show how one- and two-photon fluorescence can be considered together to produce half-adder action. These ideas are rather general,[18–21] so that all-optical half-adders abound. Additionally, this work stops tantalizingly short of a full-adder (see section 10.21.2). A unimolecular version is also available.[22]

Another all-optical case is shown by Andréasson et al.[23] They employ **9** and **10**, which use light at 1064 and 532 nm as inputs, to show both types of logic required for a half-adder. The absorption at 1000 nm (AND logic) as well as the fluorescence at 720 nm (XOR logic) are the outputs, making this system a nice addition to molecular arithmetic. AND logic gate **9** is a triad consisting of a porphyrin linked to a C_{60} electron acceptor and a dihydropyrene photochrome. XOR logic gate **10** is a molecular dyad containing a porphyrin but coupled to a dihydroindolizine photochrome instead.

How does the AND action arise? Since the two inputs are 1064 nm and 532 nm laser pulses, their simultaneous presence in an intervening third-harmonic generating crystal produces 355 nm light via frequency mixing; 355 nm is absorbed by the cyclophanediene unit of **9** and converted to the

Double Input–Double Output Systems

dihydropyrene via an electrocyclic ring closure to produce **11**. When the central porphrin unit in **11** is excited by a read laser at 650 nm, PET occurs from it to the C_{60} unit. Another thermal ET quickly follows from the dihydropyrene unit to the porphyrin radical cation. The resulting dihydropyrene radical cation spaced by the porphyrin ground state from the C_{60} radical anion lasts for microseconds. This long-lived state's absorbance at 1000 nm is the output of the AND gate. Clearly, the laser power is chosen so that the 1064 nm pulse alone will not produce 355 nm light within the third-harmonic generating crystal. The 532 nm pulse cannot do this.

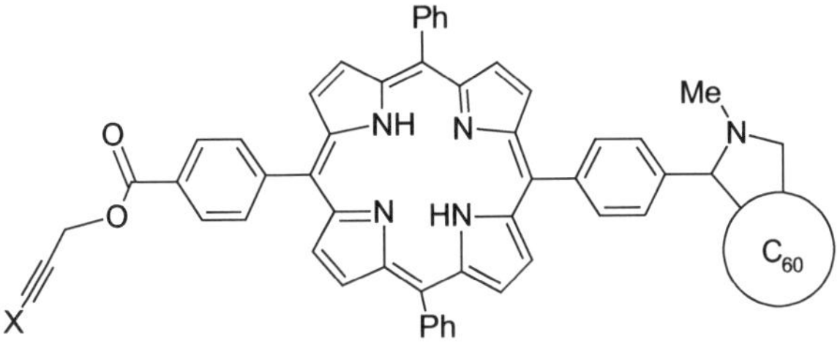

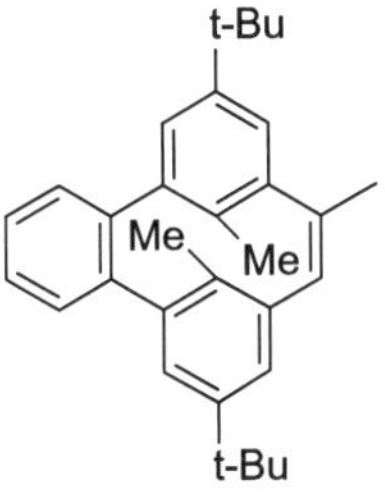

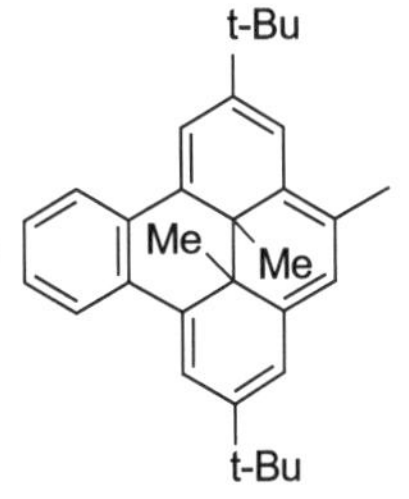

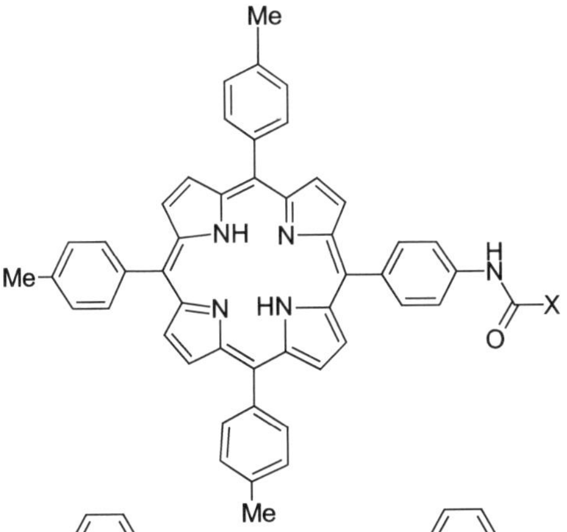

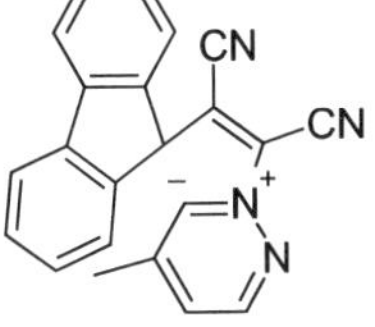

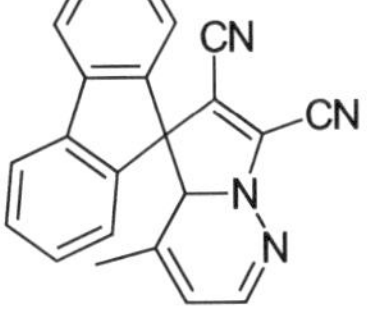

The starting state **9** can be recovered by 532 nm irradiation for absorption by the dihydropyrene unit in **11** and electrocyclic ring opening. Of course, running photochromic reactions in the presence of photoactive units of lower excited state energy generally causes EET and efficiency losses should be expected.

The action of the XOR gate **10** arises as follows. The read laser elicits a low level of fluorescence from the porphyrin unit since the betaine unit is a PET acceptor; 1064 nm illumination allows thermal ring closure of the betaine to the dihydroindolizine unit and produces **12**. With the betaine gone, **12** has a high level of fluorescence. Illumination at 532 nm produces the same result, but by photochemical ring closure. On the other hand, the two inputs of 1032 nm and 532 nm pulses mix together in the third-harmonic generating crystal to produce the 355 nm light which photoisomerizes the dihydroindolizine unit to the betaine to give **10** (the low intensity fluorescent state) again.

It is clear that the essential third-harmonic generating crystal brings in a bulk material component to what is otherwise a molecular-scale experiment. However, the all-optical nature of inputs, outputs and power supplies is to be applauded even though quantitative input–output homogeneity is not achieved. The features of multi-photon experiments[17,18] are also on show.

All-optical molecular logic gates can also be run inexpensively without laser sources.[24] Tian's **13** rotaxanated with α-cyclodextrin relies on 380 nm and 313 nm radiation inputs to isomerize *E*-azobenzene and *E*-stilbene units, respectively, to the corresponding *Z*-isomers. *Z*-isomers are less planar and allow poorer electron delocalization. The *ZZ* case shows the worst delocalization. Hence, its absorption maximum is the most blue-shifted and it shows significantly stronger absorption at 270 nm referenced to absorbance at the longer wavelength of 350 nm than *EE*-, *ZE*- and *EZ*- stereoisomers. The output measured in this way is 'high' only when the 380 nm and 313 nm radiation inputs have been applied, *i.e.* the all-optical AND logic is found.

13

The role of the rotaxane or 'bead-on-a-string' character of the complex between the **13** 'string' and the α-cyclodextrin 'bead' comes to the forefront when we note that fluorescence of either of the naphthalimide termini (at 520 nm or 395 nm) is enhanced when the 'bead' is close to it. Partial shielding of the naphthalimide excited states from solvent water could contribute to this enhancement.

The deviation from planarity of a Z-isomer mentioned above means that the α-cyclodextrin 'bead' cannot complex this unit. So the 'bead' moves to the other unit if it is of the E-form and enhances the emission of the neighbouring naphthalimide fluorophore. If both units are Z-, the 'bead' has no choice but to occupy the region around the central biphenyl moiety. Then, both fluorophores are poorly emissive. However, the EE-isomer allows the 'bead' to travel the length of the 'string', complexing any component along the way, but with an averaged position roughly at the centre. Again, both fluorophores are poorly emissive. Thus XOR logic arises if the output is chosen as the fluorescence of either naphthalimide terminal. No major complications, which could have arisen from EET in this bifluorophore case, are noted by the authors. Nevertheless, straightforward XOR logic might have arisen if the two naphthalimide termini were identical.

A receptor integrated within a rather electron-rich photochromic (see refs 125–140 in Chapter 7) **14** can be exploited by using Fe^{3+} as an input (which is also an oxidant) to demonstrate half-adder action by observing absorbance at selected wavelengths.[25] A light dose (of wavelength 365 nm) is the other crucial input. Reset is achieved by using a reductant, a complexant or 520 nm irradiation, as required by the specific situation. Siri also claims a similar instance.[26]

14

DNA-based enzymes can be adapted to produce half-adders.[27] The example hybridizes with a suitable oligonucleotide substrate and hydrolyses it at a weak point where a single ribonucleotide resides (point 'r' in Figure 9.3b and 9.3d).[28] Figure 9.3a is a schematic view of the AND gate, where the base sequence of the deoxyribonucleotide device is decided so that it folds into the enzyme motif plus two stem–loops. Notably, the enzyme heart is blocked from two sides. However, the supply of suitable oligonucleotides (input$_1$ and input$_2$) opens out these stem–loops so that the substrate can be bound and hydrolysed (Figure 9.3b). The enzymatic action is visualized by the emergence of fluorescence from a green fluorophore F once the part carrying the intramolecular quencher Q is hydrolysed off. The stem loop can be assembled from smaller components chosen from a library[29] (as described in section 5.3.1.3), so that this method becomes more flexible.

The XOR gate can be produced by running two complementary INHIBIT gates in parallel. If both have identical (red) fluorophores F', the joint emission can be read in a binary manner to result in an OR combination of the two INHIBIT gates. This can be seen, for instance, in the sum-of-products algebraic

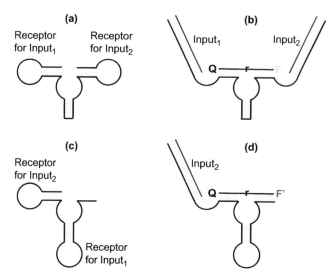

Figure 9.3 Two-input AND (a, b) and INHIBIT (c, d) logic based on deoxyribozymes. The grey rectangles are the Crick–Watson base-pairing regions. Ribonucleotide 'r' is the point of hydrolysis which separates the fluorophore F from the quencher Q.

expression for XOR logic in Chapter 3 (Figure 3.2). One of these INHIBIT gates is shown in Figure 9.3c. One side of the enzyme heart needs to be cleared by opening out the stem–loop upon hybridization with $input_2$ (Figure 9.3d). However, there is a second stem–loop which serves as a receptor for $input_1$. This stem–loop opens upon supply of $input_1$ so that the enzyme's heart is weakened. $Input_1$ is therefore the disabler.

Besides the mathematical significance, this leads to a diagnostic application where the presence of two oligonucleotides, one or none of them can be simply identified by green, red or no emissions, respectively.

Classical peptide-based enzymes also need to be mentioned here because there are cases which have been developed to show combined half-adder and half-subtractor action.[30,31] To continue with the DNA thread, DNA without enzyme action is also useful for constructing half-adders. Ghadiri's approach[32] to the AND gate component is summarized in Figure 9.4. The gate itself is composed of two strands of unequal length. Importantly, hybridization of the fluorophore-attached strand occurs at the central region of the long strand, which carries the quencher and the polymer bead. The latter is useful during protocols for washing away hybridized species. Input strands are designed so that parts of them bind to the exposed regions of the long strand of the device. Once bound to these 'toe-hold' regions, the remainder of the input strand encroaches further to push off part of the short strand of the device. However, the short strand hangs on with about half of its bases still paired with corresponding units on the longer strand. It is only when both $input_1$ and $input_2$

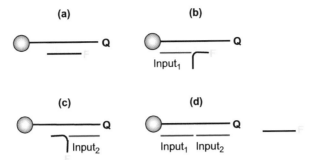

Figure 9.4 The two-input AND logic component of a half-adder based on strand displacement of oligonucleotides. The grey rectangles are the Crick–Watson base-pairing regions. The logic device is composed of a short strand carrying the fluorophore F, which is hybridized to the central part of a longer strand containing the quencher Q and attached to a polymer bead. The outer parts are the 'toe-hold' regions which the input strands can exploit, as described in the text.

attack simultaneously that the invasion is successful to the point that the short strand is pushed off completely. Once displaced into solution, the fluorophore on the short strand is free of the quencher's effects and strong emission is the result. Figure 9.4(a)–(d) illustrates this series of events. Fluorescence output therefore follows AND logic when driven by the two oligonucleotide inputs.

Notably, these two approaches use oligonucleotides as inputs and produce other oligonucleotides as output. In the former case[27] the fluorescence signal is simply a translation of the output oligonucleotide. Such a situation allows feeding of an output from one molecular device as the input of another. This[33] and related approaches[32,34–36] allow serial integration of gates, which will be discussed in Chapter 10.

Older cases of DNA molecule-based (but not molecular-scale, owing to the need for gel electrophoresis) bit addition are known.[37,38] Wild's spectral hole-burning experiments for arithmetic are even older,[39] but the logic operations were externally impressed on molecules rather than being inherently molecular properties.

9.3 Half-subtractor

Subtraction is as important as addition and is tackled similarly. The half-subtractor does this by putting two binary digits into a parallel array of INHIBIT and XOR gates where the borrow digit is outputted from the INHIBIT gate and the difference digit emerges from the XOR gate. See Figure 9.5 for an electronic symbolic representation.

In the first such molecular case,[7,40] Langford and Yann use **15** to achieve the XOR logic truth table (Table 9.2). This case was summarized in Chapter 7 (Table 7.12). Addition of H^+ (input$_1$) diprotonates **15** and shifts the absorption

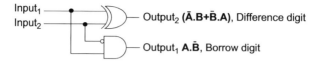

Figure 9.5 The array of XOR and INHIBIT gates corresponding to the half-subtractor.

Table 9.2 Truth table for **15**.[a]

Input₁ H⁺	Input₂ t-BuO⁻	Output₁ Borrow Fluorescence[b]	Output₂ Difference Transmittance[c]
0 (none)	0 (none)	0 (low, 5)	0 (low, 0.25)
0 (none)	1 (high, 0.1 M)	1 (high, 67)	1 (high, 47)
1 (high, 0.1 M)	0 (none)	0 (low, 0)	1 (high, 40)
1 (high, 0.1 M)	1 (high, 0.1 M)	0 (low, 4)	0 (low, 0.25)

[a]10⁻⁵ M in dimethylformamide.
[b]Relative intensities, λ_{exc} 435 nm, λ_{em} 637 nm.
[c]Percentage at 417 nm, 1 cm optical path length.

band to 447 nm (from the original value of 417 nm). The t-BuO⁻ (input₂) produces the dianionic form of **15** which shifts the absorption band to 437 nm. Such behaviour is not unlike a pH indicator with an ICT (internal charge transfer) excited state.[41] In either case, the transmittance at 417 nm increases sharply, compared with the 'low' value found originally. Of course, the application of equimolar t-BuO⁻ and H⁺ leads to neutralization and the original low transmittance is unaffected. This neutralization move was first used by Balzani and Stoddart's team, which also included Langford.[42]

15

In a subsequent correction,[43] **15** is exploited to give INHIBIT logic with the same inputs, but with fluorescence intensity at 637 nm (while exciting at 435 nm) chosen as the output. This is the normal fluorescence of porphyrins,[44] and 637 nm is also where the dianionic form of **15** happens to maximize its emission. Also, 435 nm is close to where this form absorbs the best. So, t-BuO⁻ (input₂) produces fluorescence switching 'on'. This combination of excitation and emission observation wavelengths does not elicit a strong fluorescence response from neutral **15**, *i.e.* the output is 'low' when both inputs are absent. The same 'low' output is naturally found when both inputs are present in equimolar amounts owing to their neutralization.[42] The remaining input condition that needs to be examined is the presence of H^+ (input$_1$). The diprotonated form of **15** is also moderately excited by 435 nm irradiation. Furthermore, its weak emission is very poorly collected by observing at 637 nm. So the measured output is switched 'off'.

The parallel operation of XOR and INHIBIT logic within **15** can be easily arranged by observing two outputs by using two different observation techniques – another case of logic reconfiguring. That is why it featured briefly in section 8.6. It is notable that the half-subtractor arises from commercially available **15**, acid, base and nothing else.

Structure **16**[45] is an ICT fluorophore emitting at 660 nm in CH_3CN. The dimethylamino unit is the electron pushing component in this push–pull π-electron system. The addition of t-BuO$^-$ deprotonates the phenol group and the emission is switched 'off' due to PET across the virtual spacer. Upon addition of H^+, **16** gives a blue-shifted emission at 565 nm due to protonation of the dimethylamino unit and the subsequent reduced ICT nature of the π-system. When equimolar t-BuO$^-$ and H^+ are added (both inputs 'high'), nothing happens due to neutralization[42] and the original state of 660 nm is preserved. If we choose output$_1$ emission at 565 nm, this corresponds to INHIBIT logic, with t-BuO$^-$ as the disabling input.

16

If we reconfigure **16** by choosing output$_1$ emission at 660 nm, this corresponds to XNOR logic. The latter can be quickly converted to XOR logic by using a negative logic convention for the output signals, *i.e.* by taking 'high' output signals as logic state '0' and 'low' signals as logic state '1'. The negative logic convention is sometimes invoked in semiconductor technology too. This book sticks to positive logic convention for both inputs and outputs, *i.e.* 'high' signals are taken as logic state '1' and 'low' states as logic state '0' unless otherwise specifically noted. Nevertheless, this case of **16** could have also belonged in sections 8.7.1 and 13.3.1 owing to the wavelength-based reconfigurability.

The acid–base annihilation method introduced by Balzani and Stoddart for designing XOR logic gates[42] can also be applied to an equimolar mixture of 'fluorophore–spacer–receptor' system **17** and Cu^{2+}.[46] In this instance, the absorbance at 255 nm drops seriously at pH 7 (when it exists as **17.H$^+$**) but not when an equivalent of H$^+$ or OH$^-$ is added. Naturally, when an equivalent each of H$^+$ and OH$^-$ is added, the pH value remains at 7 and the absorbance at 255 nm remains low. This is the XOR gate component and suggests aggregation of electroneutral **17$_2$.Cu^{2+}** as the underlying mechanism. Since **17** is a PET system with an amine donor, it is clear that protonation prevents binding to Cu^{2+} besides stopping PET and releasing fluorescence. In the presence of OH$^-$, Cu^{2+} is precipitated as $Cu(OH)_2$ and PET within the freed **17** switches fluorescence 'off'. At pH 7, **17$_2$.Cu^{2+}** is non-fluorescent because of EET from the fluorophore to the Cu^{2+} centre as well as PET. Here is the INHIBIT gate component. Operation of the two gate components in parallel results in the half-subtractor.

Another half-subtractor arises from a mixture of 'fluorophore–spacer–receptor' system **18** and Zn^{2+}, which is driven by inputs of H$^+$ and Et$_3$N in acetonitrile.[47] A difference from the previous cases is that the naphthalene monomer and excimer (arising from complex **18$_2$.Zn^{2+}$_2$**) emissions are the outputs from two complementary INHIBIT gates. The simultaneous observation of these INHIBIT gates through a suitable filter produces XOR behaviour according to the sum-of-products description. Thus we have half-subtractor action arising from fluorescence outputs alone when the XOR gate and either of the INHIBIT gates are considered in parallel. Depending on which of the latter is chosen, the directionality of the half-subtractor will change, *i.e.* input$_1$–input$_2$ or input$_2$–input$_1$ operation will be performed.

17; R = 9-anthrylmethyl   **18**; R = HN–CH$_2$CH$_2$–NH–(naphthyl)

A bidirectional half-subtractor is explicitly constructed by Credi[48] by modifying his previous work on XOR logic.[42] Compounds **19** and **20** are also described under XOR logic in Chapter 7, where input$_1$ and input$_2$ are H$^+$ and Bu$_3$N. These, with a small alteration in the case of **20**, play the starring roles. The XOR and complementary INHIBIT gates arise from the observation of emission at 340, 428 and 666 nm respectively when excited at an isosbestic wavelength of 264 nm. Careful observation of fluorescence spectra in several solvent systems allows distinction of XOR gate output = 0 state arising from the input$_1$ = 0, input$_2$ = 0 and input$_1$ = 1, input$_2$ = 1 states owing to significant interference of accumulated salts (especially the acid anion $CF_3SO_3^-$) with pseudorotaxane formation. Such an ability to deduce the input state from the output state can be interpreted as a reversible logic operation – something rarely seen in molecular[49] or semiconductor[3-5] computing.

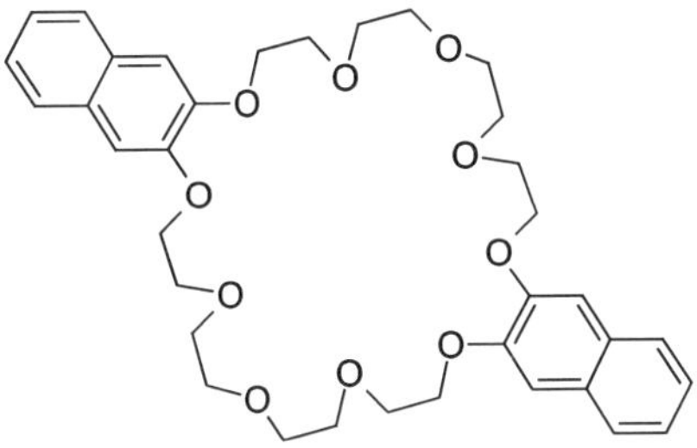

19 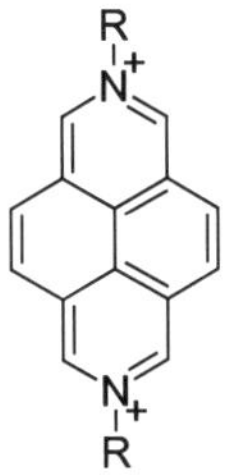 **20; R = n-decyl**

A generalization that emerges is that many compounds whose absorption and emission spectra are shifted by acid and by base, *i.e.* possessing basic and acidic groups, will exhibit half-subtractor action if operating wavelengths, output signal thresholds and logic conventions are carefully chosen. Several additional studies[49-54] support this generalization, including those involving commercially available compounds.

Examples of half-subtractors arising from input pairs that do not annihilate each other are hard to find. Some Ca^{2+}, H^+-driven half-subtractors can now be discerned in some of our older work on logic reconfiguring[55] which brought out XOR and INHIBIT gates, though we failed to recognize it at the time. There is also an example due to Jiang, where Zn^{2+} and a 254 nm light dose are applied to a Schiff base photochromic.[56] Half-subtractors can also be made from enzyme cascades but these will be taken up in Chapter 10 under the topic of combined half-adders and half-subtractors.[30,31]

9.4 1:2 Demultiplexer

A parallel integration of AND and INHIBIT gates leads to a 1:2 demultiplexer which would rightfully belong in this chapter. However, it will be convenient to discuss this device alongside its counterpart, the 2:1 multiplexer, in section 10.15.

9.5 Magnitude Comparator

Magnitude comparators can determine which of two bits is the larger. XOR gates are simpler (identity) comparators, which can only say whether two bits are equal or different. The truth table of a magnitude comparator (Table 9.3)

Table 9.3 Truth table for **21**.[a]

Input$_1$ OH$^-$	Input$_2$ H$^+$	Output$_1$ Greater Flu$_{513}$[b]	Output$_2$ Equal Flu$_{423}$[b]
0 (none)	0 (none)	0 (low, 9)	1 (high, 63)
0 (none)	1 (high, 3 × 10^{-3} M)	0 (low, 3)	0 (low, 6)
1 (high, 3 × 10^{-3} M)	0 (none)	1 (high, 100)	0 (low, 2)
1 (high, 3 × 10^{-3} M)	1 (high, 3 × 10^{-3} M)	0 (high, 9)	1 (high, 63)

[a]10^{-5} M in dimethylformamide.
[b]Relative intensities, λ_{exc} 335 nm.

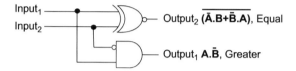

Figure 9.6 The array of XNOR and INHIBIT gates corresponding to the magnitude comparator.

shows how the bit comparison is made. If input$_1$ > input$_2$, the greater output$_1$ is true (binary 1) and the equal output$_2$ is false (binary 0). If the opposite situation of input$_1$ < input$_2$ is being examined, the greater output$_1$ is false and the equal output$_2$ is also false. A molecular implementation is Liu's **21**.[57] In its neutral form, the ICT system **21** emits strongly at 423 nm. Monoprotonated **21** is essentially non-emissive across the wavelength range. Monodeprotonated **21** emits intensely at 513 nm. The truth table (Table 9.3) corresponds to a parallel wiring of INH and XNOR gates, as shown in Figure 9.6 in terms of electronic symbols.

21

9.6 Reversible Logic

During the discussion of bidirectional half-subtractors[48] in section 9.3, the ability uniquely to deduce the input state from the output state (or *vice versa*) was mentioned as an illustration of reversible logic.[49] The rare behaviour of Pérez-Inestrosa's dual-emissive systems[58] was discussed under INHIBIT logic in section 7.6. Dual emission opens up the possibility of logic reconfiguring by changing the wavelength of observation (Chapters 6, 8 and 13). Thus, extra output channels can be made available so that the mapping between input and output states becomes more unique. An illustration is provided by the related structure **22** which has a protonatable N-oxide unit and a phenol unit which can be deprotonated. Charge transfer (CT) emission at 500 nm is seen when a substantial difference exists between the donor and acceptor abilities of the two π-systems within **22**. Phenolate–isoquinoline N-oxide or phenol-protonated N-oxide pairs can achieve this situation. This corresponds to H^+, OH^--driven XOR logic. Importantly, the CT emission bandwidth is significantly different in these two situations, *i.e.* the 'high' H^+, 'low' OH^- state displays a 'low' emission at 450 nm whereas the 'low' H^+, 'high' OH^- state produces a 'high' emission instead. Such distinguishability of the two 'high' fluorescence output signals at 500 nm by exploiting a second wavelength channel is the reversible logic aspect. However, such unique mapping of each 'high' output signal onto an input set is not extensible to each 'low' output signal of **22**. The lead from this work[58] has been developed further by the teams of Credi,[48] Gust,[59] and Pischel and Andréasson[60] for small molecules and by Turberfield for DNA hairpins.[61]

22

References

1. The British Museum, Great Russell Street, London.
2. P. S. Rudman, *How Mathematics Happened: The First 50,000 Years*, Prometheus Books, New York, 2007.
3. A. P. Malvino and J. A. Brown, *Digital Computer Electronics*, Glencoe, Lake Forest, 3rd edn, 1993.
4. J. Millman and A. Grabel, *Microelectronics*, McGraw-Hill, London, 1988.
5. A. L. Sedra and K. C. Smith, *Microelectronic Circuits*, Oxford University Press, Oxford, 5th edn, 2003.
6. A. P. de Silva and N. D. McClenaghan, *J. Am. Chem. Soc.*, 2000, **122**, 3965.
7. U. Pischel, *Angew. Chem. Int. Ed.*, 2007, **46**, 4026.
8. A. P. de Silva, H. Q. N. Gunaratne and C. P. McCoy, *Nature*, 1993, **364**, 42.
9. J. H. Qian, Y. F. Xu, X. H. Qian and S. Y. Zhang, *J. Photochem. Photobiol. A: Chem.*, 2009, **207**, 181.
10. P. R. Ashton, R. Ballardini, V. Balzani, A. Credi, K. R. Dress, E. Ishow, C. J. Kleverlaan, O. Kocian, J. A. Preece, N. Spencer, J. F. Stoddart, M. Venturi and S. Wenger, *Chem. Eur. J.*, 2000, **6**, 3558.

11. O. A. Bozdemir, R. Guliyev, O. Buyukcakir, S. Selcuk, S. Kolemen, G. Gulseren, T. Nalbantoglu, H. Boyaci and E. U. Akkaya, *J. Am. Chem. Soc.*, 2010, **132**, 8029.
12. Y. C. Zhou, H. Wu, L. Qu, D. Q. Zhang and D. B. Zhu, *J. Phys. Chem. B*, 2006, **110**, 15676.
13. L. Zhang, W. A. Whitfield and L. Zhu, *Chem. Commun.*, 2008, 1880.
14. W. Sun, C. Zhou, C. H. Xu, C. J. Fang, C. Zhang, Z. X. Li and C. H. Yan, *Chem. Eur. J.*, 2008, **14**, 6342.
15. S. Z. Kou, H. N. Lee, D. van Noort, K. M. K. Swamy, S. H. Kim, J. H. Soh, K. M. Lee, S. W. Nam, J. Yoon and S. Park, *Angew. Chem. Int. Ed.*, 2008, **47**, 872.
16. D. F. H. Wallach and D. L. Steck, *Anal. Chem.*, 1963, **35**, 1035.
17. F. Remacle, S. Speiser and R. D. Levine, *J. Phys. Chem. B*, 2001, **105**, 5589.
18. A. S. Lukas, P. J. Bushard and M. R. Wasielewski, *J. Am. Chem. Soc.*, 2001, **123**, 2440.
19. F. Remacle, R. Weinkauf and R. D. Levine, *J. Phys. Chem. A*, 2006, **110**, 177.
20. E. K. L. Yeow and R. P. Steer, *Phys. Chem. Chem. Phys.*, 2003, **5**, 97.
21. E. K. L. Yeow and R. P. Steer, *Chem. Phys. Lett.*, 2003, **377**, 391.
22. O. Kuznetz and S. Speiser, *J. Lumin.*, 2009, **129**, 1415.
23. J. Andréasson, G. Kodis, Y. Terazono, P. A. Lidell, S. Bandyopadhyay, R. H. Mitchell, T. A. Moore, A. L. Moore and D. Gust, *J. Am. Chem. Soc.*, 2004, **126**, 15926.
24. D. H. Qu, Q. C. Wang and H. Tian, *Angew. Chem. Int. Ed. Engl.*, 2005, **44**, 5296.
25. X. F. Guo, D. Q. Zhang, G. X. Zhang and D. B. Zhu, *J. Phys. Chem. B*, 2004, **108**, 11942.
26. E. A. Shilova, A. Heynderickx and O. Siri, *J. Org. Chem.*, 2010, 1855.
27. M. N. Stojanovic and D. Stefanovic, *J. Am. Chem. Soc.*, 2003, **125**, 6673.
28. R. R. Breaker and G. F. Joyce, *Chem. Biol.*, 1995, **2**, 655.
29. J. Elbaz, O. Lioubashevski, F. Wang, F. Remacle, R. D. Levine and I. Willner, *Nature Nanotechnol.*, 2010, **5**, 417.
30. R. Baron, O. Lioubashevski, E. Katz, T. Niazov and I. Willner, *Angew. Chem. Int. Ed.*, 2006, **45**, 1572.
31. R. Baron, O. Lioubashevski, E. Katz, T. Niazov and I. Willner, *J. Phys. Chem. A*, 2006, **110**, 8548.
32. N. H. Voelcker, K. M. Guckian, A. Saghatelian and M. R. Ghadiri, *Small*, 2008, **4**, 427.
33. H. Lederman, J. Macdonald, D. Stefanovic and M. N. Stojanovic, *Biochemistry*, 2006, **45**, 1194.
34. G. Seelig, D. Soloveichik, D. Y. Zhang and E. Winfree, *Science*, 2006, **314**, 1565.
35. B. M. Frezza, S. L. Cockroft and M. R. Ghadiri, *J. Am. Chem. Soc.*, 2007, **129**, 14875.
36. A. Lake, S. Shang and D. M. Kolpashchikov, *Angew. Chem. Int. Ed.*, 2010, **49**, 4459.
37. F. Guarnieri, M. Fliss and C. Bancroft, *Science*, 1996, **273**, 220.

38. B. Yurke, A. P. Mills and S. L. Cheng, *Biosystems*, 1999, **52**, 165.
39. U. P. Wild, S. Bernet, B. Kohler and A. Renn, *Pure Appl. Chem.*, 1992, **64**, 1335.
40. S. J. Langford and T. Yann, *J. Am. Chem. Soc.*, 2003, **125**, 11198.
41. E. Bishop (ed.), *Indicators*, Pergamon, Oxford, 1972.
42. A. Credi, V. Balzani, S. J. Langford and J. F. Stoddart, *J. Am. Chem. Soc.*, 1997, **119**, 2679.
43. S. J. Langford and T. Yann, *J. Am. Chem. Soc.*, 2003, **125**, 14951.
44. K. Kalyanasundaram, *Photochemistry of Polypyridine and Porphyrin Complexes*, Academic, London, 1992.
45. A. Coskun, E. Deniz and E. U. Akkaya, *Org. Lett.*, 2005, **7**, 5187.
46. G. Q. Zong and G. X. Lu, *Tetrahedron Lett.*, 2008, **49**, 5676.
47. M. Vazquez Lopez, M. E. Vazquez, C. Gomez-Reino, R. Pedrido and M. R. Bermejo, *New J. Chem.*, 2008, **32**, 1473.
48. M. Semeraro and A. Credi, *J. Phys. Chem. C*, 2010, **114**, 3209.
49. E. Pérez-Inestrosa, J.-M. Montenegro, D. Collado, R. Suau and J. Casado, *J. Phys. Chem. C*, 2007, **111**, 6904.
50. M. Suresh, D. A. Jose and A. Das, *Org. Lett.*, 2007, **9**, 441.
51. M. Suresh, D. A. Jose and A. Das, *Tetrahedron Lett.*, 2007, **48**, 8205.
52. D. Margulies, G. Melman and A. Shanzer, *Nature Mater.*, 2005, **4**, 768.
53. M. Suresh, P. Kar and A. Das, *Inorg. Chim. Acta*, 2010, **363**, 2881.
54. U. Pischel and B. Heller, *New J. Chem.*, 2008, **32**, 395.
55. A. P. de Silva and N. D. McClenaghan, *Chem. Eur. J.*, 2002, **8**, 4935.
56. S. C. Wang, L. B. Zang, L. Y. Zhao, X. L. Wang, Q. F. Hou and S. M. Jiang, *Spectrochim. Acta A*, 2010, **77**, 226.
57. Y. Liu, W. Jiang, H. Y. Zhang and C. J. Li, *J. Phys. Chem. B*, 2006, **110**, 14231.
58. J. M. Montenegro, E. Pérez-Inestrosa, D. Collado, Y. Vida and R. Suau, *Org. Lett.*, 2004, **6**, 2353.
59. J. Andréasson, U. Pischel, S. D. Straight, T. A. Moore, A. L. Moore and D. Gust, *J. Am. Chem. Soc.*, 2011, **133**, 11641.
60. P. Remon, M. Hammarson, S. M. Li, A. Kahnt, U. Pischel and J. Andréasson, *Chem. Eur. J.*, 2011, **17**, 6492.
61. A. J. Genot, J. Bath and A. J. Turberfield, *J. Am. Chem. Soc.*, 2011, **133**, 20080.

CHAPTER 10
More Complex Systems

10.1 Introduction

More complex molecular logic systems will arise when integration of simpler components is arranged. Parallel integration is not too difficult to achieve chemically because several components can be simply added into the same solution so that they share the experimental space at the same time. For instance, the earliest half-adder described in section 9.2 involved the parallel operation of carefully matched AND and XOR gates.[1]

Serial integration is harder because many molecular logic systems use inputs and outputs which are distinguishable. This is input–output heterogeneity. Hence, the output of one logic device cannot be fed as input into another. On the other hand, electronic computing thrives on quantitative input–output homogeneity. The inputs and outputs are not only both electronic but they have the same values of 'low' and 'high' voltages. Indeed, even a 2-input AND logic gate of the electronic variety consists of several wires, transistors and resistor(s) involving serial connections. Arrays involving diode–diode logic were given in Figures 3.4 and 3.5. This success in the integration of large numbers of electronic components has led many engineers and engineering-oriented scientists to demand the same quantitative input–output homogeneity of molecular systems. Actually, this is not necessary for several applications of molecular logic which are described in Chapter 14. Also, this would lead to feedback and short-circuiting unless conduits for the signals are separately arranged to/from the molecule(s). Bulk metal wires can do the job for electrical signals and optical fibres serve similarly for optical signals but such conduits are difficult to arrange for other signals, especially in a molecular context. The skeleton of a polypeptide has been proposed for charge migration, for example.[2] Conduits, whether they are for electrical or other signals, add to the bulk of the device. As will be demonstrated in later sections, conduits can be avoided

---

Monographs in Supramolecular Chemistry No. 12
Molecular Logic-based Computation
By A Prasanna de Silva
© The Royal Society of Chemistry 2013
Published by the Royal Society of Chemistry, www.rsc.org

in many molecular systems displaying chemically worthwhile levels of logic integration.

Serial integration has been tackled in five general ways up to now.

a) Functional integration. Molecules containing multiple supramolecular interactions and switching mechanisms are deliberately selected as devices. Depending on the output being observed, a relatively complicated input–output truth table can arise. As described in Chapter 3, this input–output function can be presented as a 'sum-of-products' expression and minimized according to Boolean algebraic methods.[3–9] Thus, the original logic behaviour can be functionally represented by an array of NOT, AND and OR gates. Physical linking of these component gates is not really involved. This approach produces examples of many logic types of varying levels of complexity.[10–13] These are available in many sections in Chapter 7 and in subsequent chapters, especially this one. Occasionally, each supramolecular phenomenon can represent a logic operation. The first example of functional integration,[14] which was discussed in section 7.4 as a rationally constructed NOT OR gate, is of this kind.

b) Light as a gate-to-gate linker. The direct tactic would be to use the light output from one fluorescent gate to serve as the input to the next. An early example is a close approach to a full-adder,[15] whose linking of components was outlined in section 7.7.1. Another tactic is to use a switchable absorber to modulate the light emission from a fluorophore. Although the intrinsic gate property of a fluorophore is simple, rather complex logic behaviour arises overall. This is because the absorber is controllable by several inputs and because different fluorophores produce different spectral overlaps with the absorber states. Examples of this type[16–20] will be described in section 10.13. Work described in section 10.20 will be relevant too.

c) H^+ as a gate-to-gate linker. Classical pH sensors[21] show how proton transfer, albeit involving water species, can switch absorbance signals. When such cases are coupled to acids/bases controlled by several inputs, interesting logic devices arise. For instance, some of the complex logic behaviour of **1**[22] can be reflected onto the absorbance at 556 nm (due to the protonated **2**) owing to proton transfer.[23,24] However, the intrinsic memory aspect of photochromics like **1** causes other issues which will be mentioned in Chapter 11. Replacement of **2** by **3** in this experiment produces an output in the voltammetric current at –0.6 V (*vs.* sce) instead of the absorbance signal.[25] A closely related case where a relative of **1** was coupled with a fluorescent PET sensor[26] was discussed in section 7.11.[27]

d) Cascading enzymes. Enzyme cascades are common in biochemistry[28] where a substrate is passed from one enzyme to another after each transformation. Logic devices of this kind developed by Willner, Katz and their collaborators are available in sections 10.20[29] and 10.21.1.[30,31]

e) Cascading oligonucleotides. Winfree's laboratory[32-34] uses DNA strand displacement to achieve rather complex logic arrays. The fact that oligonucleotides of different sequences and lengths serve as inputs and outputs as well as the logic devices is a notable feature. A useful solid-phase variant is also available.[35] This approach[32-36] was outlined in section 7.6.1 for simple cases.

10.2 Three-input AND

The first deliberately constructed molecular version of a 3-input AND logic gate (electronic representation: Figure 10.1) was **4** from Guo, Zhang and Zhu.[37] Its complexity was due to the presence of a photochromic spiropyran moiety and a perylenediimide unit. Connecting two photoactive switches opens up the danger of EET destroying the photoactivity of the higher-energy excited state. For example, azobenzene–porphyrin conjugates do not show significant E–Z photoisomerization of the azobenzene unit.[38] Guo et al.[37] have managed to overcome this situation with **4**, where the usual spiropyran photochromic ring opening to merocyanine[39] can be achieved by ultraviolet (uv) irradiation, in spite of the presence of the perylenediimide. The merocyanine can complex with Fe^{3+} (see refs 125–140 in Chapter 7), moving its absorption spectrum to higher energies and stopping EET acceptance from the perylenediimide fluorophore. The aniline moiety of the spiropyran can bind to H^+ and stop its PET donor action towards the perylenediimide fluorophore. Indeed, the fluorescence is switched 'on' only if a dose of uv light, H^+ and Fe^{3+} are all supplied in THF (Table 10.1). This constitutes a 3-input AND logic gate. Interestingly, the authors do not report irreversibilities arising from any Fe^{3+}-induced oxidation of the aniline moieties, and Fe^{3+}-induced oxidative electroconversion of related compounds with similar oxidation potentials is known.[40]

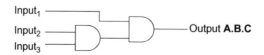

Figure 10.1 Three-input AND logic. Input$_1$, input$_2$, input$_3$,... are symbolized by variables A, B, C,... for algebraic purposes throughout.

More Complex Systems

Table 10.1 Truth table for **4**.[a]

Input$_1$ hv dose$_{365}$	Input$_2$ H$^+$	Input$_3$ Fe^{3+}	Output Fluorescence[b]
none	none	none	low (0.011)
none	high (2.5 × 10^{-3} M)	none	low (0.011)
high	none	none	low (0.011)
high	high (2.5 × 10^{-3} M)	none	low (0.011)
none	none	high (4.5 × 10^{-4} M)	low (0.034)
none	high (2.5 × 10^{-3} M)	high (4.5 × 10^{-4} M)	low (0.062)
high	none	high (4.5 × 10^{-4} M)	low (0.062)
high	high (2.5 × 10^{-3} M)	high (4.5 × 10^{-4} M)	high (0.352)

[a] 2 × 10^{-5} M in tetrahydrofuran. λ_{exc} 480 nm, λ_{em} 560 nm.
[b] Quantum yields.

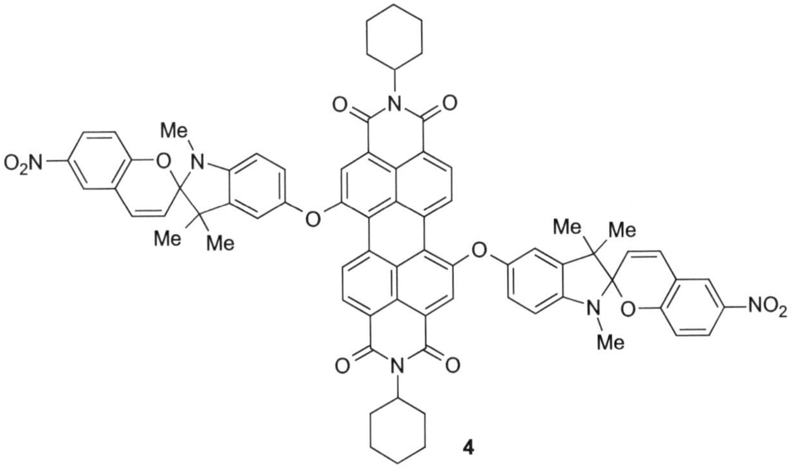

Employing all-chemical inputs to 3-input AND gates produces 'lab-on-a-molecule' systems which are of value in medical diagnostics.[41] Because of their application value, these will be analysed in Chapter 14.

Compound **5**[42] follows the previous H$^+$, Na$^+$-driven fluorescent AND cases.[43,44] It employs a 'receptor$_1$–spacer$_1$–fluorophore–spacer$_2$–receptor$_2$–spacer$_2$–fluorophore–spacer$_1$–receptor$_1$' format which allows redundancy to assist PET. Three PET processes can be identified in **5**. When the amine is bound to H$^+$, it is the binding of the second Na$^+$ to the benzocrown receptors that switches 'on' fluorescence, *i.e.* **5** shows H$^+$, (Na$^+$)$_2$-driven AND logic. Hg^{2+}, Zn^{2+}, Ca^{2+}-driven AND logic is seen in the fluorescence output of Akkaya's **6**,[45] which operates by a combination of PET and ICT switching mechanisms.

Similarly, Na$^+$, H$^+$ and PO$_4^{3-}$ can be discerned as the three inputs presented to AND gate **7**[46] which uses PET-active benzo-15-crown-5 ether and polyamine receptors to bind Na$^+$ and H$^+$ respectively. Indeed, the polyamine is partially protonated at neutral pH. The partially protonated polyamine then binds HPO$_4^{2-}$ so that PET is suppressed (as discussed in section 7.2.1.2).[47] The PET pathway originating in the benzocrown ether is suppressed by Na^+ binding. It is important to choose H^+ input levels to be 10^{-8} M ('high', where HPO_4^{2-} is dominant) and 10^{-11} M ('low', where PO_4^{3-} rules the roost) so that a fluorescence enhancement factor (at 428 nm) of 1.5 is found for the 'high' output state as compared to the highest 'low' state.

5

6; R = bis(2-pyridylmethyl)amino

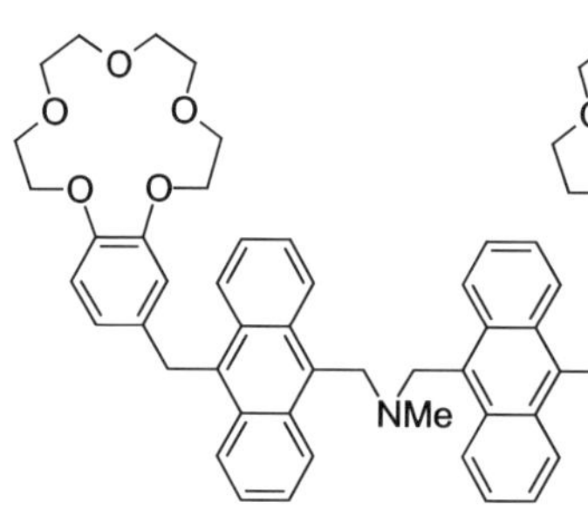

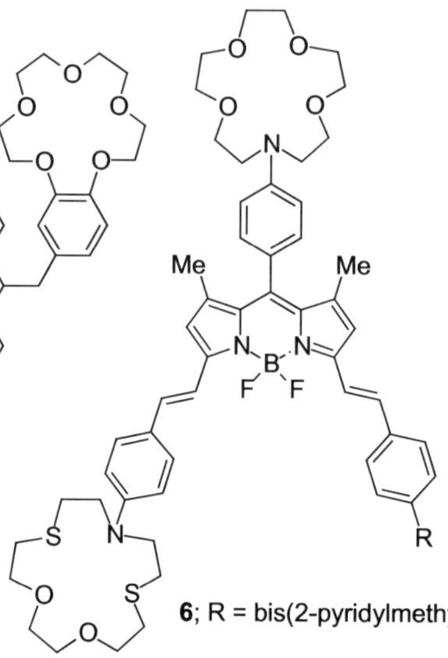

Lanthanides like Eu^{3+} absorb light poorly and require complexation to π-systems, *e.g.* aromatic β-diketones,[48] for a chance for strong luminescence to occur.[49,50] Addition of tetracycline (**8**) satisfies this condition but two water molecules in the first coordination sphere of Eu^{3+} create Born–Oppenheimer holes to convert electronic energy to vibrational quanta.[51–53] Addition of chelating anions such as citrate can displace these two water molecules to result in large enhancements of long-lived luminescence by factors as large as 22,[54] *i.e.* **8**, (citrate)$_2$-driven AND logic can be realized.

7; R = $NH(CH_2)_3N[(CH_2)_3NH_2]_2$

8

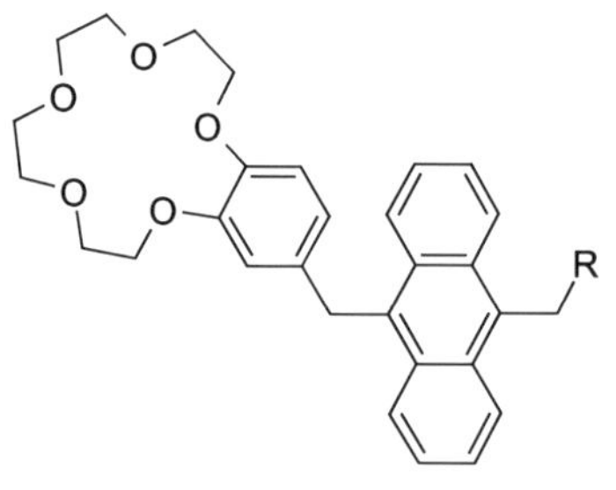

Enzyme cascades are also good candidates for multi-input AND operations.[55] Under the appropriate conditions, maltose phosphorylase produces

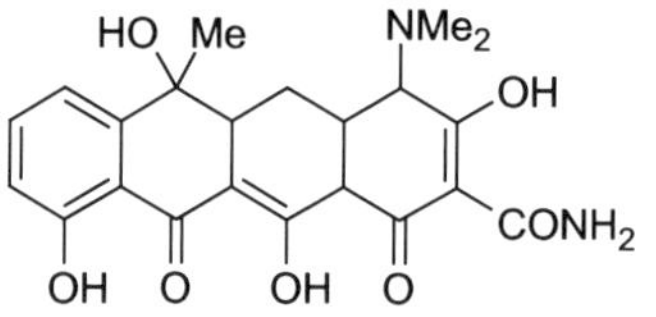

More Complex Systems

glucose and glucose phosphate from maltose and phosphate inputs. The glucose so-produced serves as input to glucose dehydrogenase along with NAD^+. Monitoring the NADH via its absorbance completes the experimental protocol. Attention is also paid to minimizing the noise accumulation in the output[55,56] of this gate array.

10.2.1 Three-input AND with Mixed Input Types

For an example of 3-input AND logic possessing a combination of connected/separate input sets, we turn to Shinkai's **9**. This is a heterobireceptor system which, once supplied with Zn^{2+}, selectively signals the presence of galacturonic acid[57,58] by employing an aminomethylphenyl boronic acid to catch a diol unit and a phenanthroline-bound Zn^{2+} centre to capture the carboxylate group. However **9** has only one PET process (going from the aminomethylphenyl boronic acid to the phenanthroline fluorophore) so that the selectivity of detection is not high. Nevertheless, galacturonic acid and Zn^{2+} produces FE = 3.5.

10.3 Three-input OR

Ion input and fluorescence output feature in the 3- (and higher-) input OR gates emerging from Bharadwaj's laboratory.[59–62] Figure 10.2 shows the corresponding electronic representation. Compound **10**[59,60] is notable because various transition metal ions serve to switch 'on' fluorescence by more or less similar amounts.[59,60] For instance, 10^{-2} M Zn^{2+}, Ni^{2+} or Cu^{2+} produces

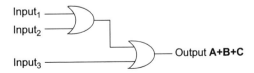

Figure 10.2 Three-input OR logic.

fluorescence quantum yields of 0.33, 0.23 and 0.23 respectively for **10** in THF solution. The surprise in this system is that transition metal ions are acting contrary to their normal behaviour. Open-shell metal ions of this type have a history of quenching fluorescence very efficiently by several mechanisms: heavy atom effects, PET, electronic energy transfers and paramagnetic effects.[63] However, the number of exceptions to this generalization is growing.[64,65] Interestingly, these metal ions are electrochemically silent when embedded in **10**. The originators of **10** have also eliminated artefacts due to protonation arising from the hydration shell of the highly charged transition metal ion.[66]

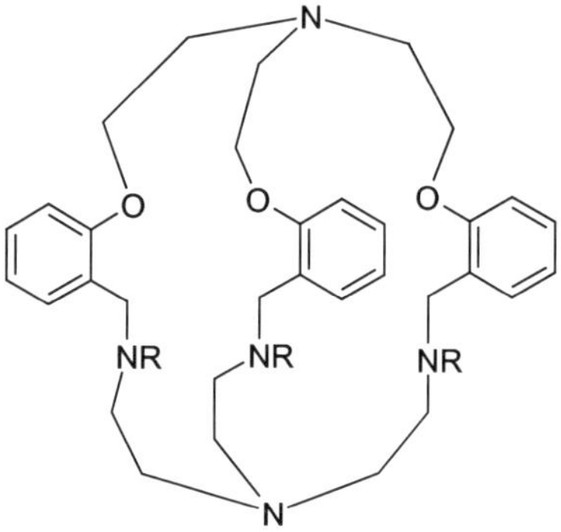

10; R = 9-anthrylmethyl

Electric voltage inputs and current as output are considered by Collier et al.[67] who use a unimolecular layer sandwiched between metallic terminals with an additional barrier at each molecular interface. The output is easily convertible to a voltage dropping across a resistor by sending the current through the latter. Compound **11** (given in section 7.2.7 as structure **82**) is used in the central layer, but the actual chemical structure does not matter in this instance provided that surface-active viologen units are available. These viologen derivatives conduct through resonant tunnelling if they are contacted with a titanium surface and interrogated with a negative voltage (from the titanium side). However, the molecules are much poorer conductors at positive voltages, *i.e.* a diode-like action.

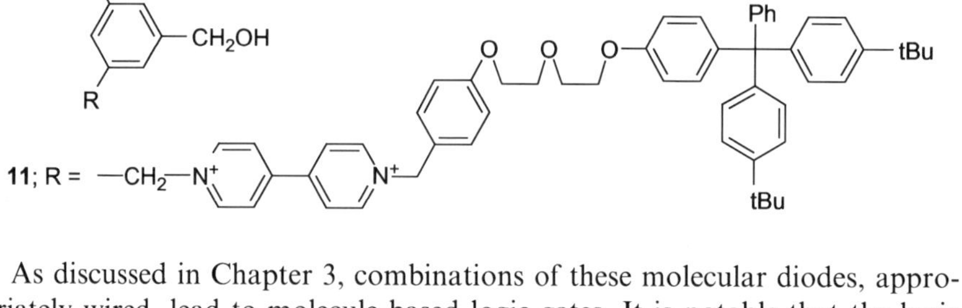

11; R = —CH$_2$—N$^+$⟨⟩—⟨⟩N$^+$—

As discussed in Chapter 3, combinations of these molecular diodes, appropriately wired, lead to molecule-based logic gates. It is notable that the logic gate action is not an intrinsic molecular property. Wiring with bulk metal conductors is essential. Collier et al.[67] illustrate this by setting up 3-input OR gates (Figure 10.2) with a small array of switches containing several million viologen molecules in each. The excellent 'on/off' discrimination in the output current can be ascribed to the diode-like nonlinearity in the switches when in their 'closed' states.

The molecular electronic system from Collier et al.[67–69] allows irreversible logic reconfiguring in the following way. Application of a positive voltage leads to irreversible oxidative damage which prevents any conduction, even upon returning to a negative voltage. The oxidative destruction of the molecular switch is exploited to electrically reconfigure individual switches in an array. For example, the 3-input OR gate (the wiring for a diode–diode logic implementation of a 2-input OR case is shown in Chapter 3, Figure 3.5) is converted to a 2-input OR gate by oxidatively destroying a switch in one of the input lines. Though also discussed in section 8.2, this protocol is reminiscent of certain programmable logic devices in silicon technology[3–9] where chosen links between individual logic gates are opened irreversibly so that the resulting device is customized according to the needs of the circuit designer.

10.4 Three-input NOR

Figure 10.3 shows the electronic representation of 3-input NOR logic. Singh and Kumar's **12**[70] shows how the general ability of d-block metal ions to quench fluorescence of neighbouring π-electron systems (*via* EET and PET mechanisms) can be exploited to emulate this logic configuration at the molecular level. The moderate fluorescence output of **12** at 402 nm is switched 'off' by Cu^{2+}, Ni^{2+} or Co^{2+} inputs or any combination of these. The residual intensity of the 'off' signal is uniformly low for all these situations, since the fluorescence process is easily overpowered over a wide range of quenching efficiencies. Thus good quality 3-input NOR logic arises.

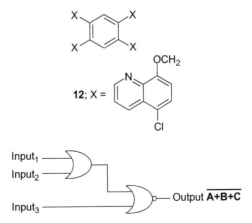

Figure 10.3 Three-input NOR logic.

Section 7.4 on 2-input NOR gates featured Chiu's crown ether–pyridinium hybrid macrocycle and tetrathiafulvalene-based clip[71] (structures **104** and **105** respectively in Chapter 7). Their complexation led to a charge transfer absorption band at 533 nm which was disrupted by K^+ and NH_4^+ inputs. The input palette can be increased by applying temperature as a new variable between 25 ('low') and 80 °C ('high').[72] The complexation of the macrocycle and clip is enthalpically controlled, so that increased temperature results in its disruption.

10.5 Three-input INHIBIT

The 3-input INHIBIT logic operation,[14] (Figure 10.4) can be demonstrated with **13** which was the first 3-input gate and the first INHIBIT gate of any kind to be published. This uses Ca^{2+} as input$_1$ to the amino acid receptor[73] to block a PET process from this receptor to the triplet excited state of the bromonaphthalene lumophore.[74] Compound **13** also uses β-cyclodextrin as input$_2$ which actually encapsulates the lumophore[75–77] to offer it protection against triplet–triplet annihilation which occurs because the triplet excited state of bromonaphthalene can easily collide with another copy of itself to cause mutual de-excitation. Importantly, β-cyclodextrin is transparent to the exciting light. The disabling input$_3$ is O_2 which, due to its paramagnetism, wrecks the phosphorescence emitted from the lumophore triplet excited state whether it is enveloped by β-cyclodextrin or not (Table 10.2). Phosphorescence quenching is

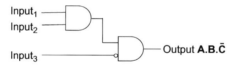

Figure 10.4 Three-input INHIBIT(input$_3$) logic.

Table 10.2 Truth table for **13**.[a]

Input$_1$ Ca^{2+}	Input$_2$ β-CD	Input$_3$ O_2	Output Phosphorescence[b]
none	none	low ($<10^{-6}$ M)	low (1.0)
none	high (5×10^{-3} M)	low ($<10^{-6}$ M)	low (1.0)
high (10^{-3} M)	none	low ($<10^{-6}$ M)	low (1.6)
high (10^{-3} M)	high (5×10^{-3} M)	low ($<10^{-6}$ M)	high (25)
none	none	high (3×10^{-4} M)	low (1.0)
none	high (5×10^{-3} M)	high (3×10^{-4} M)	low (0.5)
high (10^{-3} M)	none	high (3×10^{-4} M)	low (1.0)
high (10^{-3} M)	high (5×10^{-3} M)	high (3×10^{-4} M)	low (1.5)

[a] 10^{-5} M in water at pH 7.2.
[b] Relative quantum yields, λ_{exc} 287 nm, λ_{em} 505, 530 nm.

More Complex Systems 237

a general disabling process.[78–81] Functional, rather than physical, integration succeeds here too.

Replacement of the Ca^{2+} receptor within **13**[14] with an aliphatic amine which binds H^+ produces **14**.[82] This amine nitrogen atom serves as the PET donor to the triplet excited state of the bromonaphthalene unit. Therefore this gate is driven by H^+, β-cyclodextrin and O_2 (disabler) inputs.

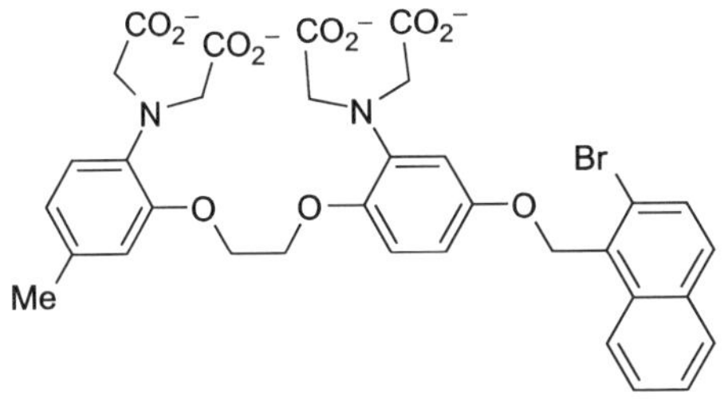

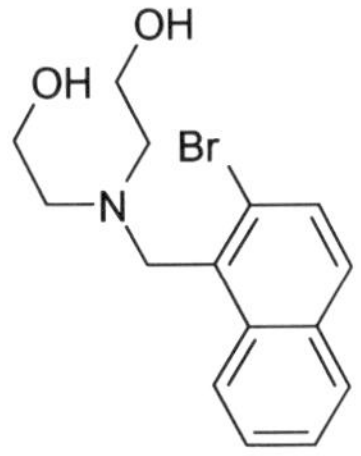

The use of phosphorescent systems in molecular-scale logic devices allows us to operate gates in a controlled time sequence. A fluorescent gate and a phosphorescent gate coexisting in solution can be interrogated at will in two different ways. Firstly, time-delayed observation can be performed following pulse excitation (over microseconds) so that the fluorescence gate responds within the excitation pulse time and is lost during the 'blind' period of observation. The millisecond duration of the phosphorescence allows it to be observed without any contamination by the fluorescence signal.[83] Second, the phosphorescent gate can be disabled by admission of oxygen so that the fluorescent gate can be observed essentially alone.

An example based on the luminescence output of a tris(bipyridyl)Ru(II) derivative **15**[84] depends on the association of its sugar terminals with a bisboronic acid **16** so that PET occurs to the viologen unit.[85] Thus **16** is the disabling input. The sugar binder Concanavalin A is the second input and a pH 7.2 buffer is the third. Concanavalin A exerts two effects: displacement of **16** so as to stop PET from **15** and displacement of water from **15**'s environment which also enhances the emission. However, the starting state of the device is **15** in unbuffered aqueous acetonitrile, which hampers further discussion concerning pH effects.

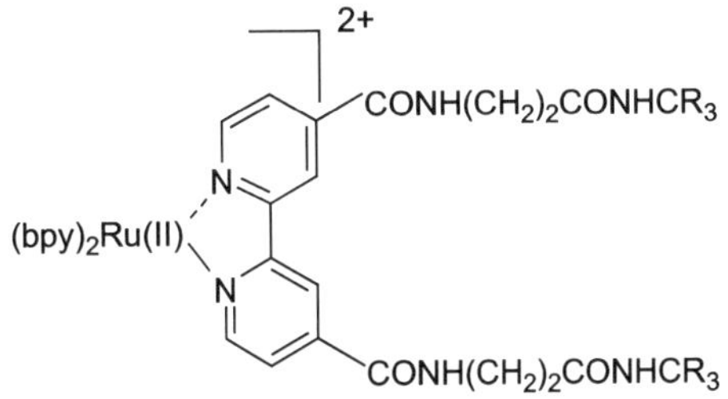

15; R = $CH_2O(CH_2)_2CONH(CH_2)_2$α-mannosyl

16

A fluorescent case arises from the 464 nm emission of Ajayaghosh's **17**,[86] whose inputs are Zn^{2+}, H^+ and ethylene diamine tetraacetic acid (EDTA) (the disabler). The blue emission arises only when Zn^{2+} is captured by the bipyridine unit and when both amines are protonated. Addition of EDTA would not only scavenge Zn^{2+} but it would also buffer the pH down to neutral. Interestingly, the logic response of **17** can be reconfigured by observing the output at 574 nm where the neutral form emits. The result is an inverted version of disabled OR logic (shown later in Figure 10.11b). Notably, the ICT nature of the fluorophore is seriously reduced when the electron-donating amines are protonated and leads to the 464 nm emission which is considerably blue-shifted as compared to the neutral form. In contrast, OH^-, a 254 nm light dose and Zn^{2+} (disabler) are the inputs for Jiang's photochromic Schiff base **18** which also contains a metal ion chelating site as well as an acidic group.[87] If the absorbance at 323 nm is monitored as the output, the gate array suggested by the authors can be reduced to a 3-input INHIBIT function. If the emission at 460 nm is chosen as output, a 2-input INHIBIT function is produced instead (after appropriate array minimization). The two inputs here are Zn^{2+} and the 254 nm light dose (disabler). An analogous result is obtained for a related Schiff base dimer adsorbed on mesoporous aminopropyl silica.[88]

17; R = N(CH₂CH₂OMe)₂

18

A wavelength-reconfigurable example **19** which uses absorbance outputs exclusively[89] is available from Wu and colleagues. This compound shows rich absorption spectra owing to the presence of different ligands and different metal centres. A Bronsted basic site and a redox-active centre allow control of absorption spectra via multiple input species. However, the use of redox inputs such as Fe^{3+} and Zn needs care owing to memory aspects coming into play. The observation of absorbance at 405 nm, due to an MLCT state involving Pt(II), fits H^+, Fe^{3+}, Zn-driven INHIBIT(Zn) logic. Protonation of the dimethylamino group creates a cationic substituent on the terpyridyl component – an electron-poor unit. Arrival of Fe^{3+} puts the ferrocenyl acetylide unit in the oxidized state which also becomes an electron-poor unit. When taken together, these two electron-poor units and the Pt(II) centre give rise to the MLCT state. Zn is the disabling input because it neutralizes H^+ by releasing the

More Complex Systems

latter as H_2, besides reducing the ferrocenyl acetylide unit back to the initial situation. Moving the observation point to 500 nm, an intra-ligand CT state signature, a simpler IMPLICATION(Zn $\Leftarrow H^+$) logic is seen instead. Protonation of the dimethylamino group arrests the intraligand CT process, whereas Zn neutralizes H^+. The longest wavelength absorption displayed by **19** is at 850 nm due to an LM'CTThough also discussed in section 8.2, this protocol state concerning the ferrocenyl acetylide unit in the oxidized state. Monitoring the absorbance here results in Fe^{3+}, Zn-driven INHIBIT(Zn) logic. The disabling nature of Zn towards the effect of Fe^{3+} is as mentioned above.

19

The intense absorbance at 670 nm arising from aggregated gold nanoparticles behaves in an INHIBIT logical fashion when the inputs are a 365 nm light dose, Cu^{2+} and EDTA (disabler) provided that the nanoparticles are decorated with spiropyran photochromics.[90] The memory effects also need to be set aside. Only the open merocyanine form produced on uv irradiation has ligating groups so that Cu^{2+}-induced crosslinking and aggregation can occur. The EDTA sequesters Cu^{2+} in order to negate this crosslinking. As discussed often in this book, the photochromic itself displays brilliant colour changes but the nanoparticles have far stronger absorption cross-sections in the visible region. The absorbance output from such nanoparticle aggregation can also display 2-input logic behaviour when combined with metal ion-dependent DNAzymes.[91]

Stojanovic and Stefanovic built a 3-input INHIBIT gate (Figure 10.5) as part of a larger study.[92] At its centre is the deoxyribozyme[93] discussed in Chapter 9 (Figure 9.3c and d). Figure 10.5a shows a schematic view of this INHIBIT gate, where the base sequence of the deoxyribonucleotide device is decided so that it folds into the enzyme motif plus three stem–loops. Notably, the enzyme heart is blocked from two sides. The supply of suitable oligonucleotides ($input_2$ and $input_3$) opens out these stem–loops so that the substrate can be grabbed and hydrolysed (Figure 10.5b). However, there is another stem–loop which serves as a receptor for another oligonucleotide ($input_1$). This stem–loop opens upon supply of $input_1$ so that the enzyme's heart is weakened. $Input_1$ is therefore the disabler. The enzymatic action is visualized by the emergence of fluorescence from a fluorophore F once the part carrying the intramolecular quencher Q has been hydrolysed off.

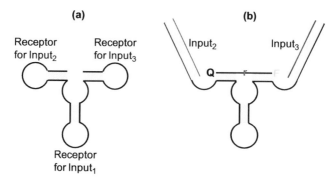

Figure 10.5 Three-input INHIBIT(input$_1$) logic based on deoxyribozymes, (a) before and (b) after presentation with input$_2$ and input$_3$. The grey regions are the Crick–Watson base-pairing regions. Ribonucleotide 'r' is the point of hydrolysis which separates the fluorophore F from the quencher Q.

10.6 Three-input IMPLICATION

Qian's team employ the effects of two detergents, sodium dodecyl sulfate (SDS) and cetyltrimethylammonium bromide (CTAB) (which possess anionic and cationic organic components respectively) and temperature on the fluorescence of **20** to demonstrate 3-input IMPLICATION(input$_2 \Leftarrow$ input$_1$) logic.[94] Figure 10.6 shows its electronic representation. The fluorescence of **20** is quenched only when it forms a pre-micellar aggregate with micromolar levels of SDS in neutral aqueous solution. Even these aggregates are broken up when equimolar amounts of CTAB are present due to the stronger association between SDS and CTAB. Higher temperatures also break up these pre-micellar aggregates. The SDS is input$_1$ in these experiments, whereas CTAB and temperature represent the other two inputs.

10.7 Three-input Enabled OR

An enabled OR logic system (Figure 10.7) can be developed from **21**,[95,96] if its memory possibilities are avoided. High proton densities can be arranged in the locality of **21** by either acidification or by adding an anionic micelle-forming detergent SDS to a solution at nearly neutral pH (5.5). Our **22** clearly showed

More Complex Systems   241

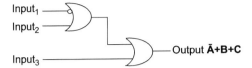

Figure 10.6 Three-input IMPLICATION(input$_2$ ⇐ input$_1$) logic.

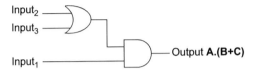

Figure 10.7 Three-input input$_1$-enabled OR logic.

Table 10.3 Truth table for **21**.[a]

Input$_1$ Light dose$_{365}$	Input$_2$ H^+	Input$_3$ SDS	Output Absorbance$_{450}$ (from **23**)
none	low ($10^{-5.5}$ M)	none	low
none	high (10^{-1} M)	none	low
high	low ($10^{-5.5}$ M)	none	low
high	high (10^{-1} M)	none	high
none	low ($10^{-5.5}$ M)	high (5×10^{-2} M)	low
none	high (10^{-1} M)	high (5×10^{-2} M)	low
high	low ($10^{-5.5}$ M)	high (5×10^{-2} M)	high
high	high (10^{-1} M)	high (5×10^{-2} M)	high

[a] 5×10^{-5} M in water.

the highly perturbed H^+ concentrations near micelles.[97,98] Thus in the case of **21**, the three inputs are acidification (to pH 1), SDS micelles and irradiation. Irradiation is essential, of course, to form **23**. So it is the enabling input (input$_1$ in Figure 10.7). Either or both of the other two inputs will guarantee production of **23** upon irradiation of **21** (Table 10.3).[95] We note that **21** and relatives, in the absence of detergents, were previously used to construct 2-input AND gates.[96–98]

A newer case of this kind is Lehn's **24** which possesses adjacent tetrahedral and octahedral sites so that interesting combinations of metal ions can be applied as inputs.[102] The OH$^-$ input prepares **24** for divalent ion binding by deprotonation, as is commonly found in many analytical reagents.[21,103,104] Zn^{2+}, say, then binds at the octahedral site to produce a complex $(24 - H^+)_2 \cdot Zn^{2+}$ absorbing at 480 nm as the output. This is Zn^{2+}, OH$^-$-driven AND logic. The Zn^{2+} can be replaced with Pb^{2+} without losing the absorbance output at 480 nm. This is a Zn^{2+}, Pb^{2+}-driven OR gate feeding its output to an AND gate whose other input receives OH$^-$, *i.e.* 3-input enabled OR.

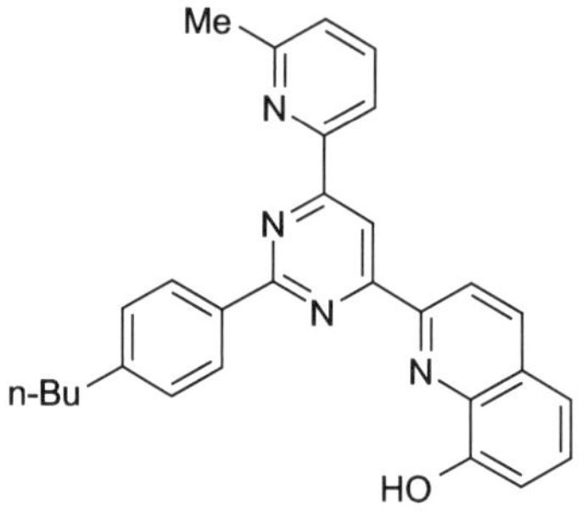

24

10.8 Three-input Enabled NOR

Rather similar to the INHIBIT gate **25**,[105] **26**[106] can sensitize Tb^{3+} luminescence. The electron donor $(C_2H_5)_3N$ (input$_1$) quenches this emission by intermolecular PET. The O_2 achieves the same end, but *via* different means. A slow EET from the phthalimide triplet to Tb^{3+} is held responsible, because triplet states of π-electron systems have long been known to be deactivated rapidly by O_2 (input$_2$). The logical result of these two inputs, either alone or together, is to control the emission output in a NOR logical manner. However, the gate properties of **26**.Tb^{3+} can be developed further by considering Cl$^-$ as another input (input$_3$). In the absence of $(C_2H_5)_3N$ and O_2, Cl$^-$ enhances the emission of **26**.Tb^{3+} by displacing trace water from the coordination sphere of Tb^{3+}. The Tb^{3+}-bound water allows H–O vibrational quanta to drain the electronic quantum in the $^5D_4$ excited state of Tb^{3+}.[49] Overall, this makes **26**.Tb^{3+} an enabled NOR gate where Cl$^-$ is the enabling input (Figure 10.8).

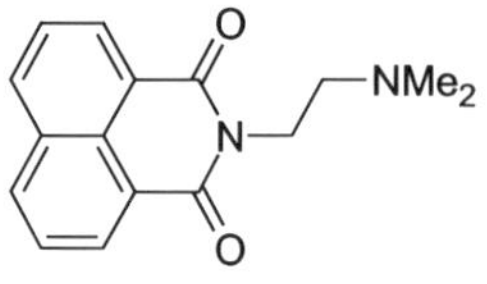

25 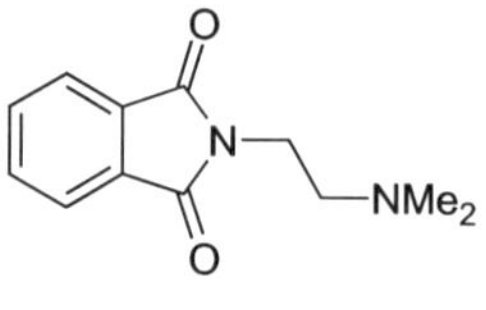   **26**

Another case of this type can be found in Perez-Inestrosa's work with **27**,[107] which is specifically a H$^+$-enabled, K$^+$, Ba^{2+}-driven NOR gate when its

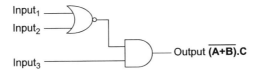

Figure 10.8 Three-input input$_3$-enabled NOR logic.

Table 10.4 Interim truth tables for pH-dependent fluorescence of **28**.[a]

Input$_1$ Cu^{2+}	Input$_2$ Ni^{2+}	Output$_1$ Flu$_{pH=1}$	Output$_2$ Flu$_{pH=4.5}$	Output$_3$ Flu$_{pH=7}$
none	none	high	high	high
none	high	high	high	low
high	none	high	low	low
high	high	high	low	low

[a]10^{-5} M in water.

emission at 550 nm is monitored. In its incarnation as an INHIBIT gate in section 7.6, we noted its possibilities for employing multiple inputs and outputs. In the current situation, the CT emission originates in the PET process from the benzocrown ether to the protonated isoquinoline N-oxide, provided that the crown cavity is not blocked by K$^+$ or Ba^{2+}.

To round off this section, let us consider the anthrylmethylpolyamine PET system **28**.[108] Protonation of **28** causes switching 'on' of emission because the critical benzylic and central amine groups are neutralized. The protonation can also compensate for the quenching effects of Cu^{2+} and Ni^{2+}, i.e. the species **28.**H$^+_3$.M^{2+} is nicely emissive, although the typically stronger binding of Cu^{2+} means that **28.**H$^+_3$.Cu^{2+} exists at lower pH values than **28.**H$^+_3$.Ni^{2+} does. Thus Cu^{2+}-induced quenching sets in at pH >4 whereas Ni^{2+}-induced quenching happens at pH >6.5. Table 10.4 shows interim truth tables where the H$^+$ input has not yet been factored in. This allows the observation of different logic behaviour depending on the 'high' level of H$^+$ chosen. The 'low' level of H$^+$ is taken as pH 10. The H$^+$-enabled, Cu^{2+}, Ni^{2+}-driven PASS 1, NOT TRANSFER(Cu^{2+}) and NOR logic emerges as a result. Multi-level logic can also be imagined in this case (Chapter 12).

27

28; R = 9-anthrylmethyl

10.9 Three-input Enabled IMPLICATION with Wavelength-Reconfigurability

Raymo and Giordani were the first to employ ion-sensitive photochromics (see refs 125–140 in Chapter 7) to achieve logic gate arrays.[109] As noted before, photochromics could be switched from their colourless state to the coloured state by uv irradiation, and returned to the colourless state by application of visible light.[110] The availability of a third state *via* protonation is the novelty. The colourless form is **1**,[22] which is ring-opened to the coloured state **29**. Additionally, **1** ring opens to **30** upon protonation and basification maintains it in the ring-opened form **29** on the timescale of the experiment (*ca.* 5 min). The doses of 254 and 524 nm light (input$_1$ and input$_2$ respectively) are chosen so that the 254 nm light dose dominates in a competition situation. The H$^+$ is input$_3$. Easily observable absorbances at the electronic absorption peak maxima at 563 nm for **29** and at 401 nm for **30** are output$_2$ and output$_1$ respectively. Table 10.5 summarizes the results. As far as output$_1$ is concerned, **1** displays enabled IMPLICATION logic or to put it more specifically, input$_3$-enabled IMPLICATION(input$_1$ ⇐ input$_2$) logic (Figure 10.9). Some readers may prefer the more concise algebraic descriptor given in Figure 10.9. Output$_2$ can be

Table 10.5 Truth table for acid- and light-induced transformations of **1**.[a]

Input$_1$ hv dose$_{254}$	Input$_2$ hv dose$_{524}$	Input$_3$ H$^+$	Output$_1$ Abs$_{401}$	Output$_2$ Abs$_{563}$
none	none	none	low	low
none	none	high	high	low
none	high	none	low	low
high	none	none	low	high
none	high	high	low	low
high	none	high	high	low
high	high	none	low	high
high	high	high	high	low

[a]10^{-4} M in acetonitrile.

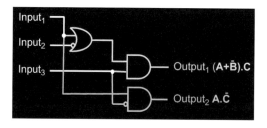

Figure 10.9 The gate arrays which are emulated by the logic behaviour of **1** when its absorbance is observed at 401 (blue) and 563 nm (green), which are taken as output$_1$ and output$_2$ respectively. a) Input$_3$-enabled IMPLICATION (input$_1$ ⇐ input$_2$) and b) input$_1$, input$_3$-driven INHIBIT(input$_3$) logic. Each array becomes visible only when interrogated at the appropriate wavelength.

More Complex Systems 245

described in an even simpler way: input$_1$, input$_3$-driven INHIBIT(input$_3$) logic (Figure 10.9). Notably, the state of input$_2$ (visible light) has no influence on output$_2$. This is clearly a case of logic reconfiguring by means of the wavelength at which the absorbance is measured, related to cases featured in Chapters 6, 8 and 13. As will be discussed in Chapter 13, wavelength-reconfiguring of logic has quantum aspects which can be discussed in terms of superposed gates. Figure 10.9, shown in reverse video with its dark background, attempts to give a flavour of this phenomenon where the particular gate array becomes visible only when interrogated at the correct wavelength. For instance, if Figure 10.9 is viewed through a suitable blue filter, only the blue and white features will be visible. Importantly, Figure 10.9 and relatives in this chapter and in Chapter 13 are not parallel arrays of gates but rather superposed gates. The quantum aspects of these Figures could perhaps be better represented by drawing all the devices/outputs on top of each other in separate colours, but issues of clarity prevent us from following this path.

29

30

10.10 Three-input Disabled OR with Wavelength-Reconfigurability

Perez-Inestrosa's **27**[107] has already been discussed in sections 7.6 and 10.8. Here it pops up again, such is its versatility. When H^+, Zn^{2+} and K^+ are fed as input$_1$, input$_2$ and input$_3$, with output being observed at 550 nm, a K^+-disabled, H^+, Zn^{2+}-driven OR logic action is recorded. This is because H^+ or Zn^{2+} unselectively ties up oxygen electron pairs on the N-oxide unit so that the isoquinoline N-oxide becomes electron deficient enough to encourage PET from the benzocrown ether, which in this case unusually produces a charge transfer emission at 550 nm. The K^+ acts as the disabling input$_3$ because it binds to the crown ether and reduces its electron donor ability so that PET is arrested. In turn, the CT emission is extinguished.

Additionally, **27** can be observed at 400 nm, the wavelength of its locally excited (LE) emission, where a H^+, Zn^{2+}-driven NOR behaviour is seen. When the N-oxide oxygen electron pairs are tied up, there is a blue-shift of the LE emission to 380 nm. Hence, observation at the latter wavelength produces H^+, Zn^{2+}-driven OR instead. The reverse video representation in Figure 10.10 with the dark background attempts to show this wavelength-configurable logic gate

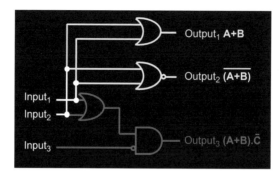

Figure 10.10 The gate arrays which are emulated by the logic behaviour of **27** when its emission is observed at 380, 400 and 550 nm, which are taken as output$_1$, output$_2$ and output$_3$ respectively. a) Input$_1$, input$_2$-driven OR, b) input$_1$, input$_2$-driven NOR logic and c) 3-input input$_3$-disabled OR logic. Each array becomes visible only when interrogated at the appropriate wavelength.

array with a hint to the quantum possibility of gate superposition. By simultaneously monitoring it at 380, 400 and 550 nm, three different logic faces of **27** can be discerned all at once. So **27** could have belonged in Chapter 13, concerning quantum aspects, as well.

Bispyrene compound **31**[111] in $CH_2Cl_2:CH_3CN$ (5:1) solution can be seen as a tetramine-disabled, H^+, Zn^{2+}-driven OR logic gate when its output is taken as monomer fluorescence observed at 376 nm. Protonation or metal-binding brings a bulky counteranion in between the fluorophores so that excimer formation is thwarted. When a linear tetramine is applied in sufficient quantity, it negates the effects of H^+ or Zn^{2+} by neutralization or complexation respectively. Naturally, the excimer fluorescence at 471 nm will produce the inverted version of disabled OR logic. The corresponding logic gate arrays are shown in Figure 10.11 after suitable array minimization. Similar minimization shows a light dose-disabled, OH^-, Zn^{2+}-driven OR logic function for Jiang's photochromic **18** immobilized on aminopropyl mesoporous silica.[112] The light dose has a wavelength of 254 nm. Notably, the behaviour of the same compound **18** in homogeneous solution fitted in section 10.5, *i.e.* logic reconfiguring occurs upon adsorption. Aminopropyl mesoporous silica features again when it carries an 8-amidoquinoline unit and leads to Cu^{2+}-disabled, Zn^{2+}, Cd^{2+}-driven OR logic in the hands of Li, Ma and their colleagues.[113] Classical literature shows that compounds like 8-amidoquinoline give emission enhancements with closed-shell dications but are quenched by open-shell analogues.[103]

31; R = 1-pyrenylmethyl

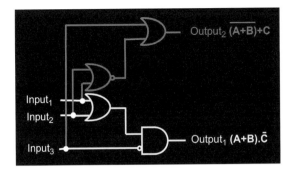

Figure 10.11 The gate arrays which are emulated by the logic behaviour of **31** when its emission is observed at 376 and 471 nm which are taken as output$_1$ and output$_2$ respectively. a) Three-input input$_3$-disabled OR logic and b) its inverted version. Each array becomes visible only when interrogated at the appropriate wavelength.

10.11 Three-input Disabled INHIBIT

During the discussion of Stojanovic and Stefanovic's deoxyribozyme-based logic gates[114] (section 5.3.1.3), it was clear that the presence of a single ribonucleotide was essential for the success of the method. A way of avoiding the use of this ribonucleotide due to the combined effort of the groups of Zhang, Fan and He[115] was briefly mentioned at that point. They depended on Breaker's Cu^{2+}-dependent DNA-cleaving deoxyribozyme.[116] Zhang, Fan and He showcase their method by building a 3-input input$_2$-disabled INHIBIT-(input$_1$) gate. The electronic representation is given in Figure 10.12 and the oligodeoxyribonucleotide implementation is outlined in Figure 10.13. Part (a) of the latter figure shows the sequences of the basic deoxyribozyme and of the fluorescently labelled substrate. The schematic picture in Figure 10.13b shows how the basic deoxyribozyme is expanded in three regions. In one of these regions, the structure of the basic deoxyribozyme is seriously perturbed by the formation of a new stem. Arrival of an input$_3$ strand opens this stem by binding to a part of it and the adjacent loop. Then the catalytic cleavage of the substrate can proceed. The other two regional expansions, one a loop and the other a 'pistol', inhibit the catalysis when they bind the strands corresponding to input$_1$ and input$_2$ respectively. Presentation of the input$_1$ strand opens this loop and the adjoining stem which is essential for catalytic activity. On the other hand, arrival of the input$_2$ strand covers up another small but essential catalytic component. So, it is only the presence of input$_3$ and the absence of input$_1$ and input$_2$ that gives rise to the cleavage product. The use of denaturing polyacrylamide gel electrophoresis to detect the cleavage of the fluorescently labelled product brings in a materials aspect to this work. However, future development with substrates carrying EET-based fluorophore–quencher pairs should permit molecular-scale logic experiments.

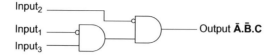

Figure 10.12 Three-input input$_2$-disabled INHIBIT(input$_1$) logic.

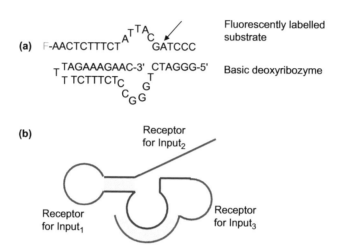

Figure 10.13 Three-input input$_2$-disabled INHIBIT(input$_1$) logic based on deoxyribozymes. (a) Basic deoxyribozyme and the fluorescently (F) labelled substrate. The cleavage site is shown with the arrow. (b) Schematic representation of the logic gate, with the basic deoxyribozyme shown in blue and receptors for the inputs shown in red. The grey regions are the Crick–Watson base-pairing regions.

10.12 Three-input Disabled XNOR

A disabled XNOR gate (Figure 10.14) at the molecular scale is obtained by Lee et al.[117] where the disabling input$_3$ is Pb^{2+}, and H^+ and OH^- are input$_1$ and input$_2$ respectively. Device **32**[117] is composed of a calix[4]arene-18-crown-6 receptor and two pyrene fluorophores linked to it via two amide linkages. In its free form in acetonitrile, normal pyrene monomer emission is seen. Upon Pb^{2+} binding to the crown and the carbonyl oxygens, fluorescence is switched 'off' owing to PET from pyrene to the electron-deficient amide. Also, addition of acid to the free **32** causes quenching of fluorescence due to PET from pyrene to the protonated amide. Addition of base, however, removes the amidic proton, resulting in PET from this electron-rich moiety to pyrene. Stoichiometric amounts of acid and base need to be maintained.[118]

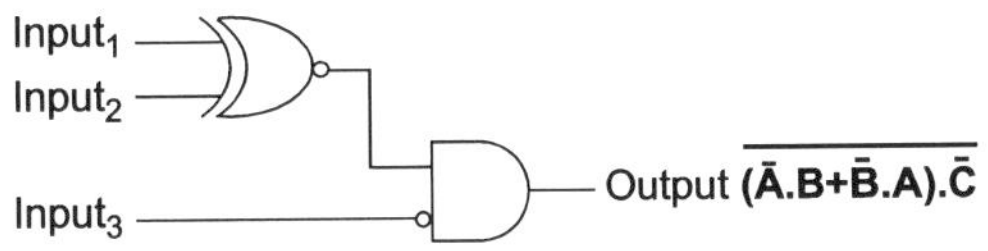

More Complex Systems 249

Figure 10.14 Three-input input$_3$-disabled XNOR logic.

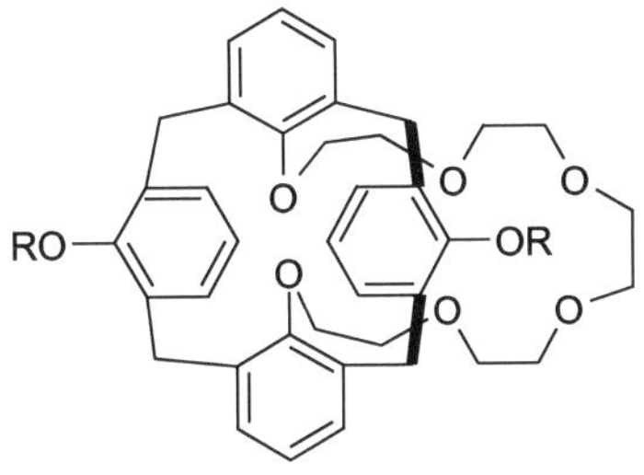

32; R = CH$_2$CONHX
X = 1-pyrenylmethyl

10.13 Three-input Disabled IMPLICATION

There is a logical situation which follows on from the experiment discussed in section 10.9, though it is logically less complex as it stands. The photochemical and H$^+$-induced transformations of the colourless **1** lie at its heart. However, instead of monitoring the absorbances of the coloured products **29** and **30**, Raymo and Giordani observed the effects on the fluorescence of an additive.[16] Commercially available pyrene (**33**) is the additive whose emission spectrum (centred at 373 nm and best excited at 336 nm) overlaps to different extents with the absorption spectra of **1**, **29** and **30**, as well as suffering competitive absorption during its excitation. Table 10.6 summarizes the truth table behaviour. It should not be surprising then that we observe a somewhat opposite logic behaviour compared with that of section 10.9. This is disabled IMPLICATION logic, or more accurately, input$_1$-disabled IMPLICATION(input$_2$ $\Leftarrow$ input$_3$) logic (Figure 10.15) where doses of 254 and 524 nm light and H$^+$ are input$_1$, input$_2$ and input$_3$ respectively. The role of 254 nm radiation dose as the disabler is particularly clear because it creates the coloured products which absorb the emission of **33**. On the other hand, a dose of 524 nm light would carry coloured products back to the colourless **1**, *i.e.* strong emission would be seen from **33**. Similarly, the absence of H$^+$ would avoid formation of the coloured **30** which absorbs strongly at 373 nm.

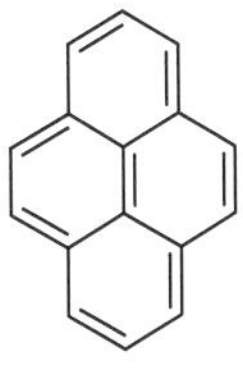

33

Table 10.6 Truth table for the fluorescence of **33** under the influence of acid- and light-induced transformations of **1**.[a]

$Input_1$ $hv\ dose_{254}$	$Input_2$ $hv\ dose_{524}$	$Input_3$ H^+	Output Fluorescence[b]
none	none	none	high
none	none	high	low
none	high	none	high
high	none	none	low
none	high	high	high
high	none	high	low
high	high	none	low
high	high	high	low

[a]10^{-4} M **1** and 10^{-4} M **33** in acetonitrile.
[b]Intensities, λ_{exc} 336 nm, λ_{em} 373 nm.

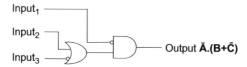

Figure 10.15 $Input_1$-disabled IMPLICATION($input_2 \Leftarrow input_3$) logic displayed by a mixture of **1** and **33** when the emission of the latter is observed at 373 nm.

This approach of applying inputs to one molecular device while the output is obtained from another can be viewed as a form of chaining two active components, *i.e.* serial integration. Given that fluorophores with different emission signatures can be substituted for **33**, each of these will suffer competitive absorption of excitation light, as well absorption of the emission by **1**, **29** and **30** in different ways. Thus different truth tables and corresponding gate arrays will result.[17] So we have another interesting method of logic reconfiguring which belongs here rather than in Chapter 8 because of its triple inputs. Though the outputs are observed at different wavelengths in the end, the primary reconfiguring variable is the additive fluorophore.

Alternatively, the fluorophore can be covalently attached to the photochromic as we find in Li's **34**.[119] When excited at 530 nm, fluorescence output at 581 nm emerges only if a base such as triethylamine ($input_1$) is absent so that rhodamine lactam formation is avoided. Also, fluorescence can only be observed if EET from the fluorophore to the coloured form of the dithienylethene[120] is avoided, *i.e.* a 312 nm light dose ($input_3$) is not provided to **34**. Another scenario is to couple the 312 nm light dose with a visible light dose ($input_2$) so that their effects cancel.

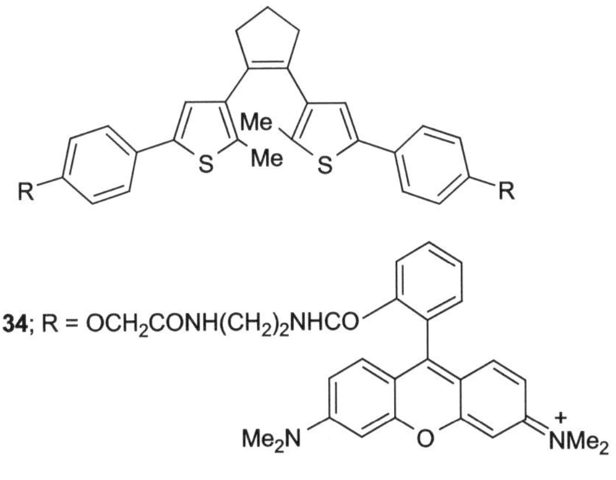

More Complex Systems 251

34; R = OCH$_2$CONH(CH$_2$)$_2$NHCO-

Li and his colleagues also target this logic type by using bifluorophoric **35**,[121] whose phenol form suffers EET so that a red emission (527 nm) from hydroxyperylenetetracarboximide is seen when the naphthoperylenetetracarboximide unit is excited at 330 nm. Under basic conditions, the phenolate form of **35** loses the spectral overlap criterion for EET (section 4.6) so that only the yellow (580 nm) emission of the naphthoperylenetetracarboximide moiety is found. Additionally, Fe^{3+} is bound adequately to the phenol group and the peptide units in the uncompetitive solvent THF so that all fluorescence is suppressed via PET/EET to the open-shell metal ion. Observation of the 580 nm emission therefore fits Fe^{3+}-disabled IMPLICATION(H$^+$ ⇐ base) logic. Perhaps this is an occasion to point out related logic work by Wang and colleagues[122] based on a Y-shaped DNA labelled with three fluorophores which interacts with three nuclease inputs. Nuclease-driven logic systems are rare. However, the result of the 3-input experiment described in this work is reducible to a single-input NOT function dependent on the nuclease Hae-III.

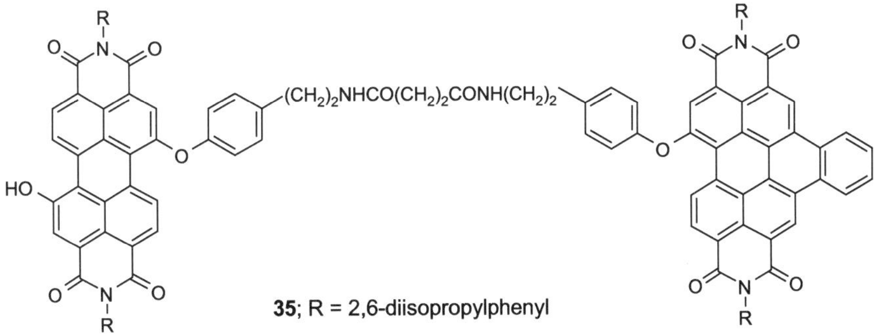

35; R = 2,6-diisopropylphenyl

A case of dansylamide covalently immobilized on silicon nanowires[123] belongs here, if we assign input$_1$ = H$^+$, input$_2$ = Cl$^-$, input$_3$ = Hg^{2+}. The output is fluorescence intensity at 530 nm, excited at 330 nm. In water, the fluorescence of **36** is quenched by H$^+$ due to removal of the amino donor electrons and subsequent reduction of the ICT character of the excited state. Hg^{2+} also serves as a quencher, but only if non-acidic Cl^- solutions are avoided. The latter situation takes Hg^{2+} out of solution as hydroxohalides.

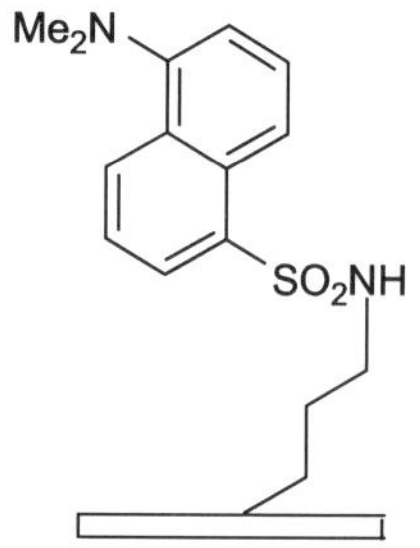

36

Gupta and van der Boom's **37**,[124] as a monomolecular layer on glass, produces disabled IMPLICATION logic as well. The inputs are NO^+, H_2O and Ce^{4+}, with the latter being the disabler. In the absence of Ce^{4+}, we can see how the IMPLICATION gate arises. NO^+ alone oxidizes **37** to its Os^{3+} state with loss of **37**'s characteristic absorption in the visible range. If NO^+ is supplied along with H_2O, they react with each other and **37** is left untouched; H_2O alone causes no change to **37** either. The non-commutative behaviour is clear. Now if Ce^{4+} is brought into the picture, it will oxidize **37** whether the other two inputs are present or not. So we have Ce^{4+}-disabled IMPLICATION logic.

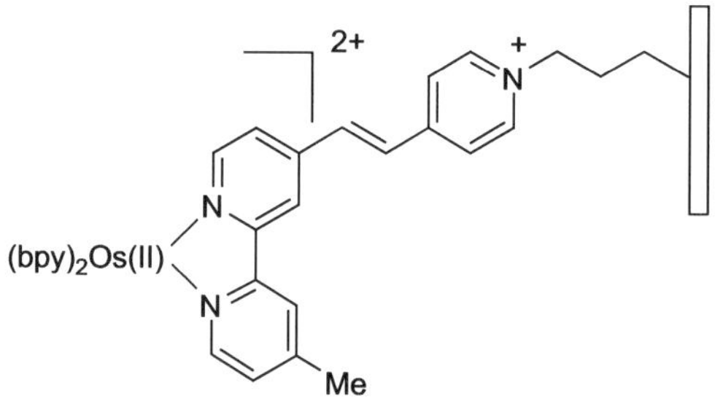

37

Because **37** is based on a bulk phase, it is also possible to let the product species from the redox reactions move into solution and diffuse across to another glass-based system which can be observed. This becomes a way of integrating two molecule-based logic systems with chemical intermediaries. Previous solution-based molecular logic integration with H^+ mediation is known.[23] Being based on redox states, **37** is also latched in a given state for some time, even when the redox agent which originally caused that state is

More Complex Systems

removed. Such memory effects need to be erased by resetting to the starting state for combinational logic development.

10.14 Three-input Inverted Enabled OR

An inverted version of the logic array in section 10.7 is available from the work of Tian.[125] Specifically, **38** is an inverted H^+-enabled T, Cu^{2+}-driven OR gate. Its electronic representation is given in Figure 10.16. Polymer **38** shares some features with a case[126] discussed in section 7.6, such as a polarity-sensitive fluorophore and a temperature-induced coil–globule transition. However, it is distinguished by possessing a Cu^{2+}-binding site within the bis(2-pyridyl)amine unit. Naturally, fluorescence quenching is seen when Cu^{2+} binds to **38**, which happens when the medium is free of OH^- which would otherwise compete for the Cu^{2+}. Fluorescence of **38** is low even when Cu^{2+} is absent, provided that the temperature is high enough to form the non-polar globular polymer so that the degree of ICT within the fluorophore is reduced.[127] Table 10.7 collects all the fluorescence responses of this kind.

38; R = bis(2-pyridylmethyl)amino

10.15 2:1 Multiplexer and 1:2 Demultiplexer

In a computing context, multiplexing is a way of economizing on the number of data lines for transmission over a distance. A 2:1 multiplexer, for instance, takes data bits arriving along two input lines and manages to send them along a single transmission line without forgetting which is which. It does so by tagging

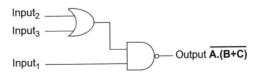

Figure 10.16 Three-input inverted input$_1$-enabled OR logic.

a given data bit with the state of a separate addressing input, *e.g.* when the addressing input is 'high', the data bit must have belonged to line 1. The 2:1 multiplexer is represented by the gate array in Figure 10.17.

While the first deliberately designed molecular 2:1 multiplexer came from Andréasson *et al.*,[128] the rare pH-switchable sensing behaviour of Brown's **39**[129] can also be seen to possess the same logic. In acid media, Na^+ causes 9-fold fluorescence enhancement while K^+ can only manage a factor of 4. In basic media, K^+ causes 8-fold fluorescence enhancement while Na^+ becomes impotent. These observations correspond to the truth table in Table 10.8.

Table 10.7 Truth table for **38**.[a]

$Input_1$ H^+	$Input_2$ Temperature	$Input_3$ Cu^{2+}	Output Fluorescence[b]
low (10^{-10} M)	low (25 °C)	none	high (1.0)
low (10^{-10} M)	high (46 °C)	none	high (1.0)
high (10^{-7} M)	low (25 °C)	none	high (1.0)
high (10^{-7} M)	high (46 °C)	none	low (0.5)
low (10^{-10} M)	low (25 °C)	high (5×10^{-5} M)	high (0.9)
low (10^{-10} M)	high (46 °C)	high (5×10^{-5} M)	high (0.9)
high (10^{-7} M)	low (25 °C)	high (5×10^{-5} M)	low (0.18)
high (10^{-7} M)	high (46 °C)	high (5×10^{-5} M)	low (0.14)

[a] 1.6 g l^{-1} in water : ethanol (5:1, v/v), weight – average molecular weight of **38** = 7026.
[b] intensities, λ_{exc} 460 nm, λ_{em} 620 nm.

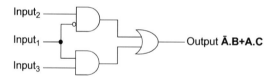

Figure 10.17 2:1 Multiplexer.

Table 10.8 Truth table for **39**.[a]

$Input_1$ H^+	$Input_2$ K^+	$Input_3$ Na^+	Output Fluorescence[b]
low[c]	none	none	low (0.048)
low[c]	high (10^{-2} M)	none	high (0.39)
high (10^{-2} M)	none	none	low (0.060)
high (10^{-2} M)	high (10^{-2} M)	none	low (0.24)
low[c]	none	high (10^{-2} M)	low (0.053)
low[c]	high (10^{-2} M)	high (10^{-2} M)	high[d]
high (10^{-2} M)	none	high (10^{-2} M)	high (0.51)
high (10^{-2} M)	high (10^{-2} M)	high (10^{-2} M)	high[d]

[a] 10^{-6} M in methanol.
[b] Quantum yields, λ_{exc} 376 nm, λ_{em} 405, 427 nm.
[c] Maintained with 10^{-2} M OH$^-$.
[d] Value not reported.

More Complex Systems

Switching the benzo-15-crown-5-ether to a calixcrown gives rise to Cs$^+$ instead of Na$^+$ being targeted.[130] Naturally, the pH-switchable sensing remains intact in **40**, with even better selectivities than those seen in **39**. Related cases where the switching variable is the nature of the pH buffer employed rather than the pH value itself are discussed by Qian's group.[131]

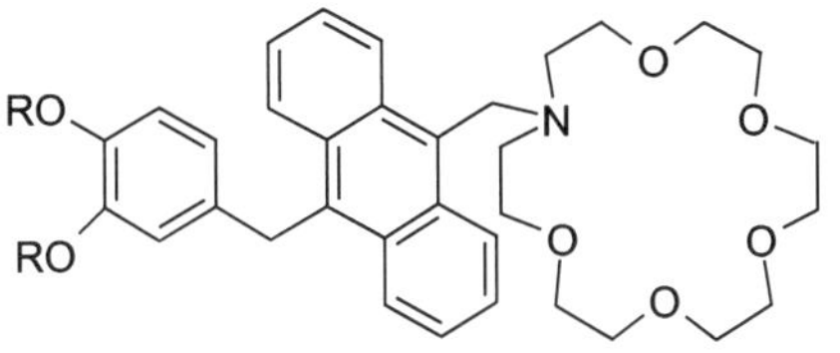

39; R,R =

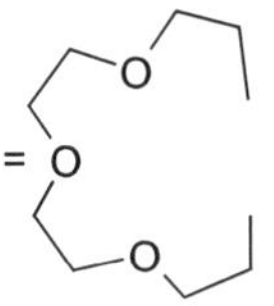

40; R,R = Pr-O- ... -OPr

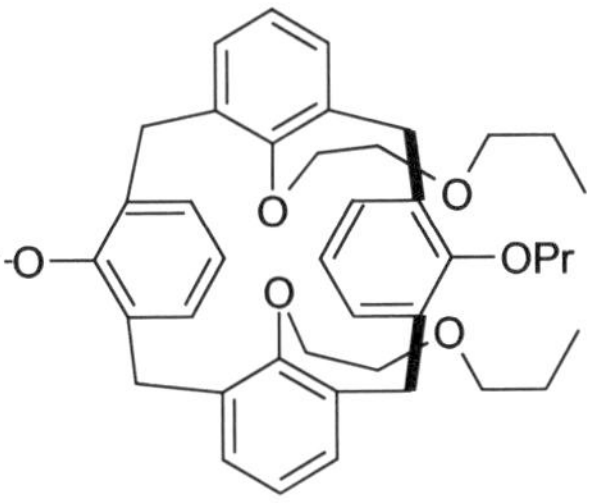

Concerning the mechanism of **39**'s behaviour, the 1,2-dioxybenzene unit of the benzocrown ether only engages in a PET process to protonated aminomethyl anthracene, *i.e.* under acidic conditions. The luminescence so suppressed can be recovered when Na$^+$ is captured by the benzo-15-crown-5 ether moiety with moderate selectivity. Under basic conditions, K$^+$ selectively enters the azacrown cavity and arrests the PET process otherwise possible from the azacrown nitrogen atom to the anthracene moiety. Closely related effects (and structural components) have previously been seen in H$^+$, Na$^+$-driven AND gate **41**[44] and K$^+$-driven YES gate **42**.[132]

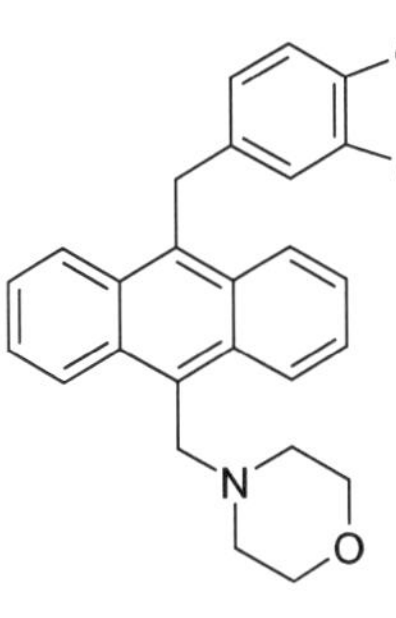

41

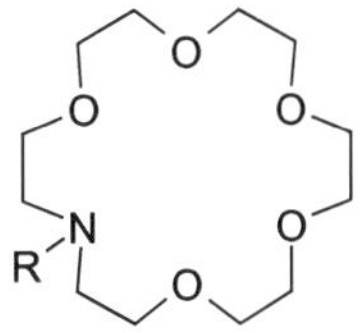

42; R = 9-anthrylmethyl

While H^+ was the addressing input for **39**, we now present cases where the addressing role is taken over by light dose or redox reagents. The general issue of memory effects arising from the latter cases is discussed in Chapter 11. When embedded in the membrane of ion-selective electrodes, Shinkai's **43**[133] is switchable between Na^+-selectivity and Li^+-selectivity by means of irradiation with a light dose of suitable wavelength. The intramolecular photodimerization of anthracene units[63] underlies this switching. However, the need for a bulk membrane sacrifices molecular-scale operation. The redox-switchable, PET-based fluorescent system **44/45**[134,135] is similarly able to pass from a Cd^{2+}-selectivity to a Hg^{2+}-selectivity. Reduction is arranged by Zn/H^+ whereas oxidation is done with I_2.

43; X = O(CH$_2$)$_2$OEt
Y = t-butyl; Z = 9-anthrylO(CH$_2$)$_2$O

44; R = 9-anthrylmethyl

45; R = 9-anthrylmethyl

The opposite of the 2:1 multiplexer is the 1:2 demultiplexer, obviously. These two are often found together in computer engineering scenarios, because a demultiplexer is needed to recover and re-route the data bits which were sent down the single transmission line from a multiplexer. The gate array underlying the 1:2 demultiplexer is shown in Figure 10.18. Molecular 1:2 demultiplexers based on photochromism,[136] photocurrent generation,[137,138] enzyme cascades[139] and fluorescence[140,141] are known, but we will discuss a recent combined multiplexer/demultiplexer from Credi's laboratory.[142]

The simplicity of structure **46** is remarkable.[142] Though simple, its push–pull π-electron system gives rise to an excited state with considerable ICT character.

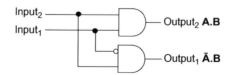

Figure 10.18 1:2 Demultiplexer.

More Complex Systems

Large H$^+$-induced red shifts are found for **46** due to the electron accepting pyridinium ring. In its 2:1 multiplexer incarnation, the two inputs are excitations close to the absorption maxima of **46** and its protonated form. The output is the emission intensity measured at a wavelength in-between the emission maxima of **46** and **46**.H$^+$, so that either form can contribute. The 1:2 demultiplexer version uses emission intensities at the two emission maxima as the two outputs. Excitation at a short-wave isosbestic point (262 nm) is the input, so that both forms of **46** would be equally excited. Naturally, the address input is H$^+$ in both these cases. We need to note that Katz's enzyme cascade based on glucose oxidase, laccase, glucose dehydrogenase and horseradish peroxidase also achieves this double of multiplexer/demultiplexer action.[139]

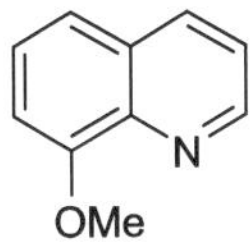

46

10.16 Other 3-Input Systems

We discussed Schmittel's linear diphenanthroline array (structure **142** in Chapter 7) in section 7.8 under XNOR logic. Here we feature the linear triphenanthroline array **47**.[143] Like its smaller cousin, **47** emits reasonably strongly only when all of its phenanthroline units are either protonated or not. The intermediate states, where only one or two units are protonated, are weak emitters. The fluorescence output can then be described in terms of the minimized Boolean algebraic expression given in Figure 10.19, which also shows the corresponding gate array in terms of electronic symbols.

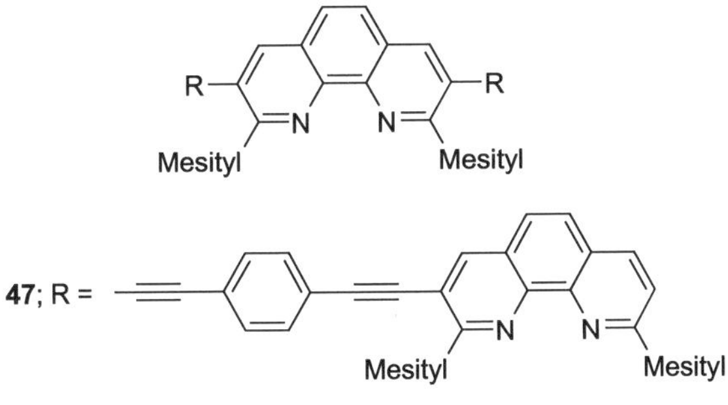

47; R =

10.17 Four-input AND

Molecular counterparts of 4-input AND logic devices (Figure 10.20) are rare,[144] but we can identify several such cases where anions are targeted by metal ions (or protons) which are themselves held by ligands (receptors).

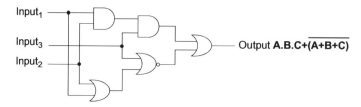

Figure 10.19 Logic gate array which arises from the fluorescence of a linear triphenanthroline array **47** upon stepwise protonation.

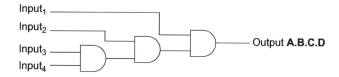

Figure 10.20 Four-input AND logic.

The end result is that fluorescence arises only when the metal ions and the anions are all present at sufficient concentrations. Vance and Czarnik's **48** goes back to 1994,[145] and represents $(H^+)_2$, pyrophosphate-driven AND behaviour. As we have seen before, the attachment of multiple PET-active receptors (*via* spacers) to a fluorophore is a general way of achieving more complex logic behaviour. For example, 2-input AND gate **49**[44] arose from the corresponding 1-input YES case **50**.[146] Czarnik expanded his older case **51**[47] to **48** so that it contained two polyamine units which could be partially protonated so each binds a phosphate motif. Crucially, **48** contains 1,8-difunctionalization on the anthracene fluorophore so that the two phosphate motifs can be fused together without bond strain. This is where the pyrophosphate selectivity comes from. As usual for fluorescent signalling systems which involve PET processes running from each free receptor, all receptors need to be bound up with their appropriate target species before fluorescence is released. So pyrophosphate and the two copies of the crucial H^+-transfer process (which was discussed earlier in section 7.2.1.2 regarding structure **25** therein) produce fluorescence switching 'on'. Related but different work from Hamachi[147–149] and Smith[150] can be recognized as $(Zn^{2+})_2$, (phosphate)$_2$-driven AND gates. These will be discussed under applications in Chapter 14.

A closely related case concerns $(Zn^{2+})_2$, pyrophosphate-driven AND logic. In the presence of Zn^{2+}, the phenol unit in **52**[151] deprotonates at pH 7.4. However, the proximity of two Zn^{2+} ions mitigates the effective electron density on the phenolate. On the other hand, this density increases upon the addition of pyrophosphate, which causes two phosphate faces to bind to the two Zn^{2+} centres. Thus the ICT excited state emits strongly at 456 nm (FE = 9.5), which serves as the output. A red shift of 20 nm is also seen.

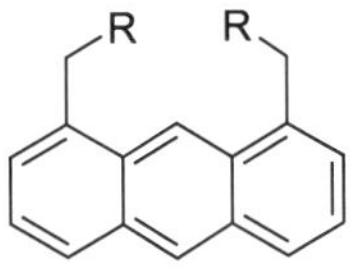

More Complex Systems 259

48; R = NH(CH$_2$)$_3$NH$^+$[(CH$_2$)$_3$NH$_3$$^+$]$_2$

50; R = N(CH$_2$CH$_2$OH)$_2$
51; R = NH(CH$_2$)$_3$NH$^+$[(CH$_2$)$_3$NH$_3$$^+$]$_2$

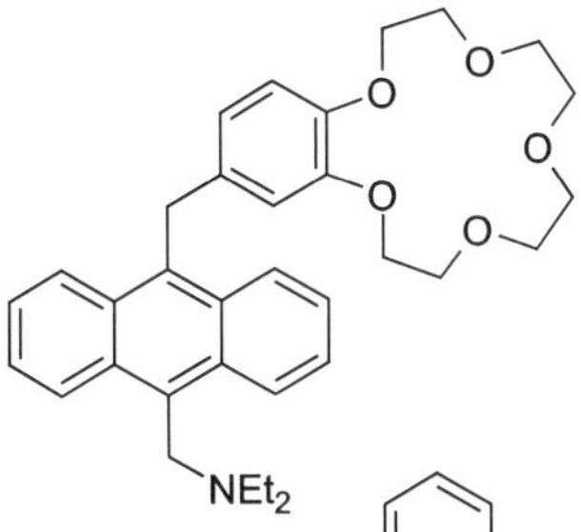

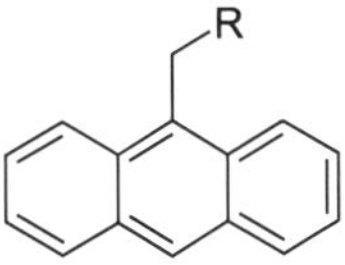

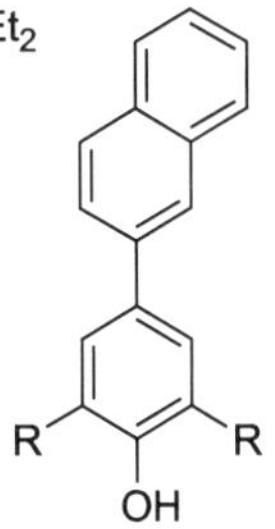

49

52; R = bis(2-pyridylmethyl)aminomethyl

10.18 Four-input Doubly Disabled AND

A (Zn^{2+})$_2$, imidazolate-driven gate (where the ligating nitrogens in imidazolate represents two disabling inputs) can be discerned in Fabbrizzi's **53**[152] where the polyamines catch a Zn^{2+} each so that they in turn trap an imidazolate unit of histidine in water at pH 9.6. Each Zn^{2+} arrests PET from the tertiary amines to the anthracene so that the fluorescence is switched 'on'. However, the imidazole unit is captured in its deprotonated form which launches a new PET process to the anthracene so that the fluorescence is switched 'off' again. The gate array corresponding to 4-input doubly disabled AND logic is given in Figure 10.21. Related works from the same laboratory have absorbance/fluorescence output which can be interpreted in similar ways.[153,154]

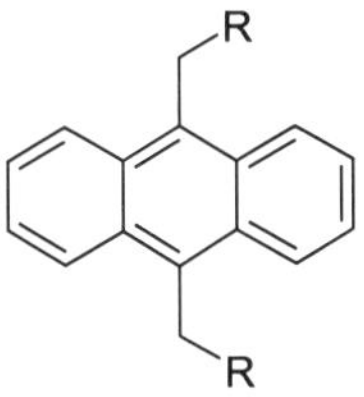

53; R = NH(CH$_2$)$_2$N[(CH$_2$)$_2$NH$_2$]$_2$

A 3-input relative of **53**'s logic behaviour can be recognized within Martínez-Máñez and Sancenón's work,[155] which was discussed in section 8.6, provided that fluorescence quenching is taken as the output. This refers to Pb^{2+}, Hg^{2+} and Fe^{3+} inputs. Pb^{2+}-disabled INHIBIT(Hg^{2+}) describes the logic seen here, but two other logic configurations are also available. When the observed parameter is altered to absorption spectral shift or voltammetric potential shift,

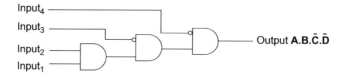

Figure 10.21 Four-input input$_3$, input$_4$-doubly disabled AND logic.

the logic type changes to TRANSFER(Pb^{2+}) or OR respectively. The interrogation of molecular switches with three (or more) techniques is rare.[155,156] Reconfigurable 3-input logic systems are still uncommon and several other examples are discussed in earlier sections of this chapter.

10.19 4-to-2 Encoder and 2-to-4 Decoder

A 4-to-2 encoder (electronic representation: Figure 10.22) compresses four unconnected bits of data into two (for long-distance transmission, for instance). This is done by reading the output as a two-bit number from two channels which permits binary–decimal number interconversions, for instance. Andreasson *et al.*[157] accomplish this compression in an all-optical manner by appreciating the rich light-induced spectral changes of **54** which contains two famous photochromes – fulgide[158] and dithienylethene[120] – while choosing the irradiating light doses carefully in order to have a workable scheme.

Two light doses at 397 (input$_2$) and 302 nm (input$_3$) were aimed at cyclizing the fulgide (signature absorption at 475 nm, output$_2$) and the dithienylethene (signature absorption at 625 nm, output$_1$) respectively. A light dose at 366 nm (input$_4$) cyclized both photochromes whereas one at >460 nm (input$_1$) caused cycloreversion of both photochromes. The latter dose is therefore also used to reset the device. Table 10.9 can be understood by these facts alone.

More Complex Systems

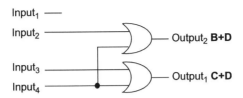

Figure 10.22 4-to-2 Encoder.

Table 10.9 Truth table for **54** operating as a 4-to-2 encoder.[a]

Input$_1$ hv dose$_{>460}$	Input$_2$ hv dose$_{397}$	Input$_3$ hv dose$_{302}$	Input$_4$ hv dose$_{366}$	Output$_1$ Abs$_{625}$	Output$_2$ Abs$_{475}$
1	0	0	0	0	0
0	1	0	0	0	1
0	0	1	0	1	0
0	0	0	1	1	1

[a]2.5×10^{-5} M in 2-methyltetrahydrofuran. The inputs and outputs are left as bits because the coding of information in this application and the binary–decimal number interconversion are clearer that way.

Table 10.10 Truth table for **54** operating as a 2-to-4 decoder.[a]

Input$_1$ hv dose$_{397}$	Input$_2$ hv dose$_{302}$	Output$_1$ Trans$_{335}$	Output$_2$ Flu$_{624}$	Output$_3$ Abs$_{393}$	Output$_4$ Abs$_{535}$
0	0	1	0	0	0
0	1	0	1	0	0
1	0	0	0	1	0
1	1	0	0	0	1

[a]2.5×10^{-5} M in 2-methyltetrahydrofuran. The inputs and outputs are left as bits because the coding of information in this application and the binary–decimal number interconversion are clearer that way.

For instance, the second line means that a dose of light at 397 nm cyclizes the fulgide alone and so the absorbance at 475 nm emerges.

It is to be noted that Table 10.9 and its counterpart Table 10.10 are operational truth tables each aimed at their particular task and do not consider all the binary combinations that are possible.

Balzani's group achieve the same mathematical truth table[159] by examining the simple, but iconic, tris(2,2′-bipyridyl)Ru(II) (**55**) as the starting state in terms of its electrochemical, luminescent and electrochemiluminescent behaviour. Electrochemical oxidation at 1.4 V, excitation at 450 nm and electrochemical reduction at −1.4 V stand for input$_1$, input$_2$ and input$_3$ respectively. Input$_4$ is represented by rapidly alternating oxidation and reduction, which is a way of inducing electrochemiluminescence.[160] Reset is arranged with electroreduction at 0.0 V. Output$_1$ is taken as absorbance at 530 nm where only reduced **55** (Ru oxidation state 1) absorbs. Output$_2$ is emission at 620 nm which only **55** can manage and only when it is excited at 450 nm ('high' input$_2$). Importantly, supply of 'high' input$_4$ produces a stationary state containing sufficient amounts of reduced **55** that output$_1$ becomes 'high'.

55

Once compressed data arrive at their destination, they need to be decoded in order to be intelligible. A 2-to-4 decoder reads data from two lines as two-bit numbers, then decompresses them into the four bits, each in its own channel. Figure 10.23 displays the corresponding gate array in electronic symbols. In its molecular equivalent, **54**, Andreasson et al.[157] employ light doses at 397 and 302 nm respectively, as input$_1$ and input$_2$ as the arriving bits. As noted in the previous paragraph, these light doses cause cyclization of the fulgide and dithienylethene moieties respectively. The outputs in Table 10.10 are assigned phenomenologically from the rich spectral information of **54** in order for the scheme to be successful, hence the mixture of transmittance, absorbance and emission employed as outputs. Output$_2$ is an exception because fluorescence at 624 nm arises from the cyclized fulgide when it is excited at 500 nm. It is worth noting that **54** can be reconfigured as several other gate arrays by choosing various wavelengths for irradiation and for observation.[161]

Balzani's group also arrange a 2-to-4 decoder[159] with **55** with a reassigned set of inputs and outputs as follows. Input$_1$ and input$_2$ are represented by oxidation at 1.4 V and reduction at −1.4 V respectively. Absorbances at 450, 310 and 530 nm are carefully chosen to represent output$_1$, output$_2$ and output$_3$ respectively. Emission at 620 nm (without optical excitation) is taken as output$_4$. Notably, the input string '11' is applied by simultaneous oxidation and reduction with separate working electrodes so that electrochemiluminescence[160] is emitted.

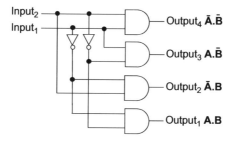

Figure 10.23 2-to-4 Decoder.

10.20 Other Four-Input (and Higher) Systems

Tian et al.[162] modify Irie's celebrated dithienylethene photochromics[120] with pyridyl groups, so as to benefit from H^+ and Zn^{2+} inputs in addition to the usual uv and visible light doses (see refs 125–140 in Chapter 7). The H^+- and Zn^{2+}-bound forms of **56** are fluorescent at 530 and 463 nm respectively, both of which disappear upon uv irradiation. An ICT behaviour can be seen, owing to the fluorophore–receptor integration. A rather complex truth table can be set up with **56**, especially when different emission and absorption wavelengths are observed,[17] to achieve four superposed logic configurations (see Chapters 8 and 13). However, memory effects need to be avoided to extract combinational logic, which may be possible by replacing the thiophene groups in **56** with pyrrole units.

56

When compared to most examples in this book, a relatively complex truth table can be found in the next example, due to Willner's group.[29] Acetylcholine, butyrylcholine, O_2 and glucose are applied as inputs to the enzyme cascade of acetylcholine esterase, choline oxidase, microperoxidase and glucose dehydrogenase, while the NADH concentration is monitored as the output. However, logic minimization with Karnaugh maps (for instance) leads to the gate array displayed in Figure 10.24. The chemistry involved in this example is built up from the discussion in section 7.2.6. Several other examples continuing in this vein, but with important differences, are due to Katz. An output, measured as a pH drop, goes 'high' if gluconic acid emerges by one of two routes.[163] The first of these involves glucose oxidase (GOx) if glucose (input$_3$) and O_2 (input$_4$) are made available to GOx. The second pathway is built around glucose dehydrogenase when it receives glucose (input$_3$) and NAD^+. The NAD^+ itself is produced from alcohol dehydrogenase when it processes NADH (input$_1$) and

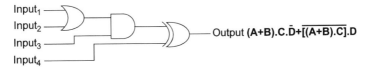

Figure 10.24 Logic array which arises from the enzyme cascade of acetylcholine esterase, choline oxidase, microperoxidase and glucose dehydrogenase operated according to Willner.[29]

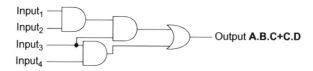

Figure 10.25 Logic array which arises from the enzyme system of glucose oxidase, glucose dehydrogenase and alcohol dehydrogenase operated according to Katz.[163]

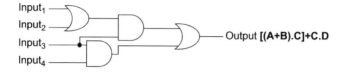

Figure 10.26 Logic array which arises from the enzyme system of horseradish peroxidase, glucose oxidase, lactate oxidase, amyloglucosidase and invertase operated according to Katz.[165]

ethanal (input$_2$). Figure 10.25 describes this gate array. The pH drop can control the current from electrochemical cells outfitted with polymer-modified electrodes,[163] similar to those described in section 7.3.1.[164] A similar analysis can be made for another gate array[165] (Figure 10.26) where the output is the chemiluminescence intensity of luminol (**57**) reacting with H_2O_2 under horseradish peroxidase (HRP) catalysis. The H_2O_2 is the product of lactate oxidase working on lactate (input$_4$) and O_2 (input$_3$) or glucose oxidase working on glucose and O_2 (input$_3$). In its turn, glucose is the result of a mixture of amyloglucosidase and invertase processing maltose (input$_1$) or sucrose (input$_2$) respectively. The substrate-selectivity of enzymes is on show in these carefully selected mixtures.

57

Some of the ideas underlying the enzyme systems in the previous paragraph are carried forward by Katz and Sokolov by adding antibody–antigen interactions:[166] 3,3′,5,5′-Tetramethylbenzidine (**58**) is oxidized to a product absorbing at 655 nm if both H_2O_2 and HRP are made available. The H_2O_2 arises only from the presentation of both glucose (input$_3$) and O_2 (input$_4$) to glucose oxidase. On the other hand, HRP arrives in the form of a HRP-labelled

More Complex Systems

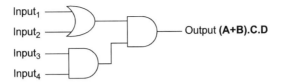

Figure 10.27 Logic array which arises from a particular antibody–enzyme cascade operated according to Katz.[163]

antigoat immunoglobulin (IgG) or a HRP-labelled antirabbit IgG. Either of these can bind to a surface prepared by treating a layer of dinitrophenyl–human serum albumin (DNP–HSA) and nitrotyrosine–bovine serum albumin (NT–BSA), with antiDNP IgG from goat (input$_1$) and antiNT IgG from rabbit (input$_2$). The absorbance output then corresponds to the gate array given in Figure 10.27. It should be noted that several components of the device are introduced in a sequence, perhaps for experimental convenience. Interestingly, an intermediate output of the logic device consisting of the antibody–antigen interaction is evaluated by atomic force microscopy.

58

We now consider other gate arrays, which have the potential to be even more complex than those considered thus far. A general approach to medium-scale integration can be achieved by using arrays of reaction pots/cells. Up to now, we distinguished inputs by their nature (*i.e.* H$^+$ or Na$^+$) alone. Now their spatial positions (or addresses) become available as a distinguishing feature. Increase in input number means that larger truth tables will result. These can be analysed in terms of bigger arrays of logic gate components.

Raymo and Giordani develop the multi-cuvet idea with photochromic compounds.[18] As mentioned in section 8.2, when a dose of ultraviolet light at 254 nm (input) hits **1**, a product absorbing at 563 nm is formed. When transmittance at 563 nm is taken as the output, a light dose-driven NOT logic action is found. If the transmittance measurement is now made through two cuvets with each running the previous experiment, formation of the product absorbing at 563 nm in either cuvet, or both, will cause a drop of the output signal.

The 254 nm light doses falling on each cuvet will now be input$_1$ and input$_2$ at 'high' levels. We now have 2-input NOR logic. Extension to a three-cuvet version will produce 3-input NOR gate action and so on.

Though these experiments are entirely optical in terms of inputs and outputs, it is to be noted that the latter are not quantitatively homogeneous. A dose of 254 nm light and a transmittance at 563 nm are not interchangeable (at two levels). So the logic integrations achievable by this approach do not arise by physically applying the output of one gate as the input to the next. We must also not lose sight of the fact that photochromic actions usually come with a memory effect (see Chapter 11). This aspect is avoided in the present discussion.

Here is an example[19,20] which is based on a classical colour test for sulfide that many of us encountered in high-school chemistry laboratories. Nitroprusside [(CN)$_5$(NO)Fe(II)]$^{2-}$ reacts with thiols to produce coloured {(CN)$_5$[N(O)SR]Fe(II)}$^{3-}$ (absorbing at 520 nm) only if the H$^+$ level is 'low' (allowing nucleophilic thiolate to be available) and the salt (KCl) concentration is 'high' (so that K$^+$ stabilizes the trianionic product by ion-pairing). Thus we have a H$^+$, K$^+$-driven INHIBIT(H$^+$) gate with absorbance output.

Unlike the zero-dimensional case where the reaction is run in a single cuvet, Figure 10.28 considers the one-dimensional case of a line of two cuvets,[20] with a weak light source (to prevent photochemical reactions) emitting at 520 nm and a photodetector as the terminus. It is clear that a high absorbance is detected if either cuvet a or cuvet b develops the coloured {(CN)$_5$[N(O)SR]Fe(II)}$^{3-}$, *i.e.* if either cuvet has 'low' H$^+$ and 'high' K$^+$. The equivalent logic gate array is composed of two H$^+$, K$^+$-driven INHIBIT(H$^+$) gates feeding an OR gate.

More complex logic gate arrays can be developed from the same thermal reaction and a given configuration of cuvets by considering photochemical (520 nm) conversion of {(CN)$_5$[N(O)SR]Fe(II)}$^{3-}$ to a secondary product which absorbs light at 700 nm, *i.e.* a photochromic process. Either the absorbance measurement at 520 nm or the photochemical experiment (followed by absorbance measurement at 700 nm) can be examined in two- or three-dimensional arrays of cuvets so that gate arrays of even higher complexity are realized. Some of these are the most complex molecular logic gate integrations

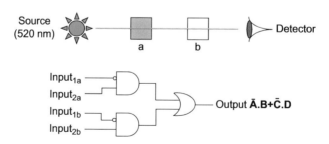

Figure 10.28 Cell arrays for developing logic gate arrays. Input$_{1a}$, input$_{2a}$, input$_{1b}$ and input$_{2b}$ are symbolized by variables A, B, C and D for algebraic purposes.

More Complex Systems 267

seen to date. This thermal reaction responds to many variables (nature of sulfide derivative, temperature, pressure, *etc.*) and these can be used to increase the logical complexity still further. However, these examples do not represent molecular-scale logic because macroscopic cells are necessary.

10.21 Higher Arithmetic Systems

The logic basis of basic arithmetic systems such as half-adders and half-subtractors has been discussed in Chapters 3 and 9 respectively. Molecular versions of these have been considered in Chapter 9. Now we use these as springboards to perform more involved arithmetic functions.

10.21.1 Combined Half-adder and Half-subtractor

Multiple logic systems loaded onto a single molecular structure[12,167] have a special charm, especially when arithmetic operations are involved. Shanzer's group developed the first unimolecular arithmetic processor **59** by incorporating a half-adder and a half-subtractor so that it allows both addition and subtraction of two bits.[168] It behaves as a reconfigurable molecular logic system, as seen in Chapters 6 and 8, by using a select few chemical inputs (H^+, $CH_3CO_2^-$ as a base, Fe^{3+} and EDTA) and fluorescence (blue and green) outputs. Compound **59** consists of a modified bacterial iron carrier chain (siderophore) linking pyrene and fluorescein fluorophores. The acid and base inputs interact with the fluorescein and siderophore component, whereas Fe^{3+} binds only with the siderophore.

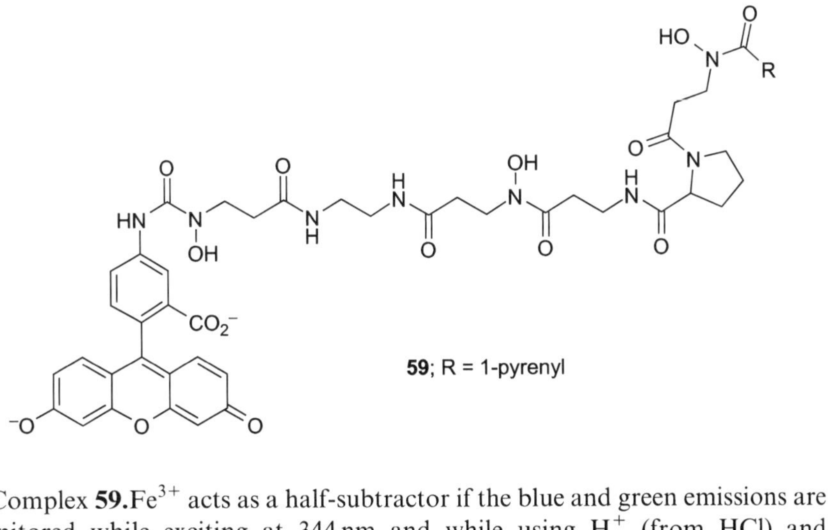

59; R = 1-pyrenyl

Complex **59**.Fe^{3+} acts as a half-subtractor if the blue and green emissions are monitored while exciting at 344 nm and while using H^+ (from HCl) and $CH_3CO_2^-$ as the inputs. In ethanol solution, no emission is observed owing to the fluorescence quenching ability of Fe^{3+} via PET/EET mechanisms.

H$^+$ addition protonates the fluorescein (rendering it a non-fluorescent and less-delocalized π-system) and hydroxamate units (causing release of Fe^{3+}). In this state, the only emission observed is from the pyrene unit at 390 nm. Addition of CH$_3$CO$_2^-$ produces fluorescein dianion emission at 525 nm (by EET from pyrene) even in the presence of Fe^{3+}. Formation of hydroxo complexes of Fe^{3+} is a probable cause of this loss of quenching ability. With equimolar H$^+$ and CH$_3$CO$_2^-$ added, neutralization occurs and the Fe^{3+} remains bound, resulting in an unchanged output. INHIBIT logic is achieved when the fluorescence is read at any one of these two wavelengths (525 or 390 nm). XOR logic requires that the fluorescence intensity is collected simultaneously at 390 and 525 nm, which can be arranged with an appropriate filter. From a computing vantage, this can be understood as the construction of the XOR logic function by summing two complementary INHIBIT functions[169,170] arising in the blue and green output channels.

Now H$^+$ (previously supplied as HCl) is replaced by EDTA, a weaker acid and a strong Fe^{3+} binder. With no inputs, no emission is observed, as before. When EDTA is added, the fluorescein is protonated (hence becoming non-fluorescent) and Fe^{3+} is extracted, causing emission from pyrene at 390 nm. If CH$_3$CO$_2^-$ is added the fluorescein is deprotonated but Fe^{3+} is not removed which leads to an increase in the 525 nm intensity, though well below the threshold. Only in the presence of CH$_3$CO$_2^-$ and EDTA can a strong output from fluorescein be observed at 525 nm. This is the AND logic that we need to combine with the XOR gate seen in the previous paragraph in order to obtain the half-adder. From a practical viewpoint, there is a difficulty in running this combination at the same time because the proton input needs to be provided in two different ways (HCl and EDTA). Except for this caveat, we have the half-adder and the half-subtractor operations within **59**.Fe^{3+}.

Newer cases of packaged half-adders and half-subtractors are available. Yan's group presents several cases,[171] including **60**.(Cu^{2+})$_2$[172] whose fluorescence is manipulated with H$^+$ and OH$^-$ (or S^{2-}). Another is Liu's **61**,[173] where the absorption and emission spectra are similarly manipulated. Starting states of **61** and **61**.H$^+$ are used as appropriate. This case is distinguished by including bidirectional half-subtractor action.[174] Its comparator capability was discussed in section 9.5. The response of the absorption spectrum of **62**.H$^+_2$ towards OH$^-$ is analysed by Song's team in a similar manner.[175] A rare example of a combined half-subtractor and comparator is unveiled by the laboratory of Tian and Zhu[176] via the analysis of Zn^{2+}, OH$^-$-induced changes in the absorption and emission spectra of **63**. The ICT nature of all these excited states contributes to the spectral diversity which is essential to yield these arithmetic functions.

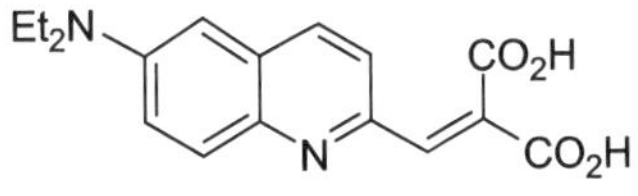

More Complex Systems   269

62

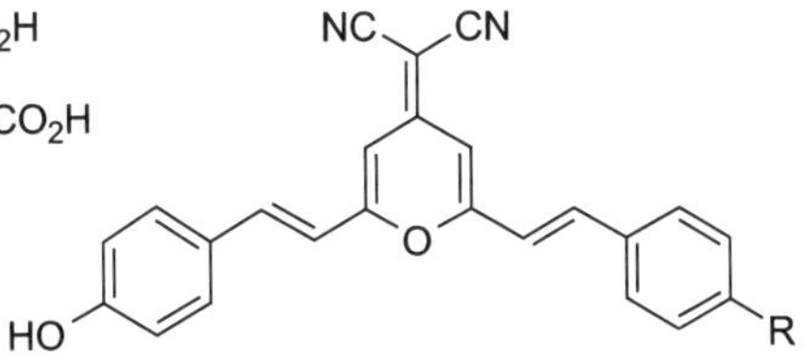

63; R = bis(2-pyridylmethyl)amino

H$^+$ and OH$^-$ serve as inputs to Qian's **64**,[177] whose absorption and fluorescence spectral outputs correspond to a half-subtractor if mixed logic conventions are used. Half-adder function is seen if two inputs of 4.2 mM SDS in neutral water are applied and the absorbances at 400 and 425 nm are taken as the outputs, along with mixed logic conventions. When free of SDS, **64** has λ_{abs} = 386 nm but it shifts to 425 nm in 8.4 mM SDS. Notably, the absorption spectra cross at 400 nm. This red-shift is due to protonation of **64** (and stabilization of its ICT excited state) by the concentrated H$^+$ population near the anionic micelle surface, given that the critical micelle concentration of SDS is 8 mM.[178] On the other hand, 4.2 mM SDS produces only pre-micellar aggregates with **64** which show a suppressed absorption spectrum maximizing at 395 nm.

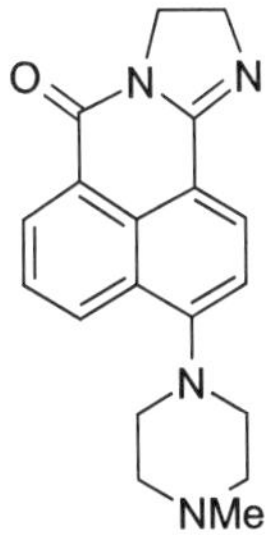

64

Though not based on a single kind of molecular structure, a single system of enzymes can also give rise to combined half-adder and half-subtractor behaviour.[30,31] Glucose, H$_2$O$_2$-driven AND logic behaviour seen in the gluconic acid output from the system of glucose dehydrogenase (GDH), HRP and NAD$^+$[179] was discussed in section 7.2.6. XOR logic can be extracted if the modulus of absorbance changes at 340 nm (where NADH absorbs) is taken as output under initial conditions where NADH and NAD$^+$ are present in equal amounts. This has the weakness of a manual mathematical operation being required in order to demonstrate a molecular mathematical action. Similar weaknesses arise when INHIBIT logic is set up *via* the same output by using a different initial condition of only NAD$^+$ (without NADH).

10.21.2 Full-adder

Full-adders allow manipulation of decimal numbers up to three. Figure 10.29 shows the gate array required, which clearly shows the concatenation of two half-adders and the OR combination of the two carry lines. Molecular versions of these have been approached with photoinduced hole transport in DNA strands[169] and with fluorogenic deoxyribozyme reactions.[180] The latter approach develops the ideas and structures discussed under half-adders in Chapter 9. We feature the former case in the next paragraphs. Several other approaches based on oligonucleotides are available.[181–184] Even a case based on silver clusters belongs here.[185]

Logic gate action is extracted from the hole-transporting ability of a short synthetic DNA oligomer comprising adenine modified with electron-rich motifs to aid the process of hole-transport along the strand (the 'logic gate strand').[169] The hole-transport efficiency is controlled by the opposite bases on the hybridizing strand (the 'input strand'). A thymine (taken as input = 1) is particularly good in this capacity whereas a cytosine (taken as input = 0) is poor, even though both hybridize successfully to the electron-rich adenine. Hole injection is done via PET to an electron-deficient thymine from an adjacent guanine following 312 nm irradiation.[186] The hole runs along the strand before being trapped at guanine triplets and causes strand scission at that location, which is detected by polyacrylamide gel electrophoresis. One set of guanine triplets is placed before the electron-rich adenine and the other is placed after. Hole transport efficiency in the logic gate strand is defined by the oxidative damage ratio of the proximal versus the distal guanine triplets. A high damage ratio is taken as output = 1.

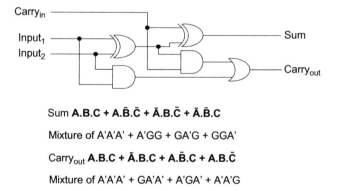

Sum $A.B.C + A.\bar{B}.\bar{C} + \bar{A}.B.\bar{C} + \bar{A}.\bar{B}.C$

Mixture of A'A'A' + A'GG + GA'G + GGA'

Carry$_{out}$ $A.B.C + \bar{A}.B.C + A.\bar{B}.C + A.B.\bar{C}$

Mixture of A'A'A' + GA'A' + A'GA' + A'A'G

Figure 10.29 Full-adder. Input$_1$, input$_2$ and carry$_{in}$ are symbolized by variables A, B and C for algebraic purposes. Schematic representation of the DNA logic gate strand mixtures required for implementation of the full-adder is also given. The electron-rich adenine is represented by A'. In practice, there are spacer bases between each of these key bases (A' and G) and guanine triplets on either side. The strands are directional, *i.e.* the 5'-terminal is on the left. The hole is injected from the left.

Table 10.11 Truth table for the full adder *via* hole-transport in DNA with electron-rich adenine derivatives.[a]

$Input_1$ T or C[b]	$Input_2$ T or C[b]	$Input_3$ $Carry_{in}$ T or C[b]	$Output_1$ $Carry_{out}$ Damage[c]	$Output_2$ Sum Damage[c]
0	0	0	0	0
0	1	0	0	1
1	0	0	0	1
1	1	0	1	0
0	0	1	0	1
0	1	1	1	0
1	0	1	1	0
1	1	1	1	1

[a] 10^{-6} M in water at pH 7, 312 nm irradiation at 0 °C, work-up and gel electrophoretic analysis.
[b] Thymine or Cytosine in the site of the input strand opposite to the electron-rich adenine in the logic gate strand. Thymine is taken as 1 and cytosine is taken as 0.
[c] Oxidative damage ratio.

If a guanine is substituted for the electron-rich adenine, a cytosine input is needed in the hybridizing strand in order to undergo base-pairing and, subsequently, efficient hole-transport. In this instance, an output of 1 is obtained for the input of 0 (which is the cytosine). Thus, YES and NOT logic output are separately obtained when the logic gate strand contains electron-rich adenine and guanine respectively.

As seen in Figure 10.29 (and as can be deduced from Table 10.11), the Sum and Carry$_{out}$ functions are composed of the sum of four triple product terms. Triple inputs are built in to the input strand by inserting a combination of three guanines or electron-rich adenines between the guanine triplets. Four logic gate strands are mixed to obtain each function. Figure 10.29 also details these requirements. The two mixtures are examined separately by irradiation and subsequent gel electrophoresis. Overall, this DNA hole-transport method appears to be rather general and exploits the 'sum of products' approach to logic analysis (section 3.3) very well.

The team responsible for the TPF-based half-adder[15] also develop simulations of full-adders. However, the authors note that experimental realization of these approaches is complex.[187,188] They also test **65**,[189,190] which is a covalently bound version of the rhodamine and azulene components used in the half-adder.[15] It is notable that the fluorescence and the TPF of each component is now observed in spite of the weakness of some of these phenomena. Then each component can be viewed as a half-adder so that fluorescence and TPF outputs follow XOR and AND logic respectively when sub-threshold laser beam intensities are used as the inputs. Indeed, Wasielewski's early work[191] can be interpreted in a similar way (section 7.2.5). Importantly, **65** shows efficient EET ($\phi_{EET} = 0.96$) from the rhodamine unit to the azulene moiety, which demonstrates the communication between the two half-adders. However, the OR combination of the two carry lines has to be done manually.

65

10.21.3 Combined Full-adder and Full-subtractor

Shanzer's team[192] use the simple absorptiometric pH sensor fluorescein (**66**) to show full-adder and full-subtractor operation, *i.e.* a molecular-scale calculator (a 'moleculator'). The gate array for a full-subtractor is given in Figure 10.30. Having already shown fluorescein to operate as a combined half-adder and half-subtractor,[193] they expand its operations by exploiting fluorescein's four ionization states (+1, 0, −1, −2) each with its own signature absorption spectrum (Figure 10.31). These signatures are all distinguishable because the neutral form **66** has limited π-conjugation when compared with the cationic (**67**) and anionic (**68**) forms. In fact, the anionic form (**68**) has a red-shifted absorption band (474 nm) as compared to that of the cationic version (**67**), whereas the neutral form absorbs minimally across the wavelength range. Mutual annihilation of inputs, degeneracy of inputs, and switching between positive and negative logic with absorbance and transmittance outputs also play essential roles.

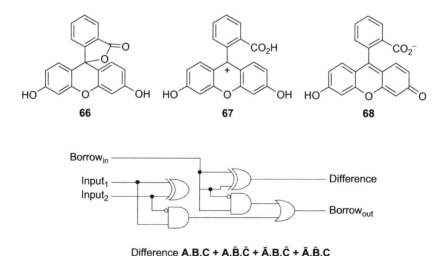

Difference **A.B.C + A.B̄.C̄ + Ā.B.C̄ + Ā.B̄.C**

Borrow_{out} **A.B.C + Ā.B.C + Ā.B̄.C + Ā.B.C̄**

Figure 10.30 Full-subtractor. Input$_1$, input$_2$ and borrow$_{in}$ are symbolized by variables A, B and C for algebraic purposes.

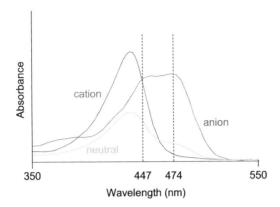

Figure 10.31 pH-Dependent absorption spectra of **66**. Each of the neutral, cationic and anionic forms becomes predominant at pH values of 3.3, 1.9 and 5.5 respectively. The dianionic form (not shown) comes to the fore at pH 12.
Adapted from D. Margulies, G. Melman and A. Shanzer, *J. Am. Chem. Soc.*, 2006, **128**, 4865 with permission from the American Chemical Society.

As generalized in section 9.3, on half-subtractors, this arithmetic device arises from molecules containing acidic and basic groups such as **66**, if an equivalent each of H^+ and OH^- are used as inputs, especially in non-aqueous solvents. In water, the arguments have to be made with regard to the protonation states rather than the actual H^+ concentrations that are applied. In the present case, the spectrum in Figure 10.31 at 447 nm shows us that the absorbance is low (output '0') for the neutral form **66** ($input_1$ '0', $input_2$ '0') found at pH 3.3. On the other hand, the absorbance is high (output '1') for the cation **67** ($input_1$ '1', $input_2$ '0') or anion **68** ($input_1$ '0', $input_2$ '1') forms which dominate the population at pH 1.9 and 5.5 respectively. Under the present regime the input condition ($input_1$ '1', $input_2$ '1') means that the neutral form of fluorescein **66** is treated with equivalent amounts of H^+ and OH^- so that the neutral form is preserved. Mutual annihilation of the two inputs can be imagined to produce the same output (output '0') as the first row ($input_1$ '0', $input_2$ '0'), *i.e.* the absorption spectrum of the fluorescein neutral form. This is XOR logic.

A similar examination of the spectra in Figure 10.31, but at 474 nm, shows us that the absorbance is high (output '1') only in the anionic form **68** ($input_1$ '0', $input_2$ '1'). All the other three sets of input conditions produce low absorbance (output '0') which satisfies the truth table for INHIBIT logic.

A third input is now introduced in the form of another equivalent of OH^-, so that the half-subtractor can be grown into the full-subtractor. The third input is the borrow-in bit. The output channel wavelengths remain the same. Note that $input_2$ and $input_3$ are only distinguishable to the operator of the 'moleculator' and not to the molecule itself. However at the end of the day, the operator is the one requiring the correct answer to the calculation.

Table 10.12 Truth table for **66** operating as a full-subtractor.[a]

Input$_1$ H$^+$	Input$_2$ OH$^-$	Input$_3$ Borrow$_{in}$ OH$^-$	Output$_1$ Borrow$_{out}$ Abs$_{474}$	Output$_2$ Difference Abs$_{447}$
0	0	0	0	0
0	1	0	1	1
1	0	0	0	1
1	1	0	0	0
0	0	1	1	1
0	1	1	1	0
1	0	1	0	0
1	1	1	1	1

[a]6×10^{-6} M in water.

Under this chemical regime, comparing the truth table for the full-subtractor (Table 10.12) with that the half-subtractor (Chapter 9, Table 9.2) shows many similarities, in spite of the initial complexity of the former. For instance any row with the third input value of zero (the top four rows) is simply carried through from the half-subtractor. Any row where equivalents of H$^+$ and OH$^-$ are both absent (0 and 0 respectively) would give the same result as another row where they appear (1 and 1 respectively) owing to their neutralization. We are only left the row with just two inputs of OH$^-$ (1 and 1) to understand. In chemical terms, this means the dianionic form of **66**, *i.e.* **69**, will be produced (at pH 12). Luckily, this form absorbs strongly at 474 nm and weakly at 447 nm which fits the outputs required quite neatly. The choice of monitoring wavelengths assists the researchers in their drive to match the observations to the truth table's requirements.

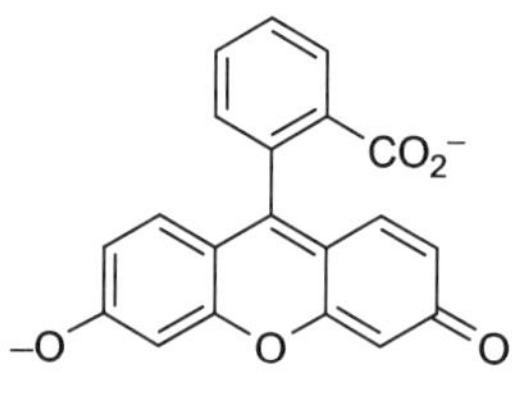

69

Now, let us see how Margulies *et al.* apply new coding schemes to the same data set employed for the half-subtractor in order to produce a half-adder. First we need a change of the starting state to the cationic state **67**. One mole of OH$^-$ (input$_1$) converts it to neutral **66** and another mole of OH$^-$ (input$_2$) produces anionic **68**. So we are working with the same three states as we did for the half-subtractor. The old set of absorption spectra is all we need, but we must carefully take the transmittance at 447 nm to attain XOR logic behaviour. Absorbance data at 474 nm still serve to produce AND logic characteristics. Such opportunistic choice of coding – switching between positive and negative

More Complex Systems 275

logic – is legal. However, we note that gates arising from opposite codings cannot be technically combined at the same time.

Similar to the discussion in the previous paragraphs concerning the full-subtractor, the full-adder truth table (Table 10.11) also has much that is familiar from the corresponding logic table for the half-adder (Chapter 9, Table 9.1), such as the first four rows. Notably, the present case uses a three-fold degeneracy of the inputs, *i.e.* input$_1$, input$_2$ and input$_3$ all use a mole of OH$^-$. The fifth row is algebraically equivalent to the second. The sixth and seventh rows are equivalent to the fourth. So only the last row remains to be justified and that is done as follows. Treatment of **67** with three moles of OH$^-$ produces **69** (again) and hence has strong absorbance at 474 nm and strong transmittance at 447 nm (as noted in different language regarding the full-subtractor).

This example is particularly remarkable because it achieves such a human-relevant computation with such a small molecule and such a simple experimental protocol. It also shows that a variety of selective receptors is not always required within a molecular device to demonstrate a rather complex logic array. Song's team[172] provide another example of a combined full-adder and full-subtractor by similarly analysing the pH-dependent fluorescence of a 1:2 mixture of **62** and **70**. Mixtures of compounds have yielded logic dividends before.[1,194,195]

70

10.22 Gaming Systems: Tic-tac-toe

All of us would have played this game in the school-yard by putting noughts and crosses on a 3 × 3 square grid. If the first player has any sense, (s)he places the cross (or nought) in the centre square because there is no losing from then on. So it is particularly apt that the first report on molecular gaming[92] takes the first move for the molecules. The human opponent has no chance. Stojanovic and Stefanovic's work featuring a set of deoxyribozyme logic gates[92,196,197] has been rightly described as a *tour de force*.[198] The planning of this system exploits an analysis detailing all possible moves in the form of a tree diagram[199] followed by pruning according to symmetry arguments. The first move by the opponent can only be to a corner or a side.

Up to six carefully selected and optimized logic gates are placed in each well of the 3 × 3 matrix, so that a competitive response is made by the molecules whatever move is made by the human opponent. Then the human has no chance of achieving three 'noughts' in a straight line while playing by the rules. When the human move is indicated, this is coded with an oligonucleotide which is then added to all wells. After 15 min of molecular activity (akin to human deliberation in this case), detectable fluorescence arises in a particular well. This is the competitive molecular response. The famed chess contests between Garry Kasparov and the Deep Blue computer now have a small molecular analogue.[200]

The most complex of these gates is a three-input INHIBIT gate which was discussed in section 10.5 and shown in Figure 10.5. These logic gates behave like molecular beacons which radiate fluorescence upon chemical command.[201] For instance, well 1 (top left corner) contains a YES gate which will respond to an input coded by an opponent move into well 4 (left side). This means that the molecular system reacts to the opponent move into well 4 by signalling its own move into well 1. Other remaining wells contain either 2-input AND, 3-input INHIBIT or a YES gate (responding to an opponent move into well 1). So it is clear that all these wells will remain at low fluorescence levels following the opponent move into well 4.

Though this is child's play, similar molecular systems can respond to disease indicators. Applications beckon in diagnostics and therapy. Games which are simpler than tic-tac-toe can even be learned by the molecular system.[202,203]

References

1. A. P. de Silva and N. D. McClenaghan, *J. Am. Chem. Soc.*, 2000, **122**, 3965.
2. F. Remacle, R. Weinkauf, D. Steinitz, K. L. Kompa and R. D. Levine, *Chem. Phys.*, 2002, **281**, 363.
3. J. Millman and A. Grabel, *Microelectronics*, McGraw-Hill, New York, 2nd edn, 1988.
4. A. P. Malvino and J. A. Brown, *Digital Computer Electronics*, Glencoe, New York, 3rd edn, 1993.
5. M. Ben-Ari, *Mathematical Logic for Computer Science*, Prentice-Hall, Hemel Hempstead, 1993.
6. J. R. Gregg, *Ones and Zeros*, IEEE Press, New York, 1998.
7. A. L. Sedra and K. C. Smith, *Microelectronic Circuits*, Oxford University Press, Oxford, 5th edn, 2003.
8. F. M. Brown, *Boolean Reasoning*, Dover, New York, 2nd edn, 2003.
9. C. Maxfield, *From Bebop to Boolean Boogie*, Newnes, Oxford, 2009.
10. E. Katz and V. Privman, *Chem. Soc. Rev.*, 2010, **39**, 1835.
11. U. Pischel, *Aust. J. Chem.*, 2010, **63**, 148.
12. H. Tian, *Angew. Chem. Int. Ed.*, 2010, **49**, 4710.
13. A. P. de Silva, *Chem. Asian J.*, 2011, **6**, 750.
14. A. P. de Silva, I. M. Dixon, H. Q. N. Gunaratne, T. Gunnlaugsson, P. R. S. Maxwell and T. E. Rice, *J. Am. Chem. Soc.*, 1999, **121**, 1393.
15. F. Remacle, S. Speiser and R. D. Levine, *J. Phys. Chem. B*, 2001, **105**, 5589.
16. F. M. Raymo and S. Giordani, *Org. Lett.*, 2001, 1833.
17. F. M. Raymo and S. Giordani, *J. Am. Chem. Soc.*, 2002, **124**, 2004.

18. F. M. Raymo and S. Giordani, *Proc. Natl. Acad. Sci. USA*, 2002, **99**, 4941.
19. K. Szacilowski, *Chem. Eur. J*, 2004, **10**, 2520.
20. K. Szacilowski and Z. Stasicka, *Coord. Chem. Rev.*, 2002, **229**, 17.
21. E. Bishop (ed.), *Indicators*, Pergamon, Oxford, 1972.
22. F. M. Raymo and S. Giordani, *J. Am. Chem. Soc.*, 2001, **123**, 4651.
23. F. M. Raymo and S. Giordani, *Org. Lett.*, 2001, **3**, 3475.
24. S. Giordani, M. A. Cejas and F. M. Raymo, *Tetrahedron*, 2004, **60**, 10973.
25. F. M. Raymo, R. J. Alvarado, S. Giordani and M. A. Cejas, *J. Am. Chem. Soc.*, 2003, **125**, 2361.
26. A. P. de Silva and R. A. D. D. Rupasinghe, *J. Chem. Soc. Chem. Commun.*, 1985, 1669.
27. X. F. Guo, D. Q. Zhang, Y. C. Zhou and D. B. Zhu, *Chem. Phys. Lett.*, 2003, **375**, 484.
28. J. M. Berg, J. L. Tymoczko and L. Stryer, *Biochemistry*, Freeman, San Francisco, 5th edn, 2002.
29. T. Niazov, R. Baron, E. Katz, O. Lioubashevski and I. Willner, *Proc. Natl. Acad. Sci. USA*, 2006, **103**, 17160.
30. R. Baron, O. Lioubashevski, E. Katz, T. Niazov and I. Willner, *Angew. Chem. Int. Ed.*, 2006, **45**, 1572.
31. R. Baron, O. Lioubashevski, E. Katz, T. Niazov and I. Willner, *J. Phys. Chem. A*, 2006, **110**, 8548.
32. G. Seelig, D. Soloveichik, D. Y. Zhang and E. Winfree, *Science*, 2006, **314**, 1565.
33. W. Fontana, *Science*, 2006, **314**, 1552.
34. G. Seelig, B. Yurke and E. Winfree, *J. Am. Chem. Soc.*, 2006, **128**, 12211.
35. B. M. Frezza, S. L. Cockroft and M. R. Ghadiri, *J. Am. Chem. Soc.*, 2007, **129**, 14875.
36. D. Y. Zhang and G. Seelig, *Nature Chem*, 2011, **3**, 103.
37. X. F. Guo, D. Q. Zhang and D. B. Zhu, *Adv. Mater.*, 2004, **16**, 125.
38. C. A. Hunter and L. D. Sarson, *Tetrahedron Lett.*, 1996, **37**, 699.
39. Y. Hirshberg, *J. Am. Chem. Soc.*, 1956, **78**, 2304.
40. F. Pragst and E. Weber, *J. Prakt. Chem.*, 1976, **318**, 51.
41. D. C. Magri, G. J. Brown, G. D. McClean and A. P. de Silva, *J. Am. Chem. Soc.*, 2006, **128**, 4950.
42. D. C. Magri, G. D. Coen, R. L. Boyd and A. P. de Silva, *Inorg. Chim. Acta*, 2006, **568**, 156.
43. A. P. de Silva, H. Q. N. Gunaratne and C. P. McCoy, *Nature*, 1993, **364**, 42.
44. A. P. de Silva, H. Q. N. Gunaratne and C. P. McCoy, *J. Am. Chem. Soc.*, 1997, **119**, 7891.
45. O. A. Bozdemir, R. Guliyev, O. Buyukcakir, S. Selcuk, S. Kolemen, G. Gulseren, T. Nalbantoglu, H. Boyaci and E. U. Akkaya, *J. Am. Chem. Soc.*, 2010, **132**, 8029.

46. A. P. de Silva, G. D. McClean and S. Pagliari, *Chem. Commun.*, 2003, 2010.
47. M. E. Huston, E. U. Akkaya and A. W. Czarnik, *J. Am. Chem. Soc.*, 1989, **111**, 8735.
48. J. P. Leonard, C. M. G. dos Santos, S. E. Plush, T. McCabe and T. Gunnlaugsson, *Chem. Commun.*, 2007, 129.
49. F. H. Richardson, *Chem. Rev.*, 1982, **82**, 541.
50. D. Parker and J. A. G. Williams, *J. Chem. Soc. Dalton Trans.*, 1996, 3613.
51. J. Michl and V. Bonačić-Koutecký, *Electronic Aspects of Organic Photochemistry*, Wiley, New York, 1990.
52. M. Klessinger and J. Michl, *Excited States and Photochemistry of Organic Molecules*, VCH, New York, 1995.
53. M. D. P. de Costa, A. P. de Silva and S. T. Pathirana, *Canad. J. Chem*, 1987, **69**, 1416.
54. Z. Lin, M. Wu, M. Schaeferling and O. S. Wolfbeis, *Angew. Chem. Int. Ed.*, 2004, **43**, 1735.
55. V. Privman, M. A. Arugula, J. Halamek, M. Pita and E. Katz, *J. Phys. Chem. B*, 2009, **113**, 5301.
56. D. Melnikov, G. Strack, M. Pita, V. Privman and E. Katz, *J. Phys. Chem. B*, 2009, **113**, 10472.
57. M. Takeuchi, M. Yamamoto and S. Shinkai, *Chem Commun.*, 1997, 1731.
58. M. Yamamoto, M. Takeuchi and S. Shinkai, *Tetrahedron*, 1998, **54**, 3125.
59. P. Ghosh, P. K. Bharadwaj, S. Mandal and S. Ghosh, *J. Am. Chem. Soc.*, 1996, **118**, 1553.
60. P. Ghosh, P. K. Bharadwaj, J. Roy and S. Ghosh, *J. Am. Chem. Soc.*, 1997, **119**, 11903.
61. B. Bag and P. K. Bharadwaj, *J. Lumin.*, 2004, **110**, 85.
62. B. Bag and P. K. Bharadwaj, *J. Phys. Chem. B*, 2005, **109**, 4377.
63. N. J. Turro, V. Ramamurthy and J. C. Scaiano, *Modern Molecular Photochemistry of Organic Materials*, University Science Books, Mill Valley, CA, 2010.
64. S. P. Wu, T. H. Wang and S. R. Liu, *Tetrahedron*, 2010, **66**, 9655.
65. S. P. Wu, Z. M. Huang, S. R. Liu and P. K. Chung, *J. Fluoresc.*, 2012, **22**, 253.
66. J. F. Callan, A. P. de Silva, J. Ferguson, A. J. M. Huxley and A. M. O'Brien, *Tetrahedron*, 2004, **60**, 11125.
67. C. P. Collier, E. W. Wong, M. Belohradsky, F. M. Raymo, J. F. Stoddart, P. J. Kuekes, R. S. Williams and J. R. Heath, *Science*, 1999, **285**, 391.
68. C. P. Collier, G. Mattersteig, E. W. Wong, Y. Luo, K. Beverly, J. Sampaio, F. M. Raymo, J. F. Stoddart and J. R. Heath, *Science*, 2000, **289**, 1172.
69. A. R. Pease, J. O. Jeppesen, J. F. Stoddart, Y. Luo, C. P. Collier and J. R. Heath, *Acc. Chem. Res.*, 2001, **34**, 433.
70. P. Singh and S. Kumar, *New J. Chem.*, 2006, **30**, 1553.
71. P. N. Cheng, P. T. Chiang and S. H. Chiu, *Chem. Commun.*, 2005, 1285.

72. P. T. Chiang, P. N. Cheng, C. F. Lin, Y. H. Liu, C. C. Lai, S. M. Peng and S. H. Chiu, *Chem. Eur. J*, 2006, **12**, 865.
73. R. Y. Tsien, *Biochemistry*, 1980, **19**, 2396.
74. R. A. Beecroft, R. S. Davidson, D. Goodwin, J. E. Pratt and X. J. Luo, *J. Chem. Soc. Faraday 2*, 1986, **82**, 2393.
75. M. Bender and M. Komiyama, *Cyclodextrin Chemistry*, Springer-Verlag, New York. 1971.
76. R. A. Bissell and A. P. de Silva, *J. Chem. Soc. Chem. Commun.*, 1991, 1148.
77. Y. L. Peng, Y. T. Wang, Y. Wang and W. J. Jin, *J. Photochem. Photobiol. A: Chem*, 2005, **173**, 301.
78. M. Gouterman, *J. Chem. Educ.*, 1997, **74**, 697.
79. V. Moshasrov, V. Radchenko and S. Fonov, *Luminescent Pressure Sensors in Aerodynamic Experiments*, Central Aerohydrodynamic Institute, Moscow, 1998.
80. X. Lu, I. Manners and M. A. Winnik, in *New Trends in Fluorescence Spectroscopy*, ed. B. Valeur and J.-C. Brochon, Springer, Berlin, 2001, p. 229.
81. J. D. Bolt and N. J. Turro, *Photochem. Photobiol.*, 1982, **35**, 305.
82. L. X. Mu, Y. Wang, Z. Zhang and W. J. Jin, *Chinese Chem. Lett*, 2004, **15**, 1131.
83. I. Hemmila, *Applications of Fluorescence in Immunoassay*, Wiley, New York, 1991.
84. R. Kikkeri, D. Grunstein and P. H. Seeberger, *J. Am. Chem. Soc.*, 2010, **132**, 10230.
85. R. Kikkeri, I. Garcia-Rubio and P. H. Seeberger, *Chem. Commun.*, 2009, 235.
86. S. Sreejith, K. P. Divya, T. K. Manojkumar and A. Ajayaghosh, *Chem. Asian J.*, 2011, **6**, 430.
87. L. Y. Zhao, D. Sui, J. Chai, Y. Wang and S. M. Jiang, *J. Phys. Chem. C*, 2006, **110**, 24299.
88. S. C. Wang, G. W. Men, Y. Wang, L. Y. Zhao, Q. F. Hou and S. M. Jiang, IEE, *Sensors J*, 2011, **11**, 137.
89. X. Y. Liu, X. Han, L. P. Zhang, C. H. Tung and L. Z. Wu, *Phys. Chem. Chem. Phys.*, 2010, **12**, 13026.
90. D. B. Liu, W. W. Chen, K. Sun, K. Deng, W. Zhang, Z. Wang and X. Y. Jiang, *Angew. Chem. Int. Ed.*, 2011, **50**, 4103.
91. S. Bi, Y. M. Yan, S. Y. Hao and S. S. Zhang, *Angew. Chem. Int. Ed.*, 2010, **49**, 4438.
92. M. N. Stojanovic and D. Stefanovic, *Nature Biotechnol.*, 2003, **21**, 1069.
93. R. R. Breaker and G. F. Joyce, *Chem. Biol.*, 1995, **2**, 655.
94. J. H. Qian, Y. F. Xu, S. T. Zhang and X. H. Qian, *J. Fluoresc.*, 2011, **21**, 1015.
95. A. Roque, F. Pina, S. Alves, R. Ballardini, M. Maestri and V. Balzani, *J. Mater. Chem.*, 1999, **9**, 2265.

96. F. Pina, M. J. Melo, S. Alves, R. Ballardini, M. Maestri and P. Passaniti, *New J. Chem.*, 2001, **25**, 747.
 97. R. A. Bissell, A. J. Bryan, A. P. de Silva, H. Q. N. Gunaratne and C. P. McCoy, *J. Chem. Soc. Chem. Commun.*, 1994, 405.
 98. M. Fernandez and P. Fromherz, *J. Phys. Chem.*, 1977, **31**, 1755.
 99. F. Pina, M. Maestri and V. Balzani, *Chem. Commun.*, 1999, 107.
100. F. Pina, M. J. Melo, M. Maestri, R. Ballardini and V. Balzani, *J. Am. Chem. Soc.*, 1997, **119**, 5556.
101. F. Pina, A. Roque, M. J. Melo, M. Maestri, L. Belladelli and V. Balzani, *Chem. Eur. J*, 1998, **4**, 1184.
102. A. Petitjean, N. Kyritsakas and J.-M. Lehn, *Chem. Eur. J*, 2005, **11**, 6818.
103. A. Fernández-Gutiérrez and A. Muñoz de la Peña, in *Molecular Luminescence Spectroscopy. Methods and Applications. Part 1*, ed. S. G. Schulman, Wiley, New York, 1985, p. 371.
104. E. B. Sandell, *Colorimetric Determination of Traces of Metals*, Interscience, London, 3rd edn, 1959.
105. M. de Sousa, M. Kluciar, S. Abad, M. A. Miranda, B. de Castro and U. Pischel, *Photochem. Photobiol. Sci.*, 2004, **3**, 639.
106. M. de Sousa, B. de Castro, S. Abad, M.A. Mirandaand and U. Pischel, *Chem. Commun.*, 2006, 2051.
107. J. M. Montenegro, E. Perez-Inestrosa, D. Collado, Y. Vida and R. Suau, *Org. Lett.*, 2004, **6**, 2353.
108. S. Alves, F. Pina, M. T. Albelda, E. Garcia-Espana, C. Soriano and S. V. Luis, *Eur. J. Inorg. Chem.*, 2001, 405.
109. F. M. Raymo, *Adv. Mater.*, 2002, **14**, 401.
110. H. Durr and H. Bouas-Laurent (ed.), *Photochromism. Molecules and Systems*, Elsevier, Amsterdam, 1990.
111. W. D. Zhou, Y. J. Li, Y. L. Li, H. B. Liu, S. Wang, C. H. Li, M. J. Yuan, X. F. Liu and D. B. Zhu, *Chem. Asian J.*, 2006, **1-2**, 224.
112. L. Y. Zhao, S. C. Wang, Y. Wu, Q. F. Hou, Y. Wang and S. M. Jiang, *J. Phys. Chem. C*, 2007, **111**, 18387.
113. Z. P. Dong, Z. H. Dong, J. Ren, J. Jin, P. Wang, J. Jiang, R. Li and J. T. Ma, *Microporous Mesoporous Mater*, 2010, **135**, 170.
114. M. N. Stojanovic, T. E. Mitchell and D. Stefanovic, *J. Am. Chem. Soc.*, 2002, **124**, 3555.
115. X. Chen, Y. F. Wang, Q. Liu, Z. Z. Zhang, C. H. Fan and L. He, *Angew. Chem. Int. Ed.*, 2006, **45**, 1759.
116. N. Carmi, L. A. Shultz and R. R. Breaker, *Chem. Biol.*, 1996, **3**, 1039.
117. S. H. Lee, S. K. Kim, J. H. Lee and J. H. Kim, *Tetrahedron*, 2004, **60**, 5171.
118. A. Credi, V. Balzani, S. J. Langford and J. F. Stoddart, *J. Am. Chem. Soc.*, 1997, **119**, 2679.
119. H. Y. Zheng, W. D. Zhou, M. J. Yuan, X. D. Yin, Z. C. Zuo, C. B. Ouyang, H. B. Liu, Y. L. Li and D. B. Zhu, *Tetrahedron Lett.*, 2009, **50**, 1588.
120. M. Irie, *Chem. Rev.*, 2000, **100**, 1685.

121. Y. J. Li, H. Y. Zheng, Y. L. Li, S. Wang, Z. Y. Wu, P. Liu, Z. Q. Gao, H. B. Liu and D. B. Zhu, *J. Org. Chem.*, 2007, **72**, 2878.
122. X. L. Feng, X. R. Duan, L. B. Liu, F. D. Feng, S. Wang, Y. L. Li and D. B. Zhu, *Angew. Chem. Int. Ed.*, 2009, **48**, 5316.
123. L. X. Mu, W. S. Shi, G. W. She, J. C. Chang and S. T. Lee, *Angew. Chem. Int. Ed.*, 2009, **48**, 3469.
124. T. Gupta and M. E. van der Boom, *Angew. Chem. Int. Ed.*, 2008, **47**, 5322.
125. Z. Q. Guo, W. H. Zhu, Y. Y. Xiong and H. Tian, *Macromolecules*, 2009, **42**, 1448.
126. S. Uchiyama, N. Kawai, A. P. de Silva and K. Iwai, *J. Am. Chem. Soc.*, 2004, **126**, 3032.
127. M. Meyer, J. C. Mialocq and B. Perly, *J. Phys.Chem.*, 1990, **94**, 98.
128. J. Andréasson, S. D. Straight, S. Bandyopadhyay, R. H. Mitchell, T. A. Moore, A. L. Moore and D. Gust, *Angew. Chem. Int. Ed.*, 2007, **46**, 958.
129. H. Xu, X. Xu, R. Dabestani, G. M. Brown, L. Fan, S. Patton and H. F. Ji, *J. Chem. Soc. Perkin Trans*, 2002, **2**, 636.
130. H. F. Ji, R. Dabestani and G. M. Brown, *J. Am. Chem. Soc.*, 2000, **122**, 9306.
131. T. Y. Cheng, T. Wang, W. P. Zhu, Y. Y. Yang, B. B. Zeng, Y. F. Xu and X. H. Qian, *Chem. Commun.*, 2011, **47**, 3915.
132. A. P. de Silva and S. A. de Silva, *J. Chem. Soc. Chem. Commun.*, 1986, 1709.
133. G. Deng, T. Sakaki, Y. Kawahara and S. Shinkai, *Supramol. Chem.*, 1993, **2**, 71.
134. G. Hennrich, H. Sonnenschein and U. Resch-Genger, *J. Am. Chem. Soc.*, 1999, **121**, 5073.
135. G. Hennrich, W. Walther, U. Resch-Genger and H. Sonnenschein, *Inorg. Chem.*, 2001, **40**, 641.
136. J. Andreasson, S. D. Straight, S. Bandyopadhyay, R. H. Mitchell, T. A. Moore, A. L. Moore and D. Gust, *J. Phys. Chem. C*, 2007, **111**, 14274.
137. S. Gaweda, G. Stochel and K. Szacilowski, *Chem. Asian J.*, 2007, **2**, 580.
138. A. Podborska and K. Szacilowski, *Aust. J. Chem.*, 2010, **63**, 165.
139. M. A. Arugula, V. Bocharova, J. Halamek, M. Pita and E. Katz, *J. Phys. Chem. B*, 2010, **114**, 5222.
140. E. Perez-Inestrosa, J.-M. Montenegro, D. Collado and R. Suau, *Chem. Commun.*, 2008, 1085.
141. M. F. Budyka, N. I. Potashova, T. N. Gavrishova and V. M. Lee, *J. Mater. Chem.*, 2009, **19**, 7721.
142. M. Amelia, M. Baroncini and A. Credi, *Angew. Chem. Int. Ed.*, 2008, **47**, 6240.
143. M. Schmittel, P. Mal and A. de los Rios, *Chem. Commun.*, 2010, **46**, 2031.
144. A. P. de Silva and S. Uchiyama, *Nature Nanotechnol*, 2007, **2**, 399.
145. D. H. Vance and A. W. Czarnik, *J. Am. Chem. Soc.*, 1994, **116**, 9397.
146. R. A. Bissell, E. Calle, A. P. de Silva, S. A. de Silva, H. Q. N. Gunaratne, J. L Habib-Jiwan, S. L. A. Peiris, R. A. D. D. Rupasinghe, T. K. S. D.

Samarasinghe, K. R. A. S. Sandanayake and J.-P. Soumillion, *J. Chem. Soc. Perkin Trans*, 1992, **2**, 1559.
147. T. Sakamoto, A. Ojida and I. Hamachi, *Chem. Commun.*, 2009, 141.
148. A. Ojida, M. Mito-oka, K. Sada and I. Hamachi, *J. Am. Chem. Soc.*, 2004, **126**, 2454.
149. A. Ojida, M. Mito-oka, M. Inouye and I. Hamachi, *J. Am. Chem. Soc.*, 2002, **124**, 6256.
150. A. V. Koulov, K. A. Stucker, C. Lakshmi, J. P. Robinson and B. D. Smith, *Cell Death Diff*, 2003, **10**, 1357.
151. D. H. Lee, S. Y. Kim and J. I. Hong, *Angew. Chem. Int. Ed.*, 2004, **43**, 4777.
152. L. Fabbrizzi, G. Francese, M. Licchelli, A. Perotti and A. Taglietti, *Chem. Commun.*, 1997, 581.
153. L. Fabbrizzi, A. Leone and A. Taglietti, *Angew. Chem. Int. Ed.*, 2001, **40**, 3066.
154. M. Bonizzoni, L. Fabbrizzi, G. Piovani and A. Taglietti, *Tetrahedron*, 2004, **60**, 11159.
155. D. Jimenez, R. Martínez-Máñez, F. Sancenón, J. Soto, A. Benito and E. García-Breijo, *Eur. J. Inorg. Chem.*, 2005, 2393.
156. M. Schmittel and H. W. Lin, *Angew. Chem. Int. Ed.*, 2007, **46**, 893.
157. J. Andreasson, S. D. Straight, T. A. Moore, A. L. Moore and D. Gust, *J. Am. Chem. Soc.*, 2008, **130**, 11122.
158. M. Badland, A. Cleeves, H. G. Heller, D. S. Hughes and M. B. Hursthouse, *Chem. Commun.*, 2000, 1567.
159. P. Ceroni, G. Bergamini and V. Balzani, *Angew. Chem. Int. Ed.*, 2009, **48**, 8516.
160. N. E. Tokel and A. J. Bard, *J. Am. Chem. Soc.*, 1972, **94**, 2862.
161. J. Andreasson, U. Pischel, S. D. Straight, T. A. Moore, A. L. Moore and D. Gust, *J. Am. Chem. Soc.*, 2011, **133**, 11641.
162. H. Tian, B. Qin, R. X. Yao, X. L. Zhao and S. J. Yang, *Adv. Mater.*, 2003, **15**, 2104.
163. M. Privman, T. K. Tam, M. Pita and E. Katz, *J. Am. Chem. Soc.*, 2009, **131**, 1314.
164. L. Amir, T. K. Tam, M. Pita, M. M. Meijler, L. Alfonta and E. Katz, *J. Am. Chem. Soc.*, 2009, **131**, 826.
165. T. K. Tam, M. Pita and E. Katz, *Sensors Actuators B*, 2009, **140**, 1.
166. G. Strack, S. Chinnapareddy, D. Volkov, J. Halamek, M. Pita, I. Sokolov and E. Katz, *J. Phys. Chem. B*, 2009, **113**, 12154.
167. A. P. de Silva, *Nature Mater*, 2005, **4**, 15.
168. D. Margulies, G. Melman, C. E. Felder, R. Arad-Yellin and A. Shanzer, *J. Am. Chem. Soc.*, 2004, **126**, 15400.
169. A. Okamoto, K. Tanaka and I. Saito, *J. Am. Chem. Soc.*, 2004, **126**, 9458.
170. M. N. Stojanovic and D. Stefanovic, *J. Am. Chem. Soc.*, 2003, **125**, 6673.
171. W. Sun, C. H. Xu, Z. Zhu, C. J. Fang and C. H. Yan, *J. Phys. Chem. C*, 2008, **112**, 16973.

172. W. Sun, Y. R. Zheng, C. H. Xu, C. J. Fang and C. H. Yan, *J. Phys. Chem. C*, 2007, **111**, 11706.
173. Y. Liu, W. Jiang, H. Y. Zhang and C. J. Li, *J. Phys. Chem. B*, 2006, **110**, 14231.
174. M. Semeraro and A. Credi, *J. Phys. Chem. C*, 2010, **114**, 3209.
175. Q. Q. Wu, X. Y. Duan and Q. H. Song, *J. Phys. Chem. C*, 2011, **115**, 23970.
176. Z. Q. Guo, P. Zhao, W. H. Zhu, X. M. Huang, Y. S. Xie and H. Tian, *J. Phys. Chem. C*, 2008, **112**, 7047.
177. J. H. Qian, X. H. Qian, Y. F. Xu and S. Y. Zhang, *Chem. Commun.*, 2008, 4141.
178. K. Kalyanasundaram, *Photochemistry in Microheterogeneous Systems*, Academic Press, Orlando, FL, 1987.
179. R. Baron, O. Lioubashevski, E. Katz, T. Niazov and I. Willner, *Org. Biomol. Chem.*, 2006, **4**, 989.
180. H. Lederman, J. Macdonald, D. Stefanovic and M. N. Stojanovic, *Biochemistry*, 2006, **45**, 1194.
181. F. Guarnieri, M. Fliss and C. Bancroft, *Science*, 1996, **273**, 220.
182. B. Yurke, A. P. Mills and S. L. Cheng, *Biosystems*, 1999, **52**, 165.
183. S. Ogasawara, Y. Kyoi and K. Fujimoto, *ChemBioChem*, 2007, **8**, 1520.
184. S. Ogasawara, T. Ami and K. Fujimoto, *J. Am. Chem. Soc.*, 2008, **130**, 10050.
185. T. H. Lee, J. L. Gonzalez, J. Zheng and R. M. Dickson, *Acc. Chem. Res.*, 2005, **38**, 534.
186. A. Okamoto, K. Tanaka and I. Saito, *J. Am. Chem. Soc.*, 2003, **125**, 5066.
187. F. Remacle, R. Weinkauf and R. D. Levine, *J. Phys. Chem. A*, 2006, **110**, 177.
188. Y. H. Yan, J. A. Mol, J. Verduijn, S. Rogge, R. D. Levine and F. Remacle, *J. Phys. Chem. C*, 2010, **114**, 20380.
189. O. Kuznetz, H. Salman, N. Shakkour, Y. Eichen and S. Speiser, *Chem. Phys. Lett.*, 2008, **451**, 63.
190. O. Kuznetz, H. Salman, Y. Eichen, F. Remacle, R. D. Levine and S. Speiser, *J. Phys. Chem. C*, 2008, **112**, 15880.
191. M. P. O'Neil, M. P. Niemczyk, W. A. Svec, D. Gosztola, G. L. Gaines III and M. R. Wasielewski, *Science*, 1992, **257**, 63.
192. D. Margulies, G. Melman and A. Shanzer, *J. Am. Chem. Soc.*, 2006, **128**, 4865.
193. D. Margulies, G. Melman and A. Shanzer, *Nature Mater*, 2005, **4**, 768.
194. S. A. de Silva, K. C. Loo, B. Amorelli, S. L. Pathirana, M. Nyakirang'ani, M. Dharmasena, S. Demarais, B. Dorcley, P. Pullay and Y. A. Salih, *J. Mater. Chem.*, 2005, **15**, 2791.
195. A. P. de Silva, C. M. Dobbin, T. P. Vance and B. Wannalerse, *Chem. Commun.*, 2009, 1386.
196. J. Macdonald, D. Stefanovic and M. N. Stojanovic, *Sci. Am.*, 2008, **299**, 84.
197. J. Macdonald, Y. Li, M. Sutovic, H. Lederman, K. Pendri, W. H. Lu, B. L. Andrews, D. Stefanovic and M. N. Stojanovic, *Nano Lett.*, 2006, **6**, 2598.

198. J. J. Tabor and A. D. Ellington, *Nature Biotechnol.*, 2003, **21**, 1013.
199. G. H. Mealy, *Bell Systems Tech. J*, 1955, **34**, 1045.
200. F. H. Hsu, *Behind Deep Blue*, Princeton University Press, Princeton, 2002.
201. S. Tyagi and F. R. Kramer, *Nature Biotechnol.*, 1996, **14**, 303.
202. R. Pei, E. Matamoros, M. Liu, D. Stefanovic and M. N. Stojanovic, *Nature Nanotechnol*, 2010, **5**, 773.
203. V. Privman, *Nature Nanotechnol*, 2010, **5**, 767.

CHAPTER 11
History-dependent Systems

11.1 Introduction

History-dependent molecular systems lead to sequential logic processors. Some of these sequential logic systems serve as memories. In fact, transistor-based memories are the workhorses of conventional computers.[1–7] While detailed discussion of molecular memories[8–11] is outside the scope of this book, we must note the great success of photochromics[12,13] in this capacity. This connection with computer science was noted as long ago as 1956.[14] Irie's dithienylethenes[13] are among the most useful photochromics. These operate down to the single molecule level[15] and, when embedded in polymer matrices, allow robust, fatigue-free operation for up to 10^4 read–write–erase cycles.[16] The coloured state is written by an ultraviolet (uv) light dose. The coloured state, which is thermally stable, can be erased with a dose of visible light. The coloured (or colourless) state can be read by interrogation with a low-intensity beam of visible (or uv) light.

11.2 R–S Latch

A semiconductor electronic analogue of photochromic memory is the R–S latch,[1–7] which can be prepared from two cross-wired and fed-back NOR gates (Figure 11.1), where the state of the inputs R (reset) and S (set) allow a memory to be held and read along the lines labelled 'output$_1$' and 'output$_2$', according to the truth table (Table 11.1). When the latch is synchronized to a clock signal, it is called a flip–flop. Given that molecular-scale experiments do not usually deal with clock signals in a computing sense, the terms 'latch' and 'flip–flop' are encountered in the chemical literature without distinction. Though we will focus on small molecule examples, even gene-based versions are known.[17]

As Table 11.1 shows, input$_R$ = 0 and input$_S$ = 1 produce one held memory state whereas swapping the inputs around (input$_R$ = 1 and input$_S$ = 0) produces

---

Monographs in Supramolecular Chemistry No. 12
Molecular Logic-based Computation
By A Prasanna de Silva
© The Royal Society of Chemistry 2013
Published by The Royal Society of Chemistry, www.rsc.org

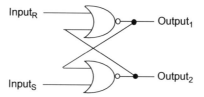

Figure 11.1 R–S flip–flop or Set–Reset latch.

Table 11.1 Truth table for the R–S latch.

$Input_R$	$Input_S$	$Output_1$	$Output_2$
0	0	holds previous state	holds previous state
0	1	1	0
1	0	0	1

the other memory state. It is notable that the condition $input_R = 0$ and $input_S = 0$ simply holds the previous memory state. The fourth condition ($input_R = 1$ and $input_S = 1$) is usually avoided by adding a small logic array before the device to act as a guard. Since $output_1$ and $output_2$ remain opposites, a given memory state can be read from the logic state of say, $output_1$ alone.

Section 10.13 concerning complex combinatorial logic featured van der Boom's electrode-immobilized electrochromic **1**, where the starting state had to be reset after each experiment in order to remove memory effects.[18] Study of **1** without such restrictions exposes R–S latch behaviour,[19] even though the persistence of written memory states is limited to 25 min. This is achieved with chemical redox agents[20] or electrode potentials[21] with mono[20]- or multi[21]-layer systems on glass. The chemical agents are $Cr_2O_7^{2-}$ in water at pH <1 for oxidation ($input_R$) and $Co(C_5H_5)_2$ in dry acetonitrile for reduction ($input_S$). Though the solvents are not mutually compatible, careful drying steps between input applications allow the system to function as required. The electrode potential approach avoids this problem anyway. Naturally, each of the redox states of **1** is stable in the absence of inputs or in the presence of the appropriate input (R or S). This is the memory aspect. Application of the other input will naturally change the existing redox state. This is the flip–flop aspect. The materials nature of van der Boom's glass-based cases goes well with Shi's flip–flops based on photochromics anchored on single ZnO nanowires.[22]

History-dependent Systems

Though not described in these terms in the primary literature, similar redox-driven fluorescent systems are known. As alluded to in section 5.2.2.7, redox-driven systems possess a memory aspect. In line with classical redox indicators,[23,24] complex **2**[25] is poorly emissive at 610 nm (when excited at 453 nm) owing to PET from the ruthenium-based lumophore to the benzoquinone moiety across the dimethylene spacer. If a potential of −0.6 V (vs. sce) is applied to it in moist acetonitrile, the benzoquinone unit is fully reduced to the hydroquinone form once two coulombs are passed per equivalent of **2**. This succeeds because the reduction potential of **2** is −0.44 V. Upon reduction, a luminescence enhancement factor of 6 is seen. However, the hydroquinone form can be re-set to **2** by application of two coulombs at a potential of +1.1 V. D'Souza's **3** shows similar fluorescence switching when it passes into its hydroquinone form.[26] However, this is achieved not with electroreduction but with catechol which launches a redox equilibrium. In contrast, Burdette's **4**[27] is a PET-based, poorly fluorescent catechol which uses Fe^{3+} as oxidant to produce the corresponding o-quinone which is highly emissive. Ascorbate achieves the reverse reaction, albeit incompletely. Interestingly, **4** is directly comparable with Rurack and Daub's benzocrown ether[28] in terms of PET design.

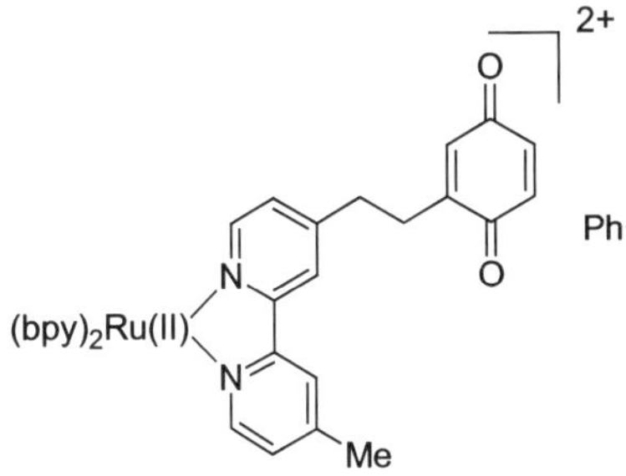

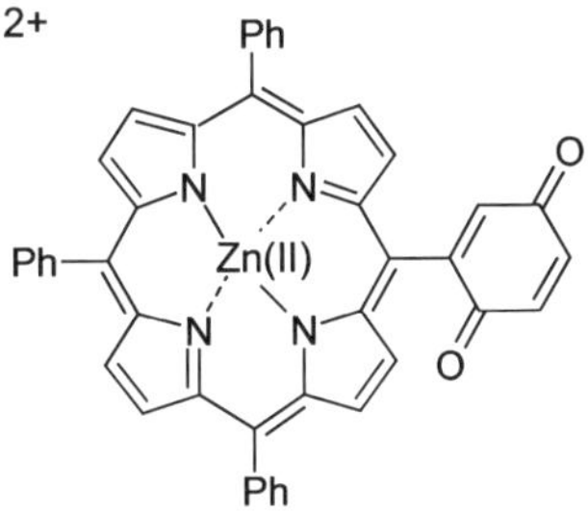

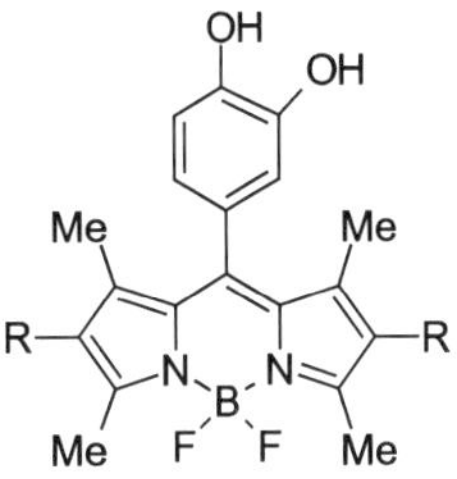

A bioinspired redox sensor **5/6**,[29] which even leads to versions capable of apparently reversible intracellular monitoring (Figure 11.2), also belongs here. Reductants such as dithionite convert highly emissive fluorescein **5** to non-absorbing and non-emissive dihydrofluorescein **6**, possibly via reduction of the disulfide and internal electron transfer. Reset is done with H_2O_2 or even air.

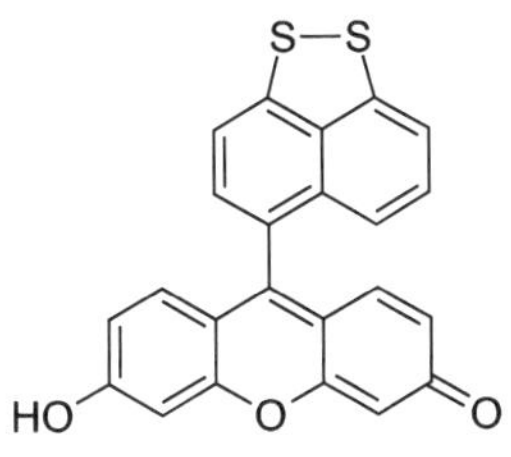

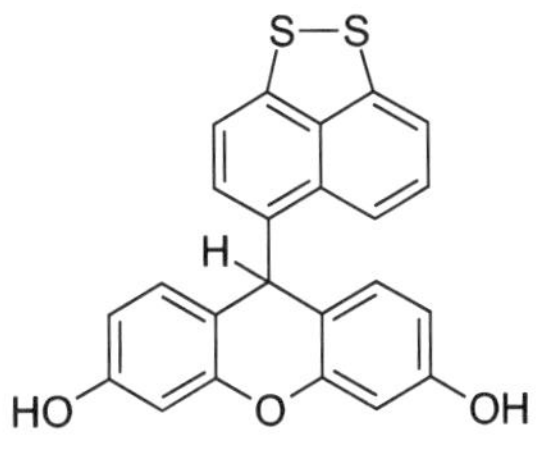

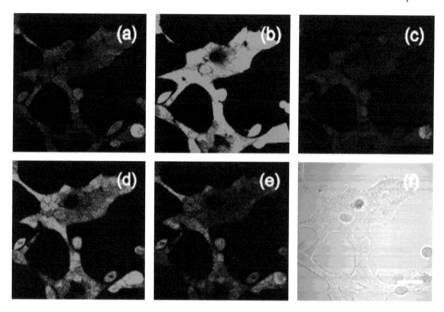

Figure 11.2 Fluorescent (a–e) and bright field (f) micrographs of live HEK cells (which have a reducing environment) loaded with 5×10^{-6} M acetoxymethyl ester of **5**. The starting state (a) is followed through two cycles of H_2O_2 stress (b,d) and recovery (c,e). Scale bar = 10 μm; $\lambda_{exc} = 488$ nm, $\lambda_{em} = 510$–550 nm.
Reprinted from E. W. Miller, S. X. Bian and C. J. Chang, *J. Am. Chem. Soc.*, 2007, **129**, 3458 with permission from the American Chemical Society.

Perylenedimide[30]-dihydropyridine system **7**[31] has its fluorescence switched 'off' by PET across the *p*-phenylene spacer. When the dihydropyridine unit is oxidized to the pyridinium form by input of tetrachlorobenzoquinone, the PET stops and the fluorescence output is switched 'on' (FE > 1000). Reset can be arranged with $NaBH_3CN$ so that **7** is recovered. A perylenedimide spaced from two ferrocene units which achieves a smaller FE value of 3.2 upon two-electron oxidation[32] is perhaps the simplest of all since no protonation steps are involved.

Earlier examples of deliberately designed R–S latches are available from Willner's laboratory.[33,34] The dithienylethene[13] **8**,[33] developed in collaboration with Tian, is immobilized on an electrode. Unusually, the application of a +0.35 V (*vs.* sce) pulse to colourless **8** produces its coloured and electro-inactive form **9**. The authors attribute this to disproportionation of the radical cation. Switching back to the colourless state **8** is achieved with irradiation with a light dose at 570 nm, as usual in the Irie systems.[13] Electrochemical set and photochemical reset is thus achieved. Reading of the output can be done by cyclic voltammetry or uv–vis spectroscopy.

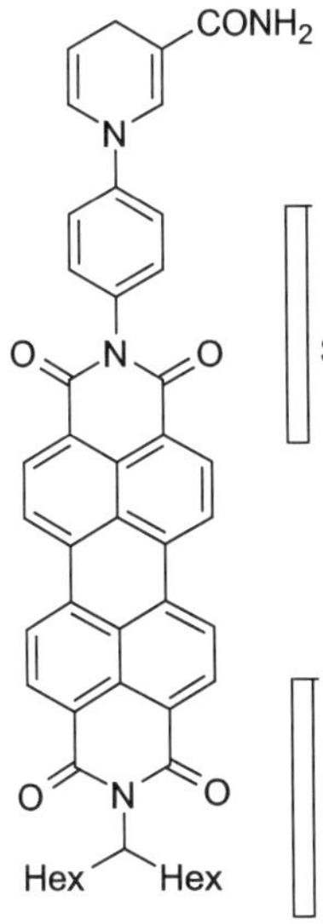

Still earlier, even simpler latch features were noted by Giordani and Raymo[35] concerning the spiropyran–merocyanine equilibrium within a silica-based sol–gel matrix. A 524 nm light dose input leads to decolorization ('low' absorbance output) which persists for 12 h. The general point that R–S latch features are found within photochromic memories has been well made by Pischel and Andréasson.[36]

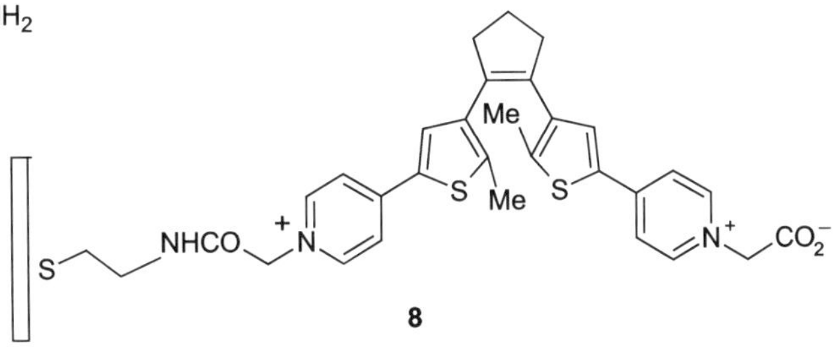

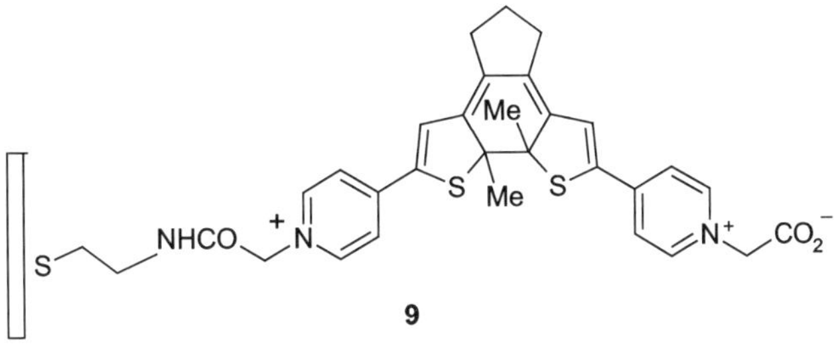

van der Boom's **1** shows a Nernstian response when studied spectroelectrochemically, *i.e.* a plateau at 'high' and 'low' absorbance (at a chosen wavelength of 496 nm) as a function of applied potential. System **1** functions well as a binary system.[20] However, he also applied accurately defined potentials to operate **1** in ternary or higher-valued logic[37,38] for memory purposes only.[21] Electrochromic polymers such as **10** coated on transparent electrodes can also replace **1** for this application.[39] Given that the number of serial logic operations needed for memory operation is very small, the often-raised objection to using ternary logic[37,38] does not apply. The same case has been made for the use of ternary logic in some aspects of molecular computational identification of objects (MCID)[40] (see Chapter 14). Case **1** builds on Raymo and Giordani's far-sighted work on logic systems with memory aspects.[41–43] Recent reviews are available.[44,45]

Just like van der Boom, who used oxidizing and reducing agents to drive a reaction forwards or backwards,[20] Katz turns to H_2O_2 with horseradish peroxidase for the oxidizing condition applied to **11**.[46] He also employs NADH with diaphorase for the reducing condition. The absorbance at 600 nm is taken as 'high' if it exceeds 0.3 after 5 min of exposure to a controlled set of concentrations, which arise only when **11** is in its oxidized state. It is notable that $input_R$ and $input_S$ (say, H_2O_2 and NADH respectively) can be substituted for by various other chemical species.

A general point is that (bio)chemical reactions, when run under irreversible regimes, naturally bring in a memory (but not flip–flop) aspect because the product will persist after the initial reaction conditions are withdrawn. Such memory aspects are also noticeable within an electrochemically driven Cu^{2+}/Cu^{+} catenate.[47] Many irreversible reactions of this general type have featured in previous chapters on combinational logic because their primary literature has put this memory aspect aside. If this primary literature had considered the history-dependent behaviour, all of these examples would have fitted in here.

To close this section, it is worth mentioning that semiconductor electronic memories can also be constructed from a ring of an even number of NOT gates with the output feeding back to the input. The corresponding case of an odd number of NOT gates is particularly interesting. This is the ring oscillator whose output oscillates between 0 and 1 with a frequency which is determined by the number of gates and the delay time of each gate. Ring oscillators are useful for prototyping and testing electronic devices. We saw in section 5.4.1 how Avouris' laboratory produced a NOT logic gate from a complementary pair of field-effect transistors built on a single-walled carbon nanotube (SWCNT).[48,49] This basic approach can be repeated five times over along the 8 µm length of a single SWCNT to give rise to a ring oscillator.[49] A readout stage is built in as well. Using two metals with different work functions (Pd and Al) as contacts proved to be an easier way to achieve the complementary field-effect transistors rather than by doping regions of the SWCNT as done previously.[48] A version based on three SWCNT-based NOT gates was known since 2001.[10]

11.3 D Latch

In the previous section, we saw the need of a guard in order to avoid a condition which could derail the correct operation of the R–S latch. The D latch provides this guard (Figure 11.3) at the front-end of the R–S latch, which shows some similarity with the 2:1 multiplexer gate array (Chapter 10, Figure 10.17). A bit of data (input$_D$) is presented to the subsequent logic array along two branches, where one branch carries an inverter. So the derailing condition of

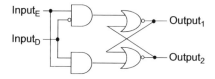

Figure 11.3 D flip–flop or Data latch.

Table 11.2 Truth table for the D latch.

$Input_E$	$Input_D$	$Output_1$	$Output_2$
0	0	holds previous state	holds previous state
0	1	holds previous state	holds previous state
1	0	0	1
1	1	1	0

two 'high' inputs is never experienced. Additionally, the state of an enabling line (input$_E$) determines whether the R–S latch component is available for communication with incoming data. If input$_E$ is held 'low', the R–S latch unit maintains its previous state. If a clock signal is used, it is sent along the enabling line and the device can be termed a D flip–flop. When so enabled, output$_1$ follows the state of input$_D$ irrespective of the previous state of output$_1$ (Table 11.2).

Andréasson, Pischel and their colleagues[50] give a nice molecular implementation of the D latch by watching the weak fluorescence of coloured **12** at 644 nm ($\lambda_{exc} = 550$ nm) as output$_1$. This is the product of the photocyclization of the photochromic fulgimide **13**, when irradiated at 365 nm. A 1064 nm laser pulse serves as input$_D$ and a 532 nm laser pulse serves as input$_E$. Since both **12** and **13** are transparent to 1064 nm radiation, it is clear that output$_1$ will remain unchanged whether exposed to the infrared pulse or not. On the other hand, coloured **12** will undergo photocycloreversion to **13** when 532 nm radiation is present by itself. The fluorescence output$_1$ from **12** thus dies with it. Thus the 'low' output$_1$ follows the absence of the infrared pulse (input$_D$). When 532 nm and 1064 nm pulses are sent through a third-harmonic generating crystal before impinging on the molecular device, 355 nm photons are created. The latter convert **13** to coloured **12** so that fluorescence output$_1$ goes 'high'. Thus 'high' output$_1$ follows the presence of the infrared pulse (input$_D$), fulfilling the requirements of a D latch. Output$_2$ of Figure 11.3 can be obtained from the absorption band of **13** (at 374 nm) if required. However, molecular-scale operation is compromised by the need for the third-harmonic generating crystal, as also found for an all-optical half-adder example in section 9.2. We note that an approach to a D latch was made by Campagna's, Gianetto's and Ziessel's laboratories[51] earlier in the same year. A comment on another type of latch – a T latch – will be reserved for section 12.2.

11.4 Molecular Keypad Lock

Keypad locks are security devices which let us into a system only when we type in the correct sequence of alphanumeric characters. No other permutation of the characters will do. The system in question could be physical, *e.g.* a room, or virtual, such as a database.

Figure 11.4 shows the electronic circuit of a 2-input keypad lock. A two-button pad can only be pressed in two ways (not counting multiple pressings of one button which cannot open locks of the type shown in Figure 11.4) to generate a two-bit string. The correct string can be guessed with a 50% probability and this weakens the security of such a keypad lock, but it will be sufficient to discuss the principles involved. Initially the lock is closed. The set/reset line supplies a 'high' input continuously unless the set/reset button is pressed. If we now momentarily supply a 'high' value for $input_1$ by pressing the appropriate button, the first AND gate provides a 'high' output which is fed back to the $input_1$ line. Thus the output line from the first AND gate will be latched in the 'high' state and will be supplied as one input to the second AND gate. Now if we press the button for the 'high' value of $input_2$ as the second part of our number sequence, the output of the second AND gate (the final output in Figure 11.4) will go 'high' and stay 'high' because of feedback. We can see a lock being actuated and opened by this 'high' signal in the output line. If the button for 'high' $input_2$ was pressed momentarily before that for 'high' $input_1$ (or if the lock was reset), no actuation would occur and lock would remain closed – clearly a case of sequence dependence or history dependence.

The complex **14.**Fe^{3+}, previously featured in section 10.21.1, has subsequently been shown to have a time-dependent response. While this can be a weakness in some respects, Shanzer's group[52] exploit it for discriminating between permutations of three inputs in their order of addition, *i.e.* ABC *vs.* CBA *etc*. This became the first molecular keypad lock.

The decomplexation of Fe^{3+} from the polydentate siderophore ligand **14** in ethanol solution displays slow kinetics under certain conditions. The carefully timed provision of EDTA, OAc^- and light intensity (344 nm) inputs (applied at 3 min intervals) in that particular order into **14.**Fe^{3+} gives a 'high' fluorescence

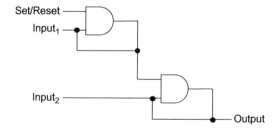

Figure 11.4 Two-input keypad lock. There should be several resistors in this circuit for its practical implementation but these have been omitted for the purpose of illustrating the general concept.

output at 525 nm whereas, say, OAc⁻, EDTA and light intensity (344 nm) inputs (in that order) does not do so. The first series of additions acts as follows. Acidic EDTA extracts Fe^{3+} to leave **14** with its fluorescein unit in its acidic lactone form which does not absorb (and emit) in the near ultraviolet. Basification with OAc⁻ takes the fluorescein unit in **14** to its basic form which is well conjugated so that its absorption spectrum maximizes at 344 nm and fluoresces strongly at 525 nm. Supply of 344 nm as an excitation now causes an intense fluorescence output. The second series of additions can also be interpreted. The initial basification with OAc⁻ takes the fluorescein unit within the complex **14**.Fe^{3+} to the basic anionic form. The subsequent addition of EDTA to this basic solution forms the tetraanion of EDTA so that the latter cannot easily approach the anionic form of **14**.Fe^{3+}. Thus the removal of Fe^{3+} from the complex does not occur over short timescales of around 10 min. The basic fluorescein unit within the complex **14**.Fe^{3+} cannot emit when excitation light input at 344 nm is applied because of the proximity of Fe^{3+} as a strong quencher via EET and PET pathways. In addition, the connection made between a specified order of addition of chemical/light inputs and an alphanumeric password is crucial to the success of this work. It is notable that Shanzer's previous experiments on the same structure but at longer timescales, *i.e.* under conditions of thermodynamic control, allowed demonstration of a combined half-adder and half-subtractor in a regime of combinational logic.[53] Generally, the existence of sequence-dependent phenomena of this kind implies that the final solutions are not at equilibrium, *i.e.* the sequence-dependence should fade over a sufficiently long timescale. The molecular keypad lock will be more valid if the kinetics of equilibration is slower.

14; R = 1-pyrenyl

The substantial body of molecular keypad locks available[54-70] prevents comprehensive discussion of them all. One of these[66] unfortunately assigns the overtone of scattered excitation light as a keypad state. However, another of these[57] is particularly interesting because two different passwords admit users into separate computing functionalities. We now consider a few representative examples.

Single-stranded DNA immobilized on silica nanoparticles can also provide examples of keypad locks, by hybridization with complementary strands carrying a substantial overhang.[71] After washing, a subsequent strand complementary to the overhang can be hybridized. This too contains an overhang so that the procedure can be repeated. Each of these added strands represent inputs and only the correct order will create the final long double-stranded moiety. The latter is detected with an intercalator which emits fluorescence only upon incorporation and whose signal increases rapidly with the lengthening of the double strand. So the strong emission arises only on addition of the correct order of input strands.

Biofuel cell current can also be an output of a keypad lock,[72,73] and we consider the demonstration by Katz and co-workers.[72] First, an enzyme cascade is set up to generate a pH decrease as the initial output. Given the famed selectivity of most enzymes, it is no surprise that the exact sequence of enzymes is crucial[55] to achieve this pH decrease. Specifically, starch is converted to maltose by β-amylase (input$_1$). Maltose is taken onwards to glucose by the duo of maltose phosphorylase and acid phosphatase (input$_2$) in the presence of phosphate. Finally, glucose is converted to gluconic acid by glucose oxidase (input$_3$) in air. So a pH decrease (6.7 to 4.2) is produced in weakly buffered solution. This acidification is linked to a fuel cell current by exploiting a layer of pH-switchable 4-vinylpyridine polymer brush on an electrode as done by these workers for OR logic gates,[74] described in section 7.3.1. Glucose is the fuel in another half-cell which also contains glucose oxidase and methylene blue as mediator.

Let us also consider **15**[64] in some detail. In THF solution, **15** possesses a strong fluorescence ($\phi_F = 0.79$) but is quenched by 20 equivalents of Hg^{2+}. The inclusion of thiaphilic Hg^{2+} in the thiacalixarene causes deprotonation of the sulfonamide units, which is a specific form of metal ion-induced proton ejection.[75,76] Deprotonated dansylamide has been long known to be non-fluorescent.[77] Addition of 20 equivalents of K^+ to this solution recovers the fluorescence because of K^+ taking up residence in the crown ether compartment of **15** and electrostatically repelling the thiacalixarene-bound Hg^{2+}. With the Hg^{2+} gone, the deprotonated dansylamides regain protons. If instead of K^+, 20 equivalents of F^- is gradually admitted to the solution of **15**.Hg^{2+}, the Hg^{2+} level is reduced (by formation of HgF_2) to the point of decomplexation from the thiacalixarene. So the deprotonated dansylamides regain protons and the fluorescence shines strongly again. However, this is only the first part of the story. As the total of 20 equivalents of F^- is approached, the dansylamides are deprotonated again by the basic F^- and the fluorescence switches 'off' again. For comparison, the analogous experiment with Cl^- leads to $HgCl_2$ and the attendant fluorescence recovery but there is no fluorescence quenching at higher Cl^- levels because its lower basicity prevents dansylamide deprotonation.

Now we reach for the molecular keypad lock. When **15** is treated sequentially with Hg^{2+}, F^- and K^+, the authors[64] find the initially strong fluorescence is quenched after Hg^{2+} input, stays low after the F^- input and recovers strongly after the K⁺ input. Evidently the deprotonated dansylamide present after the F⁻ input recovers protons upon K⁺ admission into the crown. It is not clear to this writer why this should be so. All other input sequences, *e.g.* F⁻, Hg²⁺ and K⁺, produce a low fluorescence in the end. A weakness of this system is that if 'high' fluorescence is taken as the opening of the lock, then the lock is open before the password is entered and the password becomes unnecessary.

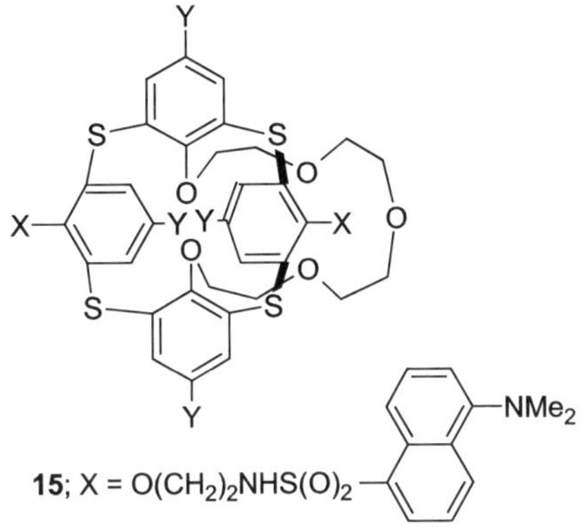

15; X = O(CH$_2$)$_2$NHS(O)$_2$—
Y = t-Bu

As work in this area grows, it might be useful to avoid keypad locks based on complex phenomena with slow and unpredictable kinetics because clear mechanistic explanations would then be difficult to achieve. Also it would be desirable to avoid encoding the device or the output as parts of the sequence of alphanumeric characters of the password. In the macroscopic world, password entries are composed of inputs only. The molecular domain would benefit by retaining aspects of this reality.

Many more such cases of molecule-based keypad locks become visible once appropriate coding schemes for inputs, outputs and devices are appreciated. Consider the following high-school classic – the multi-step synthesis of an azo dye **16**[78] (Figure 11.5). None of the steps can be swapped around if the synthesis is to be optimally successful. Part of the ingenuity of the synthetic chemist, whether computer aided[79] or not, is to present optimized conditions for application of various reagents in a carefully chosen order of steps to achieve a given target.

The distinctly and intensely coloured azo dye **16** ($\varepsilon = 26\,800$ M^{-1} cm^{-1} at $\lambda_{abs} = 462$ nm at pH 5)[80] emerges only if the particular sequence of In$_1$In$_2$In$_3$In$_4$In$_5$ is applied, with 'high' levels of each, to the starting state of the device **17** (Figure 11.5). For example, In$_2$In$_1$In$_3$In$_4$In$_5$ would not produce the azo dye. So we have a 5-input molecular keypad lock.

While absorbance in the visible range of the spectrum was used as the output for immediate recognition of the idea, we can choose any characteristic property (physical, chemical or biological)[81] of the final product as

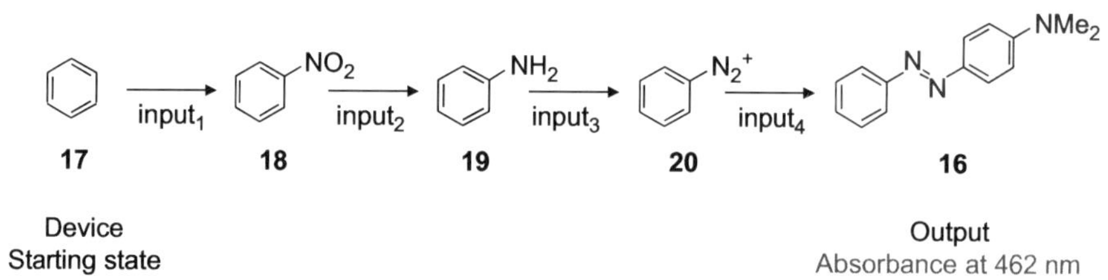

17 →input₁→ 18 (Ph-NO₂) →input₂→ 19 (Ph-NH₂) →input₃→ 20 (Ph-N₂⁺) →input₄→ 16 (Ph-N=N-C₆H₄-NMe₂)

Device
Starting state

Output
Absorbance at 462 nm

Figure 11.5 Scheme for synthesis of azo dye **16**. The device in its starting state is **17**. The output signal is the appearance of absorbance at 462 nm. The threshold between 'high' and 'low' outputs is an apparent extinction coefficient, $\varepsilon = 13\,400\,\text{M}^{-1}\,\text{cm}^{-1}$ at 462 nm. Input$_1$ = HNO$_3$, H$_2$SO$_4$. Input$_2$ = Sn, HCl. Input$_3$ = H$_2$SO$_4$. Input$_4$ = NaNO$_2$, HCl. Input$_5$ = N,N-dimethylaniline. Similar coding schemes are employed for molecular logic-based computation with deoxyribozymes,[90] for instance, except that the solvent, reagent, temperature and time conditions are less restricted during the synthesis of **16**.

the output signal. This opens up a considerable territory from the many linear multi-step synthetic sequences which are available.[82-84] Two of the most accessible of these lie in polypeptide and polynucleotide syntheses, especially those using solid-phase synthesis.[85] For instance, the nonapeptide hormone analogue deaminooxytocin is made by starting with Boc-glycine, attaching it to the solid phase and taking it though an eight-cycle linear synthesis on solid phase, followed by cleavage from solid phase, deprotection and oxidation to yield the final product.[86] The hormonal activity can be defined narrowly enough to eliminate all the other permutations of deaminooxytocin as the desired output. This would correspond to a 12-input keypad lock.

Convergent synthetic schemes[84] themselves correspond to authorized entry into high-security zones where two or more separate passwords are required consecutively. Convergent schemes can be read along the longest chain of steps and the shorter branches of steps can be called out as a complex input at that root on the main chain, in much the same way that molecular structures are read according to the IUPAC nomenclature.[87] To take a classical example, Woodward's famous chlorophyll synthesis[82,83] can be seen to use **21** as the starting material, which is converted in six steps to **22**. This is then reacted with **23** (itself prepared in six steps from **24** as a separate starting material), followed by six more steps to reach **25** with a characteristic visible spectrum [λ_{max}/nm (ε_{max}/M^{-1} cm^{-1}) 667(47 000), 610(5500), 560(2500), 529(7000) and 498(10 300)], which will serve as our defined output. So the synthetic scheme is a 6-input keypad lock, followed by a complex input (arising from a 6-input entry beginning with another device) and then another 6-input keypad lock.

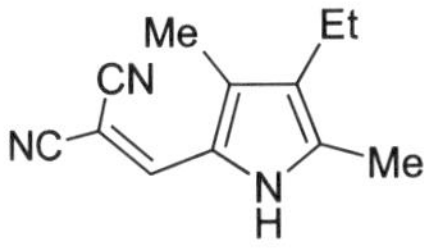

Of course, the cases discussed above are not resettable – a property shared with several important examples of molecular logic and computation with biomolecules.[88–92] Resettability is not necessary for once-only computation, which can arise in certain intelligent diagnostics applications, for instance.[93–95] Further, the best examples of keypad lock behaviour would be those schemes where the time periods required for the reaction steps are minimized.

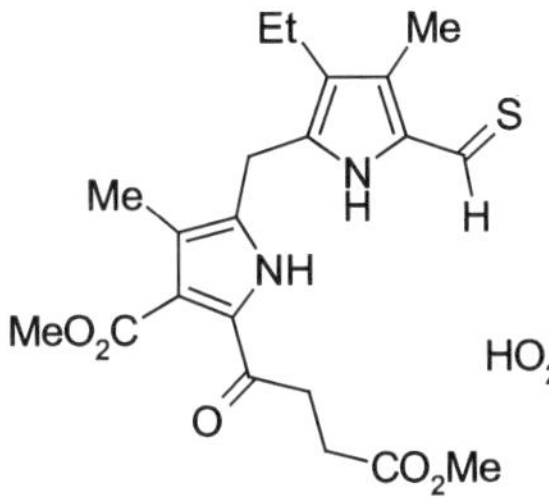

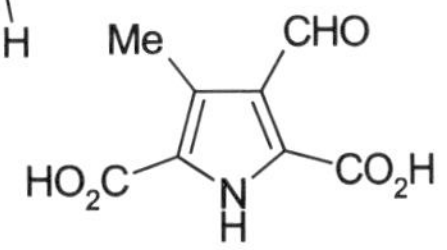

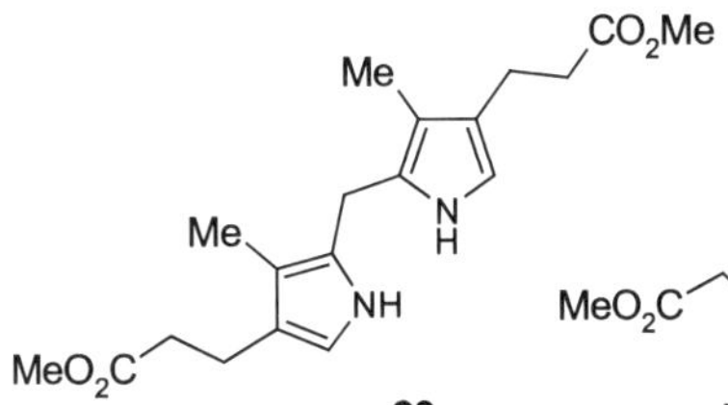

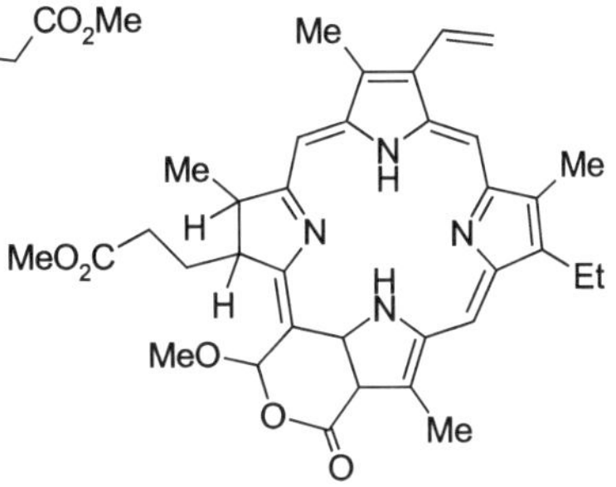

We began this section by noting the security aspect of molecular keypad locks. We close by mentioning the security-oriented use of molecules for coding alphanumeric characters.[96] Though no Boolean logic is involved, the use of chemical shift and integration of ^{1}H nuclear magnetic resonance (NMR) signals for this purpose is interesting. Sharp singlets arising from the methyl groups of compounds such as 26 and 27 are analysed as adjacent pairs. The code is developed by reading the signals in the NMR spectrum from left to right (downfield to upfield) and by noting their ratio of integrations. For instance, the ratios 1:0, 1:1, 1:1.5, 1:2 and 1:3 are assigned to A, B, C, D and E respectively. As long as the assignments are known by the sender and the receiver, different messages can be composed by choosing the right combination of compounds and their relative molar concentrations. It is clear that molecular information processing can display security possibilities beyond the old invisible inks.

26; X = CO$_2$Me, Y = H, Z = NO$_2$
27; X = OMe, Y = NO$_2$, Z = Cl

References

1. J. Millman and A. Grabel, *Microelectronics*, McGraw-Hill, New York, 2nd edn, 1988.
2. A. P. Malvino and J. A. Brown, *Digital Computer Electronics*, Glencoe, New York, 3rd edn, 1993.
3. M. Ben-Ari, *Mathematical Logic for Computer Science*, Prentice-Hall, Hemel Hempstead, 1993.
4. J. R. Gregg, *Ones and Zeros*, IEEE Press, New York, 1998.
5. A. L. Sedra and K. C. Smith, *Microelectronic Circuits*, Oxford University Press, Oxford, 5th edn, 2003.
6. F. M. Brown, *Boolean Reasoning*, Dover, New York, 2nd edn, 2003.
7. C. Maxfield, *From Bebop to Boolean Boogie*, Newnes, Oxford, 2009.
8. J. E. Green, J. W. Choi, A. Boukai, Y. Bunimovich, E. Johnston-Halperin, E. DeIonno, Y. Luo, B. A. Sheriff, K. Xu, Y. S. Shin, H. R. Tseng, J. F. Stoddart and J. R. Heath, *Nature*, 2007, **445**, 414.
9. J. Sinclair, D. Granfeldt, J. Pihl, M. Millingen, P. Lincoln, C. Farre, L. Peterson and O. Orwar, *J. Am. Chem. Soc.*, 2006, **128**, 5109.
10. A. Bachtold, P. Hadley, T. Nakanishi and C. Dekker, *Science*, 2001, **294**, 1317.
11. G. Y. Tseng and J. C. Ellenbogen, *Science*, 2001, **294**, 1294.
12. H. Durr and H. Bouas-Laurent (ed.), *Photochromism. Molecules and Systems*, Elsevier, Amsterdam, 1990.
13. M. Irie, *Chem. Rev.*, 2000, **100**, 1685.
14. Y. Hirshberg, *J. Am. Chem. Soc.*, 1956, **78**, 2304.
15. M. Irie, T. Fukaminato, T. Sasaki, N. Tamai and T. Kawai, *Nature*, 2002, **420**, 759.
16. M. Irie and K. Uchida, *Bull. Chem. Soc. Jpn.*, 1998, **71**, 985.
17. T. S. Gardner, C. R. Cantor and J. J. Collins, *Nature*, 2000, **403**, 339.
18. T. Gupta and M. E. van der Boom, *Angew. Chem. Int. Ed.*, 2008, **47**, 5322.
19. U. Pischel, *Angew. Chem. Int. Ed.*, 2010, **49**, 1356.
20. G. de Ruiter, E. Tartakovsky, N. Oded and M. E. van der Boom, *Angew. Chem. Int. Ed.*, 2010, **49**, 169.
21. G. de Ruiter, L. Motiei, J. Chowdhury, N. Oded and M. E. van der Boom, *Angew. Chem. Int. Ed.*, 2010, **49**, 4780.
22. L. X. Mu, W. S. Shi, T. P. Zhang, H. Y. Zhang, Y. Wang, G. W. She, Y. H. Gao, P. F. Wang, J. C. Chang and S. T. Lee, *Appl. Phys. Lett.*, 2011, **98**, 163101.

23. E. Bishop (ed.), *Indicators*, Pergamon, Oxford, 1972.
24. B. Kratochvil and D. A. Zatko, *Anal. Chem.*, 1964, **36**, 527.
25. V. Goulle, A. Harriman and J.-M. Lehn, *J. Chem. Soc. Chem. Commun.*, 1993, 1034.
26. G. R. Deviprasad, B. Keshavan and F. D'Souza, *J. Chem. Soc. Perkin Trans 1*, 1998, 3133.
27. D. P. Kennedy, C. M. Kormos and S. C. Burdette, *J. Am. Chem. Soc.*, 2009, **131**, 8578.
28. K. Rurack, M. Kollmannsberger, U. Resch-Genger, W. Rettig and J. Daub, *Chem. Phys. Lett.*, 2000, **329**, 363.
29. E. W. Miller, S. X. Bian and C. J. Chang, *J. Am. Chem. Soc.*, 2007, **129**, 3458.
30. L. Zang, R. C. Liu, M. W. Holman, K. T. Nguyen and D. M. Adams, *J. Am. Chem. Soc.*, 2002, **124**, 10640.
31. P. Yan, M. W. Holman, P. Robustelli, A. Chowdhury, F. I. Ishak and D. M. Adams, *J. Phys. Chem. B*, 2005, **109**, 130.
32. R. L. Zhang, Z. L. Wang, Y. S. Wu, H. B. Fu and J. N. Yao, *Org. Lett.*, 2008, **10**, 3065.
33. R. Baron, A. Onopriyenko, E. Katz, O. Lioubashevski, I. Willner, S. Wang and H. Tian, *Chem. Commun.*, 2006, 2147.
34. J. Elbaz, Z. G. Wang, R. Orbach and I. Willner, *Nanolett.*, 2009, **9**, 4510.
35. S. Giordani and F. M. Raymo, *Org. Lett.*, 2003, **5**, 3559.
36. U. Pischel and J. Andréasson, *New J. Chem.*, 2010, **34**, 2701.
37. R. W. Keyes, *Rev. Mod. Phys.*, 1989, **61**, 279.
38. B. Hayes, *Am. Scientist*, 2001, **89**, 490.
39. G. de Ruiter, Y. H. Wijsboom, N. Oded and M. E. van der Boom, *ACS Appl. Mater. Interfac.*, 2010, **2**, 3578.
40. A. P. de Silva, M. R. James, B. O. F. McKinney, P. A. Pears and S. M. Weir, *Nature Mater.*, 2006, **5**, 787.
41. F. M. Raymo and S. Giordani, *J. Am. Chem. Soc.*, 2001, **123**, 4651.
42. F. M. Raymo and S. Giordani, *Org. Lett.*, 2001, 1833.
43. F. M. Raymo and S. Giordani, *J. Am. Chem. Soc.*, 2002, **124**, 2004.
44. G. de Ruiter and M. E. van der Boom, *Acc. Chem. Res.*, 2011, **44**, 563.
45. G. de Ruiter and M. E. van der Boom, *J. Mater. Chem.*, 2011, **21**, 17575.
46. M. Pita, G. Strack, K. MacVittie, J. Zhou and E. Katz, *J. Phys. Chem. B*, 2009, **113**, 16071.
47. G. Periyasamy, J.-P. Collin, J.-P. Sauvage, R. D. Levine and F. Remacle, *Chem. Eur. J.*, 2009, **15**, 1310.
48. V. Derycke, R. Martel, J. Appenzeller and P. Avouris, *Nano Lett.*, 2001, **1**, 453.
49. Z. H. Chen, J. Appenzeller, Y. M. Lin, J. Sippel-Oakley, A. G. Rinzler, J. Y. Tang, S. J. Wind, P. M. Solomon and P. Avouris, *Science*, 2006, **311**, 1735.
50. P. Remon, M. Balter, S. M. Li, J. Andréasson and U. Pischel, *J. Am. Chem. Soc.*, 2011, **133**, 20742.
51. F. Puntoriero, F. Nastasi, T. Bura, R. Ziessel, S. Campagna and A. Giannetto, *New J. Chem.*, 2011, **35**, 948.

52. D. Margulies, C. E. Felder, G. Melman and A. Shanzer, *J. Am. Chem. Soc.*, 2007, **129**, 347.
53. D. Margulies, G. Melman, C. E. Felder, R. Arad-Yellin and A. Shanzer, *J. Am. Chem. Soc.*, 2004, **126**, 15400.
54. Z. Q. Guo, W. H. Zhu, L. J. Shen and H. Tian, *Angew. Chem. Int. Ed.*, 2007, **46**, 5549.
55. G. Strack, M. Ornatska, M. Pita and E. Katz, *J. Am. Chem. Soc.*, 2008, **130**, 4234.
56. M. Suresh, A. Ghosh and A. Das, *Chem. Commun.*, 2008, 3906.
57. W. Sun, C. Zhou, C. H. Xu, C. J. Fang, C. Zhang, Z. X. Li and C. H. Yan, *Chem. Eur. J.*, 2008, **14**, 6342.
58. W. Sun, C. H. Xu, Z. Zhu, C. J. Fang and C. H. Yan, *J. Phys. Chem. C*, 2008, **112**, 16973.
59. C. H. Xu, W. Sun, Y. R. Zheng, C. J. Fang, C. Zhou, J. Y. Jin and C. H. Yan, *New J. Chem.*, 2009, **33**, 838.
60. M. Kumar, A. Dhir and V. Bhalla, *Org. Lett.*, 2009, **11**, 2567.
61. S. Kumar, V. Luxami, R. Saini and D. Kaur, *Chem. Commun.*, 2009, 3044.
62. W. Jiang, M. Han, H. Y. Zhang, Z. J. Zhang and Y. Liu, *Chem. Eur. J.*, 2009, **15**, 9938.
63. J. Andréasson, S. D. Straight, T. A. Moore, A. L. Moore and D. Gust, *Chem. Eur. J.*, 2009, **15**, 3936.
64. M. Kumar, R. Kumar and V. Bhalla, *Chem. Commun.*, 2009, 7384.
65. J. Wang and C. S. Ha, *Analyst*, 2010, **135**, 1214.
66. P. Singh, J. Kaur and W. Holzer, *Sensors Actuators B*, 2010, **150**, 50.
67. J. Halamek, T. K. Tam, S. Chinnapareddy, V. Bocharova and E. Katz, *J. Phys. Chem. Lett.*, 2010, **1**, 973.
68. M. Zhou, X. L. Zheng, J. Wang and S. J. Dong, *Chem. Eur. J.*, 2010, **16**, 7719.
69. J. Halamek, T. K. Tam, G. Strack, V. Bocharova, M. Pita and E. Katz, *Chem. Commun.*, 2010, **46**, 2405.
70. P. Remon, M. Hammarson, S. M. Li, A. Kahnt, U. Pischel and J. Andréasson, *Chem. Eur. J.*, 2011, **17**, 6492.
71. F. Pu, Z. Liu, X. J. Yang, J. S. Ren and X. G. Qu, *Chem. Commun.*, 2011, **47**, 6024.
72. M. Privman, T. K. Tam, M. Pita and E. Katz, *J. Am. Chem. Soc.*, 2009, **131**, 1314.
73. M. Zhou, X. L. Zheng, J. Wang and S. J. Dong, *Bioinformatics*, 2011, **27**, 399.
74. X. M. Wang, J. Zhou, T. K. Tam, E. Katz and M. Pita, *Bioelectrochem.*, 2009, **77**, 69.
75. M. A. McKervey and D. L. Mulholland, *J. Chem. Soc. Chem. Commun.*, 1977, 438.
76. M. Takagi and K. Ueno, *Top. Curr. Chem.*, 1984, **121**, 39.
77. L. Prodi, F. Bolletta, M. Montalti and N. Zaccheroni, *Eur. J. Inorg. Chem.*, 1999, 455.

78. A. Vogel, *Practical Organic Chemistry*, Longman, London, 4th edn, 1978, p. 718.
79. E. J. Corey and X. M. Cheng, *The Logic of Chemical Synthesis*, Wiley, New York, 1995.
80. J. Z. Liu, T. L. Wang and L. N. Ji, *J. Mol. Catal. B Enzym.*, 2006, **41**, 81.
81. A. P. de Silva and S. Uchiyama, *Nature Nanotechnol.*, 2007, **2**, 399.
82. R. B. Woodward, G. L. Closs, E. Le Goff, W. A. Ayer, H. Dutler, W. Leimgruber, J. M. Beaton, J. Hannah, W. Lwowski, F. Bickelhaupt, F. P. Hauck, J. Sauer, R. Bonnett, S. Ito, Z. Valenta, P. Buchschacher, A. Langemann and H. Volz, *J. Am. Chem. Soc.*, 1960, **82**, 3800.
83. R. B. Woodward, W. A. Ayer, J. M. Beaton, F. Bickelhaupt, R. Bonnett, P. Buchschacher, G. L. Closs, H. Dutler, J. Hannah, F. P. Hauck, S. Ito, A. Langemann, E. Le Goff, W. Leimgruber, W. Lwowski, J. Sauer, Z. Valenta and H. Volz, *Tetrahedron*, 1990, **46**, 7599.
84. K. C. Nicolaou and E. J. Sorensen, *Classics in Total Synthesis*, VCH, Weinheim, 1996.
85. S. A. Kates and F. Albericio (ed.), *Solid Phase Synthesis: A Practical Guide*, Dekker, New York, 2000.
86. H. Takashima, V. Du Vigneaud and R. B. Merrifield, *J. Am. Chem. Soc.*, 1968, **90**, 1323.
87. J. Rigaudy and S. P. Klesney (ed.), *IUPAC Nomenclature of Organic Chemistry*, Pergamon, New York, 1979.
88. R. Baron, O. Lioubashevski, E. Katz, T. Niazov and I. Willner, *Org. Biomol. Chem.*, 2006, **4**, 989.
89. G. Strack, M. Pita, M. Ornatska and E. Katz, *ChemBioChem*, 2008, **9**, 1260.
90. M. N. Stojanovic and D. Stefanovic, *Nature Biotechnol.*, 2003, **21**, 1069.
91. G. Seelig, D. Soloveichik, D. Y. Zhang and E. Winfree, *Science*, 2006, **314**, 1585.
92. B. M. Frezza, S. L. Cockroft and M. R. Ghadiri, *J. Am. Chem. Soc.*, 2007, **129**, 14875.
93. D. C. Magri, G. J. Brown, G. D. McClean and A. P. de Silva, *J. Am. Chem. Soc.*, 2006, **128**, 4950.
94. T. Konri and D. R. Walt, *J. Am. Chem. Soc.*, 2009, **131**, 13232.
95. D. Margulies and A. D. Hamilton, *J. Am. Chem. Soc.*, 2009, **131**, 9142.
96. T. Ratner, O. Reany and E. Keinan, *ChemPhysChem*, 2009, **10**, 3303.

CHAPTER 12

Multi-level Logic

12.1 Introduction

Up to now we have been in the grip of binary logic, as it was first proposed by Boole[1] and as it was embraced by modern information technology. For good reason, because classifying signals into only 'low' and 'high' produces a robustness which current technology depends on.[2] Also, 0 and 1 are very different numbers representing nothing and something respectively, which allows their clear distinction under many imaginable circumstances. However, ternary and higher levels of logic have the strength of higher information density.[3] Ternary logic can use the numbers 0, 1 and 2 for instance. The associated weakness of being prone to error accumulation during serial operations[2] is one reason why higher-level logic systems do not play a significant role in modern information technology. This weakness can be appreciated when we see that the numbers 1 and 2 start to become indistinguishable when errors grow beyond 33% since the numbers are now 1 ± 0.33 and 2 ± 0.66. However, these higher-level logic systems can be used profitably in molecular logic and computation if we keep their strengths and weaknesses in perspective.

We will start our discussion in section 12.2 with cases where the output channel maintains two signal levels, *i.e.* a binary situation, whereas the input signal levels are divided into three. 'Off–on–off' switching systems fall into this category where ternary situations emerge. Their logic tables will be developed in a fashion similar to that used for the truth tables in the previous chapters. Binary situations are fully dispelled when we examine 'low–medium–high' switching and variants in section 12.3.

12.2 'Off–on–off' Switching Systems

We have already met many PET switching systems of the 'receptor$_1$–spacer$_1$–fluorophore–spacer$_2$–receptor$_2$' and 'fluorophore–spacer$_1$–receptor$_1$–spacer$_2$–

Monographs in Supramolecular Chemistry No. 12
Molecular Logic-based Computation
By A Prasanna de Silva

Published by the Royal Society of Chemistry, www.rsc.org

receptor$_2$' formats which show various types of binary logic. These same formats are the major contributors to fluorescent 'off–on–off' switches,[4–16] including older cases involving various mechanisms.[6,7] 'Off–on–off' switches show a 'low' output level at 'low' levels of input. As the input signal gets stronger, the output also reaches higher levels. However, the output dies away as the input is taken to yet greater heights. The dependent variable shows three plateau regions, two of which are similarly low. An instance of this for **1**[4] can be seen in Figure 12.1. PET occurs from the amine in alkaline conditions and to the pyridinium in acidic media. These 'off–on–off' systems can now be seen as ternary logic devices (Table 12.1). 'Low', 'medium' and 'high' levels of H$^+$ can be coded as digital states of 0, 1 and 2 of the input, for instance. Case **2**,[12] which arises from the combined efforts of Pina's and García-España's teams, is particularly interesting because its H$^+$-driven fluorescent 'off–on–off' response is sharpened in the presence of Zn^{2+}. The fluorescence emission of S. A. de Silva's **3**[10] takes this

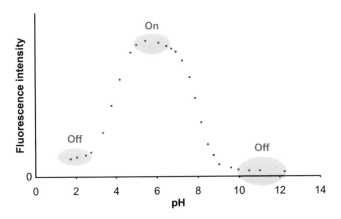

Figure 12.1 Fluorescence intensity (at 420 nm) *vs.* pH profile of **1** in methanol–water (1:4, v/v) solution. Excitation wavelength 364 nm.
Adapted from A. P. de Silva, H. Q. N. Gunaratne and C. P. McCoy, *Chem. Commun.*, 1996, 2399 with permission from the Royal Society of Chemistry.

Table 12.1 Logic table for 'off–on–off' switching behaviour of **1**.[a]

Input H$^+$	Output Fluorescence[b]
0, low (10^{-11} M)	0, low (0.009)
1, medium (10^{-6} M)	1, high (0.38)
2, high (10^{-3} M)	0, low (0.046)

[a]5 × 10^{-6} M in water:methanol (4:1, v/v); λ_{exc} 364 nm, λ_{em} 398, 420 and 443 nm.
[b]Quantum yields.

Table 12.2 Logic table for Na^+-enabled, H^+-driven 'off–on–off' switching behaviour of **3**.[a]

Input$_1$ H^+	Input$_2$ Na^+	Output Fluorescence[b]
0, none	0, none	0, low (0.04)
0, none	1, high (10^{-2} M)	0, low (0.03)
1, medium (10^{-6} M)	0, none	0, low (0.04)
1, medium (10^{-6} M)	1, high (10^{-2} M)	1, high (0.54)
2, high (10^{-2} M)	0, none	0, low (0.02)
2, high (10^{-2} M)	1, high (10^{-2} M)	0, low (0.03)

[a]10^{-6} M in methanol; λ_{exc} 377 nm, λ_{em} 407, 431 and 454 nm.
[b]Quantum yields.

sharpening action to an extreme since this case demonstrates Na^+-enabled, H^+-driven 'off–on–off' response. If Na^+ is not present in sufficient concentration, a PET process from the benzocrown ether prevents any fluorescence whatever the H^+ concentration (Table 12.2). A Ca^{2+}-enabled, H^+-driven version based on a PET basis is also available.[11]

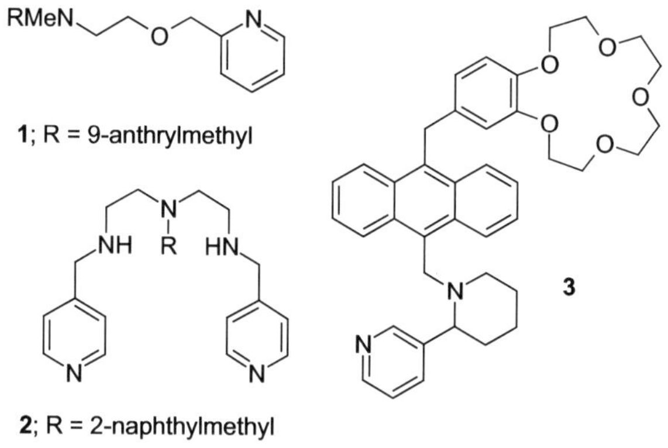

1; R = 9-anthrylmethyl

2; R = 2-naphthylmethyl

3

The operation of Perez-Inestrosa and Pischel's **4**[17] is very similar to **1** but its behaviour with added equivalents of H^+ in MeCN is interpreted as a toggle switch (T latch). However, we refrained from discussing this case in Chapter 11 because a memory function has not been demonstrated. Latching of states, while easy to achieve with photochromic or redox systems (Chapter 11), is very difficult when fast proton transfer steps are involved. Furthermore, the system needs to be reset to the neutral state **4** by treatment with base before the two consecutive applications of H^+ can be done. In other words, only two togglings can be performed before a reset is required.

4

Fluorescent 'off–on–off' systems have been interpreted as binary XOR logic devices,[18] but caution is needed because two H^+ concentrations (for example) in water cannot be chosen to serve as inputs so that their cumulative effect switches the fluorescence 'off'. pH-responsive systems in aqueous solution switch between 'on' and 'off' states over a range of 2 pH units.[19] Combination of two inputs of identical H^+ concentrations only lowers the pH value by 0.3, which means that the fluorescence output would not alter as much as desired. This point can be illustrated by examining Figure 12.1. The starting state of **1** is at pH 12, whereas the mono- and di-protonated states are found at pH 6 and 2 respectively. The difference in H^+ concentration required to produce **1**.H^+ is ca. 10^{-6} M. On the other hand, the creation of **1**.$(H^+)_2$ requires an H^+ concentration increase of not 2×10^{-6} M but rather 10^{-2} M. Similar issues were alluded to in section 10.21.3. However, non-aqueous solutions will reduce switching ranges and be more amenable to this line of thought about XOR logic. For instance, the fluorescence of **4** in MeCN switches 'on' with one equivalent of H^+ but switches 'off' again with two equivalents of H^+.[17]

Let us consider a selection of 'off–on–off' systems in some detail. Tian's **5**[20] is a polymer with a narrowly distributed molecular mass which shows H^+-induced 'off–on' action with $pK_a = 7.0$ clearly due to PET from the aliphatic tertiary amine across the spacer(s). The H^+-induced 'on–off' segment ($pK_a = 0.9$) can be attributed an ICT effect since the aromatic amine is being protonated. The poor fluorescence efficiency of this state suggests a Born–Oppenheimer hole[21] via coupling of N^+–H to water molecules.[22]

A metal-based case **6** is from the team of Ye and Ji. Two metallolumophores and two receptors can be seen within **6**.[23,24] The 'off–on–off' switching occurs via protonation and deprotonation of the imidazole receptors at the extremes of pH. The luminescence quenching can occur in several ways. At pH <4 and >8 the de/protonated imidazole units can couple to solvent water molecules[22] to give a Born–Oppenheimer hole.[21] The charged imidazole unit can control the energy of the emissive MLCT excited state so that it mixes with the non-emissive d–d excited state.

Lanthanide luminescence can also be brought into the fold as in the case of Gunnlaugsson et al.'s **7**.[25] At pH 11, the Eu^{3+} emission is 'off', since the metal-bound amide N–H is deprotonated and the amide anion launches PET to Eu^{3+}.

Upon acidification, the luminescence is switched 'on' with a pK_a of 8.1, increasing to a maximum at pH 6, because the amide anion regains a proton. On further acidification, the emission switches 'off' with a pK_a of 3.8 and levels off at pH 2, because the aminophenanthroline monoprotonates, lowering its triplet energy and losing EET efficiency towards Eu^{3+}.

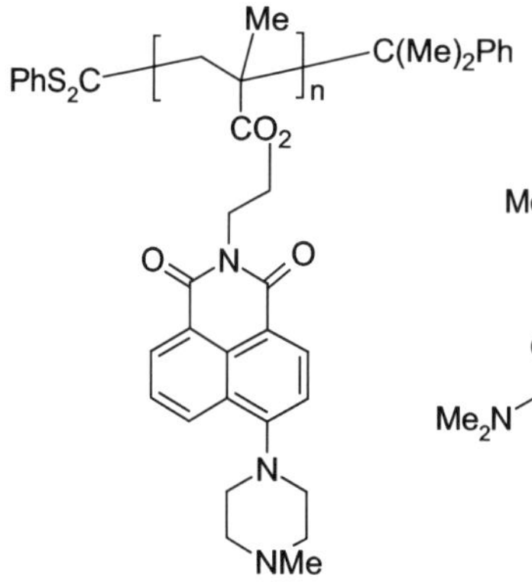

Several cases of this type arise from mixtures of compounds.[26,27] These cases tilt towards systems chemistry.[28] Compound **8**, on its own, displays fluorescence enhancement upon protonation, as seen for many 9-anthrylmethyl substituted amines. The enhancement is clearly seen at pH values below 8 and stays constant at pH <5. However, in the presence of a 3-fold excess of adenosine triphosphate (ATP), the fluorescence of **8** is strong in the pH range 5–8. The acidic branch of the fluorescence–pH profile is due to a PET-type interaction between the anthracene fluorophore in tetraprotonated **8** and the protonated adenine unit within diprotonated ATP. The registering of the tetraammonium and triphosphate moieties allows the π-stacking of anthracene and protonated adenine units for a strong quenching. Such registering is so important that looping the tetraamine chain of **8** to produce an anthracenophane completely loses the 'off–on–off' action. The genealogy of **8** can be traced back to Czarnik[29] and Lehn.[30]

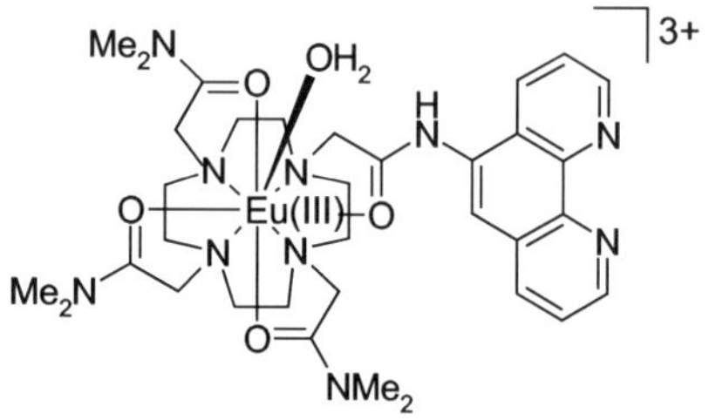

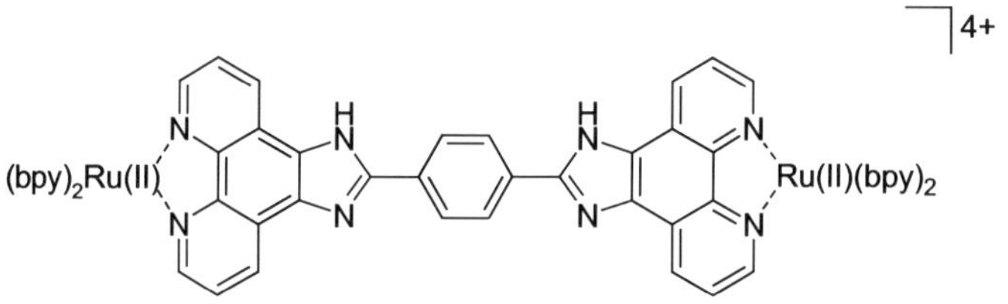

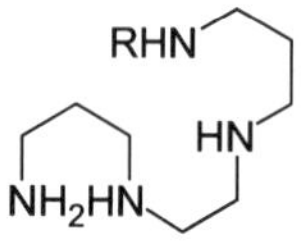

8; R = 9-anthrylmethyl

Self-assembled 'off–on–off' switching is available from Pallavicini and his team owing to the capability of detergent micelles to organize hydrophobic proton receptors *N,N*-dimethyldodecylamine and 2-dodecylpyridine and the fluorophore, pyrene.[31–33] The latter has an unusually long-lived excited state[34] which allows it to be quenched adequately by *N,N*-dimethyldodecylamine and protonated 2-dodecylpyridine according to PET within the micelle nanospace of 5 nm radius.[35] The power of the system, as opposed to an individual molecule, is even clearer here.

There are other ways to generate 'off–on–off' switching if we consider inputs other than chemical concentration, *e.g.* light intensity. As mentioned in section 7.2.5, Wasielewski and colleagues have optical switches like **9** based on PET which flip within picoseconds.[36] Optical pumping of the porphyrin absorption band at low intensities produces the perylenetetracarboxydiimide radical anion (and the porphyrin radical cation). The absorbance of the radical anion naturally increases with increasing pump intensity but falls at still higher pump intensities. So the absorbance versus pump intensity profiles shows 'off–on–off' behaviour. At these higher intensities the second porphyrin unit joins in to launch a PET process into the perylenetetracarboxydiimide radical anion which pushes the latter into the dianion state.

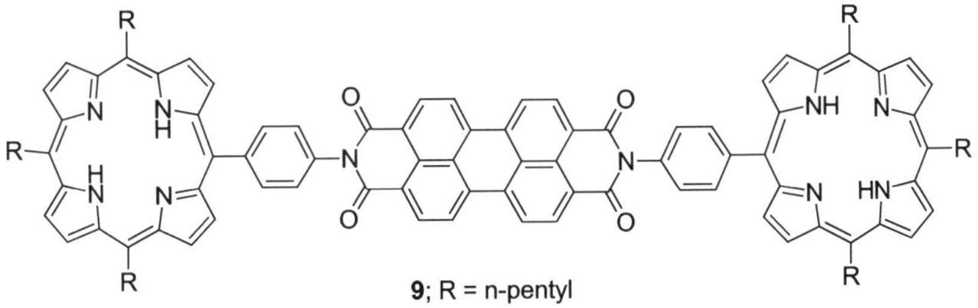

9; R = n-pentyl

A case concerning the mixture **10** and $[Co(CN)_6]^{3-}$ uses light dose as the input.[37] If irradiated by itself in acidic solution, **10** leads to **11**.[38] On its own $[Co(CN)_6]^{3-}$ photoaquates to $[Co(CN)_5(H_2O)]^{2-}$ and the liberated cyanide picks up a proton.[39] The irradiation of the mixture at pH 3.6 leads to a build-up of **11** at first but the aquation of $[Co(CN)_6]^{3-}$ gradually asserts itself and the increasing pH strangles the path to **11**. Furthermore, the increasing pH causes cycloreversion of **11** to the Z isomer of **10** which promptly photoequilibriates back to **10** itself. Thus the **11** concentration falls back to zero at long irradiation times. Fluorescence is a clear measure of the **11** concentration and serves as the output.

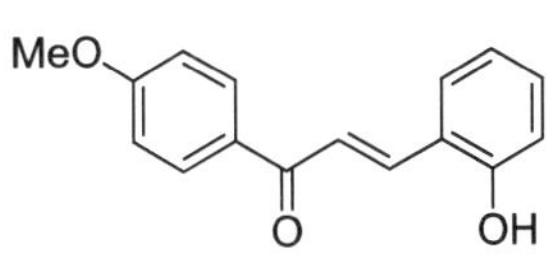

10 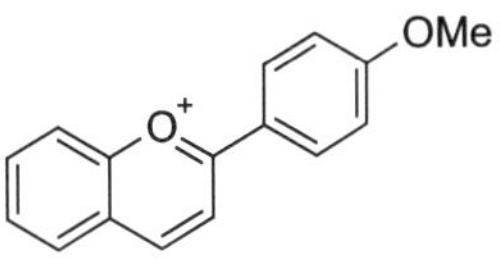   **11**

An electronic version of 'off–on–off' behaviour can be found in the molecular phenomenon of negative differential resistance (NDR). Upon application of a potential difference across a single molecule of **12**,[40,41] the current suddenly increases at a rather high potential range and then decreases catastrophically at a slightly higher voltage. However, this type of experiment has been controversial.[42–46]

The term 'off–on–off' is also used in the literature when one input causes the 'off–on' segment and a subsequent input produces the 'on–off' arm. However, these do not fit under the ternary logic umbrella. Input pairs such as Ca^{2+}/K^+ (**13**),[47] Ag^+/K^+ (**14**)[48] and DNA/Cu^{2+} (**15**)[49] have been used in this way with outputs of absorbance, fluorescence and luminescence respectively.

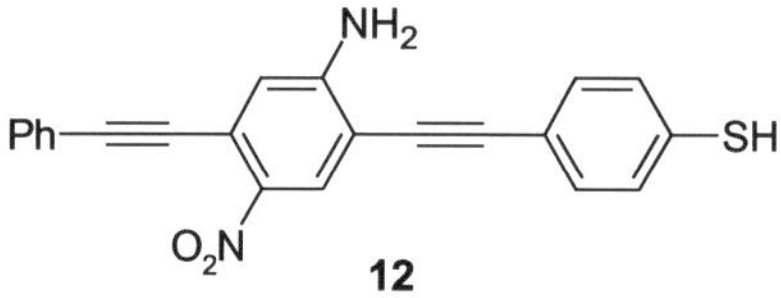

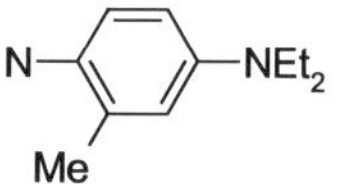

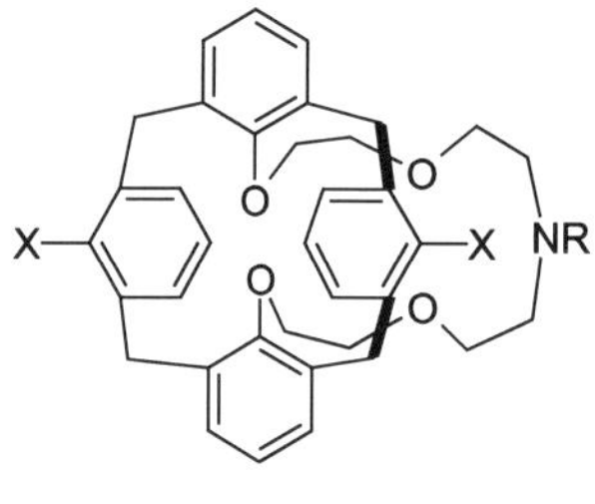

12.3 Other Variants

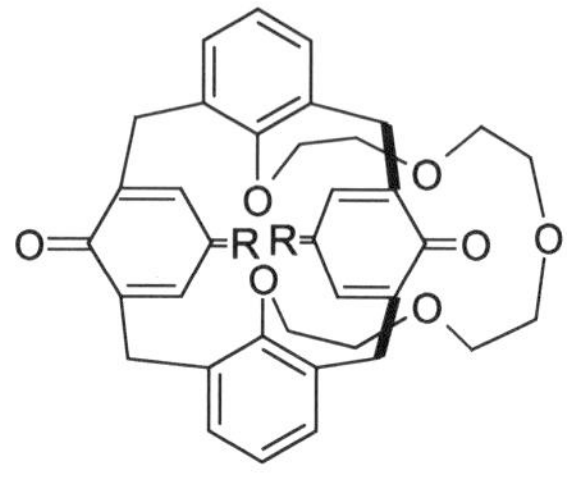

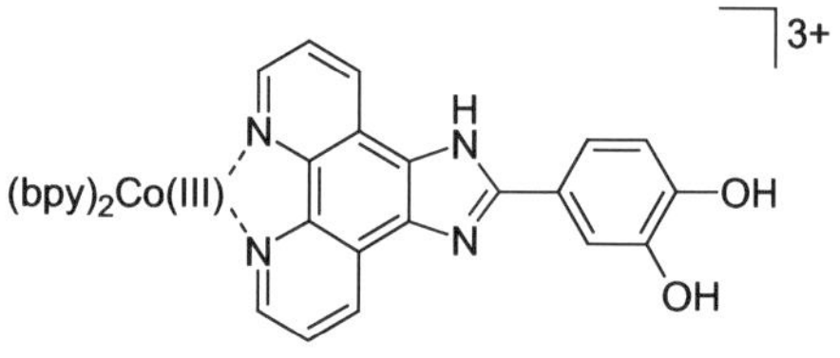

'Off–on–off' behaviour is by no means the only switching phenomenon which shows ternary logic attributes. 'Low–medium–high' output arising from gradually increasing input levels (Table 12.3) is another of these. 'Fluorophore–receptor$_1$–spacer–receptor$_2$' system **16**[50] combines PET switching and ICT switching, which occupy separate pH ranges centred at 9.3 and 6.8 respectively. Figure 12.2 clearly shows the staircase behaviour of the dependent variable. The ICT phenomenon produces strong fluorescence switching in **16** because of a Born–Oppenheimer hole[21] which develops in the excited state at the N^- of deprotonated **16** due to coupling with water molecules.[22]

Table 12.3 Logic table for 'low–medium–high' switching behaviour of **16**.^{a}

Input H^+	Output Fluorescence^b
0, low (10^{-10} M)	0, low (1.0)
1, medium (10^{-8} M)	1, medium (30)
2, high (10^{-6} M)	2, high (81)

^{a}10^{-5} M in water : methanol (4 : 1, v/v); λ_{exc} 441 nm, λ_{em} 535 nm.
^{b}Intensities.

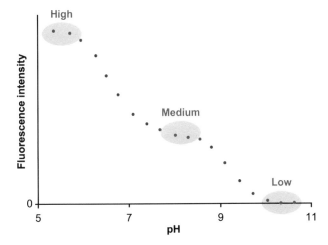

Figure 12.2 Fluorescence intensity (at 535 nm) *vs.* pH profile of **16** in methanol–water (1 : 4, v/v) solution. Excitation wavelength 441 nm.
Adapted from J. F. Callan, A. P. de Silva, J. Ferguson, A. J. M. Huxley and A. M. O'Brien, *Tetrahedron*, 2004, **60**, 11125 with permission from Elsevier Science.

The opposite situation can be seen in 'high–medium–low' output switching demonstrated by Pischel's **17**[51] in acetonitrile solution. Competition between EET and PET is involved here, as in older 'fluorophore$_1$–spacer–receptor–spacer–fluorophore$_2$' systems.[52] Introduction of F^- to monoprotonated **17** produces neutral **17** where PET from the tertiary amine to the naphthalimide fluorophore causes the fluorescence (at 520 nm and excited at 332 nm) to drop by 41%. Further F^- addition to the solution deprotonates the aromatic NH_2

unit, which reduces the fluorescence to 7% of the starting value. The fluorescence intensity–pF function shows two steps (at pF = 5.2 and 2.5) and three plateaux, *i.e.* the output shows tristable behaviour. Availability of monoprotonated, neutral and deprotonated forms of **17** is the molecular origin of tristability. The switching ranges are relatively narrow (*ca.* 1 pF unit) owing to the non-aqueous solvent.

17

The behaviour described in the prior paragraph can be developed by using **17**.H$^+$ as the starting state of a more complex logic device and by using two separate supplies of F$^-$ as degenerate inputs (Table 12.4). It is clear that the larger of the two input levels dominates in each situation. The output pattern of Table 12.4 can be assigned a ternary version of binary NOR logic (section 7.4). 'High' output arises only when both inputs are 'low'. 'Medium' output only arises when the dominant input level is 'medium'.

'Medium–low–high' switching of output is another variant. When observed at 640 nm, Akkaya's **18**[53] goes from a 'medium' emissive state to a 'low' emissive state upon addition of 10^{-6} M Zn^{2+} and then to a 'high' emissive state at much higher Zn^{2+} levels. Mechanistically, **18** has some structural features of a 'receptor–spacer–fluorophore–spacer–receptor' PET system, even though the squaraine oxygen atoms are likely to participate as ligating centres. Related ligating situations are known.[54] These lead to deviations from classical PET switching behaviour, *i.e.* the emission maxima do not remain constant. A case of H$^+$-driven fluorescence switching in a 'medium–low–high' fashion is also available,[55] having being mentioned previously in section 7.4 under NOR logic.

Table 12.4 Logic table for **17**.H$^+$.[a]

Input$_1$ F$^-$	Input$_2$ F$^-$	Output Fluorescence[b]
0, low (10$^{-6.7}$ M)	0, low (10$^{-6.7}$ M)	2, high (20.3)
0, low (10$^{-6.7}$ M)	1, medium (10^{-4} M)	1, medium (12)
0, low (10$^{-6.7}$ M)	2, high (10$^{-1.6}$ M)	0, low (1.3)
1, medium (10^{-4} M)	0, low (10$^{-6.7}$ M)	1, medium (12)
1, medium (10^{-4} M)	1, medium (10^{-4} M)	1, medium (12)
1, medium (10^{-4} M)	2, high (10$^{-1.6}$ M)	0, low (1.3)
2, high (10$^{-1.6}$ M)	0, low (10$^{-6.7}$ M)	0, low (1.3)
2, high (10$^{-1.6}$ M)	1, medium (10^{-4} M)	0, low (1.3)
2, high (10$^{-1.6}$ M)	2, high (10$^{-1.6}$ M)	0, low (1.0)

[a]10^{-5} M in acetonitrile; λ_{exc} 332 nm, λ_{em} 520 nm.
[b]Intensities.

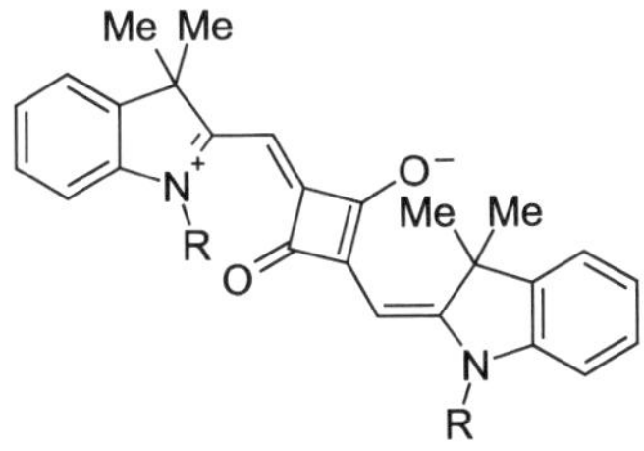

18; R = (CH$_2$)$_2$N(Me)(CH$_2$)$_2$NMe$_2$

The dendrimer–metal complex system[56] discussed under amine, H$^+$-driven XOR logic in section 7.7 also shows 'high–medium–high' behaviour of the fluorescence at 335 nm when driven by H$^+$ alone in non-aqueous solution.

The extreme version of the behaviours described in the previous two paragraphs is the 'on–off–on' response. This is seen in the fluorescence of the ICT sensor **19**[57] in the presence of anionic surfactants such as SDS in water. In detergent-free neutral water, where it is partially protonated and presumably exists as aggregates, the fluorescence quantum yield (ϕ_{Flu}) for **19** is 0.14. When the SDS concentration mounts, the ϕ_{Flu} value collapses to 0.013 owing to the formation of small premicellar aggregates between the oppositely charged species which allows much better access to water. De-excitation via Born–Oppenheimer holes (formed via hydrogen-bonding to the NH$^+$ centre) then takes place. More soluble versions of **19** do not show this effect owing to the lack of the starting aggregate. Once the SDS critical micelle concentration (8×10^{-3} M) is passed, **19** partitions into the less polar environment of the micelles so that it is shielded from water. Therefore, a ϕ_{Flu} of 0.26 is attained. A 14 nm blue-shift accompanies the partitioning of **19** into the less polar micelle.

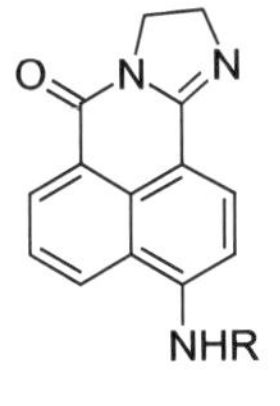

19; R = n-octyl

Another nice example comes from a micelle-based experiment due to Pallavicini, though further self-assembled structures are involved.[58] Neutral aqueous micelles of Triton X-100 serve as containers which confine **20** and **21** hydrophobically. Under alkaline conditions, **21** is axially coordinated to OH$^-$ so that **20** cannot bind. Therefore the fluorescence of **20** remains high. As we move to neutral pH, the competition from OH$^-$ ceases and **20** takes up the role of the axial ligand. Now the neighbouring open-shell Cu^{2+} centre quenches the fluorescence via PET/EET. When the H$^+$ concentration is raised still higher, the carboxylate moiety of **20** is protonated and loses its ligand ability. The fluorescence of **20**.H$^+$ is strong since it is rid of the quenching influence of Cu^{2+}.

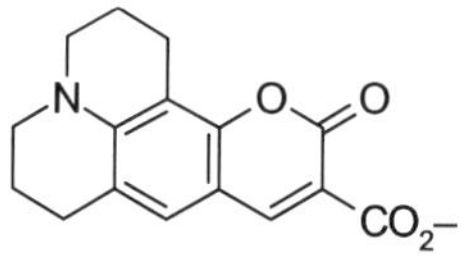

20

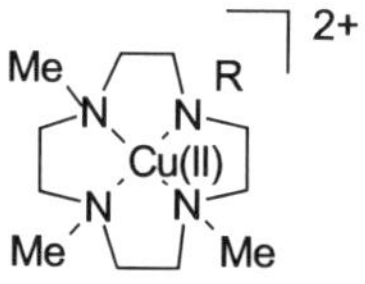

21; R = –dodecyl

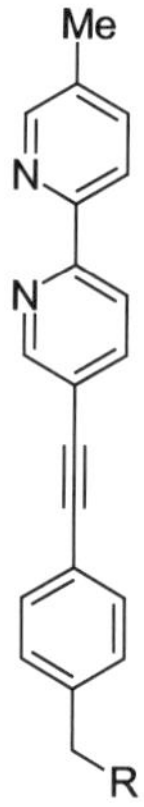

22; R = bis(2-pyridylmethyl)amino

An 'on–off–on' case, discussed in section 5.2.2.4, is also known[59] where one input causes the 'on–off' segment and a different input produces the 'off–on' portion. As mentioned in section 12.2, this case would therefore not fit ternary logic.

Another way of achieving three fluorescence output states is seen in 'off–on–on' systems where the two 'on' states possess different colours. Zhang and Zhu's **22**[60] (related to structure **6** discussed in section 9.2) shows 'off–on' switching (FE = 33, λ_{Flu} = 367 nm) as Zn^{2+} binds to the bis(2-pyridylamino) receptor thereby blocking a PET channel. If the concentration of Zn^{2+} is raised further, a second binding occurs at the 2,2'-bipyridyl receptor so that ICT nature of the excited state is increased. Thus the emission wavelength red-shifts to 415 nm, representing the second 'on' state. Other three- and higher-state systems driven by chemical inputs are also relevant here.[61–65]

To conclude, we make another mention of van der Boom's multi-state latches[66–69] described in section 11.2. By applying increasingly well-defined voltages as the independent variable, the dependent variable of absorbance can be made to specify higher and higher numbers of states. However, the robustness against the pressures of day-to-day operation will probably suffer more and more.

References

1. G. Boole, *An Investigation of the Laws of Thought*, Dover, New York, 1958.
2. R. W. Keyes, *Rev. Mod. Phys.*, 1989, **61**, 279.
3. B. Hayes, *Am. Sci.*, 2001, **89**, 490.
4. A. P. de Silva, H. Q. N. Gunaratne and C. P. McCoy, *Chem. Commun.*, 1996, 2399.

5. S. A. de Silva, A. Zavaleta, D. E. Baron, O. Allam, E. Isidor, N. Kashimura and J. M. Percarpio, *Tetrahedron Lett.*, 1997, **38**, 2237.
6. D. F. H. Wallach and D. L. Steck, *Anal. Chem.*, 1963, **35**, 1035.
7. R. F. Chen and H. Edelhoch (ed.), *Biochemical Fluorescent Concepts: Concepts*, Dekker, New York, vols 1 and 2, 1976.
8. V. Amendola, L. Fabbrizzi, P. Pallavicini, L. Parodi and A. Perotti, *J. Chem. Soc. Dalton Trans.*, 1998, 2053.
9. L. Fabbrizzi, M. Licchelli, A. Poggi and A. Taglietti, *Eur. J. Inorg. Chem.*, 1999, 35.
10. S. A. de Silva, B. Amorelli, D. C. Isidor, K. C. Loo, K. E. Crooker and Y. E. Pena, *Chem. Commun.*, 2002, 1360.
11. J. F. Callan, A. P. de Silva and N. D. McClenaghan, *Chem. Commun.*, 2004, 2048.
12. R. Aucejo, J. Alarcon, E. García-España, J. M. Llinares, K. L. Marchin, C. Soriano, C. Lodeiro, M. A. Bernardo, F. Pina, J. Pina and J. S. de Melo, *Eur. J. Inorg. Chem.*, 2005, 4301.
13. Y. Shiraishi, R. Miyamoto and T. Hirai, *Tetrahedron Lett.*, 2007, **48**, 6660.
14. G. J. Brown, A. P. de Silva, M. R. James, B. O. F. McKinney, D. A. Pears and S. M. Weir, *Tetrahedron*, 2008, **64**, 8301.
15. E. Evangelio, J. Hernando, I. Imaz, G. G. Bardaji, R. Alibes, F. Busque and D. Ruiz-Molina, *Chem. Eur. J.*, 2008, **14**, 9754.
16. J. H. Qian, Y. F. Xu, X. H. Qian and S. Y. Zhang, *J. Photochem. Photobiol. A: Chem.*, 2009, **207**, 181.
17. V. F. Pais, P. Remon, D. Collado, J. Andreasson, E. Perez-Inestrosa and U. Pischel, *Org. Lett.*, 2011, **13**, 5572.
18. K. Szacilowski, *Chem. Rev.*, 2008, **108**, 3481.
19. E. Bishop (ed.), *Indicators*, Pergamon, Oxford, 1972.
20. J. B. Jiang, B. Leng, X. Xiao, P. Zhao and H. Tian, *Polymer*, 2009, **50**, 5681.
21. M. Klessinger and J. Michl, *Excited States and Photochemistry of Organic Molecules*, VCH, New York, 1995.
22. M. D. P. de Costa, A. P. de Silva and S. T. Pathirana, *Canad. J. Chem.*, 1987, **65**, 1416.
23. H. Chao, B. H. Ye, Q. L. Zhang and L. N. Ji, *Inorg. Chem. Commun.*, 1999, **2**, 338.
24. M. J. Han, L. H. Gao, Y. Y. Lu and K. Z. Wang, *J. Phys. Chem. B*, 2006, **110**, 2364.
25. T. Gunnlaugsson, J. P. Leonard, K. Sénéchal and A. J. Harte, *J. Am. Chem. Soc.*, 2003, **125**, 12062.
26. M. T. Albelda, M. A. Bernardo, E. Garcia-Espana, M. L. Godino-Salido, S. V. Luis, M. J. Melo, F. Pina and C. Soriano, *J. Chem. Soc. Perkin Trans. 2*, 1999, 2545.
27. S. A. de Silva, K. C. Loo, B. Amorelli, S. L. Pathirana, M. Nyakirang'ani, M. Dharmasena, S. Demarais, B. Dorcley, P. Pullay and Y. A. Salih, *J. Mater. Chem.*, 2005, **15**, 2791.
28. J. N. H. Reek and S. Otto (ed.), *Dynamic Combinatorial Chemistry*, VCH, Weinheim, 2010.

29. M. E. Huston, E. U. Akkaya and A. W. Czarnik, *J. Am. Chem. Soc.*, 1989, **111**, 8735.
30. M. W. Hosseini, A. J. Blacker and J.-M. Lehn, *J. Am. Chem. Soc.*, 1990, **112**, 3896.
31. Y. Díaz-Fernández, F. Foti, C. Mangano, P. Pallavicini, S. Patroni, A. Pérez-Gramatges and S. Rodriguez-Calvo, *Chem. Eur. J.*, 2006, **12**, 921.
32. P. Pallavicini, Y. Díaz-Fernández and L. Pasotti, *Coord. Chem. Rev.*, 2009, **253**, 2226.
33. P. Pallavicini, Y. Díaz-Fernández and L. Pasotti, *Analyst*, 2009, **134**, 2147.
34. M. Montalti, A. Credi, L. Prodi and M. T. Gandolfi, *Handbook of Photochemistry*, CRC Press, Boca Raton, 3rd edn, 2006.
35. P. S. Goyal, S. V. G. Menon, B. A. Dasannacharya and P. Thiyagarajan, *Phys. Rev. E*, 1995, **51**, 2308.
36. M. P. O'Neil, M. P. Niemczyk, W. A. Svec, D. Gosztola, G. L. Gaines III and M. R. Wasielewski, *Science*, 1992, **257**, 63.
37. F. Pina, M. Maestri and V. Balzani, *J. Am. Chem. Soc.*, 2000, **122**, 4496.
38. F. Pina, M. Maestri and V. Balzani, *Chem. Commun.*, 1999, 107.
39. L. Moggi, F. Bolletta, V. Balzani and F. Scandola, *J. Inorg. Nucl. Chem.*, 1966, **28**, 2589.
40. J. Chen, M. A. Reed, A. M. Rawlett and J. M. Tour, *Science*, 1999, **286**, 1550.
41. J. Chen, W. Wang, M. A. Reed, A. M. Rawlett, D. W. Price and J. M. Tour, *Appl. Phys. Lett.*, 2000, **77**, 1224.
42. R. F. Service, *Science*, 2003, **302**, 556.
43. J. R. Heath, J. F. Stoddart and R. S. Williams, *Science*, 2004, **303**, 1136.
44. P. S. Weiss, *Science*, 2004, **303**, 1137.
45. J. L. He, B. Chen, A. K. Flatt, J. J. Stephenson, C. D. Doyle and J. M. Tour, *Nature Mater.*, 2006, **5**, 63.
46. H. Haick and D. Cahen, *Progr. Surf. Sci.*, 2008, **83**, 217.
47. Y. Kubo, S. Obara and S. Tokita, *Chem. Commun.*, 1999, 2399.
48. J. S. Kim, O. J. Shon, J. A. Rim, S. K. Kim and J. Y. Yoon, *J. Org. Chem.*, 2002, **67**, 2348.
49. Q. L. Zhang, J. H. Liu, X. Z. Ren, H. Xu, Y. Huang, J. Z. Liu and L. N. Ji, *J. Inorg. Biochem.*, 2003, **95**, 194.
50. J. F. Callan, A. P. de Silva, J. Ferguson, A. J. M. Huxley and A. M. O'Brien, *Tetrahedron*, 2004, **60**, 11125.
51. R. Ferreira, P. Remon and U. Pischel, *J. Phys. Chem. C*, 2009, **113**, 5805.
52. A. P. de Silva, H. Q. N. Gunaratne, T. Gunnlaugsson and P. L. M. Lynch, *New J. Chem.*, 1996, **20**, 871.
53. G. Dilek and E. U. Akkaya, *Tetrahedron Lett.*, 2000, **41**, 3721.
54. J. Bourson, J. Pouget and B. Valeur, *J. Phys. Chem.*, 1993, **97**, 4552.
55. Z. X. Wang, G. R. Zheng and P. Lu, *Org. Lett.*, 2005, **17**, 3669.
56. G. Bergamini, C. Saudan, P. Ceroni, M. Maestri, V. Balzani, M. Gorka, S. K. Lee, J. V. Heyst and F. Vogtle, *J. Am. Chem. Soc.*, 2004, **126**, 16466.
57. J. H. Qian, X. H. Qian and Y. F. Xu, *Chem. Eur. J.*, 2009, **15**, 319.

58. F. Denat, Y. A. Diaz-Fernandez, L. Pasotti, N. Sok and P. Pallavicini, *Chem. Eur. J.*, 2010, **16**, 1289.
59. P. P. Neelakandan, M. Hariharan and D. Ramaiah, *J. Am. Chem. Soc.*, 2006, **128**, 11334.
60. L. Zhang and L. Zhu, *J. Org. Chem.*, 2008, **73**, 8321.
61. R. Grigg and W. D. J. A. Norbert, *J. Chem. Soc. Chem. Commun.*, 1992, 1298.
62. C. Di Pietro, G. Guglielmo, S. Campagna, M. Diotti, A. Manfredi and S. Quici, *New J. Chem.*, 1998, **22**, 1037.
63. L. Fabbrizzi, M. Licchelli and P. Pallavicini, *Angew. Chem. Int. Engl.*, 1998, **37**, 800.
64. F. Pina, M. J. Melo, M. A. Bernardo, S. V. Luis and E. Garcia-Espana, *J. Photochem. Photobiol. A Chem.*, 1999, **126**, 65.
65. S. Faulkner, D. Parker and J. A. G. Williams, in *Supramolecular Science: Where It Is and Where It Is Going?*, ed. R. Ungaro and E. Dalcanale, Kluwer, Dordrecht, 1999, p. 53.
66. G. de Ruiter, L. Motiei, J. Chowdhury, N. Oded and M. E. van der Boom, *Angew. Chem. Int. Ed.*, 2010, **49**, 4780.
67. G. de Ruiter, Y. H. Wijsboom, N. Oded and M. E. van der Boom, *ACS Appl. Mater. Intterfac.*, 2010, **2**, 3578.
68. G. de Ruiter and M. E. van der Boom, *Acc. Chem. Res.*, 2011, **44**, 563.
69. G. de Ruiter and M. E. van der Boom, *J. Mater. Chem.*, 2011, **21**, 17575.

CHAPTER 13
Quantum Aspects

13.1 Introduction

We remember that the observation of particulate or wave-like behaviour from a small object depends on the experiment performed, because both properties are characteristic of the object. This is the textbook duality of quantum systems. Taking this further, we see that a bit of information in the quantum domain can exist simultaneously as 0 and as 1. This superposition of numbers can be processed with quantum logic gates.[1–3] Such a potential for parallel computation could allow certain problems to be more efficiently solved on a quantum computer than on a classical computer.[3,4] The most celebrated instance is the algorithm due to Shor,[5,6] which showed that a quantum computer should be able to factorize large numbers efficiently. This would have a large social impact because the security of many data encryption systems, *e.g.* in banks, relies on the inability of classical computers to factorize large numbers. However, it needs to be said that there is no immediate danger to our banks from this direction.

It is to be noted, however, that superposition mentioned in the previous paragraph applies to the quantum bits rather than to the logic gates. The latter instance (covered in sections 13.3 and 13.4) of logic gate superposition is a feature apparently unavailable among conventional electronic logic devices.[7–9] These quantum aspects are discussed below. Methods based on atoms, *e.g.* trapped ions,[10] are outside the scope of this book. Promising approaches based on molecular magnets also have not achieved practical computations as yet,[11–14] and so will not be covered here.

13.2 Nuclear Magnetic Resonance Spectroscopy Approach

A version of nuclear magnetic resonance (NMR) spectroscopy is an approach to quantum computation.[15–17] Nuclear spins form the bits because

Monographs in Supramolecular Chemistry No. 12
Molecular Logic-based Computation
By A Prasanna de Silva
© The Royal Society of Chemistry 2013
Published by the Royal Society of Chemistry, www.rsc.org

they can exist 'up' (0, say) or 'down' (1, say) with respect to a laboratory magnetic field. The superposition of these will be found in the quantum domain. A suitable molecule then provides a set of spins which are operated on by the spectrometer via a sequence of radiofrequency (RF) pulses. So the gates themselves are not molecular but arise from the pulse sequence. Nevertheless, a brief discussion is given because molecules are essential to the process. Importantly, nuclear spins are quite isolated from their environment so that they will maintain coherence for the duration of relaxation times, which can be tens of seconds.

An instructive case[18] uses the spins of ^{1}H and ^{13}C within ^{1}H^{13}CCl$_3$ as a two-bit system so that a simple quantum algorithm can be run. The algorithm, which determines whether a function is a constant or balanced (equal numbers of 1's and 0's), can be illustrated as follows. How do we determine if a coin is fair or fake? Classically, we would first look at one face to see if it is heads or tails. Then we would turn the coin over and inspect the other face. If the second face is different from the first, the coin is fair. Looking at one face would not be enough. However, there is a quantum algorithm[19] which allows the 'fair or fake' determination to be done with one look only.

The NMR experiment itself commences with a RF pulse sequence applied to ^{1}H^{13}CCl$_3$ so that the two bits are both, say, 'up'. This is necessary because the 'up' and 'down' states differ so little in energy that a thermal Boltzmann distribution is normally found. Such preparation of 'pure' states in a bulk sample is necessary[15-17] since single molecules (which necessarily possess pure states) are not amenable to an NMR experiment. Then a transformation RF pulse sequence is applied to the first (input) bit while the inverse transformation is applied to the second bit (called the 'work bit'). Of course, the resonant frequencies are different for the two bits. Next, the function is applied to the first bit and the result is combined with the second bit according to CNOT logic. Controlled NOT (CNOT) logic is a quantum form of XOR logic, where the state of the first bit controls whether the second bit is inverted or not.[20] There are RF pulse sequences which fit this operation. Spin–spin coupling is also involved at this stage. The second bit now contains a superposition of all possible answers. Now the inverse of the original transformations is applied so that the superposition collapses. Reading out of the first bit as 'up' would mean the function is constant. If the readout is 'down' the function is balanced.

A quantum search algorithm[21] has also received NMR spectroscopic implementations.[22-25] Such NMR methods have been taken further, *e.g.* to achieve factorization of the number 15[26] according to Shor's algorithm,[5,6] but it is becoming harder to locate molecules with larger numbers of spins of sufficiently different frequencies which couple to each other. The problem of decoherence is also growing during these longer computations. A recent review compares quantum computing by NMR spectroscopy with several other non-molecular approaches.[27]

13.3 Electronic Absorption and Emission Spectroscopy Approach

Sections 6.4 and 8.7.1 have mentioned how easy it is to observe multiple wavelengths simultaneously. In fact we do this during colour vision. Optical multi-channel analysers or even beam-splitters and filters (Figure 13.1) would permit the same parallel operation on photonic outputs in a measurable fashion. Thus the simultaneous multiplicity or superposition of molecular logic gates becomes possible, *i.e.* where a given molecular system serves as more than one logic gate at the one time simply by observing at different wavelengths. This opens interesting perspectives, when we remember the revolution in data communications caused by the move from metal to fibre-optic lines where the multiplexing opportunities offered by light were unavailable with electric voltages. However, multiplexing in an electronic logic gate context has a very different meaning[7–9] and molecular versions were described in section 10.15. The closest electronics analogy to the chemical phenomenon under discussion is perhaps found in multi-gate chip packages which will give different logic outputs from different pins, but these are parallel operations.[7–9] The molecules discussed in the following sections have applications as self-calibrating sensors owing to the availability of an internal referencing channel.

During spectroscopy experiments of this kind, inversion of a given logic operation is possible by coding for an absorbance rather than a transmittance output. Thus, if a YES operation were demonstrated through a transmittance output, a NOT operation would result from an absorbance output. Nevertheless, during our case studies, we will maintain the coding choices made by the original authors. A similar comment could have been made in Chapters 6 and 8. Conventional electronic systems also have the choice of positive or negative logic, with the former being the more common.[7–9]

13.3.1 Internal Charge Transfer (ICT) systems

As pointed out in section 4.2, 'chromophore–receptor' systems display ion-induced shifts. Unlike PET systems, where a spacer is purposely built in, the

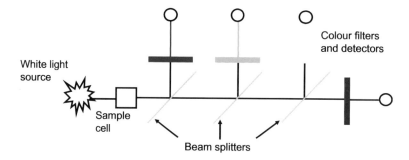

Figure 13.1 Experimental arrangement for simultaneous interrogation of a molecular logic system at four wavelengths.

Quantum Aspects 319

spacer is removed, allowing strong coupling (ICT) between the components. Such systems sacrifice the modular concept used with PET mechanisms but allow certain logic phenomena not otherwise seen, such as superposition. An analyte-induced wavelength shift of an absorption or emission spectral peak naturally causes a rise in intensity at one wavelength and a fall in another. Different logic types then arise when the intensities at several chosen wavelengths are (simultaneously) monitored as outputs.

For example, consider an ICT-based sensor **1** which shows Ca^{2+}-induced colour changes.[28] The absorption spectrum shifts from 444 to 372 nm (Figure 13.2) because of Ca^{2+} binding to the electron donor terminal of the chromophore. Observations of such target-induced transmittance changes at four clearly chosen wavelengths (Figure 13.2) are seen to exhibit all four logic types possible for a single input–single output device, *i.e.* PASS 0, PASS 1, YES and NOT. The fact, that the four wavelengths can be simultaneously observed means that four different computational operations can be run concurrently (Figure 13.3). This has a crucial consequence that any sensory molecule which shows a colour change, *e.g.* a simple pH indicator,[29] would show this quantum aspect of information processing. Since these are found in pre-university chemistry laboratories and in flower pigments such as anthocyanins, the amazing conclusion is that aspects of quantum computing are present in our schools and in our gardens. The pH-dependent emission spectra of fluorophore-appended boronic acids also come into this category, except that fructose input binds to the

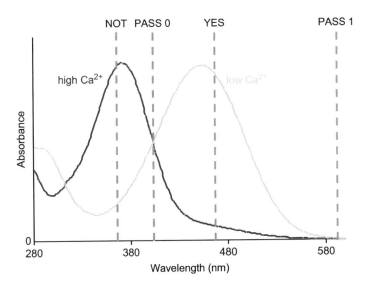

Figure 13.2 Absorption spectra of **1** with high (10^{-3} M) or low ($<10^{-9}$ M) Ca^{2+}, showing the Ca^{2+}-driven logic types found when transmittance is monitored at 363, 401, 462 and 595 nm.
Adapted from A. P. de Silva and N. D. McClenaghan, *Chem. Eur. J.*, 2002, **8**, 4935 with permission from Wiley-VCH.

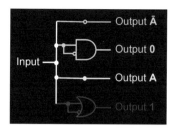

Figure 13.3 Superposition of all four single-input logic gates in Figure 13.2, where each one becomes visible only when interrogated at the appropriate wavelength.

boronic acid unit so that the operational pK_a value is shifted,[30] *e.g.* in DiCesare and Lakowicz's **2**.[31]

A double-input version of **1** is **3**,[28] where the latter possesses a H^+-binding site on the quinoline nitrogen. Protonation strengthens the electron accepting terminal of the chromophore. Sections 7.7 and 7.9 discussed the XOR and IMPLICATION logic aspects of this case respectively. In fact, Figure 13.4 extends Figure 7.11 so that five logic gates emerge (Figure 13.5). Actually, a sixth (PASS 1) can also be found if the transmittance is monitored at 600 nm, rather similar to that found in Figure 13.2 at 595 nm. Quantitative data on **1** and **3** are given in Tables 13.1 and 13.2 respectively.

The absorbance features in Upadhyay's **4** are effected with Zn^{2+} and $CH_3CO_2^-$ inputs.[32] The $CH_3CO_2^-$ causes a red-shift in the spectrum of **4** in DMSO from 407 to 505 nm, with an isosbestic point at 445 nm. Partial proton transfer from the NH group in **4** to the anion within a 2:1 complex would be the underlying cause. Addition of a 10-fold excess of Zn^{2+} reverses the phenomenon. This corresponds to IMPLICATION($CH_3CO_2^- \Rightarrow Zn^{2+}$) logic action at 407 nm and INHIBIT(Zn^{2+}) logic at 505 nm. This system is further

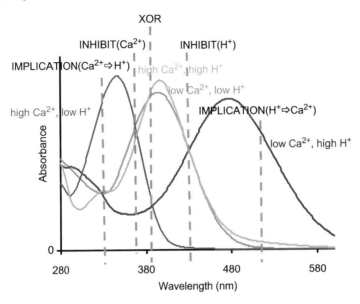

Figure 13.4 Absorption spectra of **3** with high (10^{-3} M) or low ($<10^{-9}$ M) Ca^{2+}, and high ($10^{-5.6}$ M) or low ($10^{-9.7}$ M) H^+, showing the H^+, Ca^{2+}-driven logic types found when the transmittance of **3** is observed at 330, 369, 387, 430 and 511 nm respectively. In addition, PASS 1 shows up when the transmittance is monitored at 600 nm.
Adapted from A. P. de Silva and N. D. McClenaghan, *Chem. Eur. J.*, 2002, **8**, 4935 with permission from Wiley-VCH.

reconfigurable to include **PASS 0** and **PASS 1** logic when absorbance is monitored at 700 and 445 nm respectively.

Triple-input examples of this general type are also known. Section 10.5 contained a discussion of the H^+- and redox-dependence of the rich absorption spectra of **5**.[33] The H^+, Fe^{3+}, Zn-driven INHIBIT(Zn), IMPLICATION (Zn ⇐ H^+) and Fe^{3+}, Zn-driven INHIBIT(Zn) logic types are observed at 405, 500 and 850 nm respectively.

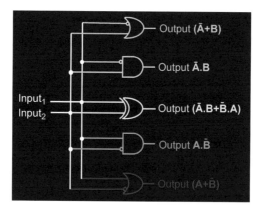

Figure 13.5 Superposition of five double-input logic gates in Figure 13.4, where each one becomes visible only when interrogated at the appropriate wavelength. A sixth (PASS 1) is also available.

Table 13.1 Set of truth tables showing superposed logic gates for **1**.[a]

Input	Output			
Ca^{2+}	$Trans_{363}$[b] NOT	$Trans_{401}$[b] PASS 0	$Trans_{462}$[b] YES	$Trans_{595}$[b] PASS 1
none	high (35)	low (5)	low (1)	high (100)
high ($10^{-2.3}$ M)	low (1)	low (5)	high (68)	high (100)

[a] 10^{-5} M **1** in water, pH 7.2.
[b] Percentage for 10 cm optical path length.

Table 13.2 Set of truth tables showing superposed logic gates for **3**.[a]

$Input_1$ Ca^{2+b}	$Input_2$ H^{+c}	Output $Trans_{330}$[d] IMPLICATION($Ca^{2+} \Rightarrow H^+$)	$Trans_{369}$[d] INHIBIT(Ca^{2+})	$Trans_{387}$[d] XOR
none	low	high (50)	low (24)	low (16)
none	high	high (50)	high (65)	high (59)
high	low	low (17)	low (25)	high (58)
high	high	high (50)	low (26)	low (15)

$Input_1$ Ca^{2+b}	$Input_2$ H^{+c}	Output $Trans_{430}$[d] INHIBIT(H^+)	$Trans_{511}$[d] IMPLICATION($H^+ \Rightarrow Ca^{2+}$)
none	low	low (32)	high (100)
none	high	low (32)	low (25)
high	low	high (96)	high (100)
high	high	low (32)	high (93)

[a] 10^{-5} M **3** in water.
[b] High value = $10^{-2.3}$ M.
[c] Low value = $10^{-9.7}$ M and high value = $10^{-5.6}$ M.
[d] Percentage for 10 cm optical path length.

Quantum Aspects 323

Baytekin and Akkaya were the first to demonstrate logic which was reconfigurable with observation wavelength.[34] Fluorescence emission was the output channel. As discussed under NAND and TRANSFER logic (sections 7.5 and 7.10), the system **6.7.8** uses a popular DNA binder **6** to interact with the matching mononucleotide pair of **7** and **8**. The output corresponds to NAND and TRANSFER(**8**) logic when viewed at 455 and 412 nm respectively (Table 13.3). INHIBIT and NOR logic results when the emission of **9**[35] is noted at 570 nm and 695 nm respectively, while interrogating with H^+ and Cu^{2+} (disabling input). The red emission is due to the strong ICT nature of the fluorophore resulting from the amino donor, whereas the orange fluorescence emerges when the amino group is removed from the π-electron system by protonation. Presentation of Cu^{2+} to **9** results in the fluorescence-quenched complex.

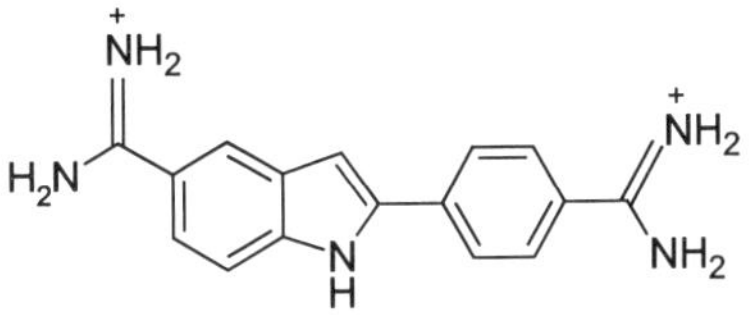

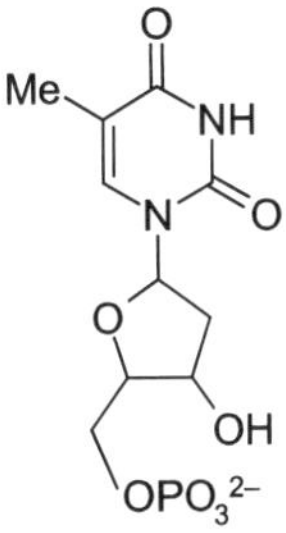

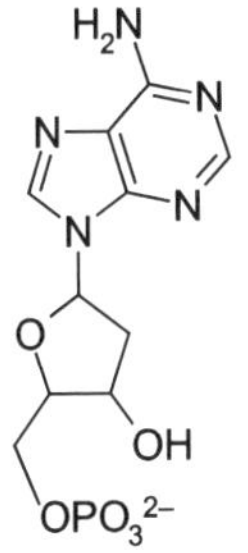

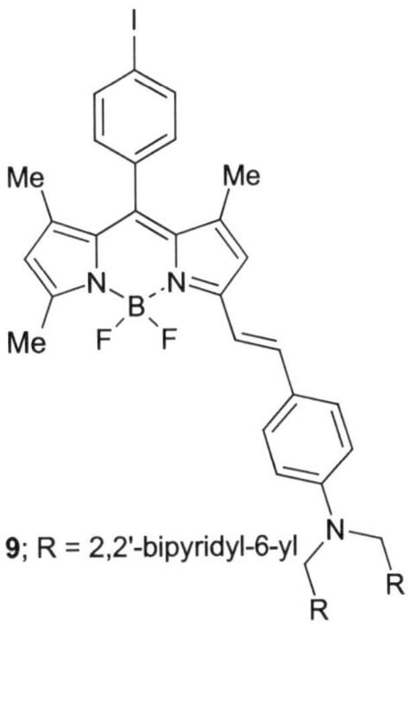

9; R = 2,2'-bipyridyl-6-yl

Table 13.3 Set of truth tables showing superposed logic gates for **6**.[a]

Input$_1$ 7	Input$_2$ 8	Output$_1$ Fluorescence$_{455}$ NAND	Output$_2$ Fluorescence$_{412}$ TRANSFER(8)
none	none	high (1.00)	low (0.28)
none	high (0.4 mM)	high (0.91)	high (1.00)
high (0.4 mM)	none	high (0.98)	low (0.32)
high (0.4 mM)	high (0.4 mM)	low (0.48)	high (1.00)

[a] 0.75×10^{-6} M **6** in dimethylsulfoxide:water (99:1, v/v), pH 7.5. Intensities, λ_{exc} 345 nm.

Yan's **10** is an example[36] where protonation shifts the fluorescence emission wavelength from 415 to 465 nm (λ_{ex} = 350 nm). Then we see an H$^+$-driven YES gate as far as the emission output at 475 nm is concerned. Similarly, NOT logic is seen in the emission at 400 nm. The wavelengths are chosen to optimize the 'high':'low' ratio of the output. Though higher logic types are discussed in this paper, we stick to the policy of regarding excitation light as a power supply rather than as an input. Although it was discussed in section 7.2.1.1, Callan's **11**[37] also needs to be mentioned here since it displays Na$^+$, K$^+$-driven AND logic when fluorescence is observed at 445 nm but has to settle for PASS 1 at 355 nm. A fluorescence output reconfigurable between NOT and INHIBIT is also known.[38]

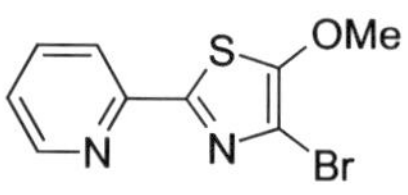

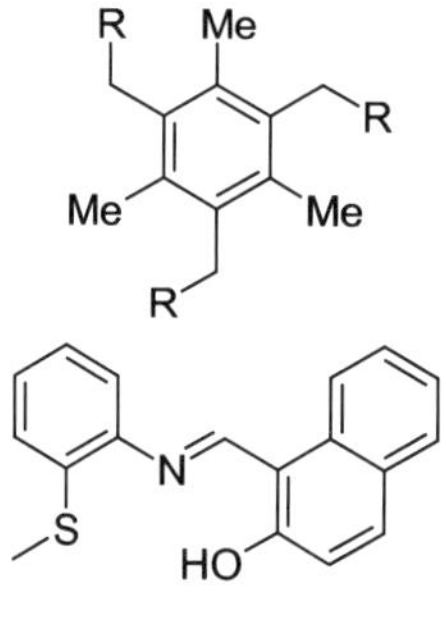

Compound **12**[39] has four states which are rather distinguishable *via* absorption and emission maxima as well as emission quantum yield and lifetime. So **12** is difficult to classify until operating conditions are restricted. For instance, if the absorbance at 440 nm is taken as the output, it fits a NAND logic classification. However if emission is taken as output after excitation at 440 nm, it becomes a 2-input INHIBIT gate. The azacrown and the thiazacrown in **12** give good selectivity for the inputs Ca^{2+} and Ag$^+$. Two ICT paths from terminal donor nitrogen atoms to the acceptor components are visible. Metal binding converts each terminal from a donor site into an acceptor.

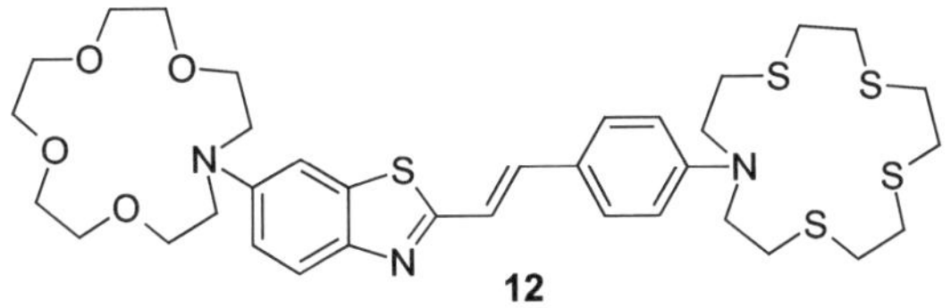

Quantum Aspects

12

Now we move to reactions concerning covalent bond formation. Amines attack the carbonyl group of **13**[40,41] reversibly to disturb the extensive delocalization of the push–pull π-electron system. Monitoring the absorbance at 468 nm displays NOT logic. Since the product **14** absorbs at 384 nm, following the absorbance there shows YES logic instead. Fluorescence (with appropriate excitation) can also be used as an output to display logic superposition.

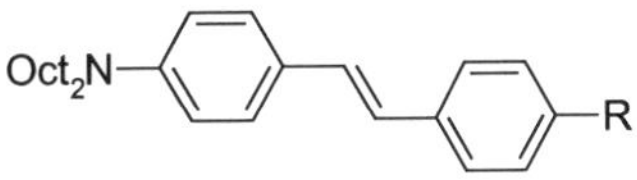

13; R = COCF$_3$
14; R = C(OH)(NHR)(CF$_3$)

The classical test for ICT character of an excited state involves solvent-dependent fluorescence wavelengths.[42] Such wavelength shifts are analysed according to the Lippert–Mataga–Bakhshiev equation[42] to yield the excited state dipole moment which quantifies the degree of ICT. Therefore solvent polarity is a useful input to reconfigure logic simultaneously while observing fluorescence. Tian's team nicely integrate an ICT fluorophore with a dithienylethene photochromic[43] within **15** so that a light dose input can also be applied profitably.[44] Table 13.4 summarizes the results. The electron-rich thienyl group at the 4-position of the naphthalimide contributes to the ICT character of the fluorophore. Indeed, the fluorescence peak shifts from 420 nm in cyclohexane to 530 nm in acetone. However, irradiation of **15** at 365 nm in either solvent converts it into the coloured photocyclized form, which is also non-fluorescent in the visible region. Hence NOR logic emerges at 420 nm whereas INHIBIT(light dose$_{365}$) logic rules at 530 nm.

Table 13.4 Set of truth tables showing superposed logic gates for **15**.[a]

Input$_1$ Light dose$_{365}$	Input$_2$ acetone	Output$_1$ Fluorescence$_{420}$ NOR	Output$_2$ Fluorescence$_{530}$ INHIBIT(light dose$_{365}$)
none	none	high (1.00)	low (0.20)
high	none	low (0.20)	low (0.07)
none	high (10% v/v)	low (0.30)	high (1.00)
high	high (10% v/v)	low (0.03)	low (0.30)

[a]2.8×10^{-5} M **15** in cyclohexane. Intensities, λ_{exc} 380 nm.

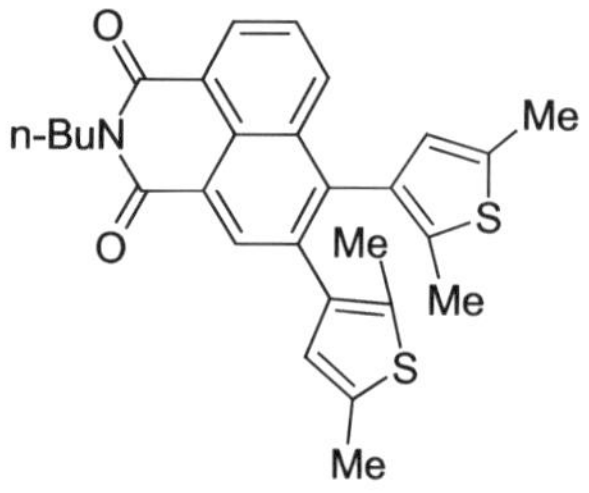

15

Other photochromic cases which display this general phenomenon are known. By its very nature, photochromism results in large wavelength shifts upon absorption of a light dose.[45] Raymo's and Giordani's work[46] was featured previously, in section 10.13. Sections 10.9 and 10.10 discussed additional cases, including non-photochromic ones.[47–50] Andreasson et al.[51] present a powerful case (structure **54** shown in section 10.19), a three-photochrome construct, which can be reconfigured into AND, XOR, INHIBIT and more complex gate arrays. This is done by changing not only the wavelength at which the absorbance is observed but the irradiation wavelength as well. Some of its features were discussed in section 10.19.

It becomes clear that any reversible chemical reaction which produces a spectral wavelength shift can be exploited in this way. Even irreversible cases have their uses in this regard, just like the irreversible single input–single output logic that was featured in sections 5.3 and 5.5.

13.3.2 Electronic Energy Transfer (EET) Systems

Differential changes of intensity in two spectral bands arise naturally in systems composed of two or more distinct fluorophores. EET occurs in such cases, as discussed in section 4.6. A selection of these follows.

Fabbrizzi and colleagues find that a mixture of **16** and **17** produces emission from **16** when it is excited directly.[52] However **16** binds Zn^{2+} to form a trigonal bipyramidal complex with one apical position being occupied by the solvent methanol.[52] Anionic **17** displaces the ligating solvent to produce a system where the EET donor is one of the three dimethylaminonaphthalene moieties and the acceptor is **17**. Now the excitation of **16** produces emission from **17**. Protonation of the aliphatic amine groups of the $16.Zn^{2+}.17$ complex causes disassembly and arrest of EET.

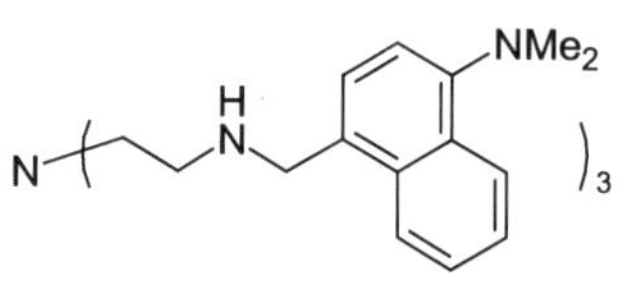

16 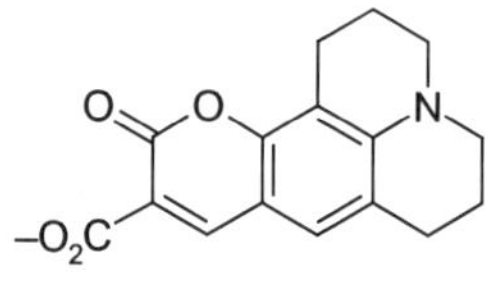   **17**

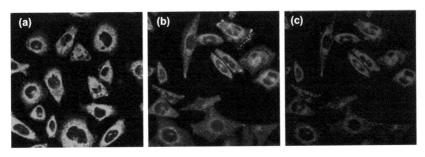

Figure 13.6 Confocal fluorescence images of MCF-7 cells (excited at 488 nm) containing (a) **18**, observed at 514 nm, (b) **18** and 5 mM Hg^{2+}, observed at 514 nm and (c) **18** and 5 mM Hg^{2+}, observed at 589 nm.
Reprinted from X. Zhang, Y. Xiao and X. H. Qian, *Angew. Chem. Int. Ed.*, 2008, **47**, 8025 with permission from Wiley-VCH.

Qian's group built in a rhodamine spirolactone close to a BODIPY unit within **18**[53] so that the initial green emission is replaced by a red emission when the spirolactone is opened[54] by Hg^{2+}. The behaviour is preserved during intracellular studies (Figure 13.6). Once the rhodamine is produced, EET occurs to it from the BODIPY moiety. Hg^{2+}-induced YES and NOT are thus seen at the observation wavelengths of 589 and 514 nm respectively (excited at 488 nm). Ratiometric measurements are clearly possible, even though the Hg^{2+}-induced reaction is irreversible.

19; R = 9-anthryl   **20**; R = 9-anthrylmethyl

Owing to good spectral overlap between donor emission and acceptor absorption (section 4.6), EET from anthracene to aminonaphthalimide fluorophores is involved in Misra's **19**[55] and our **20**.[56] Thus the anthracene emission is weak in **19**, where F^- and Hg^{2+} serve as the two inputs. The efficiency of EET in **19** drops upon admitting F^- owing to reduced spectral overlap, since the deprotonated aminonaphthalimide[57] has its absorption band red-shifted by 90 nm. The F^--induced enhancement of anthracene emission at 430 nm and concomitant loss of the aminonaphthalimide fluorescence at

530 nm is the result ($\lambda_{exc} = 380$ nm). The Hg^{2+} binds to the imino and amino groups of **19** to quench the aminonaphthalimide fluorescence *via* deprotonation of the NH moiety whereas the weak anthracene emission is unaffected. Simultaneous presence of Hg^{2+} and F^- results in complexation with each other so that **19** remains in its original state. Therefore the fluorescence output at 430 nm behaves according to INHIBIT(Hg^{2+}) logic and the output at 530 nm shows XNOR response.

PET from the amino receptor to the anthracene is additionally involved in **20**.[56] The H^+ input recovers the anthracene fluorescence at 420 nm ($\lambda_{exc} = 367$ nm) with FE = 7, *i.e.* YES logic prevails. There is no PET from the amino receptor to the aminonaphthalimide fluorophore in this case,[58] so that protonation leaves its emission intensity at 550 nm much less affected, *i.e.* PASS 1 logic is apparent. Polymer-based relatives[59] are also known.

The powerful deoxyribozyme approach to molecular logic introduced by Stojanovic and Stefanovic revolves around oligonucletides which carry the EET donor–acceptor pair of fluorescein and tetramethylrhodamine at their terminals. When discussed in section 5.3.1.3 under YES logic[60] or in section 5.5.2 under NOT logic,[60] fluorescein emission was the chosen output. Switching to the tetramethylrhodamine emission as the output would give us NOT and YES logic respectively.

Yang's team[61] have a two-fluorophore system **21** where the high-energy fluorophore can be irreversibly transformed into a third medium-energy emitter by one of two enzyme inputs. The action of β-glucosidase produces a hydroxycoumarin with an emission signature at 460 nm. Table 13.5 shows TRANSFER(β-glucosidase) logic for this case. On the other hand, phosphodiesterase I action is mirrored in the loss of 656 nm emission as the porphyrin breaks off from the coumarin moiety so that EET ceases. Importantly, the porphyrin absorbs poorly at the excitation wavelength of 340 nm. Table 13.5 shows NOT TRANSFER(phosphodiesterase) logic in this instance. Indeed, detailed studies in the analogue regime (section 2.5) should allow individual measurement of the two enzyme concentrations by way of these two emissions. These correlations persist inside liver cells (Figure 13.7). The response of alkoxycoumarin emission at 390 nm is more complicated owing to opposing effects and finally results in β-glucosidase, phosphodiesterase-driven INHIBIT(β-glucosidase) logic.

21

Quantum Aspects 329

Table 13.5 Set of truth tables showing superposed logic gates for **21**.[a]

Input$_1$ β-glucosidase	Input$_2$ phosphodiesterase I	Output$_1$ Flu$_{390}$ INHIBIT[c]	Output$_2$ Flu$_{460}$ TRANSFER[c]	Output$_3$ Flu$_{656}$ NOT TRANSFER[c]
none	none	low (1.0)	low (1.0)	high (19)
none	high[b]	low (1.0)	high (8.5)	high (14)
high[b]	none	high (5.0)	low (1.0)	low (4.0)
high[b]	high[b]	low (2.0)	high (7.0)	low (4.5)

[a]In water, pH 7.2, after incubation for 2 h; Intensities, λ_{exc} 340 nm.
[b]0.05 units ml^{-1}.
[c]See text for more details of logic type.

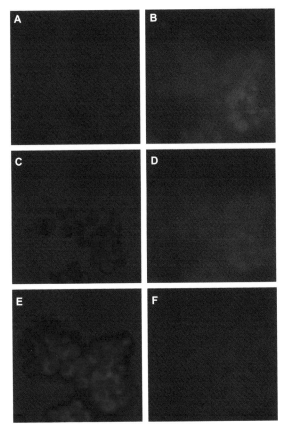

Figure 13.7 Fluorescence micrographs (in blue and red channels) of Huh7 cells incubated with **21** for 0.5 h (A, B), 1.0 h (C, D) and 2.0 h (E, F) at 37 °C in phosphate buffered saline, pH 7.2.
Reprinted from Y. H. Li, H. Wang, J. S. Li, J. Zheng, X. H. Xu and R. H. Yang, *Anal. Chem.*, 2011, **83**, 1268 with permission from the American Chemical Society.

13.3.3 Excimer and Exciplex Systems

Emission spectroscopy introduces the possibility of excited state association and, as discussed in section 4.7, this shows up as a new red-shifted band with respect to the original species. So we have conditions ripe for productive multiwavelength observation.

Modulation of monomer–excimer equilibria goes back a quarter of a century,[62,63] but we focus on di(anthrylmethyl)polyamine **22**[64] as our first illustration. Excimers can form when one of the anthracenes is excited. The difference in emission wavelength between the monomer and excimer allows observation of two fluorescence channels (at 416 nm and 520 nm respectively). Calculation of the ratio of the two band intensities gives an easily calibrated measure of the analyte. For instance, the 520 nm emission dominates at low acidity, as the hydrophobic anthracenyl terminals crowd together in aqueous solution. The 416 nm band asserts itself at moderate acidity when the nitrogen atoms pick up protons which cause repulsion between the parts of the receptor chain and prevent the anthracenyl units from approaching each other. The modulation of PET from the amines to the anthracene also plays a role. Thus the output at 416 nm indicates YES logic whereas 520 nm output displays NOT logic.

Related excimer–monomer equilbria can be seen in **23**,[65] where 2 : 1 complexes are formed with Zn^{2+}, in **24**[66] with $P_2O_7^{4-}$ and in **25**[67] with Ag^+. In the case of **25**, no excimer emission is observed with Ag^+ in aqueous solutions in excess of 70% ethanol suggesting that hydrophobic stacking[68] of the pyrene units is important. An Hg^{2+}-driven case is also available.[69] A disabled OR gate[70] discussed in Chapter 10 (as structure **31** in section 10.10) is also a member of this set.

22; R = 9-anthrylmethyl

23; R = 1-pyrenyl

24; R = 1-pyrenylmethyl

25

Excimers are also responsible for the action of **26**[71,72] when two units are encapsulated within a γ-cyclodextrin host in water. This leaves two 15-crown-5 ether moieties suitably placed to sandwich K^+ selectively, whereas a single 15-crown-5 ether is known to favour Na^+. The lining-up of fluorophores manifests itself as an intermolecular excimer signature with a concomitant decrease in the monomer band.

Quantum Aspects 331

Though exciplex-based examples have old origins,[73] we consider two from Teramae's laboratory.[74] Binding of OAc⁻ increases the electron density of the thiourea unit in **27**[74] and hence encourages PET towards the fluorophore. This gives FE = 0.12 for the monomer component of fluorescence at 378 nm in acetonitrile. Given that pyrene derivatives have relatively long-lived excited states, it is not surprising that **27** also shows an exciplex emission at 496 nm which remains unchanged upon OAc⁻ binding. This arises from the electron-rich receptor, whether complexed or not, interacting with the excited fluorophore. However the exciplex component disappears when an anthracene fluorophore with a shorter excited state lifetime is used instead. This is the case in **28**, where Gunnlaugsson finds simple PET switching (FE = 0.25) of the fluorophore emission exclusively.[75]

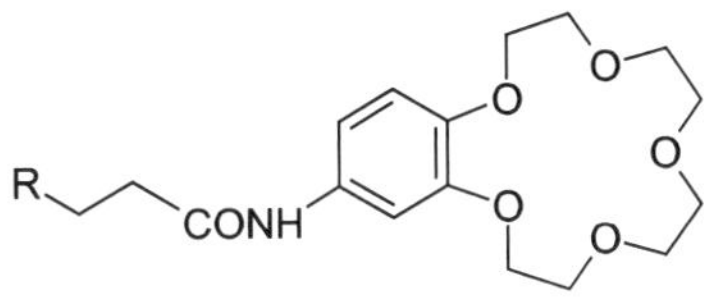

26; R = 1-pyrenylmethyl

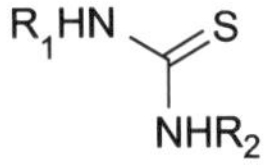

27; R_1 = 1-pyrenylmethyl, R_2 = methyl
28; R_1 = 9-anthrylmethyl, R_2 = 4-CF$_3$-phenyl

Teramae's **26**,[71] discussed above, also gives an intramolecular exciplex emission when dissolved in acetone. Addition of Na⁺ eliminates the exciplex and resurrects the monomer component. The Na⁺ binds to 15-crown-5 ether so that the electron-donor role of the benzo moiety within the exciplex is weakened. Receptors with an electron-acceptor role within exciplexes are also known.[76]

In closing this section, we note two other gate types which were not mentioned by name in the past dozen paragraphs or so. If fluorescence is the output, PASS 1 logic lurks behind every isoemissive point in the experiments discussed in this and the previous section. Conversely, PASS 0 logic is found in spectral regions where no emission is seen.

13.4 Raman Spectroscopy Approach

Raman scattering offers new ways of performing molecular logic by bringing in vibrational spectroscopy concepts and results. In contrast, many examples in this book concern designs based on electronic spectroscopy. Therefore, Raman experiments give detailed signatures which depend strongly on molecular structural features. Chemical changes can show up in the appearance and disappearance of various marker bands. The modern practice of Raman spectroscopy achieves good sensitivity of detection by adsorbing suitable molecules on roughened metal surfaces. Extra sensitivity arises when the Raman signal emerges from the electronic excited state of the molecules. Surface-enhanced resonance Raman spectroscopy (SERRS) is therefore a particularly attractive technique.[77] Flood and colleagues use the charge-transfer complex between **29** (seen before as structure **4** in Chapter 4, for instance) and **30** adsorbed on a nanostructured gold electrode.[78] Laser excitation at 785 nm allows good coupling between plasmons at the gold surface and the excited states of the molecular species. The Raman signals at 1640, 508 and 529 cm^{-1} are characteristic of **29.30**, its singly and doubly oxidized versions respectively (Figure 13.8). In fact, the singly and doubly oxidized species are simply **30^{+}** and **30^{2+}** respectively, caused by complex dissociation. Though these Raman signals are rather closely located on a wavelength (nm) scale, their separation with suitable filters can be anticipated for simultaneous viewing. If applied voltages are taken as the inputs the ambient, singly and doubly oxidized states can be seen to arise selectively from 'low, low', 'high, low' (or 'low, high') and 'high, high' input situations respectively. Therefore the Raman bands at 1640, 508 and 529 cm^{-1} show NOR, XOR and AND logic respectively.

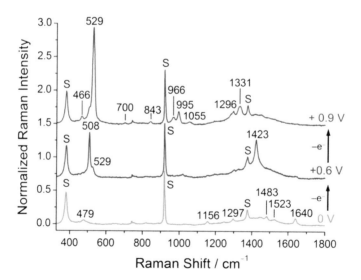

Figure 13.8 Surface-enhanced resonance Raman spectra of **29.30** in its native (green), singly oxidized (red), and doubly oxidized (blue) forms in acetonitrile (S) ($\lambda_{exc} = 785$ nm).
Reprinted from E. H. Witlicki, C. Johnsen, S. W. Hansen, D. W. Silverstein, V. J. Bottomley, J. O. Jeppesen, E. W. Wong, L. Jensen and A. H. Flood, *J. Am. Chem. Soc.*, 2011, **133**, 7288 with permission from the American Chemical Society.

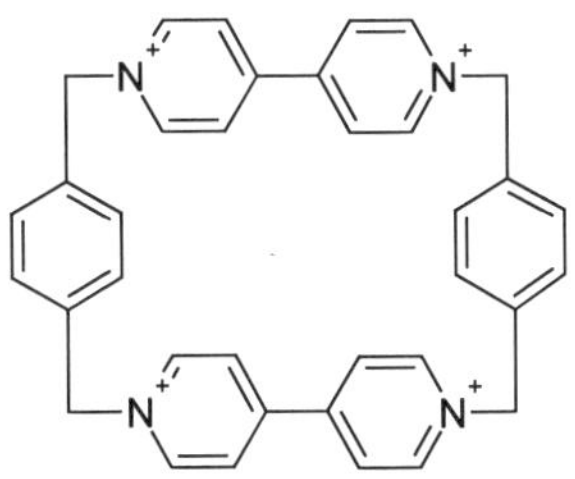

29

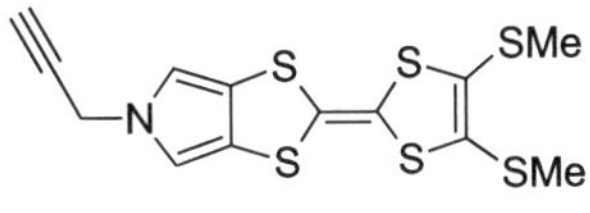

30

References

1. R. P. Feynman, *Int. J. Theor. Phys.*, 1982, **21**, 467.
2. M. A. Nielsen and I. L. Chuang, *Quantum Computation and Quantum Information*, Cambridge University Press, Cambridge, 2000.
3. D. Deutsch, *Proc. Roy. Soc. London A*, 1989, **425**, 73.
4. S. Aaronson, *Sci. Am.*, 2008, **3**, 62.
5. P. W. Shor, in *Proc. 35th Ann. Symp. Foundations Computer Sci.*, ed. S. Goldwasser, 1994, 124.
6. I. L. Chuang, R. LaFlamme, P. W. Shor and W. H. Zurek, *Science*, 1995, **270**, 1633.
7. J. Millman and A. Grabel, *Microelectronics*, McGraw-Hill, London, 1988.
8. A. P. Malvino and J. A. Brown, *Digital Computer Electronics*, Glencoe, Lake Forest, 3rd edn, 1993.
9. A. L. Sedra and K. C. Smith, *Microelectronic Circuits*, Oxford University Press, Oxford, 5th edn, 2003.
10. D. Kielpinski, C. Monroe and D. J. Wineland, *Nature*, 2002, **417**, 709.
11. M. N. Leuenberger and D. Loss, *Nature*, 2001, **410**, 789.
12. R. E. P. Winpenny, *Angew. Chem. Int. Ed.*, 2008, **47**, 7992.
13. C. F. Lee, D. A. Leigh, R. G. Pritchard, D. Schultz, S. J. Teat, G. A. Timco and R. E. P. Winpenny, *Nature*, 2009, **458**, 314.
14. G. A. Timco, S. Carretta, F. Troiani, F. Tuna, R. J. Pritchard, C. A. Muryn, E. J. L. McInnes, A. Ghirri, A. Candini, P. Santini, G. Amoretti, M. Affronte and R. E. P. Winpenny, *Nature Nanotechnol*, 2009, **4**, 173.
15. N. A. Gershenfeld and I. L. Chuang, *Science*, 1997, **275**, 350.
16. D. G. Cory, A. F. Fahmy and T. F. Havel, *Proc. Natl. Acad. Sci. USA*, 1997, **94**, 1634.
17. N. A. Gershenfeld and I. L. Chuang, *Sci. Am.*, 1998, **6**, 66.
18. I. L. Chuang, L. M. K. Vandersypen, X. L. Zhou, D. W. Leung and S. Lloyd, *Nature*, 1998, **393**, 143.
19. D. Deutsch and R. Jozsa, *Proc. Roy. Soc. Lond. A*, 1992, **439**, 553.

20. D. P. DiVincenzo, *Phys. Rev. A*, 1995, **50**, 1015.
21. L. K. Grover, *Phys. Rev. Lett.*, 1997, **79**, 325.
22. I. L. Chuang, N. A. Gershenfeld and M. Kubinec, *Phys. Rev. Lett.*, 1998, **80**, 3408.
23. J. A. Jones, M. Mosca and R. H. Hansen, *Nature*, 1998, **393**, 344.
24. L. K. Grover, *Science*, 1998, **280**, 228.
25. J. A. Jones, *Science*, 1998, **280**, 229.
26. L. M. K. Vandersypen, M. Steffen, G. Breyta, C. S. Yannoni, M. H. Sherwood and I. L. Chuang, *Nature*, 2001, **414**, 883.
27. T. D. Ladd, F. Jelezko, R. Laflamme, Y. Nakamura, C. Monroe and J. L. O'Brien, *Nature*, 2010, **464**, 45.
28. A. P. de Silva and N. D. McClenaghan, *Chem. Eur. J*, 2002, **8**, 4935.
29. E. Bishop (ed.), *Indicators*, Pergamon, Oxford, 1972.
30. J. Y. Yoon and A. W. Czarnik, *J. Am. Chem. Soc.*, 1992, **114**, 5874.
31. N. DiCesare and J. R. Lakowicz, *J. Phys. Chem.*, 2001, **105**, 6834.
32. K. K. Upadhyay, A. Kumar, R. K. Mishra, T. M. Fyles, S. Upadhyay and K. Thapliyal, *New J. Chem.*, 2010, **34**, 1862.
33. X. Y. Liu, X. Han, L. P. Zhang, C. H. Tung and L. Z. Wu, *Phys. Chem. Chem. Phys.*, 2010, **12**, 13026.
34. H. T. Baytekin and E. U. Akkaya, *Org. Lett.*, 2000, **2**, 1725.
35. F. Puntoriero, F. Nastasi, T. Bura, R. Ziessel, S. Campagna and A. Giannetto, *New J. Chem.*, 2011, **35**, 948.
36. Y. C. Bai, C. Zhang, C. J. Fang and C. H. Yan, *Chem. Asian J.*, 2010, **5**, 1870.
37. N. Kaur, N. Singh, D. Cairns and J. F. Callan, *Org. Lett.*, 2009, **11**, 2229.
38. V. S. Elanchezhian and M. Kandaswamy, *Inorg. Chem. Commun.*, 2010, **13**, 1109.
39. K. Rurack, A. Koval'chuck, J. L. Bricks and J. L. Slominskii, *J. Am. Chem. Soc.*, 2001, **123**, 6205.
40. G. J. Mohr, C. Demuth and U. E. Spichiger-Keller, *Anal. Chem.*, 1998, **70**, 3868.
41. G. J. Mohr, *Chem. Eur. J*, 2004, **10**, 1082.
42. J. B. Birks, *Photophysics of Aromatic Molecules*, Wiley, London, 1970.
43. M. Irie, *Chem. Rev.*, 2000, **100**, 1685.
44. X. L. Meng, W. H. Zhu, Q. Zhang, Y. L. Feng, W. J. Tan and H. Tian, *J. Phys. Chem. B*, 2008, **112**, 15636.
45. H. Dürr and H. Bouas-Laurent (ed.), *Photochromism. Molecules and Systems*, Elsevier, Amsterdam, 1990.
46. F. M. Raymo and S. Giordani, *J. Am. Chem. Soc.*, 2002, **124**, 2004.
47. F. M. Raymo and S. Giordani, *J. Am. Chem. Soc.*, 2001, **123**, 4651.
48. F. M. Raymo, *Adv. Mater.*, 2002, **14**, 401.
49. J.-M. Montenegro, E. Perez-Inestrosa, D. Collado, Y. Vida and R. Suau, *Org. Lett.*, 2004, **6**, 2353.
50. E. Perez-Inestrosa, J. -M. Montenegro, D. Collado, R. Suau and J. Casado, *J. Phys. Chem. C*, 2007, **111**, 6904.
51. J. Andreasson, U. Pischel, S. D. Straight, T. A. Moore, A. L. Moore and D. Gust, *J. Am. Chem. Soc.*, 2011, **133**, 11641.

52. M. DiCasa, L. Fabbrizzi, M. Licchelli, A. Poggi, A. Russo and A. Taglietti, *Chem. Commun.*, 2001, 825.
53. X. Zhang, Y. Xiao and X. H. Qian, *Angew. Chem. Int. Ed.*, 2008, **47**, 8025.
54. H. N. Kim, M. H. Lee, H. J. Kim, J. S. Kim and J. Yoon, *Chem. Soc. Rev.*, 2008, **37**, 1465.
55. M. Shahid, P. Srivastava and A. Misra, *New J. Chem.*, 2011, **35**, 1690.
56. A. P. de Silva, H. Q. N. Gunaratne, T. Gunnlaugsson and P. L. M. Lynch, *New J. Chem.*, 1996, **20**, 871.
57. T. Gunnlaugsson, P. E. Kruger, P. Jensen, F. M. Pfeffer and G. M. Hussey, *Tetrahedron Lett.*, 2003, **44**, 8909.
58. D. Yuan and R. G. Brown, *J. Chem. Res (M)*, 1994, 2346.
59. C. C. Cain and R. F. Murphy, *J. Cell. Biol.*, 1988, **106**, 269.
60. M. N. Stojanovic, T. E. Mitchell and D. Stefanovic, *J. Am. Chem. Soc.*, 2002, **124**, 3555.
61. Y. H. Li, H. Wang, J. S. Li, J. Zheng, X. H. Xu and R. H. Yang, *Anal. Chem.*, 2011, **83**, 1268.
62. H. Bouas-Laurent, A. Castellan, M. Daney, J.-P. Desvergne, G. Guinand, P. Marsau and M. N. Riffaud, *J. Am. Chem. Soc.*, 1986, **108**, 315.
63. R. Ballardini, V. Balzani, A. Credi, M. T. Gandolfi, F. Kotzyba-Hibert, J.-M. Lehn and L. Prodi, *J. Am. Chem. Soc.*, 1994, **116**, 5741.
64. Y. Shiraishi, Y. Tokitoh and T. Hirai, *Chem. Commun.*, 2005, 5316.
65. L. Prodi, R. Ballardini, M. T. Gandolfi and R. Roverai, *J. Photochem. Photobiol. A: Chem*, 2000, **136**, 49.
66. S. Nishizawa, Y. Kato and N. Teramae, *J. Am. Chem. Soc.*, 1999, **121**, 9463.
67. R. H. Yang, W. H. Chan, A. W. M. Lee, P. F. Xia, H. K. Zhang and K. Li, *J. Am. Chem. Soc.*, 2003, **125**, 2884.
68. C. Tanford, *The Hydrophobic Effect*, Wiley, New York, 2nd ed., 1980.
69. Z. Wang, D. Q. Zhang and D. B. Zhu, *Anal. Chim. Acta*, 2005, **549**, 10.
70. W. D. Zhou, Y. J. Li, Y. L. Li, H. B. Liu, S. Wang, C. H. Li, M. J. Yuan, X. F. Liu and D. B. Zhu, *Chem. Asian J.*, 2006, **1-2**, 224.
71. S. Nishizawa, M. Watanabe, T. Uchida and N. Teramae, *J. Chem. Soc. Perkin Trans*, 1999, **2**, 141.
72. A. Yamauchi, T. Hayashita, S. Nishizawa, M. Watanabe and N. Teramae, *J. Am. Chem. Soc.*, 1999, **121**, 2319.
73. G. S. Cox, N. J. Turro, N. C. Yang and M. J. Chen, *J. Am. Chem. Soc.*, 1984, **106**, 422.
74. S. Nishizawa, H. Kaneda, T. Uchida and N. Teramae, *J. Chem. Soc. Perkin Trans*, 1998, **2**, 2325.
75. T. Gunnlaugsson, A. P. Davis and M. Glynn, *Chem. Commun.*, 2001, 2556.
76. N. B. Sankaran, S. Banthia, A. Das and A. Samanta, *New J. Chem.*, 2002, **26**, 1529.
77. S. E. J. Bell and N. M. S. Sirimuthu, *Chem. Soc. Rev.*, 2008, **37**, 1012.
78. E. H. Witlicki, C. Johnsen, S. W. Hansen, D. W. Silverstein, V. J. Bottomley, J. O. Jeppesen, E. W. Wong, L. Jensen and A. H. Flood, *J. Am. Chem. Soc.*, 2011, **133**, 7288.

CHAPTER 14
Applications

14.1 Introduction

In the final analysis, a new field stands or falls depending on whether it is useful for the thinking or practice of people in the parent discipline and others further afield. If it affects the lives of these people positively, so much the better. The previous chapters have described how Boolean thinking provides a framework for slotting in many, many chemical phenomena. Thus the utility of molecular logic-based computation for the thinking of chemists has been laid out in the previous pages. However, the utility for the practice of people – scientists and the public – is the more pragmatic test. That is where hearts and minds are won and the future of a field is secured. Thankfully, molecular logic-based computation can pass this test by some margin, as this chapter will illustrate.

The field of molecular logic-based computation is now mature enough for a substantial stack of applications. Contrast this with the situation found during the formative years when sceptics abounded.[1–3] It is traditional that sceptics change their challenge from 'it is impossible' to 'OK, it is possible now but it is still useless' as a field of endeavour matures. This is another reason why new applications are so critical, especially those which address questions previously unanswered or even unasked. During this drive, it is worth remembering that DNA-based non-Boolean computation[4–6] is finding it hard[7] to live up to early claims of solving problems which are beyond semiconductor-based machines.[8] Indeed, several molecular biology laboratories are now turning their attention to developing oligonucleotide logic-based applications, as will be seen in sections 14.5, 14.6 and 14.8.

A major strength of molecular systems as compared to semiconductor electronic counterparts is their small size. The performance of even small-scale computational acts in small nanospaces would be important because semiconductor devices along with their in/out peripherals are too large to go there. To set the stage, let us recall that the operation of AND logic in the space of a

---

Monographs in Supramolecular Chemistry No. 12
Molecular Logic-based Computation
By A Prasanna de Silva
© The Royal Society of Chemistry 2013
Published by the Royal Society of Chemistry, www.rsc.org

detergent micelle[9] (including its counterion atmosphere) of about 3 nm radius[10] was discussed in section 7.2.1.1.

14.2 Optical Sensing based on YES and NOT Logic and Superpositions Thereof

As discussed in detail in Chapter 5, molecular sensors in their simple form can be claimed as rudimentary molecular logic devices.[11] Even though sensing operationally relies on small measurable responses to a small change in the stimulus property or species concentration, *i.e.* an analogue behaviour, it stands on a binary digital foundation. The sensor molecule must possess two states with measurably distinct output properties, the first when it is free of the input species or property. The second state occurs when the sensor molecule is fully engaged with the input species or environmental property. If the output signal increases with increasing stimulus we have Boolean YES logic. If the output signal decreases under the same condition, NOT logic is the result. Mass action, *i.e.* population behaviour, of the sensor molecules generates the sloping part of the response–stimulus curve. Furthermore, sensing is often exploited in an 'off–on' or binary digital manner for managerial 'go–no go' decisions, particularly in medical diagnostic situations. This area is so application-heavy that it is an established industry.

Molecular sensors extend from historical indicators for pH, metal ion and redox species[12–14] to modern intracellular monitors.[15–18] Chapter 5 contained just a selection of this huge literature. As we noted there, irreversible systems also satisfy the logic criteria though they do have weaknesses. Therefore, molecular reagents or chemdosimeters also contribute a large literature as another body of evidence for the utility of molecular logic-based computation.

14.2.1 Tracking Species and Properties within Cells and in Tissue

Being able to watch and measure the chemical species which regulate the life of cells is a dream-come-true for cellular biologists.[19] This is a *coup* for molecular logic processors, albeit of the simplest types, because semiconductor devices cannot claim similar success in this arena. A fine example is the unveiling of Ca^{2+} spiral waves by **1** inside stimulated cells (Figure 14.1).[20] Tsien pioneered this field with sensors such as **2**[21,22] which target Ca^{2+}. Even though the fluorescence emission spectrum of **2** does not show any Ca^{2+}-induced alterations, the fluorescence excitation spectrum does. Blue-shifts are seen in the excitation spectrum, so that the intensity at 340 nm shows a Ca^{2+}-induced enhancement (YES logic) and a NOT logic action is seen at 365 nm. Such blue-shifts also allow these spectra to be internally calibrated against microenvironmental variables such as sensor breakdown, cell thickness and local quencher concentrations. This is done by calculating the ratio of the intensities[23] in the excitation spectrum at 340 and 365 nm, while monitoring emission at 510 nm. Pro-drug ideas are exploited by using the acetoxymethyl ester form

Figure 14.1 Fluorescence micrograph of **1** inside *Xenopus laevis* oocytes showing spiral waves of Ca^{2+} triggered by inositol-1,4,5-trisphosphate.
Reprinted from J. Lechleiter, S. Girard, E. Peralta and D. Clapham, *Science*, 1991, **252**, 123 with permission from the American Association for the Advancement of Science.

of **2**, which is hydrophobic, to cross cell membranes.[24] Unselective esterases in the cytosol then take over to produce **2** itself, so that it can commence its sensing duties. The second-messenger role of Ca^{2+} within the cytosol was illuminated in this way. It is important to note that calculating the ratio of intensities only becomes possible because of the wavelength shifts and from the resulting superposed YES and NOT gates. Even the PASS 1 gate occurring at the isosbestic point can be employed in a similar fashion.

1; R = OH, R' = O, X = Cl
4; R = NMe$_2$, R' = N$^+$Me$_2$, X = H

Sensing of H^+,[25] Na^+,[26,27] and Mg^{2+}[28] is done analogously. All of these cases depend on ICT excited states of integrated 'fluorophore–receptor' systems where a nitrogen electron pair is decoupled from the rest of the fluorophore π-electron system once the target binds to the receptor. Cation disengagement in the excited state[29] is the reason for the usual absence of emission spectral effects. However, a few exceptions exist, such as **3**.[22,30]

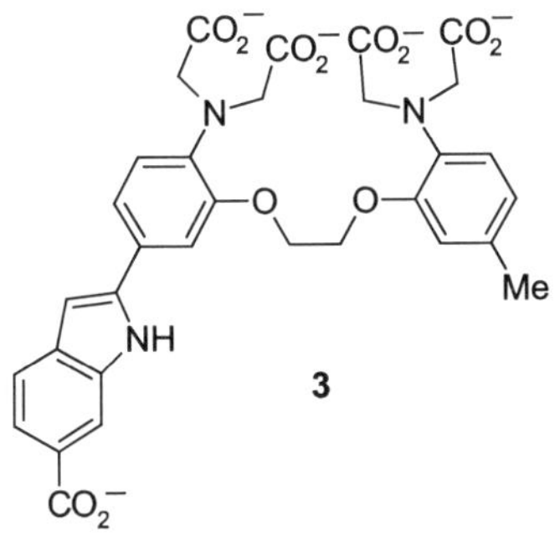

Applications 339

3

PET is also useful as a design basis[31,32] for such intracellular ion monitoring, as in the case of **4** targeting Ca^{2+}.[33] Analogously, **5** allows detection of Zn^{2+} release during cell apoptosis via fluorescence flow cytometry.[34] The emission intensity in these cases corresponds to metal ion-driven YES logic but the advantages of internal calibration are not available. The latter function can be added by employing PET systems which carry two different fluorophores.[35–38]

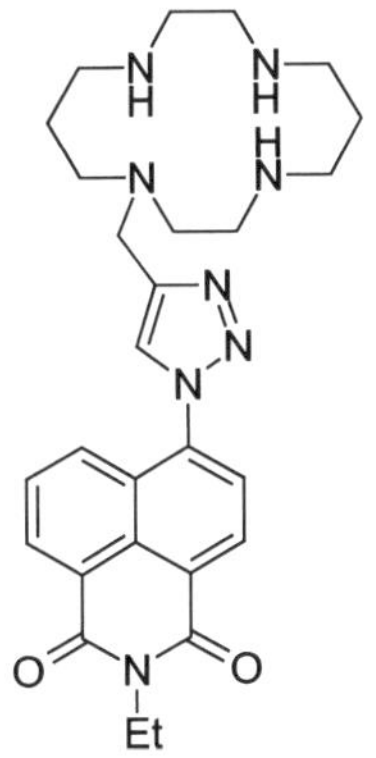

5

PET is also the design basis of **6**[39] which has been trademarked as lysosensor blue™ and detects acidic compartments in cells by emitting blue fluorescence in a YES logic fashion.[17] It helps radiologists to understand how gamma rays cause cell damage during radiotherapy (Figure 14.2).[40] Other PET sensors for monitoring cellular pH operate at longer wavelengths.[41,42] Related cases, **7** and **8**,[43] can image osteoclasts resorbing bone (Figure 14.3) because such locations are relatively acidic. The bisphosphonate group within **7** and **8**, originating in an osteoporosis drug, enables its targeting to the active sites. A non-switching fluorescent control compound (PASS 1), **9**, lights up the entire bone surface. Notably, Kikuchi's team is studying not single cells but a bone within a live mouse by two-photon excitation. Continuing on the subject of bone, Gunnlaugsson and colleagues detect microdamage sites with a Ca^{2+}-driven YES gate **10** which is also PET-based (Figure 14.4).[44] There are a number of

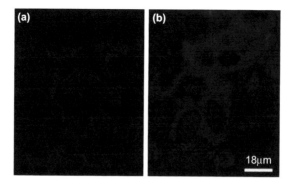

Figure 14.2 Fluorescence micrographs of mouse epithelial cells (a) before and (b) after gamma irradiation. Lysosensor blue™ allows visualization of acidic regions.
Reprinted from S. Paglin, T. Hollister, T. Delohery, N. Hackett, M. McMahill, E. Sphicas, D. Domingo and J. Yahalom, *Cancer Res.*, 2001, **61**, 439 with permission from the American Association for Cancer Research.

fluorescent sensors in use at present that allow the selective intracellular imaging of a range of species.[15–17,32,42,45]

7, R = NEt$_2$
8, R = NMe$_2$
9, R = H

Applications 341

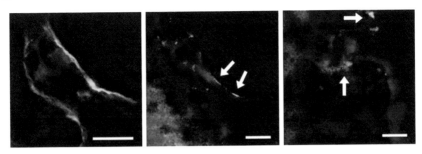

Figure 14.3 Merged two-photon excitation fluorescence micrographs of the parietal bone in red transgenic mice. The green signals are due to **9** (left), **7** (centre) and **8** (right). The red signals are due to osteoclasts. The blue signals are due to collagen. Scale bars: 40 μm.
Reprinted from T. Kowada, J. Kikuta, A. Kubo, M. Ishii, H. Maeda, S. Mizukami and K. Kikuchi, *J. Am. Chem. Soc.*, 2011, **133**, 17772 with permission from the American Chemical Society.

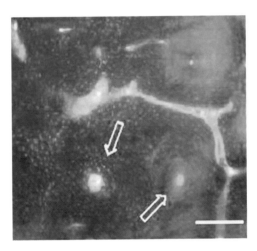

Figure 14.4 Transverse section of the bone labelled with **10** (excited at 365 nm). Arrow shows osteons with lacunae, Haversian canal and canaliculi. Bar = 100 μm.
Reprinted from R. Parkesh, T. C. Lee and T. Gunnlaugsson, *Tetrahedron Lett.*, 2009, **50**, 4114 with permission from Elsevier Science.

β-Galactosidase, a popular reporter in molecular biology, can also be visualized inside cells with the fluorescence of irreversible YES gate **11**.[46] The latter has PET arising from the alkoxyphenyl group only while the fluorescein galactosyl ether moiety is intact. Urano's team improve surgery procedures for tumours by marking them out *via* **11**.[46] The over-expression of lectins in certain tumour cells allows the latter to be bound to an avidin–β-galactosidase conjugate. Any **11** in the vicinity is then hydrolysed at the galactosyl ether bond, resulting in bright green emission.

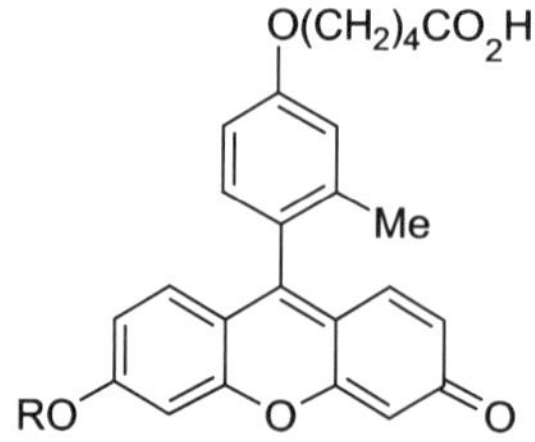

11; R = galactosyl

Intracellular properties such as temperature can also be measured by molecular thermometers, whose designs were discussed in section 5.2.3. Uchiyama uses fluorescence microscopy on one of these,[47] though in nanogel form, to demonstrate heating up of mitochondria by 0.5 °C when oxidative phosphorylation is uncoupled.[48] Excitation is done at 488 nm and observation is conducted at 515–550 nm. An elegant extension to fluorescence lifetime imaging with molecular-scale devices has also been made.[49]

Cellular properties such as membrane potentials are also open to measurement with molecular YES gates, *e.g.* Tsien's **12**.[50] Electrophysiology is the classical way of doing this but imaging is almost impossible. Also, impaling cells with electrodes is a worry. Large fluorescence enhancements are seen with **12** within microseconds when cells are depolarized (external potential becomes negative). Therefore **12** can view nerve action potentials in real time. Older fluorescent methods are less capable.[51,52] Given that **12** is an amphiphile, it embeds in cell plasma membranes when added into the cell suspension solution. The ionic and hydrophilic fluorophore of **12** protrudes from the outer leaflet of the lipid bilayer. The other, more hydrophobic, terminal is an aniline electron donor which is buried deep in the bilayer since the mid-section of **12** is an alkenic molecular wire[53] which is an excellent conduit for PET. PET can be accelerated or retarded by a molecular-scale electric field, depending on its orientation with respect to the PET path.[54] Tsien's **12** responds to the electric field created by a large fraction of the membrane potential (Figure 14.5).

Tracking species within tissue is harder owing to the requirement for deeper optical penetration,[55] but Tang's **13** (featured in section 5.2.2.1.3) yields data at a multi-cell level[56] which can be of immediate medical interest.[57,58]

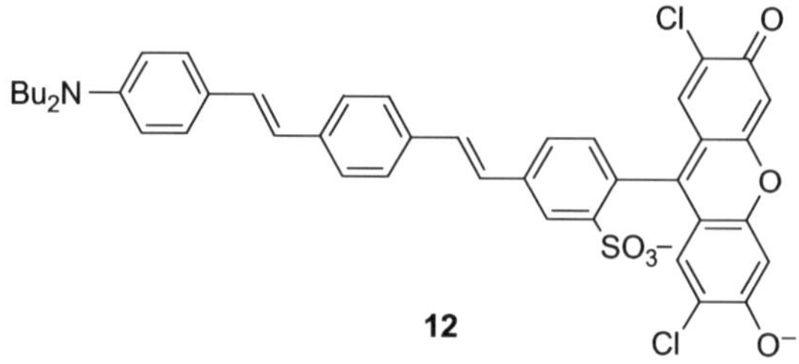

12

Applications

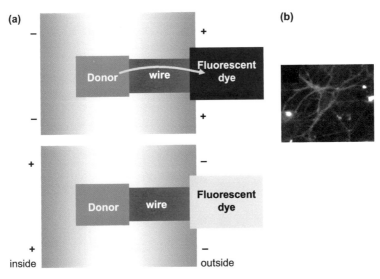

Figure 14.5 (a) Fluorescent signalling of a change in membrane potential by **12** and (b) fluorescent micrograph of a rat hippocampal neuron. The cell membranes are stained with **12**. The apolar regions and the polar head regions of the lipid bilayer membrane are shown in yellow and black respectively. The photoinduced electron transfer (PET) path is shown with the turquoise arrow.
Adapted from A. P. de Silva, *Nature Chem.*, 2012, **4**, 440 with permission from the Nature Publishing Group.

13; R = N(CH$_2$CONHOH)$_2$

14.2.2 Measuring Blood Electrolytes

How does an ambulance paramedic test the blood of a road crash victim for electrolytes? The levels of Na$^+$, K$^+$, Ca^{2+} and H$^+$ are measured in 30 s with a set of fluorescent PET-based YES gates **14–17** immobilized on an aminocellulose fibre mat (Figure 14.6).[54,59–63] The red blood cells are filtered out and the serum is presented to **14–17**. Serum Na$^+$ can be picked up by **14** so that the appropriate fraction of the population of **14** becomes Na$^+$-bound. Na$^+$-bound **14** is strongly fluorescent in the green–yellow region of the spectrum, whereas **14** itself is relatively quenched. The fluorescence intensity is directly converted into the blood Na$^+$ level.

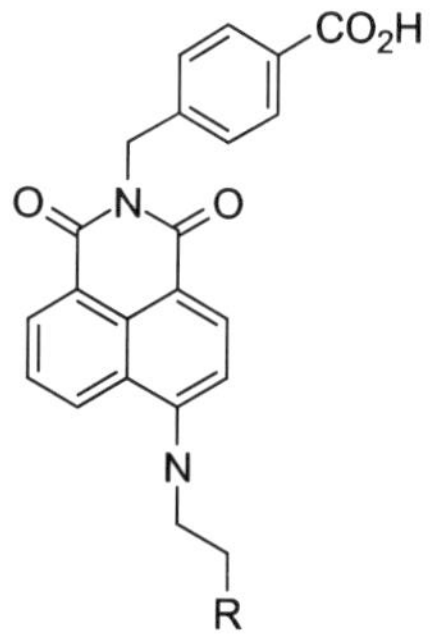

The Na$^+$-induced FE is due to PET from the azacrown receptor to the 4-aminonaphthalimide fluorophore.[64] Since the serum Na$^+$ level is 140 mM,[65] we choose the receptor so that the reciprocal of its Na$^+$ binding constant[66] is close to this figure. Selectivity considerations are also made at this stage, *i.e.* adequate Na$^+$ selectivity against K$^+$ comes from a 15-crown-5-ether receptor[67,68] augmented by a 2-methoxy lariat. Adequate selectivity against H$^+$ comes from an *N*-phenylaza-15-crown-5 ether. The latter motif is suitably electroactive (oxidation potential of 0.8 V *vs.* sce).[69] Binding of Na$^+$ to this receptor would raise its oxidation potential as a result of electrostatics[70,71] and also owing to conformational changes originating in the lariat. An absorptiometric version of **14** is available.[72]

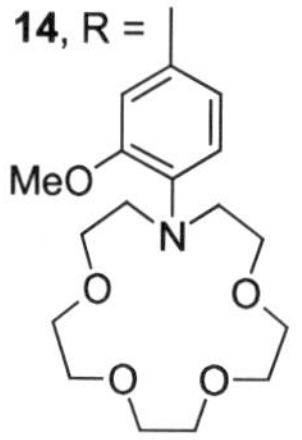

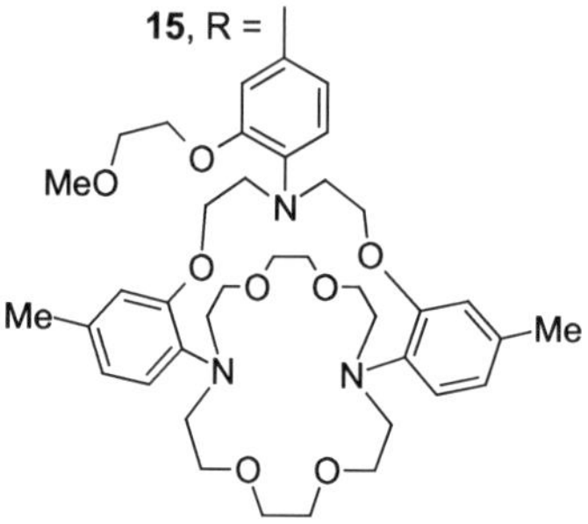

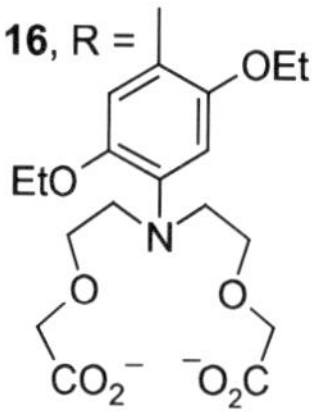

14.2.3 Monitoring Air Pressure on Aerofoils

Ever since the Wright brothers were in action, aircraft engineers have been aiming for higher performance of aerofoils. If such aims are to succeed, the ability to monitor air pressure continuously at a point on a wing with milli-second- and millimetre-resolution is essential. This is really to monitor the lift generated according to the Bernouilli principle. Given that air constantly

Applications

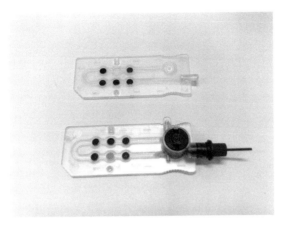

Figure 14.6 OPTI LION™ (top) and OPTI R™ (bottom) cassettes sold by Optimedical Inc. (www.optimedical.com). The black spots carry the polymer fibre mats with the appropriate sensor molecules covalently attached. Na^+, K^+, Ca^{2+}, pH and CO_2 are measured by the fluorescence of these spots. The orange spot in the bottom example is similar and responds to O_2. Reprinted from A. P. de Silva, T. P. Vance, M. E. S. West and G. D. Wright, *Org. Biomol. Chem.*, 2008, **6**, 2468 with permission from the Royal Society of Chemistry.

contains 21% O_2, the latter can serve as the entry point to this hitherto intractable problem. The excited triplet state of **18** is easily and dynamically quenched by paramagnetic O_2 via EET, even when it is applied as a polymer-based paint[73] over the wing surface. The phosphorescence output of this O_2-driven NOT gate is therefore the elegant solution, as developed by Fonov and by Gouterman.[74] Phosphorescence intensity images of the aerofoil scale-model are gathered under various flight conditions within a wind-tunnel (Figure 14.7).

18

14.2.4 Sensing Marine Toxins

Saxitoxin, **19**, is a marine toxin, which is present in the seasonal algal bloom popularly known as the red tide. Shellfish concentrate this toxin and so it is not surprising that people can fall violently ill after enjoying a bowl of seafood

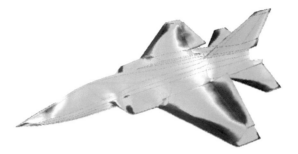

Figure 14.7 Pseudo-colour picture of a model aeroplane painted with a pressure-sensitive paint, placed in a wind tunnel and imaged *via* phosphorescence. Red: high pressure, blue: low pressure.
Reprinted from O. S. Wolfbeis, *J. Mater. Chem.*, 2005, **15**, 2657, with permission from the Royal Society of Chemistry.

chowder. Early detection of this toxin is therefore essential. Fluorescence of **20** shows saxitoxin-driven YES logic.[75] Such convenient sensing will allow deployment of these devices on coastguard boats, for instance. Mice everywhere will be relieved because samples suspected of containing this toxin are usually tested with a mouse bioassay. The binding of **19** via its guanidinium units to the azacrown ether in **20**, arrests the PET process in order to switch 'on' the fluorescence. It is interesting that **19**'s binding to Na^+-channels is the mechanism of its toxicity. Sensor **20**'s azacrown ether receptor is well known to bind alkali cations[67] and was used in the early fluorescent PET systems.[76] Simpler cations such as neopentylammonium also serve as inputs to structures related to **20**, but conformational alterations become critical.[77]

14.2.5 Monitoring Nuclear Waste Components

The nuclear industries, for energy and weapons, continually throw up difficult questions. For instance, how do we selectively monitor $^{137}Cs^+$ in nuclear waste

Applications

liquors? The early PET system **21** for sensing alkali cations such as sodium without interference from protons[78] has been evolved into excellent YES gates with Cs$^+$ input.[79] The oligooxyethylene part of **21** has been expanded into a 1,3-alternate calix[4]arene (as in **22**) or a 1,3-alternate dideoxy calix[4]arene (as in **23**). Cs$^+$-induced fluorescence enhancement factors of 12 and 54 are found for **22** and **23** respectively. Systems analogous to **22** but with only one 9-cyanoanthryl-10-methyl unit show simpler behaviour but the maximum fluorescence enhancement is similar. The special selectivity of the 1,3-alternate dideoxycalix[4]-crown shows up in improved binding and improved switching discrimination *vs.* potassium for **23**.[80] This is suitable for the convenient analysis of ^{137}Cs$^+$ in nuclear waste solutions of variable pH and salinity. The high selectivity of 1,3-alternate calixarenes towards soft alkali cations such as Cs$^+$ has been exploited within fluorescent ICT sensors, where the fluorophore is the calixarene itself.[81]

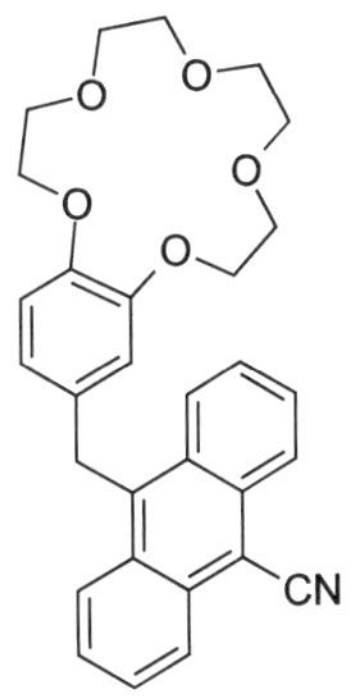

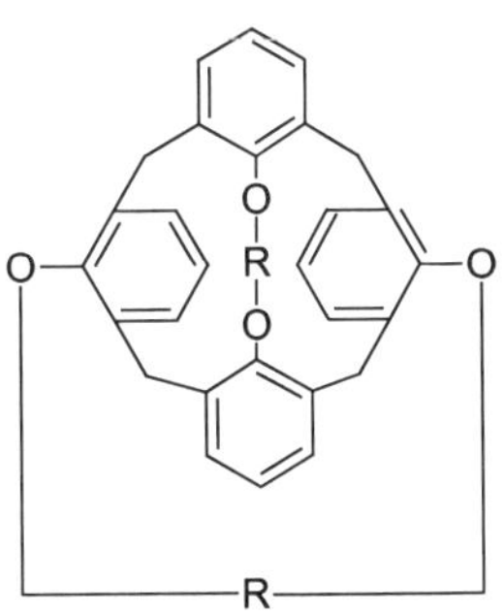

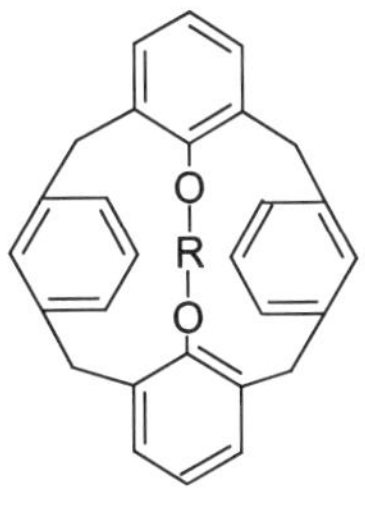

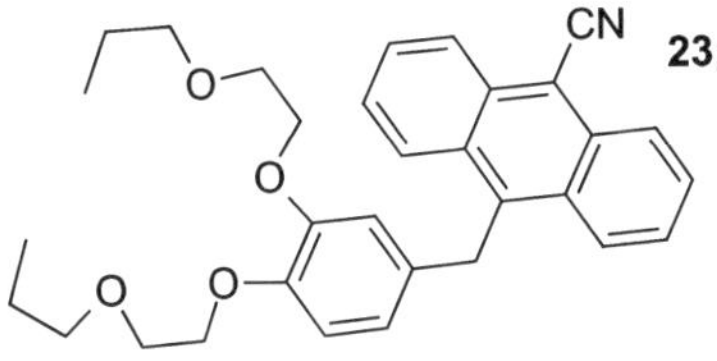

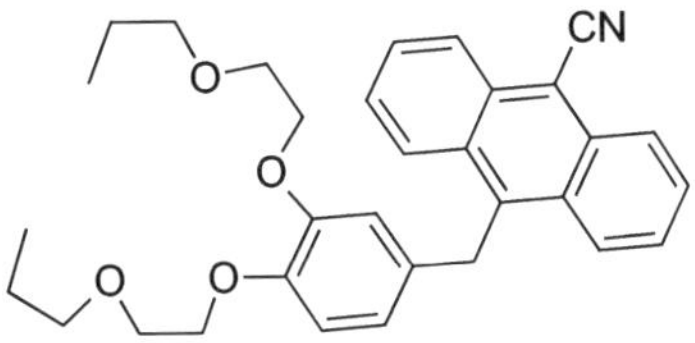

14.2.6 Screening for Catalysts

Combinatorial chemistry methods can generate many candidate structures for a particular function but a way of picking the right candidate is needed.[82,83] Fluorescent PET-based, H^+-driven YES gates are exploited by Copeland and Miller[84] for identifying transacylation catalysts from within a 'one bead–one compound'[82,83] combinatorial library. Compound **24** is bound to a polymer bead *via* its carboxylic acid terminal so that it can monitor the bead surface region for the presence of an acidic reaction product, acetic acid. The bead carries the catalyst candidate too. The reactant **25** and acetic anhydride are in toluene solution. If the catalyst candidate is effective, the bead develops a fluorescent halo at early stages of the reaction before diffusion blurs the result. A second-generation version of **24** is covalently attached to a gel[85] in which polymer beads containing catalyst candidates alone can be embedded.

24   **25**

14.2.7 Detecting Chemical Warfare Agents

Soldiers in a theatre of war are accustomed to bullets and ordnance flying towards them. Chemical warfare agents are another matter entirely and are banned by international conventions. However, the fear remains, and detectors for these agents are desperately sought. Unfortunately, no chemical receptors are known for these agents or relatives, *e.g.* diethylchlorophosphate, **26**. In these cases, detection relies on the formation of chemical bonds – rather than a reversible binding event – and changes in the fluorescence of a given reagent are monitored. Fluorescent PET[86,87] and non-PET[88] YES gates tackle this problem. For instance, the electron donor aniline unit of Walt's **27**[87] is twisted out of the xanthene fluorophore plane owing to the steric requirements of the carboxylate group. Thus it becomes a PET system with a virtual spacer.[89] pH-buffering the solution avoids protonation effects[90] and PET suppression is achieved only by nucleophilic substitution of **26** with the aniline's amine group so that a brightly fluorescent phosphoramide is produced. The first designed case of this kind targeted thiols.[91] Older examples of detecting chemical warfare agents and thiols are due to Swager[92] and Cathou[93] respectively.

As alluded to in section 14.2, chemically irreversible systems may not be appealing because such devices cannot be reset, *i.e.* once the output is produced it cannot be erased but only added to. However the case of the soldier under chemical attack is a molecular computing situation where a once-only processor is useful. Similar irreversible molecular processors allow intelligent diagnostics of illness or injury in war and peace (see section 14.5).

14.3 Improved Sensing

Throughout this book, we have seen how sensing is strongly linked to, and is indeed a part of, logic-based computation. Gathering information from a location and then preparing it for signalling to an observer is a non-trivial process. The YES and NOT logic gates involved in such operations beg the question: what can more complex logic gates bring to the sensing arena? The following sub-sections provide several answers in terms of the improvements that result.

14.3.1 Improved Sensing with AND Logic

The large literature mentioned in the sub-sections within section 14.2 refers to work by designers who, except for a few, aim for the highest selectivity of sensing possible. Various chemistry fundamentals are co-opted during this effort. For instance, the selectivity of metal ion detection benefits from appreciation of the Irving–Williams series.[94,95] However, there are strategies with a distinct Boolean basis which aid the selectivity of sensing multifunctional molecules in particular. In section 7.2.1, we discussed AND logic gates driven by two separate chemical inputs (which can, in general, even be degenerate, as seen in section 7.2.2). Now, what if these two (or more) inputs are rigidly connected together? Sections 7.2.3 and 7.2.4 tackled this question. While metal ions are not so convenient, many organic functionalities can be easily joined up in chosen geometries. For instance, even a simple sugar molecule (now detectable at the single molecule level)[96] is bristling with hydroxyl units rigidly arranged in three-dimensional space. This type of multifunctional molecule abounds in living systems. No wonder, since each of these is involved in the correct step of metabolic processes only when its specific receptor comes along. If specificity is lost, any other molecule within the compartment would be taken up by the receptor, resulting in chaos within the cell. If we build molecular logic gates carrying multiple receptors correctly positioned to capture these multi-functional molecular targets temporarily, we will have sensors with optimized selectivity, with much biorelevance. This was illustrated in section 7.2.3 with fluorescent sensor **28** selective for the protonated form of glucosamine.[97,98] Cases[99–102] showing improved selectivity for diamines (either protonated or in their neutral state) with fluorescence/absorbance output, including those discussed in section 7.2.4, can also be considered in this way. Interestingly, these diamines include cadaverine and putrescine which are the 'molecules of death'.

The specific geometric relationship between the functional groups of the target species is at its most subtle when enantioselective sensing is our goal. Discriminating different enantiomeric forms of molecules occurs, for example,[103] when the chiral AND logic gate **29** binds to the *D*-enantiomer of tartaric acid, both receptor units are occupied and a fluoresence enhancement is observed. In logic terms, both inputs are 'high' and so there is an output of 'high', manifested as an increase in fluorescence. On the other hand, *L*-tartaric acid, the mirror image form of *D*-tartaric acid, does not fit onto the AND gate as well and so one receptor site is not bound and PET prevents fluorescence. In this latter case, one of the inputs is 'low' and, therefore, the output from this AND logic gate is also 'low', *i.e.* no fluorescence enhancement occurs. This work benefited from earlier studies of this system.[104,105]

Phosphorylation reactions of peptides in water can be monitored with a 4-input AND gate in the following way. Hamachi's **30**[106–108] consists of a 1,8-difunctionalized anthracene fluorophore and a pair of dipicolylamine receptors which can capture two Zn^{2+} ions that are used to capture two organic phosphate targets so that luminescence arises in the final binding step. As mentioned in section 7.2.2, the binding of an ion into a second receptor is hindered by electrostatic repulsion by another copy of the ion held by the first receptor in a bireceptor device. A Zn^{2+} centre held by **30**'s dipicolylamine units remains coordinatively unsaturated so that it attracts an auxiliary anion in the form of a phosphorylated peptide. Such charge mitigation at the Zn^{2+} centre allows an easy binding of another Zn^{2+} at the second receptor, which finally generates fluorescence. This interplay of cations and anions, which was initially missed,[108] is perhaps involved in an older case[109] as well.

Cancer-type pathologies are monitored by Smith with **30** and **31**. When bound to two Zn^{2+} ions, these probes can detect the anionic phospholipid phosphatidylserine on the outer leaflet of cell membranes.[110] Cells hold this particular phospholipid in the inner leaflet until they are ready to die. They then signal their readiness for apoptosis so that the cell can be recycled for the common good of

Applications

the organism. Smith's group[111] extends the excitation wavelength of apoptosis detectors to 470 nm with **32**, so that it can be run on the current generation of flow cytometers. Indeed, **32** is selective for 5% incorporation of phosphatidylserine in vesicle surfaces, but not for related compounds in homogeneous solution. Other cases which only succeed in membrane media are known.[112,113] In the present case, the fluorescence enhancement is mainly due to **32** being bound to the anionic vesicle surface of lower polarity than bulk water. Its ICT fluorophore switches its emission 'on' in non-polar solvents.[114,115]

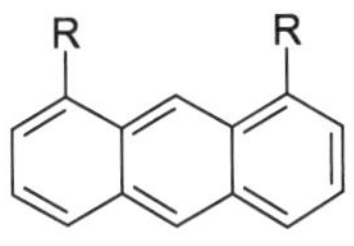

30; R = bis(2-pyridylmethyl)aminomethyl

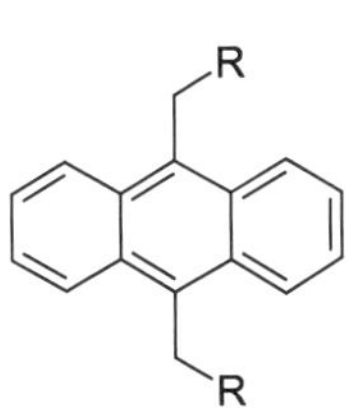

31; R = NH(CH$_2$)$_2$N[(CH$_2$)$_2$NH$_2$]$_2$

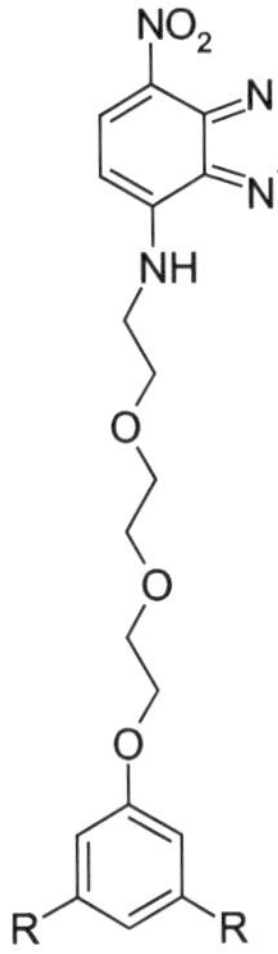

32; R = bis(2-pyridylmethyl)aminomethyl

Another nice application of fluorescent AND logic gates is selectively to detect species such as Hg^{2+} near large (bio)molecular structures, *e.g.* DNA or detergent micelles. Similar goals can be met, though with less assurance, by approximately locating the probe near the large structure.[112,113,116–118] Ihmels uses the heteroaromatic cation **33** to intercalate into double-stranded DNA[119] so that rotation about the HN–C(aryl) bond is arrested.[120] An azathiacrown unit in **33** allows binding of Hg^{2+} so as to block a PET process. Each of these two mechanisms can quench fluorescence. It is only the blocking of both mechanisms that gives 'high' fluorescence output. Interestingly, a similar AND logic function is seen in the ellipticity output at 410 nm.

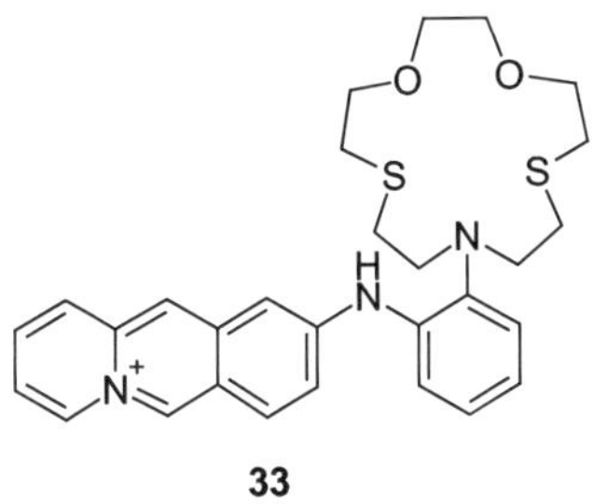

33

Visualization of tumours against a dark background would help a surgeon to remove the offending lump without leaving remnants to re-grow. Urano *et al.* enhance such visualization by building a fluorophore which is targeted to the tumour with an antibody that which becomes emissive only when encountering the higher acidity of the tumour (pH < 6).[121] Thus, the fluorescence output is driven by two inputs in an AND logical fashion. The first input is the human epidermal growth factor receptor type 2 (HER2) which is over-expressed in certain tumours, whereas the second input is H^+. The HER2 receptors can be located with the therapeutic antibody trastuzumab, so **34** contains this antibody linked to a PET system based on a BODIPY fluorophore and an aniline unit to bind H^+. Several similar PET systems were featured in Chapter 5 in the form of structures **13**, **35**, **42** and **55**. Figure 14.8 shows **34** localizing in acidic lysosomes within cells which over-express HER2 receptors. General detection of lysosomes with **6**[39] was discussed in section 14.2.1. Similar localization in peritoneal ovarian cancer metastases in live mice suggests that surgeons can perform laparoscopy with guidance from **34**.

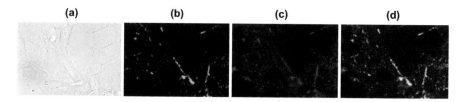

34; R = trastuzumab

DNA tiles provide a different way of using AND gates for improved sensing.[122] Dwyer's team build the tiles from nine individual strands (one core crosspiece, four struts and four sticky ends) and choose the sticky ends so that a 4 × 4 grid self-assembles by hybridization. The struts are derivatized with fluorophores

Figure 14.8 Transmission (a) and fluorescence (b–d) micrographs of NIH3T3 HER2+ cells loaded with **34** (b) and a lysosome targeting dye (c). Panel (d) is the merge of panels (b) and (d).
Reprinted from Y. Urano, D. Asanuma, Y. Hama, Y. Koyama, T. Barrett, M. Kamiya, T. Nagano, T. Watanabe, A. Hasegawa, P. L. Choyke and H. Kobayashi, *Nature Med.*, 2009, **15**, 104 with permission from the Nature Publishing Group.

so that two different EET donors locate on one side of the joined ends and one EET acceptor is located on the other. An AND logic gate is formed where the two excitations for the donors are the inputs and the emission from the acceptor is the output. The latter is taken as 'high' if its intensity exceeds an arbitrary threshold, which occurs only when both donors are being pumped. Since EET is strongly sensitive to separation distance (section 4.6), the acceptor emission intensity will collapse the moment the joined ends undergo a divorce. This happens when an analyte DNA strand arrives and hybridizes competitively with the correct sticky end of the appropriate tile. Actually, the grid needs to be partially melted for this displacement to take place but is subsequently annealed back to room temperature. Nevertheless, this method shows low nM sensitivity. There is potential for operating up to 24 wavelength-multiplexed gates on the single grid which, at $80 \times 80 \times 2\,\text{nm}^3$, is smaller than the diffraction limit.

14.3.2 Improved Sensing with Superposed AND, INHIBIT and TRANSFER Logic

The challenge of directly detecting reactive-oxygen and -nitrogen species, alone or in combination, is addressed with superposed logic in Lin's laboratory;[123] H_2O_2 and NO are the specific pair in the spotlight. A pair of fluorophores (red-emitting rhodamine and blue-emitting 7-hydroxycoumarin, which are separately excitable) is hidden within **35** in pro-drug fashion. The rhodamine is 'caged' as a *leuco* version and locked as a spirolactam. Its 1,2-diaminobenzene unit can be cut away with NO as a benzotriazole derivative[124,125] to liberate the emitter ($\lambda_{exc} = 550\,\text{nm}$, $\lambda_{em} = 580\,\text{nm}$). On the other hand, the 7-hydroxycoumarin is locked up as a benzyl ether carrying a 4-boronate substituent. Unlocking is arranged with H_2O_2 which converts the boronate into a phenylborate which promptly hydrolyses.[126] The hydroxybenzyl ether then hydrolyses further to set the blue-emitter ($\lambda_{exc} = 400\,\text{nm}$, $\lambda_{em} = 460\,\text{nm}$) free. The NO and H_2O_2 act orthogonally at the two sites, but the simultaneous presence of both species creates a blue and red fluorophore pair – a situation ripe for EET. When excited at 400 nm, the blue emission is quenched and red fluorescence is all that is seen. So we have H_2O_2, NO-driven INHIBIT(NO) logic when observing at 460 nm but H_2O_2, NO-driven AND logic when viewing at 580 nm. Similar superposition of logic gates was catalogued in section 13.3.2. Lin's **35** displays another logic type when excited at 550 nm and watched at 580 nm. Now the blue-emitter is not interrogated at all. The NO, whether alone or with H_2O_2, liberates the rhodamine and so we have H_2O_2, NO-driven TRANSFER(NO) logic. These truth tables are shown in Table 14.1.

Lin and colleagues[123] also unveil a nice colour pattern from these truth tables, which at a glance informs the observer whether H_2O_2, NO, H_2O_2 and NO, or none, are present, as shown in Table 14.2. For instance, the presence of NO alone is signalled by the pattern 'black(in blue channel)–black(in red channel)–red' when excited at 400, 400 and 550 nm respectively. Impressively, the performance of **35** translates smoothly into intracellular situations (Figure 14.9).

Table 14.1 Set of truth tables showing superposed logic gates for **35**.[a]

Input$_1$ H$_2$O$_2$	Input$_2$ NO	Output$_1$ Flu$_{460}$[b] INHIBIT(NO)	Output$_2$ Flu$_{580}$[b] AND	Output$_3$ Flu$_{580}$[c] TRANSFER(NO)
none	none	low (1.5)	low (0.5)	low (0.5)
none	high ($10^{-3.7}$ M)	low (3)	low (1)	high (38)
high ($10^{-4.3}$ M)	none	high (37)	low (0.5)	low (0.5)
high ($10^{-4.3}$ M)	high ($10^{-3.7}$ M)	low (<11)	high (7)	high (>16)

[a]10^{-6} M in water:acetonitrile (4:1, v/v), pH 7.4.
[b]Relative intensities, λ_{exc} 400 nm.
[c]Relative intensities, λ_{exc} 550 nm.

Table 14.2 Fluorescence colour pattern for **35**.

Input$_1$ H$_2$O$_2$	Input$_2$ NO	Output$_1$ (λ_{exc} 400 nm) λ_{em} 460 nm	Output$_2$ (λ_{exc} 400 nm) λ_{em} 580 nm	Output$_3$ (λ_{exc} 550 nm) λ_{em} 580 nm
none	none	black	black	black
none	high	black	black	red
high	none	blue	black	black
high	high	black	red	red

Certain macrophages can be stimulated to produce H$_2$O$_2$ endogenously by phorbol-12-myristate-13-acetate. Similarly, lipopolysaccharide stimulates formation of both H$_2$O$_2$ and NO. Figure 14.9 clearly shows the correct colour pattern as each of these stimulants operates on the macrophages.

35

14.3.3 Near-simultaneous Monitoring of Multiple Species with XOR Logic

We previously discussed a H$^+$, Ca^{2+}-driven XOR gate[127] (structure **134**, which was shown in section 7.7) designed on the basis of opposite perturbations of ICT (internal charge transfer) excited states.[128,129] Figure 7.10 schematized the behaviour of this 'receptor$_1$–chromophore–receptor$_2$' system. Suzuki's team[130] employ this approach to design Ca^{2+}, Mg^{2+}-responsive **36** which is emissive as well. Structure **36**[130] contains a coumarin fluorophore, a fluorinated BAPTA[131]

Applications

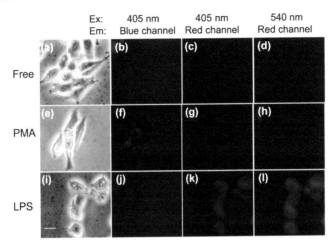

Figure 14.9 Micrographs of RAW 264.7 murine macrophages containing **35** in the absence or presence of stimulants phorbol-12-myristate-13-acetate (PMA) and lipopolysaccharide (LPS). Panels (a), (e) and (i) are a type of bright-field contrast image. All the other panels are fluorescence images under the excitation and emission conditions noted. Scale bar = 20 μm. Reprinted from L. Yuan, W. Y. Lin, Y. A. Xie, B. Chen and S. S. Zhu, *J. Am. Chem. Soc.*, 2012, **134**, 1305 with permission from the American Chemical Society.

receptor for Ca^{2+} and a ketoacid as the Mg^{2+} receptor. The BAPTA receptor is situated at the electron-donating site of the fluorophore and when Ca^{2+} is bound, a blue-shift occurs. The ketoacid receptor is located at the electron-accepting site of the fluorophore and a red-shift is found when Mg^{2+} is bound.

Fluorescence ratiometry arises naturally owing to the wavelength shifts. The fluorescence intensity output can then be given as a function of the two inputs,[132–136] and as a function of the exciting wavelength. The authors apply **36** to achieve near-simultaneous monitoring of Ca^{2+} and Mg^{2+} inside living cells by observing the fluorescence intensity while using three different exciting wavelengths of 365, 390 and 420 nm. A small time interval is needed to change the excitation wavelength of the fluorescence microscope. Perturbation of mitochondrial processes by the uncoupler **37** shows up clearly as a jump in Ca^{2+} level followed by Mg^{2+} (Figure 14.10).

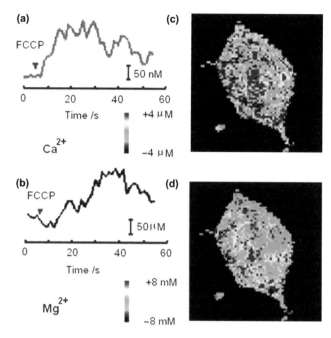

Figure 14.10 A PC12 cell containing **36** monitored for concentrations of (a) Ca^{2+} and (b) Mg^{2+} following administration of the mitochondrial uncoupler **37**. Panels (c) and (d) display images of average change of Ca^{2+} over the first 21 s and of Mg^{2+} over the first 38 s respectively.
Reprinted from H. Komatsu, T. Miki, D. Citterio, T. Kubota, Y. Shindo, Y. Kitamura, K. Oka and K. Suzuki, *J. Am. Chem. Soc.*, 2005, **127**, 10798 with permission from the American Chemical Society.

14.4 Identification of Small Objects in Populations

How do we identify a friend in a crowd at a music festival? We look for the face and for characteristic features within it. When officials, *e.g.* police, seek identification (ID), they may ask to see a passport. Micro/nanometric objects in large populations also need such ID. Such a situation arises when cells from many patients need to be screened in parallel for a disease during an epidemic. Less critically, but equally importantly, similar ID is needed for members of combinatorial chemistry libraries[82,83] bound to polymer beads. Such libraries are prepared by combining components in many different ways according to the 'split and mix' method,[137,138] for instance.

Let us consider an example briefly. Components A, B and C are attached to three separate lots of polymer beads. These lots are pooled and then divided into three lots again. Component D is attached to the first lot, E to the second and F to the third so that beads carrying AD, BD and CD are seen in lot 1, AE, BE and CE occur in lot 2 and AF, BF and CF occupy lot 3. It is clear that a few cycles build up considerable diversity in terms of the final molecules which are produced.

Semiconductor technology has a ready solution, RFID (radiofrequency identification), for identifying objects which are millimetric or larger.[139] Society is replete with these RFID chips, to prevent shoplifting and to monitor inventories for example. Approaches based on coloured magnetic particles[140,141] and stretched, dye-doped photonic crystals[142] also appear promising for tagging millimetric objects. However these approaches are impotent in the sub-millimetre domain. It is telling that commercial RFID tagging only succeeds for packages of chemically identical beads in combinatorial library development.[142,143] This is where molecular tags come in. For instance, the molecular tag can be found within beads custom-synthesized from mixtures of substituted styrenes when interrogated by Raman spectroscopy.[145] Coloured fluorescent dyes provide molecular tags for more common beads and cells.[146–148] However, colour can only code for a limited diversity owing to the width of molecular emission bands causing overlap between tag signatures. So a better solution is needed.

Such a solution arrives when the fluorescent dye molecules are enabled with logic, because the diversity of tags can be multiplied,[149] for several reasons. First, this book is testament to the many molecular logic types available and the many inputs (chemical, biochemical and physical) which are tolerated. The input level which triggers the output is also a useful variable which is adjustable *via* receptor binding strength for example. Finally, parallel arrays of the above gates create multi-valued logic systems (Chapter 12) which are very information-rich.[150]

What about implementation? Fluorescence intensities of the tagged objects are observed with a microscope after gentle exposure to the 'high' level of the chemical input (Figure 14.11a). A repeat observation follows exposure to the 'low' input level (Figure 14.11b). This simple 'wash and watch' procedure builds up logic input–output tables for many beads in parallel so that logic descriptors can be assigned to each one.

It is noteworthy that we are taking advantage of ternary logic[150] (and higher versions) – something which is discouraged in everyday semiconductor computing owing to lack of robustness. As pointed out in section 3.3, binary logic is much more robust against error accumulation over the thousands of processing steps typically occurring inside conventional semiconductor-based computers.[151] The low experimental errors accumulated in the few measurements of fluorescence intensity that are required do not cause a problem in applying ternary logic to MCID.

The specific demonstrator (Figure 14.11) shows five distinguishable bead types in terms of H^+-driven logic: YES, PASS 1, NOT, PASS 0 and PASS 1 + YES (1:1). YES, PASS 1 and NOT logic are due to tags **38**, **39** and **40** respectively. PET is the switching mechanism within **38** and **40**, as seen in many instances within Chapter 5. PASS 0 means that the bead carries no tag and is therefore silent unconditionally. Sections 5.6 and 5.7 discussed these non-switching cases. Finally, the logic array PASS 1 + YES (1:1) deserves special discussion. This is where equimolar PASS 1 and YES logic tags are fixed on one bead and fluorescence output is observed as the sum from both tags.

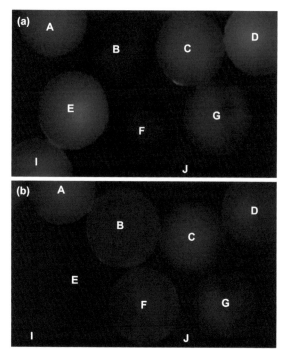

Figure 14.11 Fluorescence micrographs demonstrating molecular computational identification (MCID) of 0.1 mm polymer beads. The beads are tagged with different logic gates and treated with (a) acid and (b) alkali in aqueous methanol (1 : 1, v/v) under ultraviolet (366 nm) irradiation. The logic gate type of each bead is as follows. A; PASS 1, B; NOT, C; PASS 1, D; PASS 1 + YES (1:1), E; YES, F; NOT, G; PASS 1, I; YES, J; PASS 0.
Reprinted from A. P. de Silva, M. R. James, B. O. F. McKinney, P. A. Pears and S. M. Weir, *Nature Mater.*, 2006, **5**, 787 with permission from the Nature Publishing Group.

As an example, it is readily apparent from Figures 14.11a and 14.11b that beads B and F are silent in acid solution but shining in alkali. Thus their tags are identical copies of H^+-driven NOT logic. Each bead is identifiable in this way, except for bead D which needs special discussion. It shows significant fluorescence in alkali and double the intensity in acid. These are the two higher

Applications

states of ternary logic (2 and 1 out of 2, 1 and 0). Figure 12.2 deserves a look for those who wish to appreciate the full 2–1–0 staircase. In the present case, the 2:1 output pattern arises from the 1:1 sum of YES and PASS 1 gate outputs. Bead D is therefore displaying the PASS 1 + YES (1:1) logic tag.

It is also clear from Figure 14.11 that the current demonstrator shows tag diversity even from a single colour. The rainbow awaits. This method of molecular computational identification (MCID)[149] is ready to handle large combinatorial chemistry libraries. MCID is an early application of molecular logic-based computation which is operating in territory where semiconductor-based electronic devices fear to tread[152,153] and has been recognized as such.[154]

The discussion in the previous paragraph focussed on H^+-driven logic tags but those employing alkali cations are also available. Polymer beads **41**[149] and **42**[155] are of this type. Both involve Na^+-induced PET suppression for fluorescence output according to Na^+, H^+-driven AND and Na^+, K^+-driven OR logic respectively. The through-space PET pathway in the latter case is worth noting.

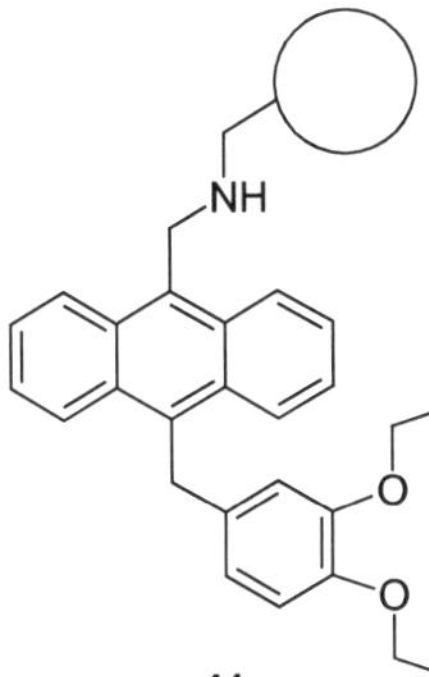

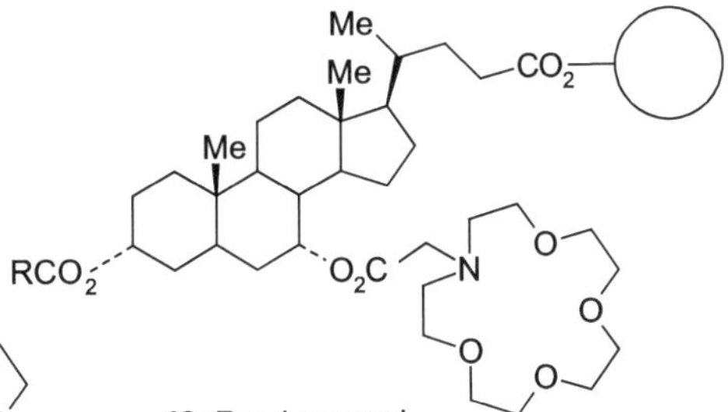

42; R = 1-pyrenyl

Many of the molecular logic gates with light output noted in the preceding pages, irrespective of their degree of complexity, can be adapted for use as MCID tags. This availability of a general application for a large number of molecular logic gates is a particular importance of MCID. Furthermore, each new molecular logic gate of this type expands the diversity of MCID tags.

14.5 Improved Medical Diagnostics

Another area where molecular logic and computation can be applied is in intelligent diagnostics for medical purposes. We have already seen how molecular logic contributes to improved sensing. Now we examine an extension of this avenue. A patient visits a doctor. After examination, some blood is drawn by a nurse and sent to a clinical laboratory for tests by a technician. When the results are available, the doctor checks the series of parameters for those which are over the normal and those which are under. Then (s)he applies education and experience to reach the diagnosis by making a logical combination of these outliers. For instance, a cardiovascular issue is indicated by a combination of 'high' cholesterol, 'high' LDL (low density lipoprotein) and 'high' CRP (C-reactive protein).[156] As discussed in section 2.5 and illustrated in Figure 2.3, such 'high' input analytes can be persuaded to exclusively elicit a fluorescence 'high' output signal. This is done by paying attention to the analyte-binding strength of the receptor unit within the molecular device. By using ideas from section 7.2 and its sub-sections to place several receptors within a multi-input AND gate, several 'high' analytes can be simultaneously detected with one fluorescence signal. It is notable that none of the analyte levels is individually quantitated.

Let us illustrate this concept with a 3-input AND gate **43**.[157] While adding another input to 2-input AND logic (section 7.2) appears evolutionary and incremental at first sight, this actually allows us to introduce the idea of a 'lab-on-a-molecule'. Indeed, new concepts can appear as we go from the number two to three, as crystallized in the old saying 'two's company, three's a crowd'.[158] Compound **43** allows the entire sequence of patient–nurse–doctor–technician interaction described in the previous paragraph to be conducted on-board and to be terminated by a fluorescent light signal to convey a 'well/ill' decision, without doctor or technician involvement. It outputs a single simple result ideal for execution by a busy doctor, who might follow up the positive hits by a thorough examination of the patients so selected. Such molecules with a bit of computing ability can help in medical diagnostic situations. Shapiro[159,160] had a related idea earlier and Stojanovic[161] later. However they aim at use within the body whereas **43** and relatives[162–164] are used externally where near-term applications following regulatory approval are more likely, *e.g.* **14**.[59]

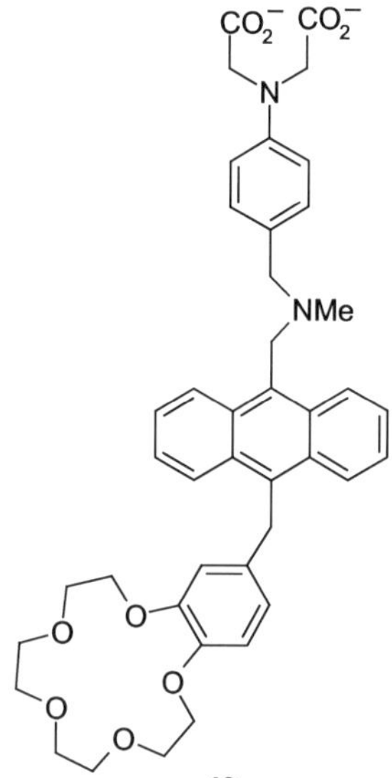

43

Compound **43** is of the 'receptor$_1$–spacer$_1$–fluorophore–spacer$_2$–receptor$_2$–spacer$_3$–receptor$_3$' format and driven by Na^+, H^+ and Zn^{2+} inputs. Benzo-15-crown-5-ether, tertiary amine and amino acid receptors capture these three

Table 14.3 Truth table for three-input AND gate **43**.[a]

$Input_1$ Na^+	$Input_2$ H^+	$Input_3$ Zn^{2+}	Output Fluorescence[b]
none	low ($10^{-9.5}$ M)	none	low (0.001)
none	high ($10^{-6.0}$ M)	none	low (0.001)
none	low ($10^{-9.5}$ M)	high ($10^{-4.8}$ M)	low (0.002)
none	high ($10^{-6.0}$ M)	high ($10^{-3.1}$ M)	low (0.003)
high (5 M)	low ($10^{-9.5}$ M)	none	low (0.006)
high (5 M)	high ($10^{-6.0}$ M)	none	low (0.007)
high (5 M)	low ($10^{-9.5}$ M)	high ($10^{-4.8}$ M)	low (0.006)
high (5 M)	high ($10^{-6.0}$ M)	high ($10^{-3.1}$ M)	high (0.020)

[a]10^{-5} M in water.
[b]Quantum yields, λ_{exc} 379 nm, λ_{em} 410, 435, 458 nm.

inputs respectively. In particular, we note that amino acid receptor$_3$ was developed by Gunnlaugsson to selectively bind Zn^{2+}.[165–167] Each of these receptors can launch a PET process to the anthracene fluorophore unless all three are blocked by binding to its analyte available at above-threshold concentration. Only then does the fluorescence switch 'on' (Table 14.3).

Konry and Walt's test is for rapid detection[163] of mass infection due to an epidemic or bioterrorism. An infected person would show evidence of the invading bacterium (as identified by a characteristic run of DNA) as well as the products of infection such as interleukin proteins. The person is referred for professional medical examination only if both these pieces of evidence are found. Such DNA, protein-driven AND logic is arranged according to the scheme in Figure 14.12a. Microspheres in microwells[146,147] are coated with antibody raised against the interleukin. If the protein is present, it can further bind to a biotinylated antibody, so that avidin can be picked up. Importantly, the microwells allow the washing steps which are essential for this scheme to work. The chain of events builds further by the bound avidin taking on a biotinylated DNA strand which then connects to a complementary run in the target bacterial DNA (if the latter is present). The scheme is visualized by picking up a green fluorescent-labelled signal DNA piece which is also complementary to a short run elsewhere in the target DNA. Emergence of green fluorescence warns the tester that the patient is doubly positive and deserves quick follow-up. Large numbers of microspheres can be examined in parallel while preserving identification information for the patient owing to the strength of the microsphere–microwell approach.[146,147] The test delivers more, because a red fluorescent-labelled signal DNA strand is picked up by the microsphere when only the protein is present, *i.e.* under DNA, protein-driven INHIBIT(DNA) logic conditions (Figure 14.12b). If both protein and DNA are absent, neither red nor green fluorescence is seen.

Katz's and Willner's enzyme-based approaches to molecular logic featured heavily in Chapters 7–11. Given that the concentrations of many enzymes or their substrates in body fluids will deviate from the norm during disease or

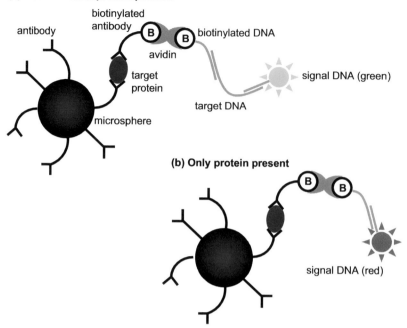

Figure 14.12 Fluorescent screening for bacterial DNA and a specific interleukin (a) when both are present and (b) when only DNA is present.
Redrawn from T. Konry and D. R. Walt, *J. Am. Chem. Soc.*, 2009, **131**, 13232 with permission from the American Chemical Society.

injury states, many diagnostic schemes can be envisaged. Katz now joins forces with Wang to diagnose injury states rapidly.[168–172]

A platoon is under fire and taking casualties. The paramedic has to take on-the-spot decisions about the nature of injuries and whether each individual soldier should be airlifted to a base hospital or treated more locally. A 'lab-on-a-molecule' will assist the paramedic greatly in such decision-making. As an example, a glucose oxidase–microperoxidase system can detect the life-threatening condition of haemorrhagic shock by clearly signalling the simultaneous increase of glucose and norepinephrine (noradrenaline) levels in 60 s.[168] The change in glucose is substantial, with 'low' and 'high' levels of glucose 4 and 26 mM respectively. On the other hand, norepinephrine jumps from a 'low' of 2 nM to a 'high' of 3500 nM. When glucose and norepinephrine are present at 'high' levels, a rapid increase of absorbance at 487 nm is seen, corresponding to AND logic (Figure 14.13).

The aqueous enzyme system operates in the following way. Glucose oxidase processes glucose and air to produce H_2O_2 which is required for microperoxidase to oxidize norepinephrine to norepiquinone. The latter absorbs at 487 nm. If the different observation technique of chronoamperometry is employed after 60 s, the AND logic action is perhaps not as clear-cut (Figure

Applications

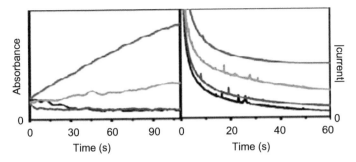

Figure 14.13 Direct detection of haemorrhagic shock by measurement of absorbance at 487 nm (left panel) and chronoamperograms at –0.4 V (*vs.* Ag/AgCl, carbon working electrode) (right panel) for the glucose oxidase–microperoxidase system. States of the norepinephrine and glucose inputs (0,0), (0,1), (1,0) and (1,1) are shown by black, blue, green and red traces respectively.
Reprinted from J. Halamek, J. R. Windmiller, J. Zhou, M. C. Chuang, P. Santhosh, G. Strack, M. A. Arugula, S. Chinnapareddy, V. Bocharova, J. Wang and E. Katz, *Analyst*, 2010, **135**, 2249 with permission from the Royal Society of Chemistry.

14.13). Nevertheless, the use of screen-printed carbon electrodes in the latter case simplifies device construction for battlefield deployment.[168] Extensions to 3-input situations and reconfigurable logic schemes are also available.[170]

Given the intensity of battlefield situations, multiple injuries are more than likely. A range of battlefield injuries can be rapidly assessed in parallel with a suite of enzyme systems so that a code can be built up from the string of outputs.[168] However, the success of such codes requires the enzyme systems to be lined up in an agreed order within the parallel array.

Pattern-based protein sensing ensembles can also be outfitted with a binary logic readout.[162,164] Guanosine-rich oligonucleotides which fold into G-quadruplexes can serve as receptors.[173] A mixture of these, each labelled with a different fluorophore emitting at a distinguishable wavelength, is used as the sensor array. Different proteins will produce different patterns. For instance, myelin basic protein (an important marker of multiple sclerosis) at concentrations of 500 nM or above produces a NOT logic output at 525 nm (fluorescein emission) and a PASS 1 logic output at 590 nm (tetramethylrhodamine carboxylate emission) with appropriate thresholding of fluorescence intensity. In contrast, >125 nM avidin gives NOT and YES logic outputs at these two wavelengths respectively. In either case, the logic output at 480 nm (pyrene emission) is essentially PASS 0.

A double-input AND logic output is seen at 480 nm when the ubiquitous protein calmodulin and Ca^{2+} are presented to the sensor ensemble in the previous paragraph. This can be understood according to the pyrene hydrocarbon fluorophore interacting with the exposed hydrophobic patch of the Ca^{2+}–calmodulin complex.

Protein sensing also includes the nuclease detection method of Wang and colleagues,[174] which was discussed in section 10.13. Oligonucleotide sensing in a diagnostic context according to Stojanovic and Stefanovic's approach[175] was noted in section 9.2.

14.6 Improved Therapy

Pro-drug activation[176] is a very useful logic output especially because it involves an actuation. A few cases of logic-controlled actuation have featured in, for example, section 7.2.7, though more will follow in this section and the next. In the present instance, OR logic gates which release drugs can improve therapeutic efficacy by allowing pro-drug metabolism *via* two triggers in a diseased tissue rather than one. Then the drug release will be successful even if one of the triggers is overcome by the disease agents. Shabat's **44** contains two different biocatalyst substrates and protected 4-nitrophenol as a model pro-drug. The two substrates are a phenylacetamide moiety that is targeted to an amidase and a unit aimed at a retro-aldol, retro-Michael-active catalytic antibody. Either the enzyme or the catalytic antibody stimulates the activation of **44** to release 4-nitrophenol, though not at equal rates. Corresponding experiments with a version of **44** containing protected doxorubicin (an anti-cancer drug) in leukaemia cell lines are similarly successful in an OR logical manner.

The two concepts of pro-drug activation and 'lab-on-a-molecule' come together in Shapiro's hands.[159,160,177] In section 14.5, we met a patient interacting with a doctor, technician and a nurse. Now an additional interaction with a pharmacist is included. As before, a series of parameters are checked as to whether they are 'high' or 'low' compared with norms. Then, instead of activating a light signal as in section 14.5, a drug molecule is unmasked and released. Having such a molecular computing system patrolling inside a patient will face considerable regulatory hurdles, partly due to technical issues of long-term stability *in vivo*. Nevertheless, aspects of this approach can end up in intelligent therapies.

Specifically, a DNA drug for prostate cancer forms the loop of a stem–loop. We met similar motifs several times in Chapter 5 and in sections 7.7.2, 9.2, 10.5 and 10.11. Once folded into the loop, the drug is veiled and unavailable. Crucially, the long stem has a sticky end. A restriction enzyme, FokI, recognizes the sequence GGATG (or CCTAC) in a double-stranded oligonucleotide

Applications

and then cuts it at the 9th and 13th bases further along the two strands so that a sticky end is produced. FokI is offered a short double-stranded oligonucleotide (a transition molecule) which contains the recognition sequence and a sticky end of its own. The latter locks onto the terminal of the long stem so that FokI can operate on the stem. The transition molecule is produced by a messenger RNA (mRNA) which is the disease indicator. A competing transition molecule is made available when the mRNA is absent. Four different mRNAs are handled in this way, where under-expression of two and over-expression of the others strongly suggest the disease. Once FokI has finished working with all four transition molecules, the long stem is so degraded that the loop opens out to unveil the drug. In some ways, the nett result is reminiscent of 4-input doubly disabled AND logic, which was discussed in section 10.18. Actually, the present molecular case is a device (a type of Turing machine) where each input is applied in turn in a particular order and the output state changes each time according to a given algorithm.[178] Aspects of such history-dependence featured in section 11.4. Though our brief discussion stops here, Shapiro's work here[159] contains other refinements such as control of the effective drug dosage *via* suppressors. His work elsewhere handles targets besides mRNAs[179] and develops logic programs[180] from the same starting point. In a nice extension, Keinan's design[181] produces a plasmid instead of the cancer drug as the output of a similar computation. Actually, the direct product is a linear double-stranded (ds)DNA whose sticky ends ligate to give the circular plasmid. This plasmid contains the lacZ gene so that it can express β-galactosidase in *Escherichia coli* cells.

Kolpashchikov and Stojanovic apply Shapiro's logic-directed oligonucleotide release approach[159] to non-nucleotides.[182] An RNA-based aptamer which bound malachite green (MG)[183] was featured in section 7.2.1.4. The relative employed here binds MG by itself and renders it fluorescent at 650 nm (Figure 14.14). For the present purpose, the aptamer carries a DNA extension which can disrupt the MG binding by self-complexation to produce a stem–loop. The latter event is prevented by hybridizing the extension with a separate oligonucleotide. This oligonucleotide contains a single key ribonucleotide which allows hydrolysis by a deoxyribozyme, as seen in section 5.3.1.3. In the current work, a deoxyribozyme of the AND logic type which is driven by two oligonucleotide inputs (as described in Chapter 9, Figure 9.3a) causes the hydrolysis. AND logic-directed MG release is the upshot, even though NAND logic behaviour is seen in the 650 nm emission. Since aptamers can be raised to bind various drug species including proteins, Kolpashchikov and Stojanovic's approach[182] is important for therapies targeted to sites expressing particular combinations of oligonucleotide markers of disease.

Logic-directed enzyme release can be similarly useful. Gianneschi and Ghadiri achieve this by employing enzyme–single-stranded (ss)DNA_1 conjugate to undergo base-pairing with a suitable inhibitor–ssDNA_2 construct.[184] The run of bases adjacent to the enzyme are left unhybridized to hang as a loop. Naturally, the resulting complex will have little enzymatic activity because of

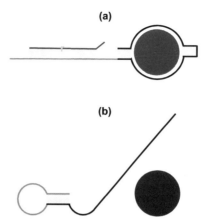

Figure 14.14 (a) An aptamer (black) binds malachite green (bright red) and renders it fluorescent. The aptamer has an extension which is hybridized to the substrate (blue) of an upstream logic gate containing the hydrolysable ribonucleotide (r in green). (b) When the substrate is hydrolysed, the freed extension forms a stem–loop and opens the aptamer binding site in order to release malachite green in its normal non-emissive form (dark brown).

inhibitor binding to the active site (owing to high effective molarity). Presentation of an input$_1$ consisting of a longer DNA single strand to hybridize with DNA$_2$ will displace the enzyme and recover the enzyme performance. Similar presentation of an input$_2$ DNA strand which hybridizes with the loop will rigidify it so that the inhibitor's effective molarity at the enzyme active site is severely reduced. Again, the enzyme recovers its activity. These two ways of activating the enzyme are independent, so that both inputs can be applied simultaneously to achieve the same activation. OR logic is the result. Therapeutic enzyme activation responding to oligonucleotide disease markers thus comes into view.

14.7 Photodynamic Therapy

Photodynamic therapy (PDT) is a useful cancer treatment where singlet O_2 is generated by a photosensitizer so that neighbouring cells are oxidatively killed.[185–187] It is clear that such photosensitizers are AND logic devices which require a dose of light and O_2 (or air) as the two inputs. Singlet O_2 would be the output. Visudyne^{TM} (Novartis), **45**, which is used to treat a type of macular degeneration, is one example among several marketed as photosensitizers. The need for O_2 in PDT is easy to forget because of the near ubiquity of air. Our memory rapidly returns when hypoxic solid tumours are targeted because PDT is weakened under such O_2-depleted conditions.[188] Since the photosensitizer acts in a catalytic fashion, this is also an opportunity to note AND-type logic avenues in generalized photocatalysis.[189]

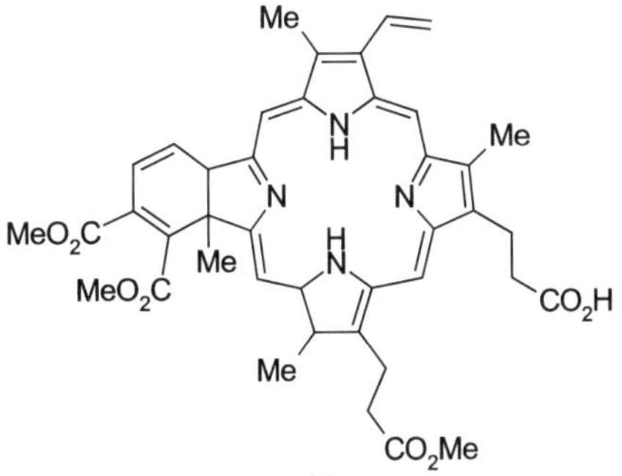

45

14.7.1 Targeted Photodynamic Therapy

Targeted PDT is a clear improvement over the conventional version because cell destruction requires specific local conditions. Some cancer cell interiors can have elevated H^+ and Na^+ levels compared with those in normal cells.[190–192] Akkaya arranges for targeted photodynamic therapy (PDT) since 'high' Na^+ and 'high' H^+ are required for success besides a light dose.[193] Compound **46** is therefore a 3-input AND logic gate where singlet O_2 is the output (Table 14.4); O_2 could also have been used as a fourth input variable. Compound **46** thus would be expected to attack cancer cells selectively.

The triplet excited state of the diiodoborondipyrromethene moiety within **46** is responsible for producing $^1O_2$ in aerated solution *via* EET to $^3O_2$.[128,129] However, this only happens if **46** is excited by the incoming 660 nm radiation and if there are no other competing processes such as PET.[128,129,194] H^+-controlled and PET-based PDT is known in a structure closely related to **46** due to O'Shea,[195] who also reports its clear function within cells. Very similar experiments concerning an aminoalkyl phthalocyanineSi(IV) derivative are also available from Ng's laboratory.[196] The absorption band of **46** red-shifts from 620 nm to 650 nm at high H^+ concentrations because the singlet excited

Table 14.4 Truth table for **46**.[a]

$Input_1$ hv $dose_{660}$	$Input_2$ H^+	$Input_3$ Na^+	Output $^1O_2$ yield[b]
none	none	none	low (0.0)
none	high (10^{-4} M)	none	low (0.0)
high	none	none	low (1.0)
high	high (10^{-4} M)	none	low (1.0)
none	none	high (10^{-3} M)	low (0.0)
none	high (10^{-4} M)	high (10^{-3} M)	low (0.0)
high	none	high (10^{-3} M)	low (2.9)
high	high (10^{-4} M)	high (10^{-3} M)	high (6.1)

[a] 2.0×10^{-5} M in acetonitrile.
[b] Relative quantum yields.

state of **46** has significant ICT character[128,129] owing to the pyridinium acceptor. So **46** is excited significantly in acidic solution only. PET is thermodynamically allowed from the benzo-15-crown-5 ether to the diiodoborondipyrromethene moiety. PET thermodynamics become unfavourable when Na^+ binds to the crown receptor.[64,89,197] PDT requires the 660 nm light dose to be applied (Table 14.4).

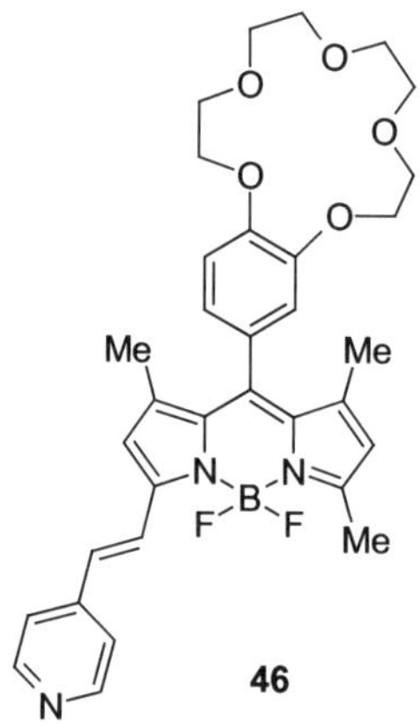

46

Though not as old as O'Shea's case[195] mentioned in the previous paragraph, Aida's chaperonin AND gate[198] responding to light dose and ATP is of similar vintage. GroEL is a chaperonin originating in *Escherichia coli* bacteria which has a double doughnut structure composed of seven subunits each. GroEL is now genetically engineered so that there is only one cysteine residue per subunit and it is positioned at the mouths of the doughnut. Michael addition of the cysteine thiol to **47** follows. Denatured green fluorescent protein (GFP) is swallowed by the derivatized GroEL and imprisoned. ATP is known to open up the doughnut mouths but the *E*-azobenzene units are still blocking the orifices. A dose of ultraviolet (uv) light ($\lambda = 350$ nm) alters the latter to *Z*-azobenzenes which are folded back so as to be less of a block. Thus the simultaneous presence of ATP and a dose of uv light permits the denatured GFP to escape so that it can re-fold in solution to the natural fluorescent form. Photo-release of a therapeutic protein cargo at a location of heightened ATP levels can thus be imagined.

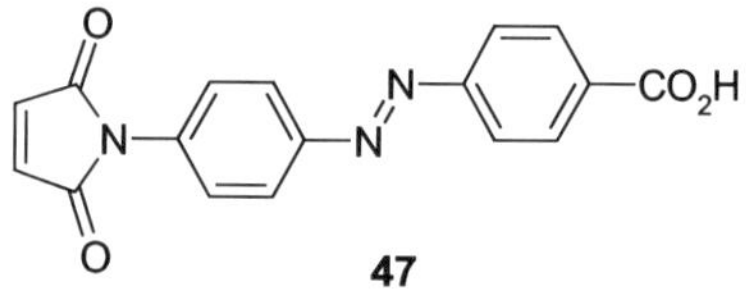

47

Photochromic spiropyrans have featured several times in this book, especially those which were also susceptible to protonation. Andréasson and colleagues[199] now take these down a new, potentially therapeutic avenue by

Applications 369

exploiting the fact that protonated ring-opened merocyanines are cationic and rather flat. They augment the basic spiropyran system (*e.g.* that seen in structures **1**, **29** and **30** which were shown in Chapter 10) by appending an extra cation in the form of a quaternary ammonium. Protonated merocyanine **48** therefore intercalates into double-stranded DNA polyanion whereas neither the unprotonated version **49** nor the ring-closed spiropyran **50** do. DNA effects on the uv–vis spectrum of these species clearly demonstrate this behavioural difference. Flow-oriented linear dichroism experiments also support the pH-dependent part of this difference. Light dose, H^+-driven AND logic-controlled DNA-binding is a useful starting point for selective photocytotoxicity, especially because cells of solid tumours tend to be acidic relative to healthy counterparts.[190,191]

The related case of light dose, OH^--driven AND logic-controlled release of cargo from mesoporous SiO_2 nanoparticles[200] (discussed in section 7.2.7 and illustrated in Figure 7.6) is also relevant here, especially if the cargo is chosen to be a drug. Hamachi's team[202] logically control gels[202–203] (section 7.2.7) in order to store/release insulin and other bioactive agents. Gel monomer **51** is amphiphilic owing to polar phosphate and apolar cyclohexyl termini. Importantly, it also carries a rigidly flat fumaramide centrepiece which permits stacking of monomers needed for fibre formation on the way to gel formation.[204] Nevertheless, the gel state is found only at reasonably high monomer concentration and low enough temperature. Reducing the hydrophilicity of the phosphate terminal by protonation or by binding to Ca^{2+} naturally encourages the gel state. However, distortion of the centrepiece by uv irradiation to produce the maleamide destroys the gel. Our starting state is **51** laced with H^+ and Ca^{2+} so that a gel is formed. The gel is loaded with insulin (fluorescently labelled for visualization). Insulin release is seen if both H^+ and Ca^{2+} are stripped by the application of EDTA (input$_1$) and NH_3 (input$_2$) respectively. Similar release is also found if the gel is irradiated with broad-band uv light (input$_3$). The corresponding logic gate array is given in Figure 14.15.

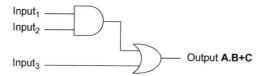

Figure 14.15 Logic gate array corresponding to Hamachi's cargo-releasing gel based on **25** pre-treated with Ca^{2+} and H^+. $Input_1$, $input_2$ and $input_3$ correspond to EDTA, NH_3 and broad-band ultraviolet light respectively. For algebraic purposes, these inputs are symbolized by variables A, B and C respectively.

51; R = cyclohexyl

14.8 Intracellular Computation

As indicated on several previous occasions, *e.g.* Chapter 2, molecules come into their own when they process information in a compatible way inside living cells. After all, they do it on a daily basis to keep us alive and well. In contrast, semiconductor processors are hampered by their larger size and incompatibility with cellular components. This section focuses on relatively complex logic operations whereas intracellular single-input logic was covered in section 14.2.1. The fact that molecular biology is a rich source of logic systems has emerged only in recent decades, even though Monod was prophetic in this regard.[205] Genes express proteins as output. The genes themselves can be up- or down-regulated by various agents. Regulation by more than a single agent opens the way to multi-input logic.[206–208] Within gene expression, it is RNA translation which produces proteins as output. RNA can be influenced by input agents *via* RNA interference[209] or *via* riboswitches[210] to create logic.[177] Some proteins also accept input species in order to switch their own regulatory activity in a Boolean manner.[211] Several approaches are therefore available to accomplish designed intracellular computation with various levels of logical complexity. The fact that sigmoidal stimulus–response curves (see Figure 2.3) are found for gene regulation and other intracellular processes shows some conceptual similarity of these biological processes with the chemical equilibria discussed in earlier chapters and in previous sections of this chapter.

Commanding protein–protein association in yeast cells leads to rather complex logic.[206,212] A protein with a DNA-binding domain as well as a receptor site for methotrexate, and another protein carrying a transcription activator as well as a receptor site for dexamethasone can be brought together by offering them a covalently linked 'methotrexate–spacer–dexamethasone' supermolecule. Even a 'trimethoprim–spacer–dexamethasone' version will suffice. Under these conditions, yeast cells produce the *lacZ* gene efficiently. Adding methotrexate alone would competitively inhibit this supramolecular protein–protein linking so that *lacZ* formation is suppressed.[213] This corresponds to 3-input disabled OR logic, where methotrexate is the disabling input.[212] β-Galactosidase activity, of the type used in sections 14.2.1 and 14.6, can monitor the presence of *lacZ*. This system can be built-up into a 5-input logic array by further regulating the DNA binding domain and the transcription activator.[212] Cases of this broad type can perform computations as complex as counting.[214,215] Another early approach, due to Fussenegger,[216] also employs small drug molecules as inputs. This case succeeds with mammalian cells and thus augurs well for clinical uses down the line.

Orthogonal ribosomes, made by Rackham and Chin,[217] illustrate yet another way of performing computing inside cells.[218] The orthogonal ribosomes do not interact with the mRNAs which work on the wild-type ribosome. Furthermore, each orthogonal ribosome does not interact with the other's mRNA. The orthogonal mRNAs for synthesizing the α- and ω-sections of β-galactosidase can be built into a single plasmid which serves as the logic gate. Only when both the necessary orthogonal ribosomal RNAs are offered as inputs do the newly produced α- and ω-sections join up and tetramerize to give the active enzyme. This can be signalled by hydrolysis of non-emissive fluorescein di-β-galactopyranoside to give strongly emissive fluorescein. So we have AND logic. Notably, the wild-type ribosome is always present to keep the cells viable.

Gene logic operating in *Escherichia coli* cells achieves an edge detection program in the hands of Voigt and collaborators.[219,220] Since edges in images are light–dark boundaries, the first trick is to get the bacteria in the dark to produce a diffusible compound while the lit-up cousins do not. When the compound reaches the illuminated bacteria, they respond to it in order to produce a dye as the second trick. Thus all the edges are visualized in parallel by the bacterial lawn (Figure 14.16). A darkness sensor from the same collaboration[221] is now adapted to release a cell–cell communication compound **52** by inserting the *LuxI* gene from *Vibrio fischeri*. The darkness sensor is also engineered to contain the *cI* gene (from phage λ), a repressor of the activity of *luxR* (also from *Vibrio fischeri*) which is present downstream. However, *luxR* is activated by **52** binding. The nett result for dark cells is moderate *luxR* activity. In contrast, the illuminated cells have no *cI*-induced repression of *luxR* but they do receive **52** by diffusion across the dark–light edge. So they have **52**-induced activation of *luxR*. Downstream of *luxR* is the common reporter gene *lacZ*. As seen in sections 14.2.1 and 14.6, the expression of β-galactosidase from *lacZ* allows production of a dye, which is black in the present case. Figure 14.16a shows the level of success achieved in edge visualization. The equivalent logic array is given in Figure 14.16b. The darkness sensor is a light dose-driven NOT gate outputting **52** and also *cI*. A **52**, *cI*-driven INHIBIT(*cI*) gate is the other half of the logic array and produces the black dye as output. It is important to note the unconventional element of the

Figure 14.16 (a) Output of edge-detecting gene arrays within a lawn of *Escherichia coli* cells (right panel) when illuminated through a mask (left panel). (b) Corresponding logic gate array. The dotted horizontal line represents diffusion of **52** across the dark–light edge; cI is a crucial repressor (see text). Part (a) is adapted from J. J. Tabor, H. M. Salis, Z. B. Simpson, A. A. Chevalier, A. Levskaya, E. M. Marcotte, C. A. Voigt and A. D. Ellington, *Cell*, 2009, **137**, 1272 with permission from Elsevier Science.

array which is the spatial diffusion of **52** across the dark–light edge. Indeed, cell–cell communication combines with the genetic machinery within each cell to produce this parallel computation. While the darkness sensor is basically bacterial photography,[221] edge detection goes further towards image-processing which is important in artificial intelligence schemes. Light arrests expression of the genetic apparatus in the darkness sensor.

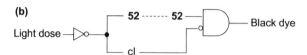

52

Section 14.6 featured Shapiro's 4-input doubly disabled AND logic system but now Benenson's and Weiss' teams take a 5-input triply disabled AND logic array inside cells so that a specific cervical cancer cell type (HeLa) can be identified and attacked.[222] All of these input species are microRNAs and one of the inputs is actually a composite of two others. The final attack is arranged by expressing the apoptosis-inducing protein hBax[223] as output. Logic processing[177,224–226] occurs as follows. Small interfering RNAs bind to chosen parts of the untranslated regions of a mRNA which encodes for the protein. The microRNA inputs hybridize with the small interfering RNAs to block them so as to remove the RNA interference. All the inputs need to be present in order

Applications

that the output protein will be expressed, *i.e.* AND logic prevails. NOT logic elements are introduced *via* microRNA–small interfering RNA interactions which activate the RNA interference. Though not employed in the case under discussion, OR logic can be seen to arise when two mRNAs which express the same protein are chosen. Even complex developmental processes, *e.g.* the formation of muscle cells from undifferentiated cousins in chicken embryos, can be addressed by protein-driven logic operating in a somewhat similar way.[227] However, ternary logic (Chapter 12) is required for an adequate description.

Riboswitches are present within untranslated regions of mRNA and bind small molecules. Since regulation of mRNA activity is the result, logic possibilities arise. Win and Smolke arrange computing inside yeast cells. Their logic work starts with a well-engineered riboswitch containing a RNA aptamer (to bind theophylline) integrated with a ribozyme which undergoes self-cleavage[210] (Figure 14.17). The riboswitch is positioned in the untranslated region of a mRNA which encodes for green fluorescent protein (GFP). When the theophylline input is present at 5 mM, the aptamer binding frees up a pentamer unit to launch a strand displacement which distorts the ribozyme active site (Figure 14.17). Then the self-cleavage stops and gene expression (output) is enhanced 25-fold, *i.e.* theophylline-driven YES logic. Similar designs yield NOT gates too.[210]

Win and Smolke proceed logically to place two well-spaced YES-logic riboswitches in the untranslated region of a mRNA.[228] One switch responds to theophylline as before but the other operates with tetracycline, *i.e.* carrying the corresponding RNA aptamer. Now both ribozymes need to be distorted by providing 'high' levels of both drug inputs so that all self-cleavage activity ceases. Then the gene is expressed in a theophylline, tetracycline-driven AND logical fashion. Indeed, the GFP gene is expressed for all to see. Similar placement of two NOT-logic riboswitches in series produces NOR logic

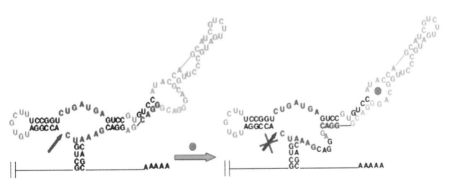

Figure 14.17 Structural change of a riboswitch upon binding theophylline (red circle) within the aptamer unit (brown). The resulting strand displacement inactivates the hammerhead ribozyme unit (purple, blue and black). The horizontal lines are untranslated regions of the messenger RNA. Reprinted from M. N. Win and C. D. Smolke, *Proc. Natl. Acad. Sci. USA*, 2007, **104**, 14283 with permission from the National Academy of Sciences USA.

behaviour. The approach is cleverer still, because extra units can be placed at sites other than the mRNA untranslated regions, *e.g.* the ribozyme core, stems or even the ribozyme–aptamer linker. It is worth noting the scale of this achievement because many previous efforts concerning simple riboswitches had failed *in vivo* owing to unknown interactions. The success of Win and Smolke's approach will allow a degree of designed interventions to be developed in yeast cell biology. Then, systems biology will benefit. Even when the designs come up against unanticipated interactions, an education surely awaits.

14.9 Conclusion

Here we are, at the end of our journey as far as this book is concerned. Now is the time to happily close with a summary of the sights seen and the experiences gained. Even as we wrap up, it is good to know that the journey will go on in the hands and minds of molecular logic enthusiasts around the world.

We have seen how chemical reactions, whether reversible or not, generally depend on prior actions such as the addition of reagents. That recognition allowed much chemistry to be viewed as input–output phenomena of molecular devices which could then be cast in logical terms. The realization that similar molecular logic exists within biology allowed natural and suitably modified biomolecules to occupy the Boolean arena. Though molecular logic systems are still elementary when compared to semiconductor machinery, we saw how they are already performing a number of useful tasks which silicon devices cannot. We saw that small and/or living spaces are fertile grounds for molecular logic-based computation. It was clear that the available techniques of chemical synthesis and molecular biology allow the construction and evaluation of many, many nanometric logic devices with a variety of functions. The ability to view the molecular/biomolecular devices in terms of components allowed engineering-like designs. Some of these designs, such as the fluorescent PET switching principle, proved to be quite robust and should serve designers well into the future.

During our journey we stepped from the simplest single-input systems to cases with two inputs or more. Each of the possible logic types within the Boolean scheme was met along the way. The diversity of chemistry was on show with a variety of molecular species, observation techniques and design principles on offer. We saw how general molecular-scale logic has become, even to the point of developing sequential, multi-valued and quantum facets. We met molecular manifestations of computational ideas such as integration, reconfigurability, multiplexing, numeracy and superposition. We had to learn some elements of physiology, molecular biology, biochemistry, device physics, computer science, mathematics and aeronautical engineering (besides the several sub-disciplines of chemistry) in order to appreciate the gamut of molecular logic systems that we encountered.

In terms of applications of molecular logic systems, we found that the monitoring and imaging of intracellular players is growing nicely. Planned interventions in bioprocesses were also noticeable. Nonetheless, it was good to see that living cells were not the only places where monitoring of chemical species is valuable. Clinical tests for electrolytes remain a clear commercial success and others are bound to follow suit. It was apparent that multi-input logic systems are pushing further forward to improve sensing, diagnosis and therapy. We saw how even elementary molecular logic could empower the identification of small objects in large populations. This and molecular keypad locks showed value in the security area.

Finally, it would be fair to say that molecular logic-based computation has come of age. If its adult life is half as interesting and educational as its baby phase has been, we can all be pleased.

References

1. A. J. Bard, *Chem. Eng. News*, 1999, September 6, 5.
2. J. M. Tour, *Acc. Chem. Res.*, 2000, **33**, 791.
3. R. S. Williams, quoted in P. Ball, *Nature*, 2000, **406**, 118.
4. L. M. Adleman, *Science*, 1994, **266**, 1021.
5. M. Amos, *Theoretical and Experimental DNA Computation*, Springer, Berlin, 2005.
6. C. L. Dwyer and A. LeBeck, *Introduction to DNA Self-Assembled Computer Design*, Artech House, Norwood, MA, 2007.
7. J. C. Cox, D. S. Cohen and A. D. Ellington, *Trends Biotechnol.*, 1999, **17**, 151.
8. L. M. Adleman, *Sci. Am.*, 1998, **279**, 54.
9. S. Uchiyama, G. D. McClean, K. Iwai and A. P. de Silva, *J. Am. Chem. Soc.*, 2005, **127**, 8920.
10. K. Sumaru, H. Matsuoka, H. Yamaoka and G. D. Wignall, *Phys. Rev. E*, 1996, **53**, 1744.
11. A. P. de Silva, D. B. Fox and T. S. Moody, in *Stimulating Concepts in Chemistry*, ed. F. Vogtle, J. F. Stoddart and M. Shibasaki, Wiley-VCH, Weinheim, 2000, p. 307.
12. A. Fernández-Gutiérrez and A. Muñoz de la Peña, in *Molecular Luminescence Spectroscopy. Methods and Applications. Part 1*, ed. S. G. Schulman, Wiley, New York, 1985, p. 371.
13. E. Bishop (ed.), *Indicators*, Pergamon, Oxford, 1972.
14. E. B. Sandell, *Colorimetric Determination of Traces of Metals*, Interscience, London, 3rd edn, 1959.
15. D. W. Domaille, E. L. Que and C. J. Chang, *Nature Chem. Biol*, 2008, **4**, 168.
16. E. L. Que, D. W. Domaille and C. J. Chang, *Chem. Rev.*, 2008, **108**, 1517.
17. *The Molecular Probes Handbook*, 11th edn, (www.invitrogen.com/site/us/en/home/References/Molecular-Probes-The-Handbook.html)
18. T. Ueno and T. Nagano, *Nature Meth*, 2011, **8**, 642.
19. J. G. McGeown, *Exp. Physiol.*, 2010, **95**, 1049.
20. J. Lechleiter, S. Girard, E. Peralta and D. Clapham, *Science*, 1991, **252**, 123.
21. R. Y. Tsien, *Am. J. Physiol.*, 1992, **263**, C723.

22. G. Grynkiewicz, M. Poenie and R. Y. Tsien, *J. Biol. Chem.*, 1985, **206**, 3440.
23. K. Kano and J. H. Fendler, *Biophys. Biochim. Acta*, 1978, **509**, 289.
24. B. Rotman and B. W. Papermaster, *Proc. Nat. Acad. Sci. USA*,1966, **55**, 134.
25. N. Klonis and W. H. Sawyer, *J. Fluoresc.*, 1996, **6**, 147.
26. A. Minta and R. Y. Tsien, *J. Biol. Chem.*, 1989, **264**, 19449.
27. A. T. Harootunian, J. P. Y. Kao, B. K. Eckert and R. Y. Tsien, *J. Biol. Chem.*, 1989, **264**, 19458.
28. B. Raju, E. Murphy, L. A. Levy, R. D. Hall and R. E. London, *Am. J. Physiol.*, 1989, **256**, C540.
29. M. M. Martin, P. Plaza, Y. H. Meyer, F. Badaoui, J. Bourson, J. P. Lefebvre and B. Valeur, *J. Phys. Chem.*, 1996, **100**, 6879.
30. G. A. Smith, T. R. Hesketh and J. C. Metcalfe, *Biochem. J.*, 1988, **250**, 227.
31. A. P. de Silva, H. Q. N. Gunaratne, A. T. M. Kane and G. E. M. Maguire, *Chem. Lett.*, 1995, 125.
32. X. H. Qian, Y. Xiao, Y. F. Xu, X. F. Guo., J. H. Qian and W. P. Zhu, *Chem. Commun.*, 2010, 6418.
33. A. Minta, J. P. Y. Kao and R. Y. Tsien, *J. Biol. Chem.*, 1989, **264**, 8171.
34. E. Tamanini, A. Katewa, L. M. Sedger, M. H. Todd and M. Watkinson, *Inorg. Chem.*, 2009, **48**, 319.
35. A. P. de Silva, H. Q. N. Gunaratne, T. Gunnlaugsson and P. L. M. Lynch, *New J. Chem.*, 1996, **20**, 871.
36. C. C. Woodroofe and S. J. Lippard, *J. Am. Chem. Soc.*, 2003, **125**, 11458.
37. R. Ferreira, P. Remon and U. Pischel, *J. Phys. Chem. C*, 2009, **113**, 5805.
38. S. Banthia and A. Samantha, *J. Phys. Chem. B*, 2006, **110**, 6437.
39. A. P. de Silva and R. A. D. D. Rupasinghe, *J. Chem. Soc. Chem. Commun.*, 1985, 1669.
40. S. Paglin, T. Hollister, T. Delohery, N. Hackett, M. McMahill, E. Sphicas, D. Domingo and J. Yahalom, *Cancer Res.*, 2001, **61**, 439.
41. M. J. Hall, L. T. Allen and D. F. O'Shea, *Org. Biomol. Chem.*, 2006, **4**, 776.
42. B. Tang, F. B. Yu, P. Li, L. L. Tong, X. Duan, T. Xie and X. Wang, *J. Am. Chem. Soc.*, 2009, **131**, 3016.
43. T. Kowada, J. Kikuta, A. Kubo, M. Ishii, H. Maeda, S. Mizukami and K. Kikuchi, *J. Am. Chem. Soc.*, 2011, **133**, 17772.
44. R. Parkesh, T. C. Lee and T. Gunnlaugsson, *Tetrahedron Lett.*, 2009, **50**, 4114.
45. E. M. Nolan and S. J. Lippard, *Acc. Chem. Res.*, 2009, **42**, 193.
46. M. Kamiya, H. Kobayashi, Y. Hama, Y. Koyama, M. Bernardo, T. Nagano, P. L. Choyke and Y. Urano, *J. Am. Chem. Soc.*, 2007, **129**, 3918.
47. S. Uchiyama, Y. Matsumura, A. P. de Silva and K. Iwai, *Anal. Chem.*, 2003, **75**, 5926.
48. C. Gota, K. Okabe, T. Funatsu, Y. Harada and S. Uchiyama, *J. Am. Chem. Soc.*, 2009, **131**, 2766.

49. K. Okabe, N. Inada, C. Gota, Y. Harada, T. Funatsu and S. Uchiyama, *Nature Commun*, 2012, **3**, 705.
50. E. W. Miller, J. Y. Lin, E. P. Frady, P. A. Steinbach, W. B. Kristan and R. Y. Tsien, *Proc. Natl. Acad. Sci. USA*, 2012, **109**, 2114.
51. L. M. Loew, *Pure Appl. Chem.*, 1996, **68**, 1405.
52. J. E. Gonzalez and R. Y. Tsien, *Chem. Biol.*, 1997, **4**, 269.
53. W. B. Davis, W. A. Svec, M. A. Ratner and M. R. Wasielewski, *Nature*, 1998, **396**, 60.
54. A. P. de Silva, H. Q. N. Gunaratne, J. -L. Habib-Jiwan, C. P. McCoy, T. E. Rice and J.-P. Soumillion, *Angew. Chem. Int. Ed. Engl.*, 1995, **34**, 1728.
55. P. Li, X. Duan, Z. Z. Chen, Y. Liu, T. Xie, L. B. Fang, X. R. Li, M. Yin and B. Tang, *Chem. Commun.*, 2011, **47**, 7755.
56. A. P. de Silva, J. Eilers and G. Zlokarnik, *Proc. Natl. Acad. Sci. USA*, 1999, **96**, 8336.
57. H. Kobayashi, M. Ogawa, R. Alford, P. L. Choyke and Y. Urano, *Chem. Rev.*, 2010, **110**, 2620.
58. R. Weissleder and M. J. Pittet, *Nature*, 2008, **452**, 580.
59. J. K. Tusa and H. He, *J. Mater. Chem.*, 2005, **15**, 2640.
60. H. He, M. Mortellaro, M. J. P. Leiner, S. T. Young, R. J. Fraatz and J. Tusa, *Anal. Chem.*, 2003, **75**, 549.
61. H. He, M. Mortellaro, M. J. P. Leiner, R. J. Fraatz and J. Tusa, *J. Am. Chem. Soc.*, 2003, **125**, 1468.
62. H. R. He, K. Jenkins and C. Lin, *Anal. Chim. Acta*, 2008, **611**, 197.
63. www.optimedical.com
64. R. A. Bissell, A. P. de Silva, H. Q. N. Gunaratne, P. L. M. Lynch, G. E. M. Maguire, C. P. McCoy and K. R. A. S. Sandanayake, *Top. Curr. Chem.*, 1993, **168**, 223.
65. S. Klahr, in *Textbook of Primary Care Medicine*, ed. J. Noble, Mosby, St. Louis, MO, 1996.
66. R. A. Schultz, B. D. White, D. M. Dishong, K. A. Arnold and G. W. Gokel, *J. Am. Chem. Soc.*, 1985, **107**, 6659.
67. G. W. Gokel, *Crown Ethers and Cryptands*, Royal Society of Chemistry, Cambridge, 1991.
68. C. J. Pedersen, *Science*, 1988, **241**, 536.
69. H. Siegerman, in *Techniques of Electroorganic Synthesis. Part II*, ed. N. L. Weinberg, Wiley, New York, 1975.
70. P. D. Beer and P. A. Gale, *Angew. Chem. Int. Ed.*, 2001, **40**, 487.
71. S. Kenmoku, Y. Urano, K. Kanda, H. Kojima, K. Kikuchi and T. Nagano, *Tetrahedron*, 2004, **60**, 11067.
72. T. Gunnlaugsson, M. Nieuwenhuyzen, L. Richard and V. Thoss, *J. Chem. Soc. Perkin Trans*, 2002, **2**, 141.
73. X. Lu, I. Manners and M. Winnik, in *New Trends in Fluorescence Spectroscopy*, ed. B. Valeur and J.-C. Brochon, Springer, Berlin, 2001, p. 229.
74. M. Gouterman, *J. Chem. Educ.*, 1997, **74**, 697.
75. P. Kele, J. Orbulescu, R.E. Gawley and R.M. Leblanc, *Chem. Commun.*, 2006, 1494.

76. A. P. de Silva and S. A. de Silva, *J. Chem. Soc. Chem. Commun.*, 1986, 1709.
77. P. Kele, K. Nagy and A. Kotschy, *Angew. Chem. Int. Ed. Engl.*, 2006, **45**, 2565.
78. A. P. de Silva and K. R. A. S. Sandanayake, *J. Chem. Soc. Chem. Commun.*, 1989, 1183.
79. H. F. Ji, G. M. Brown and R. Dabestani, *Chem. Commun.*, 1999, 609.
80. H. F. Ji, R. Dabestani, G. M. Brown and R. A. Sachleben, *Chem Commun.*, 2000, 833.
81. L. Prodi, F. Bolletta, M. Montalti, N. Zaccheroni, A. Casnati, F. Sansone and R. Ungaro, *New J. Chem.*, 2000, **24**, 155.
82. S. R. Wilson and A. W. Czarnik (ed.), *Combinatorial Chemistry, Synthesis and Application*, Wiley, New York, 1997.
83. G. Jung (ed.), *Combinatorial Peptide and Nonpeptide Libraries*, VCH, Weinheim, 1996.
84. G. T. Copeland and S. J. Miller, *J. Am. Chem. Soc.*, 1999, **121**, 4306.
85. R. F. Harris, A. J. Nation, G. T. Copeland and S. J. Miller, *J. Am. Chem. Soc.*, 2000, **122**, 11270.
86. T. J. Dale and J. Rebek, *J. Am. Chem. Soc.*, 2006, **128**, 4500.
87. S. Bencic-Nagale, T. Sternfeld and D. R. Walt, *J. Am. Chem. Soc.*, 2006, **128**, 5041.
88. T. J. Dale and J. Rebek, *Angew. Chem. Int. Ed.*, 2009, **48**, 7850.
89. R. A. Bissell, A. P. de Silva, H. Q. N. Gunaratne, P. L. M. Lynch, G. E. M. Maguire and K. R. A. S. Sandanayake, *Chem. Soc. Rev.*, 1992, **21**, 187.
90. C. Munkholm, D. R. Parkinson and D. R. Walt, *J. Am. Chem. Soc.*, 1990, **112**, 2608.
91. A. P. de Silva, H. Q. N. Gunaratne and T. Gunnlaugsson, *Tetrahedron Lett.*, 1998, **39**, 5077.
92. S. W. Zhang and T. M. Swager, *J. Am. Chem. Soc.*, 2003, **125**, 3420.
93. J. K. Weltman, R. P. Szaro, A. R. Frackelston, R. M. Dowben, J. R. Bunting and R. E. Cathou, *J. Biol. Chem.*, 1973, **218**, 3173.
94. F. A. Cotton, G. R. Wilkinson, C. A. Murillo and M. Bochmann, *Advanced Inorganic Chemistry*, Wiley, New York, 6th edn, 1999.
95. K. Rurack, *Spectrochim. Acta A, Mol. Biomol. Spectrosc*, 2001, **57**, 2161.
96. M. Elstner, K. Weisshart, K. Mu?llen and A. Schiller, *J. Am. Chem. Soc.*, 2012, **134**, 8098.
97. C. R. Cooper and T. D. James, *Chem. Commun.*, 1997, 1419.
98. C. R. Cooper and T. D. James, *J. Chem. Soc. Perkin Trans*, 2000, **1**, 963.
99. F. Fages, J.-P. Desvergne, K. Kampke, H. Bouas-Laurent, J.-M. Lehn, J.-P. Konopelski, P. Marsau and Y. Barrans, *J. Chem. Soc. Chem. Commun.*, 1990, 655.
100. A. P. de Silva and K. R. A. S. Sandanayake, *Angew. Chem. Int. Ed. Engl.*, 1990, **29**, 1173.
101. S. Misumi, *Top. Curr. Chem.*, 1993, **165**, 163.

102. K. Secor, J. Plante, C. Avetta and T. Glass, *J. Mater. Chem.*, 2005, **15**, 4073.
103. J. Z. Zhao, T. M. Fyles and T. D. James, *Angew. Chem. Int. Ed.*, 2004, **43**, 3461.
104. T. D. James, K. R. A. S. Sandanayake and S. Shinkai, *Nature*, 1995, **374**, 345.
105. C. W. Gray Jr. and T. A. Houston, *J. Org. Chem.*, 2002, **67**, 5426.
106. T. Sakamoto, A. Ojida and I. Hamachi, *Chem. Commun.*, 2009, 141.
107. A. Ojida, M. Mito-oka, K. Sada and I. Hamachi, *J. Am. Chem. Soc.*, 2004, **126**, 2454.
108. A. Ojida, M. Mito-oka, M. Inouye and I. Hamachi, *J. Am. Chem. Soc.*, 2002, **124**, 6256.
109. D. H. Vance and A. W. Czarnik, *J. Am. Chem. Soc.*, 1994, **116**, 9397.
110. A. V. Koulov, K. A. Stucker, C. Lakshmi, J. P. Robinson and B. D. Smith, *Cell Death Diff*, 2003, **10**, 1357.
111. C. Lakshmi, R. G. Haneshaw and B. D. Smith, *Tetrahedron*, 2004, **60**, 11307.
112. Y. Nakahara, T. Kida, Y. Nakatsuji and M. Akashi, *Chem. Commun.*, 2004, 224.
113. S. Bhattacharya and A. Gulyani, *Chem. Commun.*, 2003, 1158.
114. S. Mukherjee, A. Chattopadhyay, A. Samanta and T. Soujanya, *J. Phys. Chem.*, 1994, **98**, 2809.
115. S. Uchiyama, K. Takehira, S. Kohtani, K. Imai, R. Nakagaki, S. Tobita and T. Santa, *Org. Biomol. Chem.*, 2003, **1**, 1067.
116. M. S. Fernandez and P. Fromherz, *J. Phys. Chem.*, 1977, **81**, 1755.
117. C. J. Drummond, F. Grieser and T. W. Healy, *J. Phys. Chem.*, 1988, **92**, 2604.
118. R. A. Bissell, A. J. Bryan, A. P. de Silva and C. P. McCoy, *J. Chem. Soc. Chem. Commun.*, 1994, 405.
119. M. Q. Tian, H. Ihmels and K. Benner, *Chem. Commun.*, 2010, **46**, 5719.
120. A. Granzhan, H. Ihmels and G. Viola, *J. Am. Chem. Soc.*, 2007, **129**, 1254.
121. Y. Urano, D. Asanuma, Y. Hama, Y. Koyama, T. Barrett, M. Kamiya, T. Nagano, T. Watanabe, A. Hasegawa, P. L. Choyke and H. Kobayashi, *Nature Med.*, 2009, **15**, 104.
122. C. Pistol, V. Mao, V. Thusu, A. R. Lebeck and C. Dwyer, *Small*, 2010, **6**, 843.
123. L. Yuan, W. Y. Lin, Y. A. Xie, B. Chen and S. S. Zhu, *J. Am. Chem. Soc.*, 2012, **134**, 1305.
124. H. Kojima and T. Nagano, *Adv. Mater.*, 2000, **12**, 763.
125. M. J. Plater, I. Greig, M. H. Helfrich and S. H. Ralston, *J. Chem. Soc. Perkin Trans*, 2001, **1**, 2553.
126. A. R. Lippert, G. C. Van De Bittner and C. J. Chang, *Acc. Chem. Res.*, 2011, **44**, 793.
127. A. P. de Silva and N. D. McClenaghan, *J. Am. Chem. Soc.*, 2000, **122**, 3965.

128. A. P. de Silva, H. Q. N. Gunaratne, T. Gunnlaugsson, A. J. M. Huxley, C. P. McCoy, J. T. Rademacher and T. E. Rice, *Chem. Rev.*, 1997, **97**, 1515.
129. B. Valeur and M. N. Berberan-Santos, *Molecular Fluorescence*, Wiley-VCH, Weinheim, 2nd edn, 2012.
130. H. Komatsu, T. Miki, D. Citterio, T. Kubota, Y. Shindo, Y. Kitamura, K. Oka and K. Suzuki, *J. Am. Chem. Soc.*, 2005, **127**, 10798.
131. R. Y. Tsien, *Biochemistry*, 1980, **19**, 2396.
132. K. P. Zauner and M. Conrad, *Biotechnol. Prog.*, 2001, **17**, 553.
133. P. L. Gentili, *Chem. Phys. A*, 2007, **336**, 64.
134. P. L. Gentili, *J. Phys. Chem. A*, 2008, **112**, 11992.
135. V. Privman, M. A. Arugula, J. Halamek, M. Pita and E. Katz, *J. Phys. Chem. B*, 2009, **113**, 5301.
136. D. Melnikov, G. Strack, M. Pita, V. Privman and E. Katz, *J. Phys. Chem. B*, 2009, **113**, 10472.
137. K. S. Lam, S. E. Salmon, E. M. Hirsh, V. J. Hruby, W. M. Kazmierzki and R. J. Knapp, *Nature*, 1991, **354**, 82.
138. Á. Furka, F. Sebestyen, M. Asgedom and G. Dibo, *Int. J. Pept. Prot. Res*, 1991, **37**, 487.
139. S. Shepard, *RFID: Radio Frequency Identification*, McGraw-Hill, New York, 2005.
140. H. Lee, J. Kim, H. Kim, J. Kim and S. Kwon, *Nature Mater*, 2010, **9**, 745.
141. K. Braeckmans and S. C. De Smedt, *Nature Mater*, 2010, **9**, 697.
142. Z. Y. Xie, Y. J. Zhao, L. G. Sun, X. W. Zhao, Y. Shao and Z. Z. Gu, *Chem. Commun.*, 2009, 7012.
143. K. C. Nicolaou, X. Y. Xiao, Z. Parandoosh, A. Senyei and M. P. Nova, *Angew. Chem. Int. Ed. Engl.*, 1995, **34**, 2289.
144. E. J. Moran, S. Sarshar, J. F. Cargill, M. M. Shahbaz, A. Lio, A. M. M. Mjalli and R. W. Armstrong, *J. Am. Chem. Soc.*, 1995, **117**, 10787.
145. J. Raez, D. R. Blais, Y. Zhang, R. A. Alvarez-Puebla, J. P. Bravo-Vasquez, J. P. Pezacki and H. Fenniri, *Langmuir*, 2007, **23**, 6482.
146. D. R. Walt, *Science*, 2000, **287**, 451.
147. www.illumina.com
148. B. J. Battersby, G. A. Lawrie, A. P. R. Johnston and M. Trau, *Chem. Commun*, 2002, 1435.
149. A. P. de Silva, M. R. James, B. O. F. McKinney, D. A. Pears and S. M. Weir, *Nature Mater*, 2006, **5**, 787.
150. B. Hayes, *Am. Sci*, 2001, **89**, 490.
151. R. W. Keyes, *Rev. Mod. Phys.*, 1989, **61**, 279.
152. R. Webb, *Nature*, 2006, **443**, 39.
153. M. Inman, *New Sci.*, 2006, **191**(2568), 28.
154. D. Gust, T. A. Moore and A. L. Moore, in *From Non-covalent Assemblies to Molecular Machines*, ed. J.-P. Sauvage and P. Gaspard, Wiley-VCH, Weinheim, 2011, p. 321.
155. S. Nath and U. Maitra, *Org. Lett.*, 2006, **8**, 3239.

156. M. Mancini, J. M. Ordovas, G. Riccardi, P. Rubba and P. Strazullo (ed.), *Nutritional and Metabolic Bases of Cardiovascular Disease*, Wiley-Blackwell, Chichester, 2011.
157. D. C. Magri, G. J. Brown, G. D. McClean and A. P. de Silva, *J. Am. Chem. Soc.*, 2006, **128**, 4950.
158. E. D. Hirsch, J. F. Kett and J. Trefil (ed.), *The New Dictionary of Cultural Literacy*, Houghton Mifflin, Boston, 3rd edn, 2002.
159. Y. Benenson, B. Gil, U. Ben-Dor, R. Adar and E. Shapiro, *Nature*, 2004, **429**, 423.
160. E. Shapiro and Y. Benenson, *Sci. Am.*, 2006, **294**(5), 44.
161. J. Macdonald, D. Stefanovic and M. N. Stojanovic, *Sci. Am.*, 2008, **299**(5), 84.
162. D. Margulies and A. D. Hamilton, *J. Am. Chem. Soc.*, 2009, **131**, 9142.
163. T. Konry and D. R. Walt, *J. Am. Chem. Soc.*, 2009, **131**, 13232.
164. D. Margulies and A. D. Hamilton, *Curr. Opin. Chem. Biol.*, 2010, **14**, 705.
165. T. Gunnlaugsson, T. C. Lee and R. Parkesh, *Org. Biomol. Chem.*, 2003, **1**, 3265.
166. T. Gunnlaugsson, T. C. Lee and R. Parkesh, *Org. Lett.*, 2003, **5**, 4065.
167. T. Gunnlaugsson, T. C. Lee and R. Parkesh, *Tetrahedron*, 2004, **60**, 11239.
168. J. Halamek, J. R. Windmiller, J. Zhou, M. C. Chuang, P. Santhosh, G. Strack, M. A. Arugula, S. Chinnapareddy, V. Bocharova, J. Wang and E. Katz, *Analyst*, 2010, **135**, 2249.
169. D. Melnikov, G. Strack, J. Zhou, J. R. Windmiller, J. Halamek, V. Bocharova, M. C. Chuang, P. Santhosh, V. Privman, J. Wang and E. Katz, *J. Phys. Chem. B*, 2010, **114**, 12166.
170. J. Halamek, V. Bocharova, S. Chinnapareddy, J. R. Windmiller, G. Strack, M. C. Chuang, J. Zhou, P. Santhosh, G. V. Ramirez, M. A. Arugula, J. Wang and E. Katz, *Mol. BioSyst.*, 2010, **6**, 2554.
171. M. Pita, J. Zhou, K. M. Manesh, J. Halamek, E. Katz and J. Wang, *Sensors Actuators B*, 2009, **139**, 631.
172. J. Zhou, J. Halamek, V. Bocharova, J. Wang and E. Katz, *Talanta*, 2011, **83**, 955.
173. D. Margulies and A. D. Hamilton, *Angew. Chem. Int. Ed.*, 2009, **48**, 1771.
174. X. L. Feng, X. R. Duan, L. B. Liu, F. D. Feng, S. Wang, Y. L. Li and D. B. Zhu, *Angew. Chem. Int. Ed.*, 2009, **48**, 5316.
175. M. N. Stojanovic and D. Stefanovic, *J. Am. Chem. Soc.*, 2003, **125**, 6673.
176. J. R. Amir, M. Popkov, A. R. Lerner, C. F. Barbas III and D. Shabat, *Angew. Chem. Int. Ed.*, 2005, **44**, 4378.
177. Y. Benenson, *Mol. BioSyst.*, 2009, **5**, 675.
178. Y. Benenson, T. Paz-Elizur, R. Adar, E. Keinan, Z. Livneh and E. Shapiro, *Nature*, 2001, **414**, 430.
179. B. Gil, M. Kahan-Hanum, N. Skirtenko, R. Adar and E. Shapiro, *Nano Lett.*, 2011, **11**, 2989.
180. T. Ran, S. Kaplan and E. Shapiro, *Nature Nanotechnol*, 2009, **4**, 642.

181. E. Kossoy, N. Lavid, M. Soreni-Harari, Y. Shoham and E. Keinan, *ChemBioChem*, 2007, **8**, 1255.
182. M. N. Stojanovic and D. M. Kolpashchikov, *J. Am. Chem. Soc.*, 2004, **126**, 9266.
183. D. M. Kolpashchikov and M. N. Stojanovic, *J. Am. Chem. Soc.*, 2005, **127**, 11348.
184. N. C. Gianneschi and M. R. Ghadiri, *Angew. Chem. Int. Ed.*, 2007, **46**, 3955.
185. B. W. Henderson and T. J. Dougherty (ed.), *Photodynamic Therapy*, Dekker, New York, 1992.
186. T. Patrice (ed.), *Photodynamic Therapy*, Royal Society of Chemistry, Cambridge, 2003.
187. M. Hamblin and P. Mroz (ed.), *Advances in Photodynamic Therapy*, Artech House, Boston, 2008.
188. S. Stern, R. J. Hodgkiss and M. Guichard, *Radiother. Oncol.*, 1996, **39**, 129.
189. R. S. Stoll and S. Hecht, *Angew. Chem. Int. Ed.*, 2010, **49**, 5054.
190. P. Montcourrier, P. H. Mangeat, C. Valembois, G. Salazar, A. Sahuquet, C. Duperray and H. Rochefort, *J. Cell Sci.*, 1994, **107**, 2381.
191. J. A. Goode and D. J. Chadwick, *The Tumor Microenvironment*, Wiley, New York, 2001.
192. I. L. Cameron, N. K. R. Smith, T. B. Pool and R. L. Sparks, *Cancer Res.*, 1980, **40**, 1493.
193. S. Ozlem and E. U. Akkaya, *J. Am. Chem. Soc.*, 2009, **131**, 48.
194. R. A. Bissell and A. P. de Silva, *J. Chem. Soc. Chem. Commun.*, 1991, 1148.
195. S. O. McDonnell, M. J. Hall, L. T. Allen, A. Byrne, W. M. Gallagher and D. F. O'Shea, *J. Am. Chem. Soc.*, 2005, **127**, 16360.
196. X. J. Jiang, P. C. Lo, S. L. Yeung, W. P. Fong and D. K. P. Ng, *Chem. Commun.*, 2010, **46**, 3188.
197. A. J. Bryan, A. P. de Silva, S. A. de Silva, R. A. D. D. Rupasinghe and K. R. A. S. Sandanayake, *Biosensors*, 1989, **4**, 169.
198. S. Muramatsu, K. Kinbara, H. Taguchi, N. Ishii and T. Aida, *J. Am. Chem. Soc.*, 2006, **128**, 3764.
199. M. Hammarson, J. Andersson, S. M. Li, P. Lincoln and J. Andréasson, *Chem. Commun.*, 2010, **46**, 7130.
200. S. Angelos, Y. W. Yang, N. M. Khashab, J. F. Stoddart and J. I. Zink, *J. Am. Chem. Soc.*, 2009, **131**, 11344.
201. H.-J. Schneider, T. J. Liu, N. Lomadze and B. Palm, *Adv. Mater.*, 2004, **16**, 613.
202. H. Komatsu, S. Matsumoto, S. I. Tamaru, K. Kaneko, M. Ikeda and I. Hamachi, *J. Am. Chem. Soc.*, 2009, **131**, 5580.
203. K. Gawel and B. T. Stokke, *Soft Matter*, 2011, **7**, 4615.
204. J. H. Fuhrhop and J. Koening, *Membranes and Molecular Assemblies: The Synkinetic Approach*, RSC Publishing, Cambridge, 1994.
205. J. Monod, *Chance and Necessity*, Knopf, New York, 1971.

206. J. Hasty, D. McMillen and J. J. Collins, *Nature*, 2002, **420**, 224.
207. S. Istrail, S. Ben-Tabou De-Leon and E. H. Davidson, *Developmental Biol*, 2007, **310**, 187.
208. S. Istrail and E. H. Davidson, *Proc. Natl. Acad. Sci. USA*, 2005, **102**, 4954.
209. J. W. Gaynor, B. J. Campbell and R. Cosstick, *Chem. Soc. Rev.*, 2010, **39**, 4169.
210. M. N. Win and C. D. Smolke, *Proc. Natl. Acad. Sci. USA*, 2007, **104**, 14283.
211. J. E. Dueber, B. J. Yeh, K. Chak and W. A. Lim, *Science*, 2003, **301**, 1904.
212. J. E. Bronson, W. W. Mazur and V. W. Cornish, *Mol. BioSyst.*, 2008, **4**, 56.
213. H. Lin, W. M. Abida, R. T. Sauer and V. W. Cornish, *J. Am. Chem. Soc.*, 2000, **122**, 4247.
214. A. E. Friedland, T. K. Lu, X. Wang, D. Shi, G. Church and J. J. Collins, *Science*, 2009, **324**, 1199.
215. C. D. Smolke, *Science*, 2009, **324**, 1156.
216. B. P. Kramer, C. Fischer and M. Fussenegger, *Biotechnol. Bioeng.*, 2004, **87**, 478.
217. O. Rackham and J. W. Chin, *Nature Chem. Biol*, 2005, **1**, 159.
218. O. Rackham and J. W. Chin, *J. Am. Chem. Soc.*, 2005, **127**, 17584.
219. J. J. Tabor, H. M. Salis, Z. B. Simpson, A. A. Chevalier, A. Levskaya, E. M Marcotte, C. A. Voigt and A. D. Ellington, *Cell*, 2009, **137**, 1272.
220. T. Ratner and E. Keinan, *ChemBioChem*, 2010, **11**, 947.
221. A. Levskaya, A. A. Chevalier, J. J. Tabor, Z. Booth Simpson, L. A. Lavery, M. Levy, E. A. Davidson, A. Scouras, A. D. Ellington, E. M. Marcotte and C. A. Voigt, *Nature*, 2005, **438**, 441.
222. Z. Xie, L. Wroblewska, L. Prochazka, R. Weiss and Y. Benenson, *Science*, 2011, **333**, 1307.
223. S. L. Lowe, S. Rubinchik, T. Honda, T. J. McDonnell, J. Y. Dong and J. S. Norris, *Gene Ther*, 2001, **8**, 1363.
224. K. Rinaudo, L. Bleris, R. Maddamsetti, S. Subramanian, R. Weiss and Y. Benenson, *Nature Biotechnol.*, 2007, **25**, 795.
225. Z. Xie, S. J. Liu, L. Bleris and Y. Benenson, *Nucl. Acids Res*, 2010, **38**, 2692.
226. M. Leisner, L. Bleris, J. Lohmueller, Z. Xie and Y. Benenson, *Nature Nanotechnol*, 2010, **5**, 666.
227. R. Piran, E. Halperin, N. Guttmann-Raviv, E. Keinan and R. Reshef, *Development*, 2009, **136**, 3831.
228. M. N. Win and C. D. Smolke, *Science*, 2008, **322**, 456.

List of Abbreviations and Glossary

A	Adenine
Abs	Absorbance
AMP	Adenosinemonophosphate
Aptamer	DNA or RNA or peptide strand with receptor properties
Aptazyme	DNA or RNA or peptide strand with enzyme properties
ATP	Adenosinetriphosphate
β	Binding constant
bpy	2,2'-Bipyridine
BSA	Bovine serum albumin
C	Celsius
C	Cytosine
CB	Cucubituril
CD	Cyclodextrin
CD	Circular dichroism
cm	centimetre
CT	Charge transfer
DMF	Dimethylformamide
DMSO	Dimethylsulfoxide
ε	Extinction coefficient
ECL	Electrochemiluminescence
EDTA	Ethylenediaminetetraacetic acid
EET	Electronic energy transfer
em	Emission
EPR	Electron paramagnetic resonance
ESIPT	Excited state intramolecular proton transfer

Monographs in Supramolecular Chemistry No. 12
Molecular Logic-based Computation
By A Prasanna de Silva
© The Royal Society of Chemistry 2013
Published by the Royal Society of Chemistry, www.rsc.org

List of Abbreviations and Glossary

exc	Excitation
φ	Quantum yield
Fc	Ferrocene (reference electrode)
Flu	Fluorescence
FE	Fluorescence enhancement factor
G	Guanine
GDH	Glucose dehydrogenase
GOx	Glucose oxidase
GTP	Guanosinetriphosphate
HEK	Human embryonic kidney (cells)
HRP	Horseradish peroxidase
hv dose	Light radiation dose
ICT	Internal charge transfer
IgG	Immunoglobulin
IUPAC	International Union of Pure and Applied Chemistry
λ	Wavelength
LE	Luminescence enhancement factor
LMCT	Ligand-to-metal charge transfer
Lum	Luminescence
M	Molar
MC	Metal-centered (excited state)
MeCN	Acetonitrile
MLCT	Metal-to-ligand charge transfer
μm	micrometre
mm	millimetre
NAD^+	Nicotinamide adenine dinucleotide, oxidized form
NADH	Nicotinamide adenine dinucleotide, reduced form
nir	Near infra-red
nm	nanometre
NMR	Nuclear magnetic resonance
ns	nanosecond
n-type	Negative majority charge carriers (in semiconductor)
OAc^-	Acetate
ORD	Optical rotatory dispersion
PCR	Polymerase chain reaction
PDT	Photodynamic therapy
PET	Photoinduced electron transfer
phen	1,10-Phenanthroline
Pho	Phosphorescence
p-type	Positive majority charge carriers (in semiconductor)
RF	Radiofrequency
sce	Standard calomel electrode (reference electrode)
SDS	Sodium dodecylsulfate
T	Thymine
%T	Percentage transmittance
Trans	Percentage transmittance

THF	Tetrahydrofuran
TICT	Twisted internal charge transfer
TPF	Two-photon fluorescence
uv	Ultra-violet
V	Volts

Subject Index

Note: page numbers in *italic* refer to figures or tables.

absorption spectroscopy *see* electronic absorption spectroscopy
acetate ion input 68, 85, 268
acetone input 79
acid–base annihilation 222–3
Adams, M. 51
adenosine triphosphate (ATP) *71*, 124, 156, 306, 368
Adleman, L. 7
aerofoils and air pressure 344–5, *346*
Ajayaghosh, A. 238
Akkaya, E. 158, 212, 231, 323
α,ω-alkanediammonium ion 134–5, 197, 198
alphanumeric characters, coding for 297
aminomethylphenylboronic acid 133
analogue–digital relationship 19–20
AND logic 27, *28*, 203, 332
  associated collected data *118*
  distinguishable connected inputs 132–4
  distinguishable separate inputs 115
    cation and anion 122–4
    cation and biomolecule 126–7
    cation and neutral 124–6
    cation and redox 122–4
    two cations 115–22, 202
    two neutral 128–9
  indistinguishable connected inputs 134–6
  indistinguishable separate inputs 129–32
  light dose inputs 136–41
  molecular materials 143–5
    biopolymeric gates 141–3
  multi-input systems
    four-input 257–60
    three-input 230–3
  semiconductor arrays 29–31, *32*
  uses in
    half-adders 210
    improved sensing applications 349–53
Andréasson, J. 205, 214, 225, 254, 260, 262, 289, 291, 326, 368
anthrylmethylpolyamines 199
applications
  improving sensing/selectivity
    AND logic 349–53
    AND, INHIBIT and TRANSFER logic 353–6
    XOR logic for multiple species 354–6
  intracellular computation 370–4
  medical uses 350–1, 352
    diagnostics 350–1, 352, 359–64
    improved and photodynamic therapies 359–70
  optical sensing
    air pressure on aerofoils 344–5
    blood electrolytes 343–4
    catalyst screening 348
    cells and tissues 337–43

applications (*continued*)
  chemical warfare agent
    detection 348
  marine toxins 345–6
  nuclear waste monitoring
    346–7
  tagging/identification 356–9
aptamers 72, 89, 127, 366–7
arithmetic *see* number manipulation
Ashkenasy, G. 141, 160
Asoh, T. 168
ATP *see* adenosine triphosphate
Aviram, A. 1
Avouris, P. 83
azo dye 295–6

Bag, B. 120
Ballester, P. 68
Balzani, V. 169, 173, 220, 222, 261, 262
BAPTA receptor 354–6
Baytekin, H. 323
Belousov–Zhabotinskii reaction 151
Benenson, Y. 372
benzoquinone 87, *88*
Bharadwaj, P. 233
Bielawski, C. 165
bifunctional input species 132–4
biomolecules 126–8
  see also enzyme; oligonucleotide; protein
biopolymeric gates 141–3
blood electrolytes, measuring 343–4
bone 229–40
Boole, G. 24
Boolean algebra 25, 26, *28*, *32*
boronic acids 88, 133, 237, 319–20
Brown, G. 254
Burdette, S. 287

caffeine 69
calixpyrrole 165
Callan, J. 120, 122, 324
Campagna, S. 291
cancer 350–1, 352, 364, 372
  photodynamic treatment 366–70

carotene–porphyrin–fullerene
  triad 140–2
cascades
  enzyme 141–4, 204, 229–30, 232–3, 263–4
  oligonucleotide 167, 230
catalyst screening 348
cell biology 16–17
  β-galactosidase 341, 371
  gene expression 370–4
  intracellular ion monitoring 337–9
  lysosensor blue™ 339–40
  membrane potentials 342, *343*
  molecular thermometers 342
Chae, M. 78
Chang, J. 66
chaperonin 368
charge transfer 35–7, 154
chemical warfare agents 79, 348
chemidosimeters 78
chemistry, computational aspects
  of 12–15
  digital–analogue
    relationships 19–20
  indicators and sensors 18
  interdisciplinary aspects 21
  molecular device
    characteristics 20–1
  molecules *vs*
    semiconductors 15–18
Chen, C. 126
Chiu, S. 154, 236
chlorophyll synthesis 296–7
chromophore–receptor systems 35, 318–19
α-chymotrypsin 142
circular dichroism 17
CNOT logic 317
cocaine 89, 174
COINCIDENCE logic 27, *28*
Collier, C. 151, 195, 234–5
computational chemistry 12
concanavalin A 88, 237
conduits 228–9
Conrad, M. 141
Cooper, C. 133

Copeland, G. 348
Cram, D. 63–4
Credi, A. 222, 225, 256
cyclodextrin encapsulation 129, 137, 236
cyclophane 121–2
cytochrome c 142, 199
cytosine triphosphate *71*
Czarnik, A. 87, 123, 258, 306

D latch 290–8
Daniel, E. 89
dansylamide 251–3, 294–5
de Shayes, K. 122
de Silva, A. *13, 14, 43,* 343, *345*
  double input systems *198,* 210–12
  multi-level systems 303, *309, 319, 321*
  single input systems *59, 60,* 64, *75, 91,* 134, 171
decoders 262
demultiplexers 223, 256–7
deoxyribozymes 80, 217–18, 239–40, 247–8, 270, 275
  electronic energy transfer systems 328
detergents *see* surfactants
diagnostics 16, 359–64
diarylpyrazoline fluorophore 147
DiCesare, N. 320
Dielferich, F. 137
digital–analogue relationship 19–20
diodes 30, *31*
  molecular 144, 151, 195–6, 234–5
dithienylethene photochrome 165, 202, 260–1, 262, 263, 325
  history-dependent systems 285, 288
DNA 7, 142–3
  detecting species near 351
  in full-adder 270–1
  irreversible YES logic 80
  and keypad locks 294
  OR logic 150, 158–60
  strand displacement 167, 230
  tiles 352–3
Dong, S. 176, 206

double input–double output systems 210
  demultiplexer 223
  half-adder *see* half-adder
  half-subtractor 222–3
  magnitude comparator 223–4
  reversible logic 224–5
double input–single output systems 26–9
  AND logic 114–45
  IMPLICATION logic 176–9
  INHIBIT 161–9
  NAND logic 156–61
  NOT TRANSFER logic 180–1
  OR logic 145–55
  PASS 0 and PASS 1 181
  TRANSFER logic 27, *28,* 179–80
  XOR logic 169–76
  see also reconfigurable double input
D'Souza, F. 79, 87, 124, 154
dual emission systems 163, 224–5
Dwyer, C. 352

edge-detecting gene arrays 371–2
EET *see* electronic energy transfer
electrochemical systems 174–5
electrochemiluminescence 151, 203
electron paramagnetic resonance 165
electronic absorption and emission spectroscopy 17–18, 202–3, 318
  quantum aspects of
    electronic energy transfer systems 326–33
    excimer and exciplex systems 330–1
    internal charge transfer systems 318–26
    Raman spectroscopy 331–3
electronic energy transfer (EET) 41
  quantum aspects of 326–33
electronic logic devices 13, 15
  D latch 290–1
  keypad lock 292
  ring oscillators 290

electronic logic devices (*continued*)
  R–S latch 285–6
  truth tables and logic gates 25–31
emission spectroscopy *see* electronic absorption and emission
enantioselectivity 70–1, 88–9, 135–6, 198, 350
encoders 260–2
enzyme assays 82
enzyme systems
  AND logic 141–2
  cascades 141–2, 204, 229–30, 232–4, 263–5
  combined half-adder/half-subtractor 269
  DNA-bases in half-adder 217–18
  four-input system 263–5
  NAND logic 160
  OR logic 151, 202
  three input AND 232–3
  XOR logic 173
EPROMs 109
EQUIVALENCE logic 26
ESIPT 42
excimers 41–2, 330–1
exciplexes 41–2, 330–1
excited states
  charge transfer 35–7
  intramolecular proton transfer 42
  metal-centred 37
  $n\pi^*$ and $\pi\pi^*$ 37–8, 63

Fabbrizzi, L. 156, 259, 326
Fages, F. 153
Fahrni, C. 59
Fallis, I. 149
Fan, C. 247
ferrocene derivatives 68, 149, 175, 288
Flood, A. 332
fluoresceins 272, 287
fluorescence
  excitation *4*
  output 15–16
  photochemical foundation of *see* photochemistry
fluorescence microscopy 34–5

fluorophore–receptor system *35*, 318–19
fluorophore–receptor interactions *43*
fluorophore–receptor–spacer–receptor system 308–9
fluorophore–spacer–receptor systems 38, *39*, 147, 206, 222
fluorophore–spacer–receptor–spacer–receptor system 196, 302–3
four-input logic
  AND 257–9
  doubly disabled AND 259–60
  encoders and decoders 260–2
  other high-input systems 263–7
FPGAs 109
FRET 41
fuel cells 151, 206–7, 294
Fujita, M. 129, 149, 200
Fukuzumi, S. 165
fulgide photochrome 260–2, 262
full-adder 270–2
  combined with full-subtractor 272–5
full-subtractor 272–5
fullerine derivatives 124, 154
functional integration 229

β-galactosidase 341, 371
gaming systems 275–6
Garcia-Espana, E. 156, 303
gate-to-gate linkers 229
Gawel, K. 145
gel-based systems 144–5, 168–9, 369–70
gene expression 370–4
Ghadiri, M. 167, 168, 200, 218, 365
Gianetto, A. 291
Gianneschi, N. 365
Giordani, S. 181, 197, 244, 249, 265, 289, 326
glucosamine 133
glucose
  enantioselectivity 70–1, 135–6, 198
  glucose dehydrogenase 141–2, 204, 233
glutathione 88

Gouterman, M. 345
guanosine triphosphate 69, *71*, 87
Guchhait, N. 65
guest-induced fluorescence 38–9
Gunnlaugsson, T. 37, 64–5, 154, 162, 305, 331, 339
Guo, X. 181, 230
Gupta, T. 252
Gust, D. 140, 203, 225

half-adder 32, 210–22
  all-optical 214–16
  cation input 210–14
  combined with half-subtractor 267–9
  enzyme systems 217–19
  truth table *32*
half-subtractor 165, 222–3
  combined with half-adder 267–9
Hamachi, I. 144, 258, 350, 369, 370–1
Harima, Y. 73
He, L. 247
Heath, J. 143
heavy atom effects 44
historical overview
  early proposals 1–5
  photochemical approach 5–7
history-dependent systems 285
  D latch 290–8
  molecular keypad lock 292–8
  R–S latch 285–90
Holdt, H. 66
horseradish peroxidase 141–2, 204, 289
hydrogen peroxide 353–4
hydrophobic effects 44
hydroquinone 79

ICT *see* internal charge transfer
identification, small object 356–9
IDENTITY logic 26
IF THEN operation 27
Ihmels, H. 351
IMPLICATION logic 320–1
  and internal charge transfer 320–1
  molecular 176–8, 202, 204
    collected data *177*
  polymer-based 179
  three-input systems 249–53
indicators 18–19
INHIBIT logic 27, *28*
  electronic energy transfer systems 328
  internal charge transfer systems 320, 323, 324
  molecular 161–9, 199, 202–3
    collected data *163*
    in half-adders 224–5, 268
    in half-subtractors 219–23, 268
  molecule-based materials 167–9
  three-input systems 236–40, 247–8
Inouye, M. 137
inputs, types of 13–15
internal charge transfer (ICT) 35–7
  quantum aspects of 318–26
intracellular computation 370–4
intracellular monitoring 287–8
intramolecular proton transfer 120–1
ion-driven luminescent signalling 5
ion-pair detectors 122–4
ionic input 15–16
Irie, M. 263
irreversible NOT logic 92
  anion input 93
  oligonucleotide input 93–4
  protein input 94–5
irreversible reconfigurable logic 235
irreversible YES logic 77–8
  chemical input
    cation 78
    oligonucleotides 80–2
    organic molecules 78–80
    proteins 82–3
  light dose input 83
isomerization 216–17
Iwata, S. 156

James, T. 122, 133
Jiang, S. 223, 246

Kandish, K. 165
Katz, E. 151, 160, 264, 289, 294
Keinan, E. 365

keypad lock 292–8
Kim, K. 151
Kolpashchikov, D. 94, 365
Konermann, L. 142, 199
Kosower, E. 89

lab-on-a-molecule 231, 360, 362, 364
Langford, S. 219, 220
Langmuir–Blodgett films 144
lanthanides 37, 42, *43*, 61, 232
   and INHIBIT logic 164–5
   and luminescence 131–2
   off–on–off switching 305–6
   OR logic 157
latch systems
   D latch 290–8
   R–S latch 285–90
Lee, S. 248
Lehn, J. 242, 306
Leigh, D. 91
Levine, R. 138, 140
Li, C. 132, 250–1
Licchelli, M. 65
ligand-to-metal charge transfer 154
light dose input
   AND logic 136–41, 136–43
   irreversible YES logic 83
   NOT logic 91–2
   XOR logic 173–4
   YES logic 76–7
Lin, W. 353
linkers, gate 229
Lippert–Mataga–Bakhshiev equation 325
Liu, Y. 268
logic gates, electronic devices 29–31
   double input–single output 27–9
   single input–single output 25–6
luminescence 129–32
   see also photochemistry
lumophore–spacer–receptor system 130
lumophore–spacer–receptor–spacer–receptor systems 115–17
lysosensor blue™ 339–40

Ma, D. 150
macrocycle and clip 154–5, 236
magnitude comparator 223–4
Magri, D. 128, 155
malachite green 127–8, 365, *366*
malate dehydrogenase 202
Maligaspe, E. 124, 154
maltose phosphorylase 232–3, 294
Margulies, D. 274
marine toxins 345–6
Martinez-Máñez, R. 111, 203, 257
materials *see* molecule-based materials
Matsui, J. 152
medical applications 350–1, 352
   improved diagnostics 16, 359–64
   therapies 359–70
Mello, J. 66
membrane potentials, cell 342, *345*
memory systems *see* history-dependent
merocyanine *see* spiropyran–merocyanine pair
mesoporous silicon nanoparticles 145, *146*, 294, 369
metal-to-ligand charge transfer (MLCT) 36–7
methotrexate 371
Miller, S. 348
Misra, A. 327
Miyoshi, D. 150
module connectivity 43
molecular abacus 212
molecular beacon 80
molecular biology *see* cell biology
molecular computational identification (MCID) 120, 357–9
molecular electronic switch 1–2
molecular logic devices
   characteristics 20–1
   classified by input/output *14*, 18
   digital–analogue relationships 19
   electronic approaches to 1–3, 50–1, 83
   history of 1–7
   world map of laboratories 7
molecular tags 356–9
molecular thermometers 74–6, 91, 342

molecule input *see* organic molecule
molecule-based materials
   AND logic 143–5
   IMPLICATION logic 179
   INHIBIT logic 167–9
   OR logic 151–2, 155
   XOR logic 174–6
Molina, P. 68
molybdate ion 85
Monod, J. 370
monolayer *see* unimolecular layer
multi-level logic 302
   see also switching systems
multiplexer 253–6, 318

Nagamura, T. 39
Nagano, T. 94
NAND logic 27, *28*
   internal charge transfer systems 323, 324
   molecular 4, 156–61, 203, 206, 323
Nandi, A. 145
nanocrystalline titanium oxide 155, 174, 207
nanoparticles
   gold 239
   silica 145, *146*, 294, 369
nanosheets 144, 152
nanospaces 67
nanotubes 83, *84*, 144, 151
nanowires
   silicon 251–2
   zinc oxide 286
naphthalenediimides 138–9
negative differential resistance 308
nerve action potentials 324
nerve gases 79, 348
neutral inputs 124–6, 128–9
Ng, D. 367
nitric oxide 353–4
nitroprusside reaction 266–7
NMR
   and alphanumeric coding 297
   and quantum computing 316–17
Nocera, D. 129
Nojimax, T. 150

Nolan, E. 65
NOR logic 27, *28*, 29
   molecular 152–5, 323, 332
      collected data *153*
   three-input systems 235–6, 242–4
NOT IMPLICATION 199
NOT logic 25–6, 83
   chemical input
      anions 85–7
      cations 84–5
      organic molecules 87–9
      polarity 89–90
   in electronic arrays 29–31
   electronic energy transfer system 327, 328
   excimer/exciplex systems 330
   internal charge transfer systems 319, 325
   in reconfigurable system 110
   temperature input 90–1
   see also irreversible NOT
NOT TRANSFER logic 27, *28*
   molecular 180–1, 205, 328
$n\pi^*$ excited states 37–8, 63
nuclear magnetic resonance *see* NMR
nuclear waste, monitoring 346–7
nuclease-driven logic systems 251
number manipulation 31–3, 210
   see also full adder; half-adder; half-subtractor

observation technique-reconfigurable behaviour 202–5
observation wavelength 204–5, 318
off–on–off switching 302–8
oligonucleotides 7, 142, 230
   cascades 167, 230
   half-adders 218–19
   INHIBIT logic 167–8
   irreversible NOT logic 93–4
   irreversible YES logic 80–2
   OR logic 150
   three-input INHIBIT 239
   YES logic 71–2
optical inputs 14
optical outputs 13

optical rotary dispersion 17–18
optrode membrane 169
OR logic 27, 28
  in electronic arrays 29–31
  molecular 145–50, 207
    collected data 148
  molecular-based materials 151–2
  three-input systems 233–5, 240–3
ORD 17–18
organic molecule input
  AND logic
    biomolecules and cations 126–8
    neutral and cation 124–6
    neutral inputs 128–9
  bifunctional species 132–4
  irreversible YES logic 78–80
  NOT logic 87–9
  YES logic 69–71
O'Shea, D. 367, 368
output observation technique 202–4
output, types of 13–15

Pallavicini, P. 307, 311
parallel integration 228
paramagnetic effects 44
Parker, D. 42
Pasparakis, G. 167
PASS 0 logic 25–6, 27, 319, 321, 331
  double input–single output systems 181
  in electronic arrays 30
  single input–single output 95–6, 111
PASS 1 logic 25–6, 27, 319, 321, 331
  double input–single output systems 181
  in electronic arrays 30
  single input–single output 95
peptides 127, 160
Perez-Inestrosa, E. 227, 242, 245, 304
perylene tetracaboxydiimide moiety 139–40, 307
PET 38–41
pH dependent fluorescence 5, 36, 130, 202
pH sensors 229, 272
  switchable 243, 254–5, 305, 307

phenanthrolines 175–6, 257, 258, 306
1,4-phenylenediacetate 135
phosphate ion 123–4, 231–2, 258
  pyrophosphate 136, 258–9
phosphines 69, 78
phosphorescent systems 237
phosphorylation reactions 350
photochemistry, overview of 3, 5–7, 34–5
  electronic energy transfer (EET) 41
  excimers and exciplexes 41–2
  excited states
    charge transfer 35–7
    and intramolecular proton transfer (ESIPT) 42
    metal-centred 37
    $n\pi^*$ and $\pi\pi^*$ 37–8
  photoinduced electron transfer (PET) 38–41
  switching/photochemistry relationships 43–4
photochromics 76–7, 260, 263
  arrays of solutions 197, 265–6
  in internal charge transfer systems 325–6
  ion sensitive switches 158, 244
  memory 285, 286
  spiropyran–merocyanine pair 137–8, 201–2, 230, 368–9
photodynamic therapy 366–70
photoinduced electron transfer (PET) 38–41
photosensitized reactions 136
$\pi\pi^*$ excited states 37–8, 63
Pimental report 3
Pina, F. 138, 173, 303
Pischel, U. 205, 225, 289, 304, 309
poly(acrylamide) chain folding 165–7, 179, 180–1
porphyrins 3, 4, 124–5, 139–41, 154
pro-drug activation 364–6
PROM semiconductor 196
protein input
  irreversible NOT logic 94–5
  irreversible YES logic 82–3
  YES logic 72–3

Subject Index

protein kinase C 126–7
pseudorotaxane 169, 175
Pu, L. 70
pyrophosphate ion 136

Qian, H. 65, 66, 212, 240, 255, 269, 327
Qin, W. 62
quantum computing 316
  NMR approaches 316–17
  see also electronic absorption and emission spectroscopy

radiology 229
Raman spectroscopy 331–3
Raymo, F. 181, 197, 244, 249, 265, 289, 326
reactive oxygen/nitrogen species 353–4
receptor–chromophore–receptor system 171
receptor–spacer–fluorophore–spacer–receptor system 196, 197, 302
receptor–spacer–lumophore–spacer–receptor system 129–30
reconfigurable double input–single output systems 5
  inputs and connectivity 195–202
  output observation technique 202–4
    observation wavelength 204–5
  starting state of device 205–7
  voltage-reconfigurable behaviour 207–8
reconfigurable irreversible logic 235
reconfigurable single input–single output 5
  inputs 109–11
  output observation 111–12
reconfigurable three-input systems 244–7, 250, 260
rectification, monolayer 2–3
redox potential 38, 207
  AND logic 122–4
  YES logic 73
reversible logic 224–5
RFID chips 357
Rhind papyrus *211*
rhodamine spirolactone unit 327

riboswitches 373–4
rigidification 44, 127, 132
ring oscillators 290
RNA 7
  and intracellular computation 370–4
  YES logic 71–2
robustness, molecular device 17–18
rotaxanes 61, 91–2, 179
  bead on a string 216
  pseudorotaxane 169, 175
Rurack, K. 66, 158, 176
R–S latch 285–90

Sames, D. 37
Samoc, M. 202, 285
Savage, P. 66
saxitoxin 345–6
Schmittel, M. 109, 111, 175, 203, 257
Schneider, H. 121, 128, 144
self-replicating peptide 141
semiconductor-based computing 13, 15
  see also electronic logic devices
sensitizers 136
sensors 18–19
sequential logic 285
serial integration 228–30, 250
Sessler, J. 165
Shanzer, A. 207, 267, 272, 292, 293
Shapiro, E. 364, 372
Shi, W. 286
Shinkai, S. 70, 135, 233, 256
Shor, P. 316, 317
Singh, P. 148, 235
single input–single output systems 25–6
  NOT logic systems *see* NOT logic
  reconfigurable systems 109–12
  YES logic systems *see* YES logic
Siri, O. 26
Sivan, S. 142
Smith, B. 110, 119, 258
sodium dodecyl sulfate 212, 240–1, 269, 311
solvent polarity 73, 325
Song, Q. 268, 275
Sparano, B. 72

spectroscopy 17–18
  Raman 331–3
  see also electronic absorption and emission
spiropyran–merocyanine pair 138, 201–2, 230, 239, 289, 368–9
starting state, device 205–7
Stefanovic, D. 41, 80, 194, 239, 247, 275, 328
stimulus–response *13*
  curves 19–20
Stoddart, J. 128, 143, 145, *146*, 169, 220, 222
Stojanovic, M. 41, 80, 194, 239, 247, 275, 328, 365
structural rigidification 44, 127, 132
superposition 316
surfactants 195, 212, 240–1, 269, 307, 311
switching systems
  high–medium–high 311–12
  high–medium–low 309–10
  low–medium–high 308–9
  medium–low–high 310–11
  off–on–off 302–8
  off–on–on 312
  photochemical principles 43–4

Tàrraga, A. 124
tagging, molecular 356–9
Tang, B. 66, 342
tartaric acid enantiomers 136, 350
technique–reconfigurable logic behaviour 203
temperature input
  IMPLICATION logic 179
  INHIBIT logic 166
  NOT TRANSFER logic 180–1
  three-input IMPLICATION 240
  three-input NOR 236
  YES logic 74–6
Teramae, N. 331
ternary logic 302
  see also switching systems
therapies
  improved 364–6
  photodynamic 366–70

thermometer, molecular 74–6, 91, 342
three-input logic
  AND 230–3
  diphenanthroline array 257, *258*
  disabled IMPLICATION 249–53
  disabled INHIBIT 247–8
  disabled XNOR 248–9
  enabled NOR 242–4
  enabled OR 240–3
  INHIBIT 236–40
  inverted enabled OR 253
  NOR 235–6
  OR 233–5
  wavelength-reconfigurable disabled OR 245–7
  enabled IMPLICATION 244–5
Tian, H. 60, 91, 93, 216, 253, 288, 305, 325
tic-tac-toe 275–6
Tomizaki, K. 201
Tour, J. 2
Toyo'oka, T. 61
TRANSFER logic 27, *28*
  electronic energy transfer systems 328
  internal charge transfer systems 323, 326
  molecular 179–80, 205, 207
tris(2, 2'-bipyridyl)Ru(II) moiety 237–8, 261–2
truth tables
  D latch 291
  double input–single output 26–7
  half-adder 32
  R–S latch 286
  single input–single output 25–6
tryptophan 160, *161*, 199
Tsien, R. 17, 171, 196, 337, 342
tumours 352
twisted internal charge transfer (TICT) 36
two-photon absorption 165, 203, 214

unimolecular layers 2–3, 143–4, 195, 234
Upadhyay, K. 320
Urano, Y. 341